Computational Civil Engineering (CCE 2021)

IOP Conference Series: Materials Science and Engineering
Volume 1141

Iasi, Romania
27 - 29 May 2021

Editors:

Gabriela Covatariu
Vasile-Mircea Venghiac
Petru Mihai
Ioana Olteanu-Dontov
Gabriel Barbat
Daniel Covatariu

ISBN: 978-1-7138-4618-5
ISSN: 1757-8981

Printed from e-media with permission by:

Curran Associates, Inc.
57 Morehouse Lane
Red Hook, NY 12571

Some format issues inherent in the e-media version may also appear in this print version.

Institute of Physics
Dirac House, Temple Back
Bristol BS1 6BE UK

Phone: 44 1 17 929 7481
Fax: 44 1 17 920 0979

techtracking@iop.org

Additional copies of this publication are available from:

Curran Associates, Inc.
57 Morehouse Lane
Red Hook, NY 12571 USA
Phone: 845-758-0400
Fax: 845-758-2633
Email: curran@proceedings.com
Web: www.proceedings.com

TABLE OF CONTENTS

2021
Computational
Civil
Engineering
Conference
CCE2021
May 27th - 29th Iasi - Romania
Empower the
digital transition
in the construction
world
Co-organizers:
UT CB
construcții
Partners:
Unitec
LEEDS
www.cce.ci.tuiasi.ro

IOP Conf. Series: Materials Science and Engineering 1141 (2021) 011001 doi:10.1088/1757-899X/1141/1/011001

Proceedings of the
"Computational Civil Engineering" CCE2021
Iasi, Romania - May 27th – 29th, 2021

Editors:
Gabriela Covatariu
Vasile-Mircea Venghiac
Petru Mihai
Ioana Olteanu-Donțov
Gabriel Barbat
Daniel Covatariu

Organizers
Faculty of Civil Engineering and Building Services Iasi, Romania
CIMNE - International Centre for Numerical Methods in Engineering (CIMNE), UPC Barcelona, Spain
"Anton Sesan" Civil Engineering Academic Society, Romania
"Matei Teiu Botez" Academic Society Iasi, Romania

Co-Organizers
"G.M. Cantacuzino" Faculty of Architecture Iasi
Technical University of Civil Engineering Bucharest
Faculty of Civil Engineering Cluj-Napoca
Faculty of Civil Engineering Timisoara

Partners
Technische Universität Berlin
Odessa State Academy of Civil Engineering and Architecture
Unitec Institute of Technology Auckland
University of Leeds

IOP Conf. Series: Materials Science and Engineering 1141 (2021) 011001 doi:10.1088/1757-899X/1141/1/011001

The Committees of *CCE 2021 Conference*

Honorary Committee

Prof. Constantin Ionescu	"Gheorghe Asachi" Technical University of Iasi, Romania
Prof. Eugenio Oñate	International Centre for Numerical Methods in Engineering CIMNE Barcelona, Spain
Prof. Mihai Budescu	"Gheorghe Asachi" Technical University of Iasi, Romania
Prof. Nicolae Ţăranu	"Gheorghe Asachi" Technical University of Iasi, Romania
Prof. Alex H. Bărbat	International Centre for Numerical Methods in Engineering CIMNE Barcelona, Spain
Prof. Daniela Manea	Technical University of Cluj-Napoca, Romania
Prof. Raul Zaharia	"Politehnica" University of Timisoara, Romania
Prof. Radu Văcăreanu	Technical University of Civil Engineering Bucharest, Romania
Assoc.prof. Tania Hapurne	"Gheorghe Asachi" Technical University of Iasi, Romania

Scientific committee

Prof. Dorina-Nicolina Isopescu	"Gheorghe Asachi" Technical University of Iasi, Romania
Prof. Ioan Nistor	University of Ottawa, Canada
Prof. Konstantinos Daniel Tsavdaridis	University of Leeds, UK
Prof. Wynand Steyn	University of Pretoria, South Africa
Prof. Thanasis C. Triantafillou	University of Patras, Greece
Prof. João Carlos Gonçalves Lanzinha	University Beira Interior, Portugal
Prof. Cristina Cosma	Wentworth Institute of Technology, USA
Prof. Cristina Mihaela Câmpian	Technical University Cuj Napoca, Romania
Prof. Cedric D'Mello	City University of London, UK
Prof. Anatolii Kovrov	Odessa State Academy of Civil Engineering and Architecture, Ukraine
Prof. Chris Roth	University of Pretoria, South Africa
Prof. Fideliu Păuleţ-Crăiniceanu	"Gheorghe Asachi" Technical University of Iasi, Romania
Prof. Anatolii Popov	Odessa State Academy of Civil Engineering and Architecture, Ukraine
Prof. Irina Lungu	"Gheorghe Asachi" Technical University of Iasi, Romania
Assoc. prof. Piero Colajanni	Università degli Studi di Palermo, Italy
Assoc. prof. Viorel Popa	Technical University of Civil Engineering of Bucharest, Romania
Assoc. prof. Mihail Iancovici	Technical University of Civil Engineering of Bucharest, Romania
Assoc. Prof. Tiberiu Cătălina	Technical University of Civil Engineering of Bucharest, Romania
Assoc. prof. Stefan Gutiu	Technical University of Cluj Napoca, Romania
Assoc. prof. Ferdinand-Zsongor Gobesz	Technical University of Cluj Napoca, Romania
Assoc. prof. Ligia Moga	Technical University of Cluj Napoca, Romania
Assoc. prof. Carmen Ema Panaite	"Gheorghe Asachi" Technical University of Iasi, Romania
Assoc. prof. Carlos Chastre	Universidade NOVA de Lisboa, Portugal
Assoc. Prof. Vasilică Ciocan	"Gheorghe Asachi" Technical University of Iasi, Romania
Assoc. Prof. Andrei Burlacu	"Gheorghe Asachi" Technical University of Iasi, Romania

IOP Conf. Series: Materials Science and Engineering 1141 (2021) 011001 doi:10.1088/1757-899X/1141/1/011001

Assoc. Prof. Gabriel Oprişan	"Gheorghe Asachi" Technical University of Iasi, Romania
Assoc. Prof. Tania Mariana Hapurne	"Gheorghe Asachi" Technical University of Iasi, Romania
Assoc. Prof. Mebarek-Oudina Fateh	University of Skikda, Algeria
Assoc. Prof. Anastasiia Pandas	Odessa State Academy of Civil Engineering and Architecture
Assoc. prof. Nelu-Cristian Cherecheş	"Gheorghe Asachi" Technical University of Iasi, Romania
Sen.lect. Patricia Alonso Alonso	University of Coruña, Spain
Sen.lect. Luşa Tuleasca	Unitec Institute of Technology Auckland , New Zealand
Sen.lect. Sergiu Băetu	"Gheorghe Asachi" Technical University of Iasi, Romania
Sen.lect. Anca-Mihaela Ciupală–Costin	University of East London, UK
Sen.lect. Daniel Covatariu	"Gheorghe Asachi" Technical University of Iasi, Romania
Sen.lect. Ioana-Sorina Enţuc	"Gheorghe Asachi" Technical University of Iasi, Romania
Sen.lect. Sorin Codrut Floruţ	Politehnica University Timisoara, Romania
Sen.lect. Arya Assadi Langroudi	University of East London, UK
Sen.lect. Eugen Lozinca	Technical University of Civil Engineering of Bucharest, Romania
Sen.lect. Aliz Éva Máthé	Technical University of Cluj Napoca, Romania
Sen.lect. Ionuţ-Ovidiu Toma	"Gheorghe Asachi" Technical University of Iasi, Romania
Sen.Lect. Radu Andrei	"Gheorghe Asachi" Technical University of Iasi, Romania
Sen.lect. Wei Loo	Unitec Institute of Technology Auckland , New Zealand
Sen.lect. Valeriu-Sebastian Hudişteanu	"Gheorghe Asachi" Technical University of Iasi, Romania
Sen.lect. Michal Krajcik	Slovak University of Technology in Bratislava, Slovakia
Sen.Lect. Rodica Dorina Cadar	Technical University of Cluj Napoca, Romania
Sen.Lect. Rozalia Melania Boitor	Technical University of Cluj Napoca, Romania
Sen.Lect. Mihai Liviu Dragomir	Technical University of Cluj Napoca, Romania
Assist.prof. Bertha Santos	University of Beira Interior, Portugal
Dr. Melito A. Baccay	Technological University of the Philippines
Dr. Jan Červenka	Červenka Consulting, Czech Republic
Dr. Lucian-Constantin Ungureanu	Technical University Berlin, Germany
Dr. Ionut Cristian Scurtu	Mircea cel Batran Naval Academy Constanta, Romania
Dr. Gabriel Bărbat	International Centre for Numerical Methods in Engineering CIMNE Barcelona, Spain

Steering Committee from "Gheorghe Asachi" Technical University of Iasi, Romania
Prof. Dorina-Nicolina Isopescu - **president**
Prof. Irina Lungu
Assoc.Prof. Petru Mihai
Assoc. Prof. Vasilică Ciocan
Sen.Lect. Vasile Mircea Venghiac
Sen.Lect. Gabriela Covatariu
Sen.Lect. Ionuţ Ovidiu Toma
Sen.Lect. Daniel Covatariu
Sen.Lect. Ioana-Sorina Enţuc
Sen.Lect. Radu Andrei
Sen.Lect. Sebastian Hudişteanu
Sen.Lect. Ioana Olteanu-Donţov
Sen.Lect. George Ţăranu
As. Arch. Tiberiu Teodor Stanciu
As. Arch. Ramona Costea

Preface

The digitalisation determines a high level of increased productivity in any domain that was applied before. Digital transformations in Civil Engineering are thriving today on gaining precedent to traditionally dedicated computation to construction design. The need for the remote control of the construction process to meet quality exigencies, budget constraints, and duration, for balancing risk evaluation and risk mitigation, for new accuracy levels in predicting soil-structure interaction in every unique environment, for sustainability and circular economy viable in the construction sector require a multiverse approach in our Conference - Computational Civil Engineering CCE2021.

This year more than ever, we become open to all issues addressing the future of our profession as civil engineers, architects, structural engineers, MEP engineers a.s.o. Specialization in knowledge by following the separation of interests is falling apart in the world where connectivity is available for everyone. The subjects of interest for this conference may look singular items on the list, still they belong to each construction and the key ingredient is team work and the shared value is cooperation.

The pandemic, climatic and economic crises are converging to trigger a massive competition to find answers and fierce determination to share knowledge and forge alliances between all stakeholders in the construction sector. We invite architects, engineers with interests in material, constructions, chemical, computer, energy, building services, management, and economy domains to join this conference with their scientific contributions and competence on 27th-29th May 2021, in Iasi – Romania.

The Conference Computational Civil Engineering become traditional (since 2003).

The 16th edition CCE2021 was organized by Faculty of Civil Engineering and Building Services of Iasi in partnership with the International Centre for Numerical Methods in Engineering (CIMNE) Barcelona Spain, the "Anton Sesan" Civil Engineering Academic Society Iasi and the "Matei Teiu-Botez" Academic Society Iasi, Romania. We were very honoured to collaborate for conference organizing with the "G.M. Cantacuzino" Faculty of Architecture Iasi, the Technical University of Civil Engineering Bucharest, the Faculty of Civil Engineering Cluj-Napoca and the Faculty of Civil Engineering Timisoara. Also, we were glad for support from our partners Technical University of Berlin, Odessa State Academy of Civil Engineering and Architecture, Unitec Institute of Technology Auckland and University of Leeds.

The scientific event was held on May 30th-31st on line using Zoom platform.

CCE2021 aimed to bring together leading academic scientists, researchers, practitioners, numerical analysis programs developers to exchange / share their experiences to the Computational Civil Engineering domain to share their research findings to the theoretical and practical levels.

The proposed topics for CCE2021 edition of the conference was the following:

- *Computer Aided Design and Engineering*
- *Computational Methods in Civil Engineering*
- *Computational Fluid Dynamics*
- *Computer-Assisted Management in Civil Engineering*
- *Numerical Analyses of Civil Engineering Structures*
- *Computer Aided Training*
- *Civil Engineering Database*
- *Software for Civil Engineering*

IOP Conf. Series: Materials Science and Engineering 1141 (2021) 011001 doi:10.1088/1757-899X/1141/1/011001

In plenary sessions were presented very interesting and up-to-date exposures of the keynote speakers, well-known researchers or practicians in civil engineering domain:

- Prof.Dr. **Pere Roca**, Technical University of Catalonia (UPC), Spain – *Studies of the structure of Mallorca Cathedral. The contribution of numerical methods*
- Prof.Dr. **Wynand Steyn**, University of Pretoria, South Africa - *Improving livelihood through Civil Engineering research*
- Dr. **John Cribbs**, Wentworth Institute of Technology, U.S.A. - *Building Information Modeling (BIM) Workflows to Enable Cross-Disciplinary, Stakeholder Collaboration*
- Prof. Dr. Ir. **Pieter de Wilde**, University of Plymouth, UK - *The Energy Performance Gap - Challenges in Computational Building Engineering*
- Arch. **Karol Argasiński**, BIMfaktoria, buildingSMART Polish Chapter, Poland - *Pointcloud workflows in architectural HBIM practice*
- Dr. **Gabriel Barbat**, International Centre for Numerical Methods in Engineering, Spain - *Finite Element Modelling of Quasi-brittle Fracture with Enhanced Accuracy*

This year, the online organization offered an opportunity for all the participants make it easy for the researchers and experts from 12 countries to attend CCE2021.

The organizers want to express gratitude to all members of the Scientific Committee and the reviewers, and welcomes the speakers, papers presenters and participants of this conference and wish all of you a fruitful event.

Finally, but not at the end, thanks all sponsors and individuals who have contributed to the organization and success of this special event CCE2021. Hope to meet them again to the next edition CCE2023 at the end of May.

More information about this scientific event can be found on the conference website www.cce.ci.tuiasi.ro.

The editors

Peer review declaration

All papers published in this volume of IOP Conference Series: Materials Science and Engineering have been peer reviewed through processes administered by the proceedings Editors. Reviews were conducted by expert referees to the professional and scientific standards expected of a proceedings journal published by IOP Publishing.

- **Type of peer review:** Single-blind

To ensure quality and professional ethics a peer review mechanism has been adopted. The reviewers (at least two for each paper) received the manuscript with all the details of the authors and a form, prepared for the conference, with certain fields for comments and recommendations. The authors, at the next step, received anonymous reviews with the following possible recommendations: accept as is/minor revisions/major revisions/reject. The papers were revised by the authors accordingly and then forwarded to the same reviewers for final recommendation. The final decision was taken by the editors.

- **Conference submission management system:**

The papers were submitted to Easychair platform, handled by the editors.
https://easychair.org/my/conference?conf=cce20210
Two of the editors managed the correspondence between reviewers and authors.

- **Number of submissions received:** 89

- **Number of submissions sent for review:** 63

- **Number of submissions accepted:** 48

- **Acceptance Rate (Number of Submissions Accepted / Number of Submissions Received X 100):** 53.39

- **Average number of reviews per paper:** 3

- **Total number of reviewers involved:** 66

- **Any additional info on review process:**

  A pre-screening process has been adopted to comply with the quality of technical content under the following steps:

  1. Screening of the papers has been done by the editors on the originality of the content and technical area. If the paper was related to the theme of the conference, reviewers have been assigned by the editors.

  2. If the technical area of the submitted paper is not related to the theme of the conference, the paper has been sent back to the authors by the editors.

3. If the similarity index was more than 15%, the paper has also been sent back to the authors for revision.

4. The similarity check was done in paralell with the review process. The similarity report was sent together with the reviews to the authors. If the similarity index was greater than 15% the authors were asked to review their paper.

5. After authors revison the paper was submitted again for similarity check.

- **Contact person for queries:**
Gabriela Covatariu, PhD, PMP
Faculty of Civil Engineering and Building Services, "Gheorghe Asachi" Technical University of Iași, Romania,
Email: *gabriela.covatariu@tuiasi.ro*

The influence of the architecture form in the performance analysis of a passive house in Moldavia

A Cojocaru[1], D N Isopescu[1]
[1]"Gheorghe Asachi" Technical University of Iasi, Faculty of Civil Engineering and Building Services, Department of Civil and Industrial Engineering, 1, Prof. Dr. docent Dimitrie Mangeron Blvd., Iasi, 700050, Romania

E-mail: alexandra.cojocaru@student.tuiasi.ro

Abstract. Over time, a building's energy consumption in the use phase was not decisive in the design process. However, in recent years, energy efficiency and energy autonomy have become a defining criteria in buildings' performance. In 2020, residential buildings became the space for rest, recreation, personal activities, but also for professional activities. Thus, it is essential to ensure optimal temperature, air, humidity and psychological comfort due to architectural quality. The passive house concept sets the maximum allowable energy consumption to heat the building and limit the total primary energy consumption. Passive houses improve energy-efficient buildings by ensuring a comfortable indoor environment in all seasons without requiring a conventional heat distribution system. Based on local characteristics, creating and analysing model study cases is important and could solve cities' fast development. The shape of a building influences the solar input and the total energy consumption through the total surface exposed to the outside and subject to heat loss. In order to gain knowledge about the influence of the buildings' form on energy balance, this paper presents a case study of simple shape buildings with numerical analysis according to the passive house principles in the Moldavia area of Romania. The case study of a parallelepiped model with various dimensions, varying aspect ratio and storey, was made with the planning tool Passive House Planning Package. The conclusions suggest further case studies with various roof shapes and windows' areas.

1. Introduction

One of the basic needs of life, shelter or housing, is closely connected with the natural environment. From the perspective of Le Corbusier, a French architect, the home is "a cover that meets certain requirements and establishes fair relationships between the cosmic environment and human biological phenomena"[1]. The concept of housing is a complex system of functional structures arranged according to certain relationships (individual - family - social), which function and evolve.

People spend most of their time in a home, where safety, well-being, peace, comfort are the essential elements we are looking for. Along with the components of image, perception and philosophy itself, psychological and sociological, somatic and sensory factors are equally important, defining the state of the comfort of the human body: light intensity, air temperature and humidity, air quality itself (freshness, concentration of various substances and odours), movement of air currents, noise and vibration, factors that must it fall within specific value ranges in order to maintain our well-being and hygiene at high standards. The human body, as a biological organism, physically feels a state of comfort. The air temperature should be 20-25° Celsius in an indoor space, the relative humidity 45-55%, the carbon dioxide concentration below 500ppm and 30mc of fresh air per hour. The natural environment must not influence the constancy of these elements. One of the critical roles of any construction, besides the aesthetic and functional, is the guarantor of comfort and well-being at the physical level.

The first step in designing an energy-efficient house should be a rational passive design, focused on climate conditions, the building's form and orientation, offering good thermal and living comfort, even without active systems. Implementing passive solar design principles in an architectural project determines a significant reduction of energy consumption throughout the life cycle of a building without a significant impact on the initial investment or maintenance costs. Simultaneously, strategies specific to passive design can be easily integrated into a contemporary architectural solution. The building's form influences the energy balance of the building. The compact form is preferred in many cases, defined by a small shape factor (defined as the ratio between the envelope surface (A) and the building volume (V)). Also, another parameter important in the energy balance is the aspect ratio defined by the ratio between length and width of the building.

The residential field is well documented and studied from the point of view of energy efficiency. Energy consumption of residential buildings increases worldwide by about 30% annually [2]. Because of this, the residential buildings had priority in implementing energy efficiency concepts.

2. Passive House

Passive House is a concept of energy-efficient buildings, presented by Dr Wolfgang Feist from the Passive House Institute in Germany, which has had the most outstanding impact on architects, customers and researchers. The concept sets the maximum allowable energy consumption for heating the building and limits the total primary energy consumption. The passive house standard is met if a building has a heating energy consumption under 15kWh/m²yr and a total energy requirement of a maximum 120kWh/m²yr [3]. Passive houses improve low-energy houses by ensuring a comfortable indoor environment in all seasons without requiring a conventional heat distribution system [3]. According to J Schnieders, W Feist and L Rongen [4], passive houses use 80% - 90% less energy for heating than conventional buildings for the same interior space conditions for an additional cost of up to 10%. Unfortunately, most new constructions in Romania have heating energy consumption over 150kWh/m²yr, ten times higher than Passive Houses. According to the Passive House Database, there are 22 certified passive house buildings in Romania, built from 2008 till 2021. The number increases as energy-efficient buildings become a mandatory standard.

Passive House Planning Package, an Excel-based calculation program, was first launched in 1998 and is constantly evolving to verify the buildings' compliance with the Passive House standards. Calculation sheets for the energy balance (annual and monthly), for the distribution and supply of heating and the need for electricity and primary energy, are the main elements of this calculation tool. Also, calculations are analysed regarding the windows' characteristic values, the impact of shading, the heating load and the demand for cooling and dehumidification.

3. Review of studies for the influence of architecture form in energy efficiency balance

The shape of a building influences the solar input and the total energy consumption through the total surface exposed to the outside and subject to heat loss. In the design of a building, the ratio between its external surface and the total volume must be minimal, the ideal case being a hemisphere. Due to the problems of designing the interior space of a hemisphere, many researchers have conducted studies of the performance of parallelepiped buildings, followed by optimizations related to the planimetric shape (square, rectangle, hexagon, octagon, oval).

Studies on the shape of buildings and the outdoor climate have been made: research on the shape of the building based on passive strategies that provide the comfort of the interior [5], research on the use of the sun for heating [6] and research on the effect of urbanization on urban climate - the effect of the island of urban heat [7]. In the study conducted by M Premrov, M Žigart and V Žegarac Leskovar [8], 216 models of box-type buildings were analysed parametrically, with wooden frame structure, with variations on thermal transmittance, size of glazed areas, orientation, the ratio of facades, horizontal and vertical volumetric extension in several cities: Ljubljana, Helsinki, Munich, London, Athens and Madrid. The size of the glazed surfaces and their positioning linearly influence the energy balance, and the shape of the building has an impact on the building's energy efficiency.

The rapid development of cities requires a detailed study of buildings' shape, from micro to macro scale. Okeil A., in his article "A holistic approach to energy-efficient building forms" [9], highlights the lack of sets of design guidelines for achieving energy-efficient forms of construction throughout the year, forms that can create building blocks to increase the performance of the whole city. The analysis compares three types of residential neighbourhoods at a latitude of 48 degrees, as shown in figure 1: Linear, block (conventional) and RSB (Residential Solar Block - energy efficient). The residential solar block (RSB) is an energy-efficient form of construction, optimized to avoid winter shading, functional and aesthetic. Several types of housing can be organized in RSB: apartments, garden apartments (on the roof of other apartments), terraced houses. The analysis results highlight the advantages of the RSB form: functional, versatile, social, aesthetic, high sun exposure, increased airflow, the possibility of green roofs in direct connection with housing [9].

The main output of the current study and these previous studies is to offer architects general information on buildings' energy-efficient design parameters.

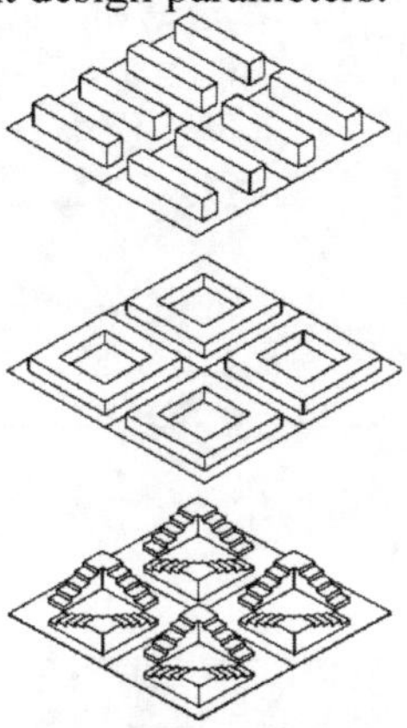

Figure 1. Three types of residential neighbourhoods: linear, block, RSB [9]

4. Numerical study

The presented numerical research is based on a case study of a parallelepiped model with various dimensions (square plan vs linear plan, one-story house vs two-story house) made of good thermal envelope elements (U=0,10W/ m²K), with windows on the south façade of the buildings but also on west and east. The study analyses the energy performance in the Moldavian Area of Romania, in the North-East of Romania, with characteristics of temperate climate, hill area, detailed in Table 1.

Table 1. Climate conditions in the Moldavian Area of Romania

	UM	Values
Average annual temperature	°C	-10 / +30
Maximum positive outdoor temperature	°C	+28
Maximum negative outdoor temperature	°C	-13
Length of the heating period	days/year	190
Total solar radiation on the south vertical surface in the heating period	kWh/ m²	380
Total solar radiation on the west/east vertical surface in the heating period	kWh/ m²	200

The study was made with the planning tool Passive House Planning Package to calculate energy flows and energy needs for heating and cooling. The planning tool PHPP 9 is based on EN ISO 13790, which shows high levels of precision and accuracy in calculating energy balances.

The first comparison is between two single-storey models with a constant occupied floor area of 64 m², a constant heated volume of 192 m³ and a varying aspect ratio (the ratio between length and width): a square plan (1A) and a linear plan (1B). Both models have a constant window surface of 50% of the south façade. The first model, with the square plan, has the heating energy consumption 22,7kWh/m²yr and the thermal load of 16,2W/ m², while the second model, with the south façade larger than the western and eastern one, has the heating energy consumption 22,9kWh/m²yr and the thermal load remain 16,2W/ m². The aspect ratio of the first model (1) is slightly preferred rather than the aspect ratio of the second one (1,52), as shown in Table 2.

Table 2. First comparison – model 1A and model 1B

Nr.	Shape	Storey	Area (m²)	Volume (m³)	Aspect ratio (L/W)	South Windows (%)	West Windows (%)	East Windows (%)	Heating Energy Consumption (kWh/m²yr)	Thermal Load (W/ m²)	Frequency of the overheating period (%)
1A	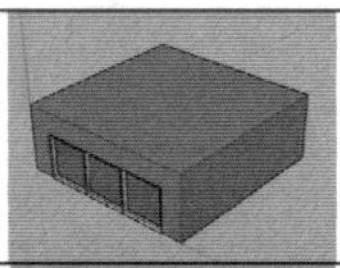	1	64	192	1	50	0	0	22,7	16,2	40,4
1B	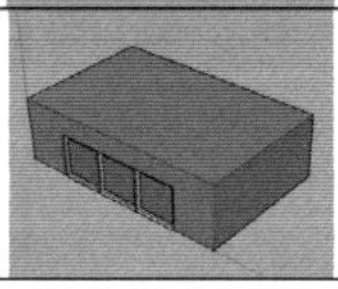	1	64	192	1,52	50	0	0	22,9	16,2	40,2

The second comparison is between two models: a single-storey model (1A) and a two-storey model (2A) with a constant occupied floor area of 64 m² and a constant heated volume of 192 m³. The previous calculations indicated that the first model, a single-storey with a square plan, has a heating energy consumption 22,7kWh/m²yr and a thermal load of 16,2W/ m². In comparison, the two-storey model, with the same occupied area floor and the same window surface, has the heating energy consumption 20,5kWh/m²yr and the thermal load of 14W/m², as shown in Table 3. The results show that the two-storey houses are preferred rather than single-storey ones. Besides, if choosing to design a single-storey house to gain solar radiation, the area of the south-oriented window should be larger than in two-storey house designs.

Table 3. Second comparison – model 1A and model 2B

Nr.	Shape	Storey	Area (m²)	Volume (m³)	Aspect ratio (L/W)	South Windows (%)	West Windows (%)	East Windows (%)	Heating Energy Consumption (kWh/m²yr)	Thermal Load (W/ m²)	Frequency of the overheating period (%)
1A	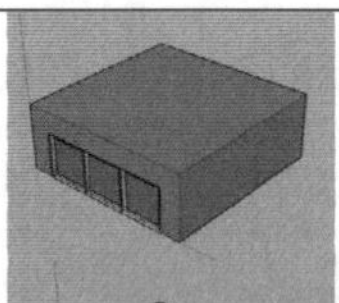	1	64	192	1	50	0	0	22,7	16,2	40,4
2A		2	64	192	1	50	0	0	20,5	14	40,7

The third comparison is between four single-storey models with a constant occupied floor area of 64 m², a constant heated volume of 192 m³ and a varying window's position: only on the south façade of the building (1A), on south and east façades (3A), on the south and west facades (3B) and on south, east and west facades (3C). The previous calculations indicated that the first model, a single-storey with square plan and windows only on the south façade, has a heating energy consumption 22,7kWh/m²yr and a thermal load of 16,2W/ m². The second model, with the same area and volume but with windows also on the west façade (16% of the west façade), has the heating energy consumption 24,4kWh/m²yr and the thermal load of 17,4W/ m². The frequency of the overheating period also increased by 3,6%. The third model, with the same area and volume, but with windows on the south façade (50% of the south façade) and on east façade (16% of the east façade), has the same heating energy consumption (24,4kWh/m²yr) and thermal load (17,4W/ m²). The fourth model, with the same area and volume, but with windows on the south façade (50% of the south façade), on east façade (16% of the east façade) and west façade (16% of the west façade), has the heating energy consumption 26,1kWh/m²yr and the thermal load of 18,7W/ m². The frequency of the overheating period also increased by 6,3%. The results of this third comparison conclude that the glazing areas are better placed on the south-oriented building envelope to obtain an increased share of solar gains useful for the building's passive heating. However, a good and evenly distributed natural illumination is also influenced by space's comfort factor, gained by positioning some windows on other building facades (west and east facade).

Table 4. Third comparison – model 1A and model 2B

Nr.	Shape	Storey	Area (m²)	Volume (m³)	Aspect ratio (L/W)	South Windows (%)	West Windows (%)	East Windows (%)	Heating Energy Consumption (kWh/m²yr)	Thermal Load (W/ m²)	Frequency of the overheating period (%)
1A		1	64	192	1	50	0	0	22,7	16,2	40,4
3A	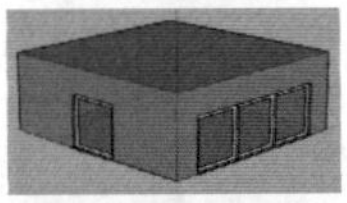	1	64	192	1	50	0	16	24,4	17,4	43,6
3B	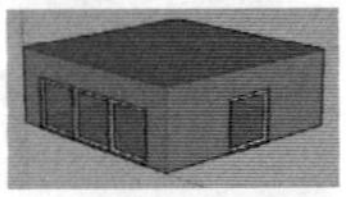	1	64	192	1	50	16	0	24,4	17,4	43,6
3C	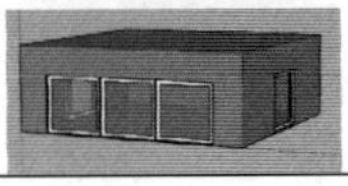	1	64	192	1	50	16	16	26,1	18,7	46,3

The essential heat source is solar energy, which heats the indoor air through incident solar radiation. The percentage of glazing area in the south façade was kept constant in the first two comparisons. The windows' characteristics were chosen to assure a high level of heat insulation and light transmission: Ug=0,50W/ m²K, Uf=0,73W/ m²K, g=0,52. If the overheating period's frequency is over 10%, the solution is the shading systems, such as vegetation, roof consoles, light exterior window shading systems, that reduce by 10% to 40% of the energy required for cooling (Maleki, 2011). The use of vegetation (trees, intensive vegetation, green facades, green roof) offers increased comfort to building users through shading, protection against wind, pollution and noise. Planting deciduous trees and shrubs, which shade the southern side of buildings, reduces the temperature by 5 degrees Celsius on the thermal envelope's surface (Kamal, 2012). Evergreen trees are planted on the west and east sides to protect the building from wind and sun at sunrise and sunset.

IOP Conf. Series: Materials Science and Engineering 1141 (2021) 012001 doi:10.1088/1757-899X/1141/1/012001

5. Conclusions

The general conclusion from the presented study and the reviewed studies is that the building shape influences the energy balance. The presented study results can lead to optimal design solutions for single and two-storey houses with a constant floor area and a varying building aspect ratio and storey. The general results' benefit is to set basic guidelines for architects for a quick estimation of buildings' energy performance, designed in different shapes.

The recommendations for the climate area of the Moldavian Area in Romania are to design compact forms with an aspect ratio near 1, with at least two storeys and small windows on west and east facades. Generalizing, those guidance lines could work on almost all zones of cold temperate climate areas.

There are still a series of parameters which is recommended to be further analysed. The south-oriented window ratio could vary as the house's interior space is designed, and it can influence the energy balance. Furthermore, in Romania's climate condition, a flat roof is not recommended. It would also be helpful to analyse a similar case with a slope roof and roof windows, as the zenithally light makes a significant contribution to well-being.

6. References

[1] Le Corbusier 1963 *Maniere de penser l'urbanisme* Gonthier

[2] Allouhi A, El Fouih Y, Kousksou T, Jamil A, Zeraouli Y, Mourad Y 2015 *Energy consumption and efficiency in buildings: current status and future trends* Journal of Cleaner Production 109 118-130

[3] Passive House Planning. Package 2007 – specifications for quality approved passive houses. Passive House Institute

[4] Schnieders J, Feist W, Rongen L 2015 *Passive Houses for different climate zones* Energy and Buildings 105 71-87

[5] Olgyay V 1963 *Design with Climate: Bioclimatic Approach to Architectural Regionalism* Princeton University Press

[6] Knowles RL 1981 *Sun Rhythm Form* MIT Press Cambridge MA

[7] Oke TR 1978 *Boundary Layer Climates* Methuen&CO Ltd

[8] Premrov M, Žigart M, Žegarac Leskovar V 2018 *Influence of the building shape on the energy performance of timber-glass buildings located in warm climatic regions* Energy 149 496–504

[9] Okeil A 2010 *A holistic approach to energy efficient building forms* Energy and Building 42 1437-1444

Characteristics of bituminous shales used in asphalt mixtures

D-Dr Hanganu[1)]
[1)] Gheorghe Asachi Technical University of Iasi, Faculty of Civil Engineering and Building Services, Bd. Dimitrie Mangeron, no 43, 700050, city of Iasi, Romania

E-mail: hanganu.dragos@gmail.com

Abstract. Bituminous shales, bitumolites or pyrosists are sedimentary rocks impregnated with an insolvent bitumen. In Romania, studies have been carried out on bituminous shales by almost all Romanian geologists, starting with Cobâlcescu in 1877, continuing with Mrazec in 1907, Athanasiu in 1926, Popescu - Voiteşti in 1943, etc. In 1956, Nicolae Grigoraş was the one who made the first studies in detail on bituminous shales. Considering that the only norm on the basis of which the technological recipes for asphalt mixtures with bituminous sand are made, it underwent the last modifications in 1985, one of the specific objectives of the research was to update the technical specifications contained in the regulations in order to streamline the technological process associated with these mixtures and for a better implementation at national level of these practices. Without this long-awaited review, the use of this unconventional material would become more difficult, although it has both technical, economic and environmental benefits.

1. Introduction
Several studies were carried out in Romania by scientists regarding bituminous shales [1, 4, 8]. Considering that the only regulation which details the technological recipes for asphalt mixtures with bituminous sand are made, underwent numerous modifications, the last one occurred in 1985 [7]. Without this long-awaited review, the use of this unconventional material would become more difficult, although it has both technical, economic and environmental benefits [2, 6].

The bituminous shales were previously ground, after which the crushing sand was used to prepare the types of asphalt mixture type AC 22.4 BIN 50/70 SI AC 16 SURF 50/70. After the preparation of the asphalt mixtures, the physical-mechanical characteristics were determined.

Through her study, Ionita G, experimental and theoretical activities were carried out, undertaken, in order to acquire new knowledge on the specified materials, respectively the crushing sand used to prepare asphalt mixtures and based on observations, tests and interpretation of laboratory results, which, based on some knowledge already existing, will lead to a significant improvement of the knowledge of applicability of these materials in a specific field, respectively that of construction materials [9].

This research is not carried out for the purpose of direct commercial use but for the purpose of enriching the consciousness in the scientific sphere in terms of a topical material, but poorly capitalized in road construction.

2. Design
In order to design an asphalt mixture recipe, respectively to establish the best proportions of the component materials so as to obtain its best characteristics, it is necessary to analyse the component materials considering that the properties of both mineral aggregates and binder influences differently the properties of the mixture. The components of the asphalt mixture will be analysed first, in order to take into account all the premises for the elaboration of the study [5].

2.1. Natural mineral aggregates
The mineral aggregates used to obtain the asphalt mixtures come from stone quarries or gravel pits. Their destination may differ depending on the destination of the respective asphalt mixture. In general,

the quarry aggregates are used on technical class roads I-III (high-traffic roads such as highways, express roads, national roads) and the gravel roads are used in the construction of lower technical class roads, IV-V (county roads, communal, local roads) [3].

In the present study, mineral aggregates from the Suseni quarry - Harghita county and natural sand from the Tupilați gravel pit - Neamț county will be used.

2.2. Bitumen

In order to obtain an asphalt concrete, this being a mixture of agglomerated mineral materials with a bituminous binder, plain petroleum bitumen for roads, non-paraffin, class 50/70 was used.

2.3. The filler

The granularity of the filler required to determine the weight in the dosing study is shown in table 1.

Table 1. The granularity of the limestone filler

Nr. crt.	Characteristic	Limestone filler
1	Granularity (passes% through sieve of ... mm)	Average test
	2	
	0,125	100
	0,063	87,5
	Average	75,4

2.4. The share of mineral aggregates in asphalt mixtures.

The aim is to obtain high-performance asphalt mixtures, in which the mineral skeleton consists of sorts of natural mineral aggregate quarry (Suseni quarry) and crushing sand from bituminous shale, useful for making asphalt layers of wear and bonding.

AC 16 SURF 50/70 and AC 22.4 BIN 50/70, of asphalt concrete type with chipping - hereinafter referred to as asphalt mixtures with bituminous crush sand for wear and connection layers of road systems.

The preparation of the asphalt mixtures was performed in the laboratory, by manually mixing all the components, Figure 1, at the temperatures specified in table 2 until homogeneous mixtures were obtained, completely coated, which were then processed.

Table 2. Temperatures of components and asphalt mixture in preparation

Nr. crt.	Bitumen type	Bitumen temperature, °C	Mineral aggregate temperature, °C	Asphalt concrete temperature at preparation, °C
1	D 50/70	150 - 170	140 - 190	140 - 180

Individually constituted batches were made for each proposed dosage, a load that initially had a set weight of 9500 g, the weight required to obtain 8 Marshall cylinders from each type of asphalt mixture, cylinders that were subsequently tested.

Figure 1. Mixing aggregates and manual bitumen in the laboratory

2.5. Bitumen dosage

Asphalt mixture samples, Figure 2, consisting of asphalt mixture are:

I. AC 16 SURF 50/70 with bituminous shale crushing sand for wear layers, series: 300-307; 310-317; 320-327, at which the bitumen dosage differs: 5.7%, 5.9% respectively 6.1%.

II. AC 22.4 BIN 50/70 with bituminous shale crushing sand for connecting layers, series: 600-607; 610-617; 620-627, at which the bitumen dosage differs: 4.3%, 4.5% and 4.7% respectively

Marshall cylinders were made from the asphalt mixture thus prepared, applying 50 ram blows on each face. The specimens were stored in laboratory conditions, at room temperature, until completely cooled, Figure 3.

Figure 2. Marshall cylinders made of AC 16 SURF 50/70 asphalt mixture with bituminous shale crushing sand from Village Nechit, Neamt county

Figure 3. Marshall cylinders made of AC 22.4 BIN 50/70 asphalt mixture with crushed sand from bituminous shales from Nechit Village, Neamt County

3. Tests performed:

3.1. Bulk density

The ratio between the mass of the unit volume of the compacted asphalt mixture specimen, including the gaps filled with air, and the volume of the specimen, expressed in g / cm^3, determined on specimens kept for one hour in water at 20 °C.

Tables 3 and 4 show the values obtained of the apparent densities on the series of cylinders from the asphalt mixture with crushing sand from bituminous shale. The variation of the apparent densities of the three dosage variants is shown in the diagrams in figures 4 and 5.

Table 3. Results for bulk density on AC 16 SURF 50/70 bituminous sand crushing mixture

Asphalt mixture type	Bitumen dosage, %	Cylinder series	Apparent density, g / cm^3	AC 16 SURF 50/70 Standard
AC 16 SURF 50/70 with bituminous shale crushing sand	5,7	300-307	2,351	2,376
AC 16 SURF 50/70 with bituminous shale crushing sand	5,9	310-317	2,370	2,376
AC 16 SURF 50/70 with bituminous shale crushing sand	6,1	320-327	2,376	2,376

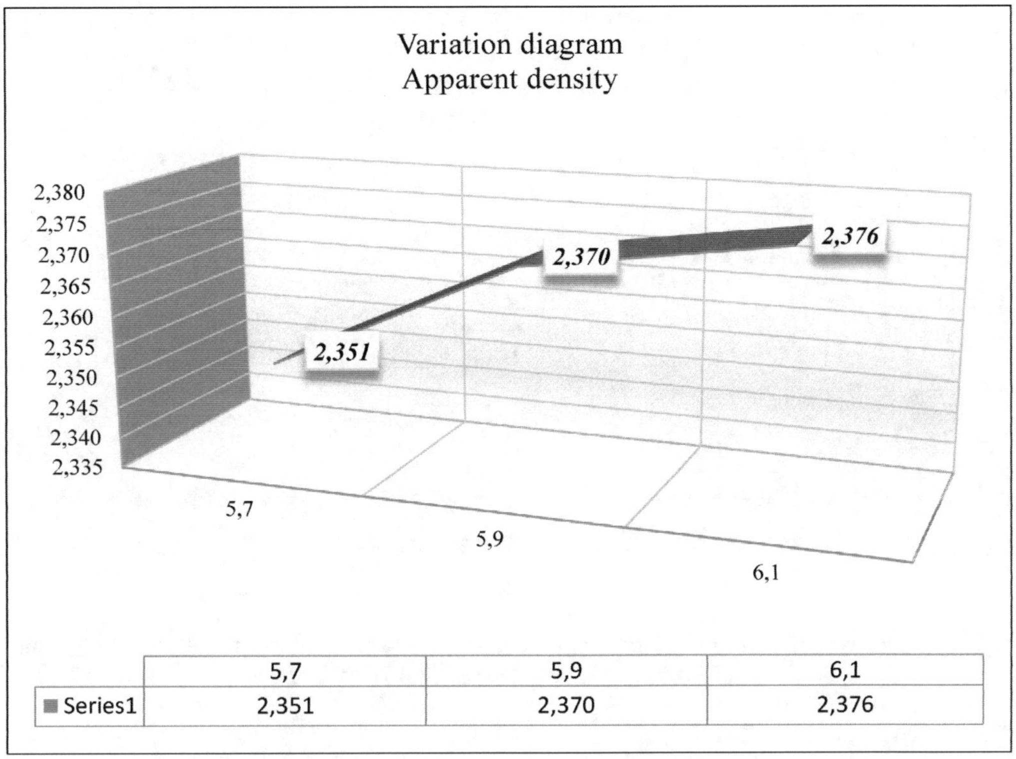

Figure 4. Diagram of variation of apparent densities for asphalt mixture with crushing sand from bituminous shale type AC 16 SURF 50/70

Table 4. Results obtained for the apparent density on the bituminous shale sand crushing mixture type AC 22.4 BIN 50/70

Asphalt mixture type	Bitumen dosage,%	Cylinder series	Apparent density, g / cm3	AC 22.4 BIN 50/70 Standard
AC 22.4 BIN 50/70 with bituminous shale crushing sand	4,3	600-607	2,339	2,368
AC 22.4 BIN 50/70 with bituminous shale crushing sand	4,5	610-617	2,365	2,368
AC 22.4 BIN 50/70 with bituminous shale crushing sand	4,7	620-627	2,399	2,368

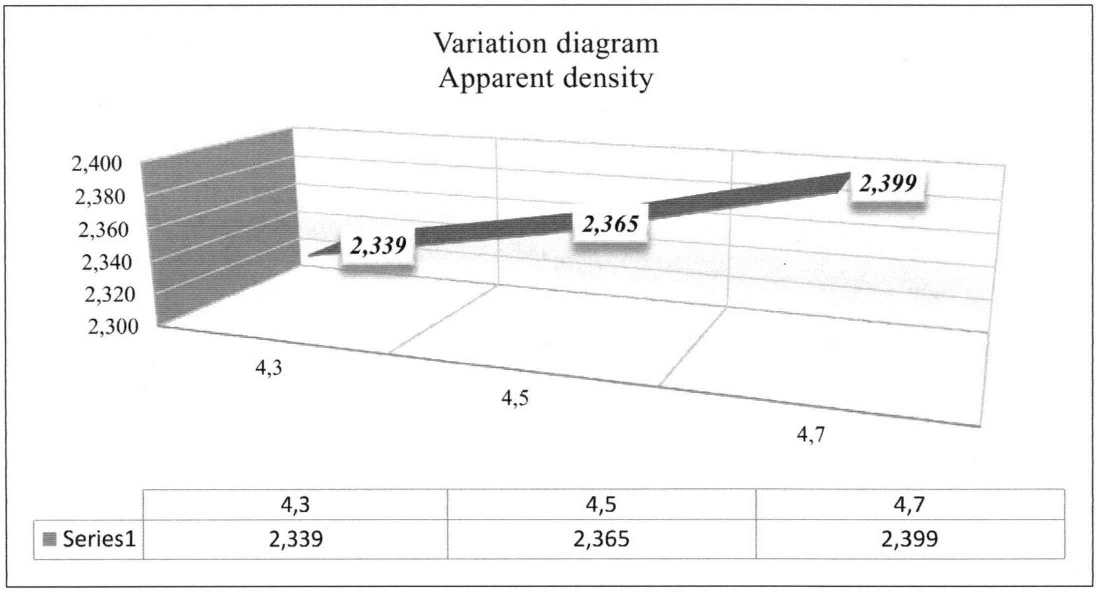

Figure 5. Diagram of variation of apparent densities for asphalt mixture with crushing sand from bituminous shale type AC 22.4 BIN 50/70

3.2. Water absorption

The amount of water absorbed by the specimens is measured after the sample is immersed in water for 3 hours at a vacuum of 15 ... 20 mmHg, followed by a time of 2 hours maintained in water at atmospheric pressure, expressed in % vol.

Tables 5 and 6 give the obtained values of the water absorption on the series of cylinders from the asphalt mixture with crushing sand from bituminous shale. The variation of the apparent densities of the three dosage variants is shown in the diagrams from figures 6 and 7.

Table 5. Results obtained for water absorption on asphalt mixture with crushed sand from bituminous shale AC 16 SURF 50/70

Asphalt mixture type	Bitumen dosage, %	Cylinder series	Water absorption, g / cm3	AC 16 SURF 50/70 Standard
AC 16 SURF 50/70 with bituminous shale crushing sand	5,7	300-307	1,065	1,80
AC 16 SURF 50/70 with bituminous shale crushing sand	5,9	310-317	0,686	1,80
AC 16 SURF 50/70 with bituminous shale crushing sand	6,1	320-327	0,202	1,80

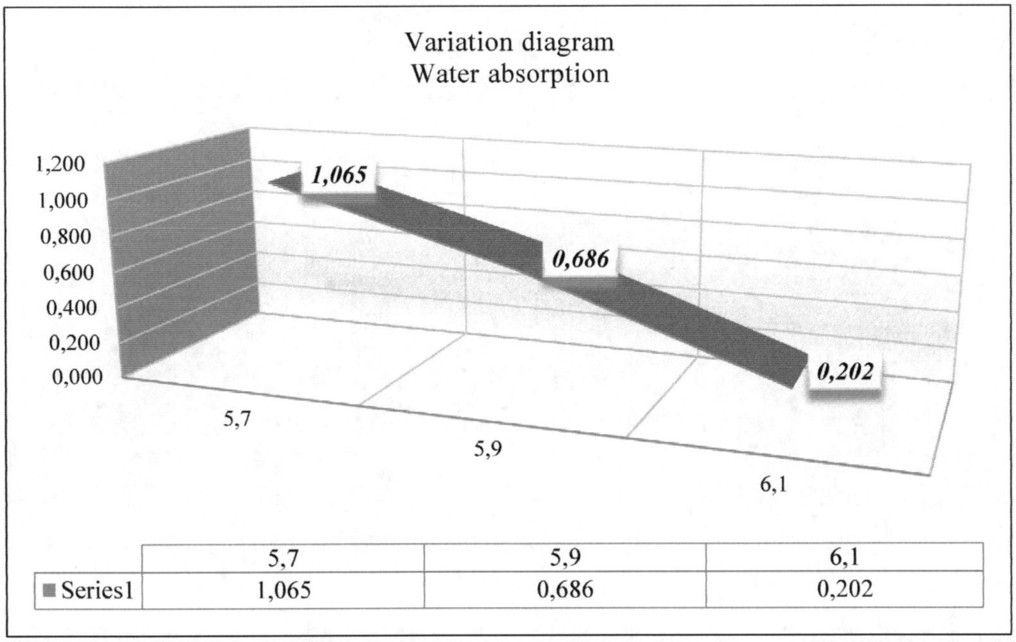

Figure 6. Diagram of variation of water absorption for asphalt mixture with crushing sand from bituminous shale BA16

Table 6. Results obtained for water absorption on asphalt mixture with crushing sand from bituminous shale AC 22.4 BIN 50/70

Asphalt mixture type	Bitumen dosage,%	Cylinder series	Water absorption, g / cm3	AC 22.4 BIN 50/70 Standard
AC 22.4 BIN 50/70 with bituminous shale crushing sand	4,3	600-607	2,900	2,90
AC 22.4 BIN 50/70 with bituminous shale crushing sand	4,5	610-617	2,154	2,90
AC 22.4 BIN 50/70 with bituminous shale crushing sand	4,7	620-627	1,915	2,90

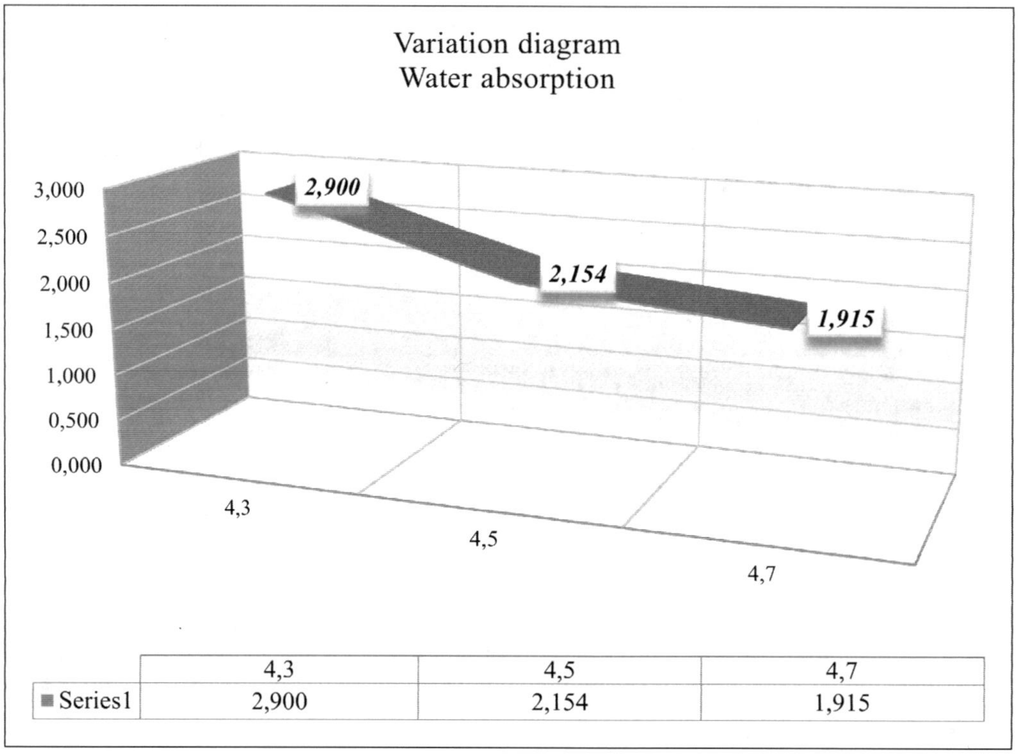

Figure 7. Variation diagram of water absorption for asphalt mixture with crushing sand from bituminous shale type AC 22.4 BIN 50/70

3.3. Marshall stability

Is assessed by the force reached when the test specimen breaks, expressed in KN.

Tables 7 and 8 give the obtained values of the Marshall stability on the series of cylinders from the asphalt mixture with crushing sand from bituminous shale. The variation of the Marshall stability of the three dosage samples is shown in the diagrams presented in figures 8 and 9

Table 7. Results for Marshall stability on AC 16 SURF 50/70 asphalt mixture with bituminous shale crushing sand

Asphalt mixture type	Bitumen dosage, %	Cylinder series	Stability Marshall, KN	AC 16 SURF 50/70 Standard
AC 16 SURF 50/70 with bituminous shale crushing sand	5,7	300-307	8,8	10,30
AC 16 SURF 50/70 with bituminous shale crushing sand	5,9	310-317	8,0	10,30
AC 16 SURF 50/70 with bituminous shale crushing sand	6,1	320-327	7,4	10,30

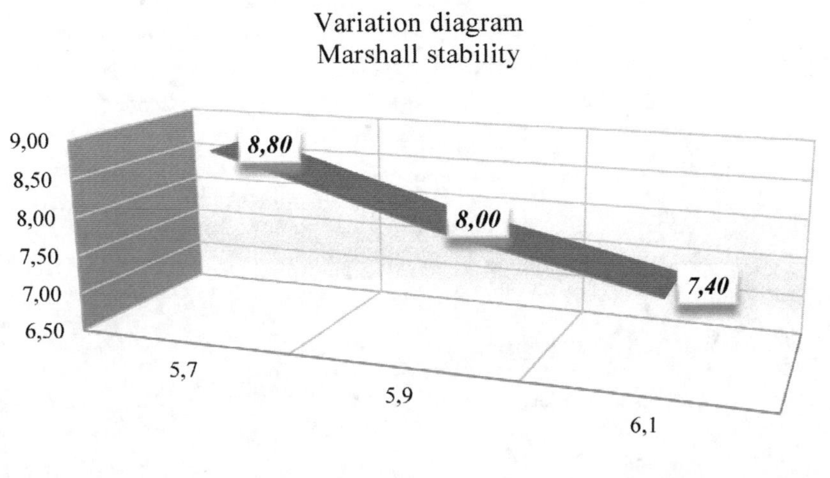

	5,7	5,9	6,1
Series1	8,80	8,00	7,40

Figure 8 .Marshall stability variation diagram on asphalt mixture AC 16 SURF 50/70 with bituminous shale crushing sand.

Table 8. Results obtained for Marshall stability on AC 22.4 BIN 50/70 asphalt mixture with bituminous shale crushing sand

Asphalt mixture type	Bitumen dosage,%	Cylinder series	Stability Marshall, KN	AC 22.4 BIN 50/70 Standard
AC 22.4 BIN 50/70 with bituminous shale crushing sand	4,3	600-607	8,0	8,30
AC 22.4 BIN 50/70 with bituminous shale crushing sand	4,5	610-617	8,4	8,30
AC 22.4 BIN 50/70 with bituminous shale crushing sand	4,7	620-627	9,2	8,30

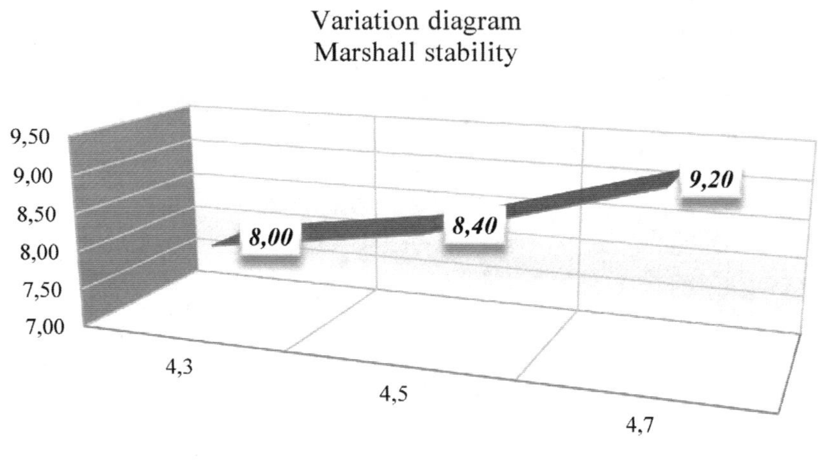

	4,3	4,5	4,7
Series1	8,00	8,40	9,20

Figure 9. Marshall stability variation diagram for asphalt mixture AC 22.4 BIN 50/70 with bituminous shale crushing sand

3.4. Marshall flow

The Marshall flow refers to the deformation reached when the test specimen breaks and is expressed in mm.

Tables 9 and 10 give the obtained values of Marshall stability on the series of cylinders from the asphalt mixture with crushing sand from bituminous shale. The variation of the Marshall stability of the three dosage variants is shown in the diagrams from figures 10 and 11

Table 9. Results for Marshall flow on asphalt sand mix with bituminous shale type AC 16 SURF 50/70

Asphalt mixture type	Bitumen dosage, %	Cylinder series	Flowing, mm	AC 16 SURF 50/70 Standard
AC 16 SURF 50/70 with bituminous shale crushing sand	5,7	300-307	3,77	3,70
AC 16 SURF 50/70 with bituminous shale crushing sand	5,9	310-317	4,46	3,70
AC 16 SURF 50/70 with bituminous shale crushing sand	6,1	320-327	4,68	3,70

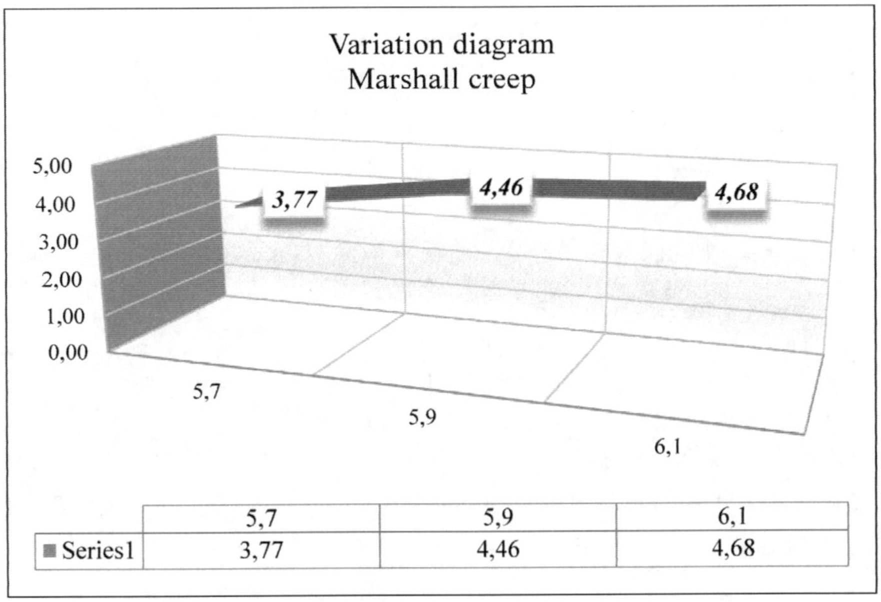

Figure 10. Marshall flow variation diagram for asphalt mixture with crushing sand from bituminous shale type AC 16 SURF 50/70

Table 10. Results obtained for Marshall flow AC 22.4 BIN 50/70 asphalt mixture with bituminous shale crushing sand

Asphalt mixture type	Bitumen dosage,%	Cylinder series	Flowing, mm	AC 22.4 BIN 50/70 Standard
AC 22.4 BIN 50/70 with bituminous shale crushing sand	4,3	600-607	4,15	3,30
AC 22.4 BIN 50/70 with bituminous shale crushing sand	4,5	610-617	4,27	3,30
AC 22.4 BIN 50/70 with bituminous shale crushing sand	4,7	620-627	4,41	3,30

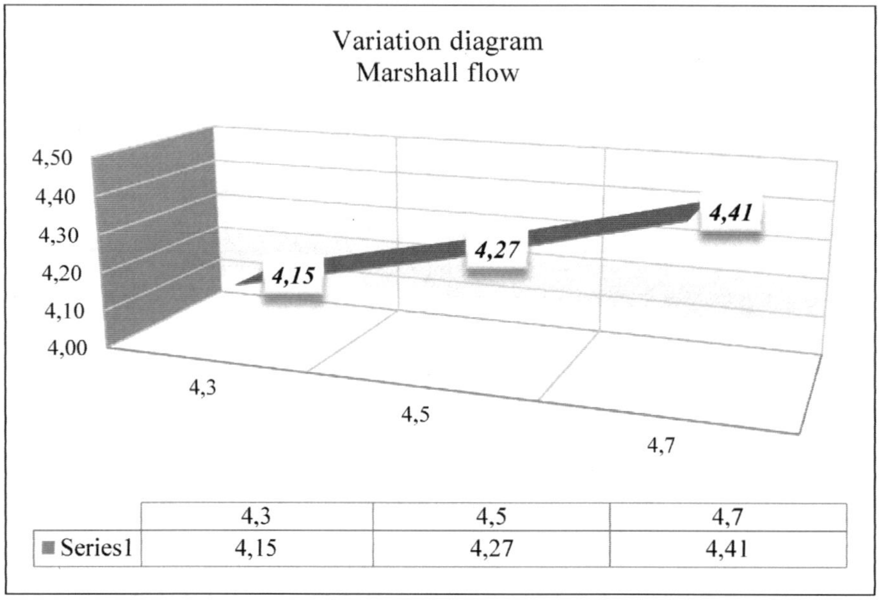

Figure 11. Marshall flow variation diagram for AC 22.4 BIN 50/70 mixture with bituminous shale crushing sand

3.5. Swelling in water

The swelling in water phenomenon establishes the harmful influence of constituent components of mineral aggregates that are sensitive to swelling - clay, by keeping them in water for 28 days [6]. Swelling was also determined at regular intermediate intervals of 7, 14, 21 and including 28 days. The results obtained at regular intermediate intervals of 7, 14, 21 and even at 28 days showed that the swelling in the water is 0.

4. Conclusions

This report shows solutions that could be put into practice regarding the preparation of asphalt mixtures using bituminous shale crushing sand from Romania. These solutions can be adopted in order to achieve significant savings in bitumen.

For all technological recipes of asphalt mixtures prepared in the laboratory of roads with bituminous shale crushing sand, the values of physical-mechanical characteristics were determined by static tests: bulk density, water absorption, stability at 60 ° C, flow index, swelling at 28 days.

The results obtained from the studies performed in the road laboratory are favourable for asphalt mixtures prepared with bituminous shale crushing sand from Nechit Village, Neamt County.

Based on the results obtained from the tests made on the asphalt mixtures with crushing sand from bituminous shales from Village Nechit, it is recommended to use them in the composition of asphalt mixtures, in order to significantly reduce the costs related to the production of asphalt mixtures. This solution is very advantageous especially for areas near natural deposits.

Acknowledgments

Regards to professor Gugiuman Gheorghe and professor Crina Miclăuş, for offering technical documentation, support and making this study possible.

5. References

[1] Bancila I 1958 Geology of the Eastern Carpathians. Scientific Publishing House, Bucharest
[2] Bolis B, Di Renzo A 1959 Road pavements. Publisher Ulrico Hoelpi Milan

[3] Cososchi B, Gugiuman Gh, Vlad V N 1981, Road Laboratory Guide, „Gheorghe Asachi" Polytechnic Institute from Iași, Faculty of Constructions – course guide
[4] Grasu C, Catana C, Grinea D, Ionesi L 1975 Geological and geochemical considerations on Oligocene bituminous rocks between Pârâul Calu and Tazlau. Work. State. "Oak", geol. - grogr., VI, Pângarați
[5] Matasaru Tr, Craus I, Dorobanțu St 1966, Drumuri. Ed Tehnica București
[6] Vrtis C M 2013 Creating a performance – based asphalt mix design to incorporate Uinta basin oil sands. Master thesis, Departament of Civil and Environmental Engineering, The University of Utah
[7] Nițulescu I, Spiroiu P, David M, Butucescu D 1977 *Contribuții la cunoașterea compoziției și structurii șisturilor bituminoase de la Anina și a transformarilor suferite în procesele termice în vederea valorificarii lor complexe* (in Romanian), D.S. Inst. de Geol. și Geofiz., București
[8] Ionița G 2017 The physical – mechanical properties of asphalt mixtures containing bituminous sand, *The XXV Intern Scientific Conf "trans&MOTAUTO'17"*, year XI, iss **5**/2017, ISSN 1313-0226, Technical Union of Mechanical Engineering „INDUSTRY 4.0", Bulgaria, pp 246 – 249
[9] Ionița G 2017 Bituminous sand road asphalt pavement, *Bulletin of the Polytechnic Institute of Jassy, Construction, Architecture Section*, vol. **63** (**67**), Nr 3, ISSN 2068-4762, pp 55 – 63

IOP Conf. Series: Materials Science and Engineering 1141 (2021) 012003 doi:10.1088/1757-899X/1141/1/012003

High-early strength concretes modified with polycarboxylate admixture on different cement types

V Kryzhanovskyi[1], S Kroviakov[2] and M Zavoloka[3]
[1,2,3] Odessa State Academy of Civil Engineering and Architecture, 4 Didrihsona str., *Ukraine*

E-mail: skrovyakov@ukr.net

Abstract. Early and grade strength of modified concrete on different cement types (CEM II/A-S 42,5 and pozzolanic cement CEM IV/A(P) 42,5 R-SR) for concrete pavements repair was explored. Application range of the polycarboxylate modifier was 0.8-1.4% by weight of cement. All concrete compositions had equal workability (slump 5…9 cm). The aim of the investigation was to determine the optimal volume of the polycarboxylate modifier to achieve high strength indicators for repair material at the early and grade age.
High-strength concrete of grades C 50/60 and C 40/50 were obtained, with early strength at the 3 days age above 35 MPa, which allows them to be used as a repair concrete material for rigid road pavements.
X-ray phase analysis confirmed the positive effect of the polycarboxylate admixture in the structure formation of concrete. With an increase in the dosage of the BASF MasterGlenium SKY 608 admixture to 1.2...1.4%, the intensity of the diffractogram lines, which corresponds to the metastable hydration phase in the cement-sand matrix, decreased for concretes on both cement types.

1. Introduction

Growth of the transport and air flows traffic demands an enhancement in the road pavement features. The global trend in the transport industry provides for the development of cement concrete pavements technology for highways and airfield pavements. The annual increase in passenger and freight traffic leads to the destruction of existing hard road surfaces. It will be necessary to rehabilitate and repair these pavements with fast opening of traffic [1-3]. Full-depth repair technology with high-early strength concrete compositions are increasingly used as the major repair strategy [4-8].

Our research is aimed at choosing the optimal binder type and rational amount of polycarboxylate type admixture for high-early strength repair concrete mixtures.

2. Analysis of recent researches and problem statement

Flexible road pavements predominate in Ukraine [9]. However, it is necessary to note the vector of road network development in favour of concrete pavement technology [10]. The lifespan of rigid pavement is 3-6 times longer than flexible asphalt concrete pavement, and the crude oil reserves for bitumen production are limited compared to the raw materials for cement production [2].

According to the 21st World Road Congress [2] and data [1,3,9], the length of highways with rigid pavement is as follows: USA – 60%, Austria – 46%, Belgium – 41%, Germany – 31%, France – 20%, Netherlands – 15%, Portugal – 10%, Spain, Italy, Canada, Switzerland, Great Britain <5 %, Ukraine – 1,92%. Most of them were built over 40 years ago and need major repairs or reconstruction.

Repair concrete mixtures must have high early compressive and flexural strength (1-3 days), which is regulated depending on the place of work [1,3,4,6]. The technology of high-early concrete is actively used, that is specified by national and other specifications [4-6]. According to [11], strength of repair concrete at 28 days of age must be no less than design strength of existing concrete pavement, for a road of the 1st category - this is concrete grade C30/35. Based on the world experience of concrete pavements, the strength characteristics of the repair concrete must be changed taking into account the necessary conditions for opening the road section for traffic and the region where the repair is being carried out.

Early-strength concrete is based on the following principles [12-15]: reduction W/C due to the use of admixtures-superplasticizers affecting on the hydration processes and structure formation; use of rapid-hardening binders and an optimal selection of grain size using high-quality aggregates is also required. Today, the most effective are polycarboxylate-type superplasticizers, which have a steric mechanism of action. Having a long chain of molecules, cement particles are dispersed at a greater distance from each other, therefore, more surfaces of contact with water are formed [16,17].

In Ukraine, rigid pavement repair technology is underdeveloped and there is practically no standards base. Our research will make it possible to formulate technical requirements and recommendations for the use of high-early strength concrete compositions as the main repair material for rigid road pavements.

3. Research objective

Selection the optimal volume of polycarboxylate type modifier for concrete mixture repairing on different cement types. Consideration of the admixture effect on the early and grade compressive strength of repair concrete for rigid road pavements.

4. Materials and research methods

Two cement types were used in the research: CEM II/A-S 42,5 and pozzolanic cement CEM IV/A(P) 42,5 R-SR. Polycarboxylate type modifier was used in volume of 0,8-1,4% by weight of cement. Granite crushed stone with nominal maximum size 20 mm and quartz sand were used as aggregates. Sieve analysis of aggregates is presented in figure 1. Conventional concrete compositions were also investigated.

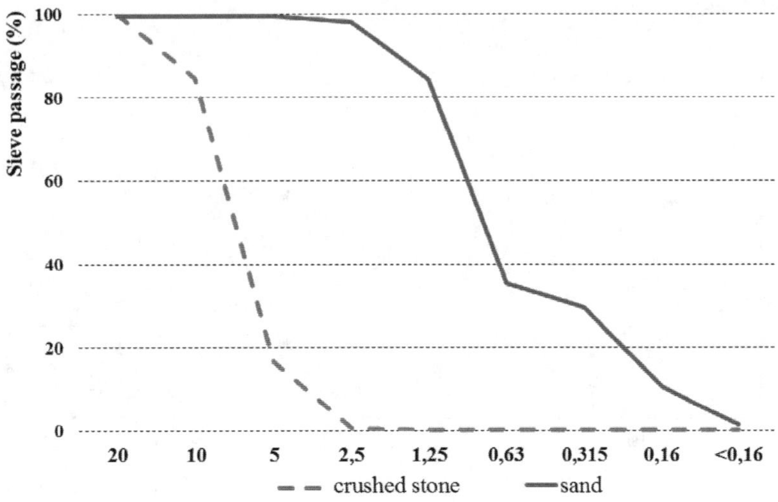

Figure 1. Coarse and fine aggregates gradation

5. Research results

10 batches of concrete samples were fabricated (in each batch 6 cubes 10x10x10 cm). Slump of concrete mixtures was (5…9 cm), that corresponds to the most common flowability of concrete mixtures for the rigid pavements repair. The compositions of the investigated concrete mixtures and W/C are shown in table 1.

Table 1. Concrete mixtures and W/C ratio

№	Cement type	Cement, kg/m^3	Breakstone, kg/m^3	Sand, kg/m^3	Admixture MasterGlenium SKY 608, %	Water, l/m^3	W/C
1	CEM IV/A(P) 42,5 R-SR	400	1140	705	-	216,8	0,542
2				735	0,8	188,4	0,471
3				740	1,0	181,6	0,454
4				755	1,2	148,4	0,371
5				750	1,4	160,8	0,402
6	CEM II/A-S 42,5	400	1140	720	-	200,8	0,502
7				765	0,8	148,4	0,371
8				785	1,0	125,2	0,313
9				800	1,2	116,8	0,292
10				795	1,4	119,2	0,298

The use of the plasticizer significantly reduced the W/C of concrete mixtures in comparison with unmodified mixtures № 1 and 6 and retained their flowability. Effect of the volume admixture and the cement type on mixtures W/C ratio are shown in figure 2.

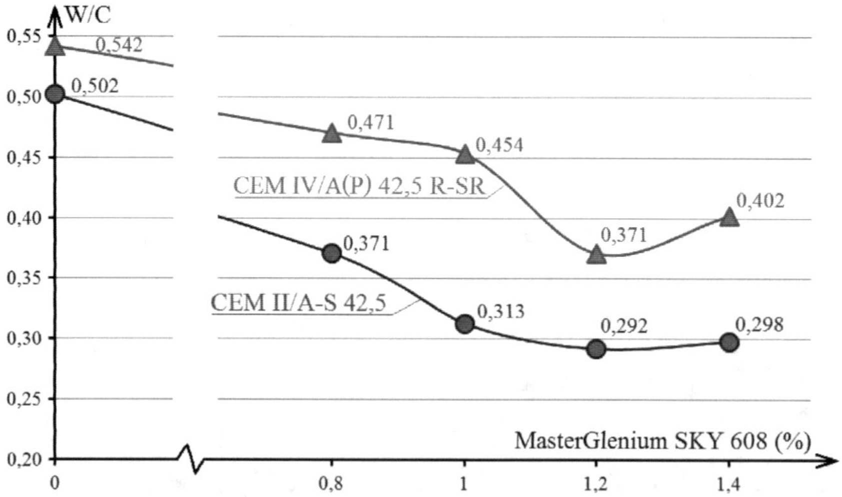

Figure 2. Effect of the volume of polycarboxylate type admixture on W/C ratio of equal flowability mixtures

The greatest decrease in the W/C ratio is observed with the 1,2% volume of the admixture by weight of cement. In this case, the W/C of concrete mixtures based on pozzolanic Portland cement CEM IV/A(P) 42,5 R-SR was significantly higher than the W/C of similar flowability mixtures based on Portland cement CEM II/A-S 42,5.

The average density of the investigated concretes ranged from 2430 to 2510 kg/m^3. Strength characteristics of the obtained concrete are shown in figure 3. Compression strength tests were carried out for each batch at the age 3 and 28 days.

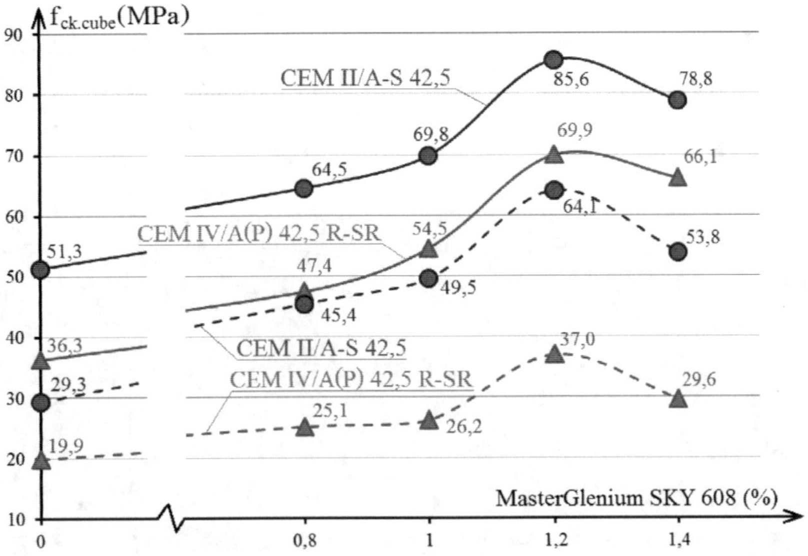

Figure 3. Compressive strength of the investigated concretes at the age 3, 28 days

As can be seen from figure 3 early and grade strength of conventional concrete based on CEM II/A-S 42,5 are higher than of similar concrete based on pozzolanic cement CEM IV/A(P) 42,5 R-SR. For mixtures № 2-5 and №7-10 there is also a tendency towards greater concrete strength based on the Portland cement CEM II/A-S 42,5. This is due to the increased water demand of pozzolanic cement and, consequently, concrete mixtures based on it.

Concretes on different cement types with the admixture in amount of 0.8% by weight of cement increased the compressive strength by almost 2 grades. With the addition of 1% modifier, the strength increased by another 1 grade, and with 1,2% - by 3 grades for concrete based on Portland cement CEM IV/A(P) 42,5 R-SR and by 2 grades with Portland cement CEM II/A-S 42,5. With a further increase in the admixture to 1.4%, an increase in the W/C of concrete mixtures is observed, respectively, strength of the investigated concrete decreases.

Thus, the rational amount of admixture for concrete on both cement types is 1.2% of the binder mass, which provides maximum compressive strength by minimizing W/C.

High-strength concrete was obtained based on Portland cement CEM II/A-S 42,5 with compressive strength 85,6 MPa (grade no less than C 50/60). Concrete based on Portland cement CEM IV/A(P) 42,5 R-SR had compressive strength 64,1 MPa (grade no less than C 40/50).

Strength of modified concrete at 3 days age on Portland cement CEM II/AS 42.5 - 68...75%, and concrete based on pozzolanic cement CEM IV/A (P) 42.5 R-SR was 45...53% depending on the amount of admixture. Compressive strength required for most concrete for rigid pavements is achieved at the 3 days age, which is especially important for repair materials.

According to the data of X-ray phase analysis, it was found that on the diffraction patterns (Fig. 4. a, b) samples № 1-4 of the cement matrix of concrete based on Portland cement CEM IV/A (P) 42.5 R-SR and CEM II/AS 42.5, the lines of natural quartz are recorded (d/n = 0.424; 0.334; 0.245; 0.228 nm) and minor reflections of portlandite (d/n = 0.493; 0.263 nm).

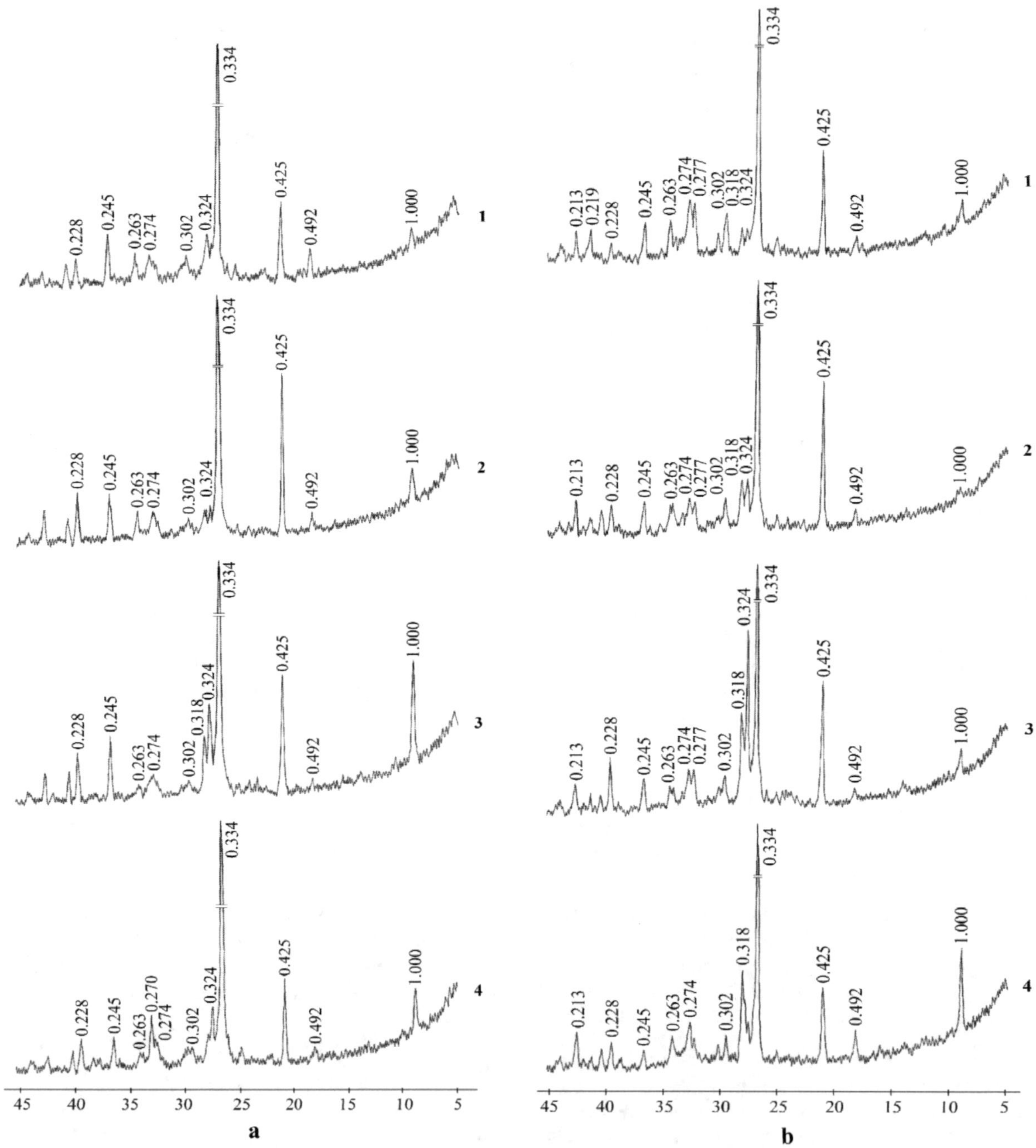

Figure 4. Diffraction patterns of the concrete cement matrix: a – based on Portland Cement CEM IV/A (P) 42.5 R-SR; b – based on Portland Cement CEM II/AS 42.5

a - 1 - CEM II/AS 42.5 (1,4% polycarboxylate admixture); 2 - CEM II/AS 42.5 (1,2% polycarboxylate admixture); 3 - CEM II/AS 42.5 (1,0% polycarboxylate admixture); 4 - CEM II/AS 42.5 (0,8% polycarboxylate admixture)

b - 1 - CEM IV/A (P) 42.5 R-SR (1,4% polycarboxylate admixture); 2 - CEM IV/A (P) 42.5 R-SR (1,2% polycarboxylate admixture); 3 - CEM IV/A (P) 42.5 R-SR (1,0% polycarboxylate admixture); 4 - CEM IV/A (P) 42.5 R-SR (0,8% polycarboxylate admixture)

The d/n = 1,000 nm line also appears, which corresponds to the metastable hydration phase in the cement-sand matrix - sodium containing calcium hydrosulfoaluminate $4CaO \cdot 0.9Al_2O_3 \cdot 1.1SO_3 \cdot 0.5Na_2O \cdot 16H_2O$. (according to Larionov Z. M.). It is important to note that the line intensity d/n = 1,000 nm at high dosages of the MasterGlenium SKY 608 polycarboxylate admixture (1.2% and 1.4% of the cement mass) both for concretes based on CEM II/AS 42.5 and for concretes based on CEM IV/A (P) 42.5 R-SR is less than the intensity of this line at a lower dosage of the admixture (0.8% and 1%). This testifies to the positive effect of the polycarboxylate admixture in the structure formation of concrete.

6. Conclusions and further researches

Using the optimal amount of BASF MasterGlenium SKY 608 admixture based on polycarboxylate ethers makes it possible to obtain high-strength and high-early-strength concrete. The use of such concrete for the rigid highway and airfield pavements repair contributes to the early start traffic. The rational amount of MasterGlenium SKY 608 admixture is 1.2% of the binder mass for concrete on both cement types - CEM II/AS 42.5 and CEM IV/A (P) 42.5 R-SR. The use of the obtained concrete as repair material will extend the service life of the pavement by 5-10 years.

Further research will be aimed at determining the flexural strength of such concrete and fiber concrete, as the main characteristic for road pavements. Also, the indicators of wear resistance and frost resistance of these concrete will be determined, as the main ones in assessing the durability of rigid road pavements.

7. References

[1] Till R D, VanPortfliet R 1992 European *Concrete Pavement Tour* (Michigan Department of Transportation, Research Report № R-1462) p 20

[2] Afinogenov O P 2004 *Proektirovanie zhestkih dorozhnyih odezhd* (Kemerovo: Kuzbassvuzizdat) p 227

[3] Hall K, Dawood D, Vanikar S, Tally R 2007 *Long-life Concrete Pavements in Europe and Canada* (Wisconsin Department of Trapsportation, Report № WHRP 0092-15- 08) p 81

[4] Cramer S M, Pham L T, Snyder M B 2007 *Better Concrete Mixes for Rapid Repair in Wisconsin* (Federal Highway Administration, Report № FHWA-PL-07-027) p 80

[5] Official Site of the Federal Highway Administration, *Full-Depth Repairs* - https://www.fhwa.dot.gov/pavement/concrete/full.cfm

[6] Smith K, Harrington D, Pierce L, Ram P, Smith K 2014 *Concrete Pavement Preservation Guide, Second Edition* (Federal Highway Administration, Report № FHWA-HIF-14-014) p 304

[7] Bassim R, Issa M F 2020 *Maturity-Based Estimates of Concrete Strength for Portland Concrete Cement Pavements and Patches at Early Age of Opening to Traffic* (Aci Materials Journal) pp 197-211

[8] Porras Y A 2018 Durable High-Early Strength Concrete. B. S., Kansas State University, p142.

[9] Mishutin A V, Smolyanets V V, Krovyakov S A 2013 *Perspektiva ispolzovaniya zhestkih dorozhnyih pokryitiy dlya gorodskih ulits i magistrali «Sever-Yug» v Odesse (*Vestnik Odesskoy gosudarstvennoy akademii stroitelstva i arhitekturyi) pp 176-182

[10] Official Site of the Highway Department of Ukraine - https://ukravtodor.gov.ua/

[11] GBN B.2.3-37641918-557: 2016. *Highways. Road rigid pavement. Designing.*

[12] Dvorkin L Y, Babich E M, Zhitkovskiy V V, Bordyuzhenko O M 2017 *Visokomitsni shvidkotverdnuchi betoni ta fibrobetoni: monografiya (*Rivne: NUVGP) 331 p

[13] Sanytskyi M A, Marushchak U D, Mazurak T A 2016 *Nanomodyfikovani portlandtsementni kompozytsii z vysokoiu mitsnistiu u rannomu vitsi (*Budivelni materialy i vyroby. Vypusk 1) pp 34-37

[14] Halim J G, Kusuma O C, Hardjito D 2017 Optimizing polycarboxylate based superplasticizer dosage with different cement type (Sustainable Civil Engineering Structures and Construction

Materials, SCESCM 2016, *Procedia Engineering* 171) pp 752-759 https://doi.org/10.1016/j.proeng.2017.01.442
[15] Kosmatka S H, Kerkhoff B, Panarese W C 2005 *Design and Control of Concrete Mixtures, Fourteenth Edition* (Portland Cement Association, Skokie, Illinois, USA) p 358
[16] Puertas F, Santos H, Palacios M, Martinez-Ramirez S 2005 Polycarboxylate superplasticizer admixtures: effect on hydration, microstructure and rheological behavior in cement pastes. *Advances in Cement Research*, **17** (2) pp 77-89 doi: 10.1680 / adcr.17.2.77.65044
[17] Janowska-Renkas E 2015 The influence of the chemical structure of polycarboxylic superplasticizers on their effectiveness in cement pastes, *Procedia Engineering*, vol **108**, p. 575-583 https://doi.org/10.1016/j.proeng.2015.06.180

Experimental and finite element modelling of prestressed hollow-core slabs

I F Moldovan[1], M Nedelcu[1], B Hegheş[1], H Constantinescu[1], Ş M Buru[1]
[1] Technical University of Cluj-Napoca, Faculty of Civil Engineering, 15 Constantin Daicoviciu Street, *Romania*

E-mail: ilinca.lungu@mecon.utcluj.ro

Abstract. This paper presents the results of a study conducted on prestressed hollow-core slabs (HCS), both experimentally and numerically. The research has been carried out on real scale HCS manufactured at a local precast factory. The concrete specimens were prestressed using ten 12.9 mm diameter steel tendons placed both at the bottom and the top flange of the cross-section. They were tested under concentrated load until shear failure. Validation of the numerical results undertaken using ABAQUS 6.13 software demonstrated that the values obtained from the numerical simulation match the experimental ones in terms of force-displacement relationship.

1. Introduction

Prestressed hollow core slabs (HCS) are structural elements predominantly used in the construction of residential, industrial buildings and bridges also. Hollow core slabs are floor and wall elements made from prestressed concrete with voids in the cross-section. The presence of the voids leads to a significant reduction of the element's weight in comparison to solid slab units. As in the construction industry nowadays the main characteristics are energy and cost effectiveness together with sustainability, and because prestressed concrete elements accomplish all the aforementioned aspects, HCS represent a future-oriented choice. The development of hollow core HCS dates back from the 1950s, when long-line pre-tensioning systems emerged [1] [2]. HCS are generally produced either through extrusion, when a low-slump concrete is pushed through the casting machine or by using a higher slump concrete with the voids being obtained by slip forming with long tubes attached to the casting machine [3].

Shear behaviour of HCS and other concrete structures too is a sophisticated issue and a problem studied for decades. The traditional techniques of adjusting the shear capacity of these units consisted in the increase of slab thickness and or filling the voids. Nevertheless, it has been confirmed that these practices had basically a negative effect on the shear capacity and the HCS particularities self-weight and low cost. Consequently, the process of retrofitting the shear capacity of HCS remains the most notable measure to prevent shear failure [4].

2. Experimental work

Experimental investigation represents a significant and efficient tool for a better understanding of HCS behaviour. Two full-scale precast prestressed HCS were tested to failure under monotonic load. The specimens were supplied by a local precast factory. The experimental program was carried out in the Test Laboratory of the Faculty of Civil Engineering. The specimens had a length of 5000mm, a depth of 320mm and a width of 1200mm. Figure 1 shows the cross-section of the HCS tested:

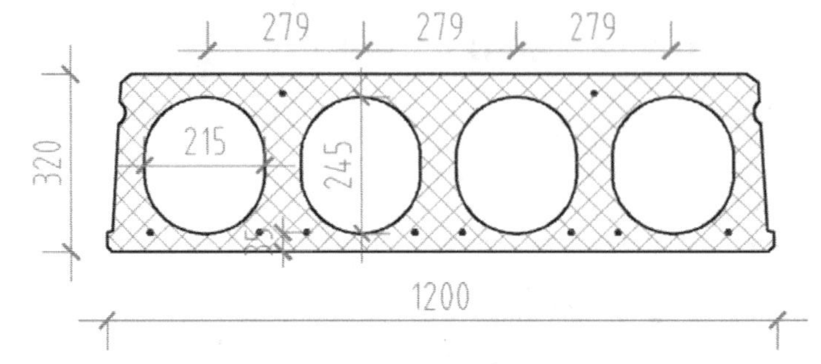

Figure 1. Cross-section of HCS.

Each test specimen has ten prestressing tendons, composed of 7-wire, low relaxation strands with a 12.9 mm diameter and a cross-sectional area of 100 mm^2. Moreover, they present high prestressing level at the bottom of the slab (eight longitudinal tendons with a 110 KN value of the prestressing force) and a low prestressing level at the top flange (two longitudinal tendons with a 50 KN value of the prestressing force).

Materials

Concrete samples tests were undertaken in the Structures Laboratory of the Faculty of Civil Engineering and resulted in an average value of 64.87 N/mm^2 for the compressive strength. Splitting tensile tests performed in the same Structures Laboratory obtaining a mean tensile strength of 2.02 N/mm^2. The modulus of elasticity has also been determined and the value obtained was 35240.55 N/mm^2. The results obtained were in concordance to the ones supplied by the precast factory's laboratory for a C 50/60 concrete class.

The mechanical properties of the prestressing tendons provided by the strand manufacturers included the values of the nominal tensile strength of 1860 N/mm^2 and modulus of elasticity of 195000N/mm^2. However, after tests performed in the Structures Laboratory on these prestressing strands resulted a value of 195558.74 N/mm^2 for the Young's modulus of elasticity.

Test setup

The HCS elements were simply supported at each end. In addition, the point load acted along the width of the element through a 250mm H steel beam. In order to satisfy the value of 2.5h of the shear span-to-depth ratio the point load was placed at a distance of 850mm from the hinged support. Figure 2 depicts the unit static scheme. The supports of the tested units consisted of two steel H profiles with the dimensions 250x1500mm and a third one having the dimensions 100x1500mm, which were fixed on the laboratory's floor.

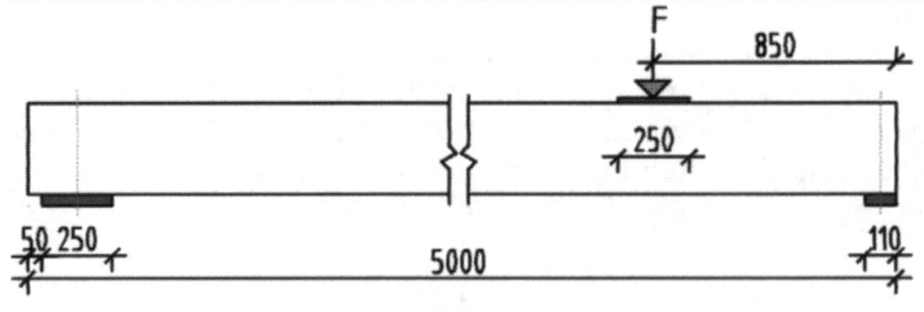

Figure 2. Static scheme.

The tested specimens were instrumented at the loaded end, as displayed in figure 3. Two linear variable differential transformers (LVDTs) were installed both on the left and the right side of the element to measure the vertical displacements along the significant location on the slab's length. Furthermore, electronic dial gauges were positioned on the specimen's loaded end for a manually strain measurement. The force was applied with a hydraulic press of 1200kN and an extension of 40 cm.

Figure 3. Experimental setup.

Experimental observations

The main purpose of this section is to outline the experimental results for the specimens tested in this study. These observations are going to be emphasized in terms of failure load and crack appearance.

The two tested specimens presented a typical shear failure, as it can be noticed from figures 4 and 5. The crack appeared at a distance of approximately 110mm from the edge of the support close to the point load and extended until the loading plate. Regarding the failure load, the first specimen reached up to 313 kN while the second one failed at 345 kN.

As noticed from figure 6, the second unit exhibited a longitudinal crack which developed suddenly on the top surface of the first void.

Figure 4. Test 1 failure results.

Figure 5. Test 2 failure results.

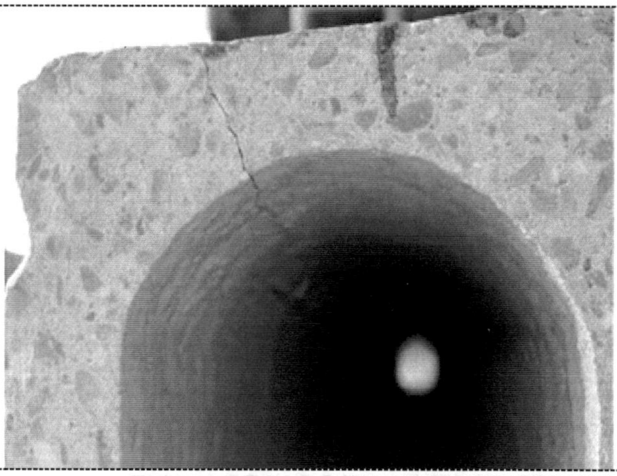

Figure 6. Test 2 failure results.

3. Numerical studies and calibration versus the experimental program

Finite element modelling represents an effective means for a better understanding of structural elements with a complex nonlinear behaviour. Finite element programme Abaqus 6.13 was used during the numerical analysis carried out in this study [5]. The details of the adopted model will be described and then the calibration of the numerical model with the data obtained from the experimental investigation will be highlighted.

The concrete damage plasticity model has been chosen for the simulation of the precast HCS[6]. This model is distinguished for the fact that crushing in compression and cracking in tension causes the principal failure mechanism of concrete [7]. The parameters used have been detailed in table 1.

Table 1. Concrete damage plasticity model parameters

Dilation angle	Eccentricity	f_{b0}/f_{c0}	k	Viscosity parameter
30°	0.1	1.16	0.667	0.0001

The constitutive model for the prestressing tendons is expressed by the idealised bilinear stress-strain curve [8].

The element type adopted for modelling concrete volumes was C3D8 (8-node linear brick, fully integrated) and T3D2 (2-node, three-dimensional truss element) for the prestressing tendons modelling.

It has been demonstrated that the accuracy of the results obtained in numerical analysis is highly dependent on the mesh size of the elements used. Therefore, a mesh sensitivity analysis has been undertaken and three different mesh sizes were studied: a fine (25mm), a medium (50mm) and a coarse mesh (75mm). The comparison of the three mesh sizes is emphasized in figure 7. Thus, it has been adopted the optimum size of 25mm for the finite elements modelling.

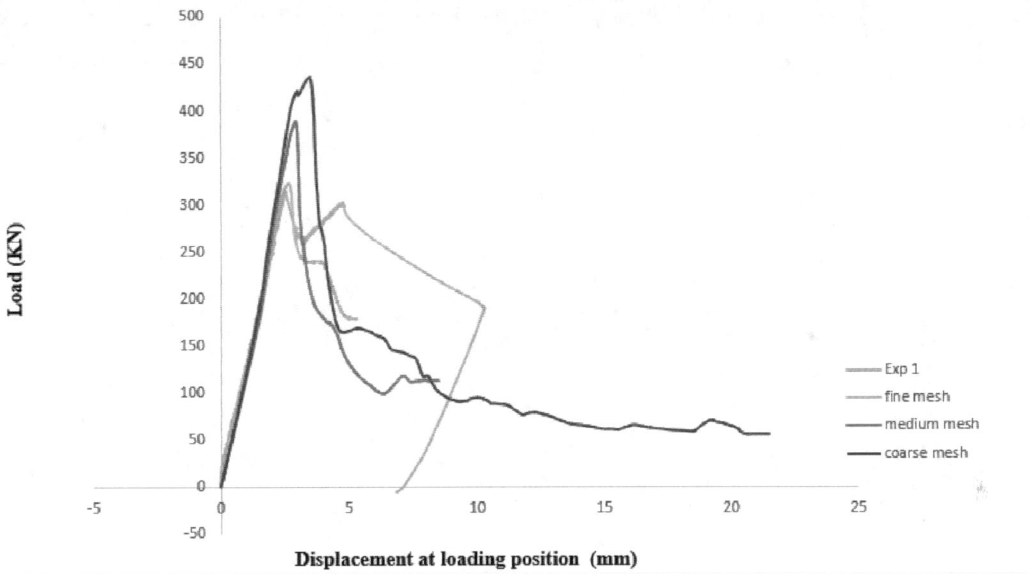

Figure 7. Mesh sensitivity analysis.

Figure 8 illustrates the studied finite element's geometry for hollow core slab.

Regarding the boundary conditions, the support plates were represented with 3D discrete rigid elements which were then connected to the hollow-core slab with a 'surface-to-surface' contact interaction option. The contact property consisted in a 'hard' contact normal behaviour and a frictionless tangential behaviour. The loading plate has been defined as a 3D discrete rigid element also, but the Tie constraint was used for connecting it to the concrete part, an option which allows the fusion of the two surfaces.

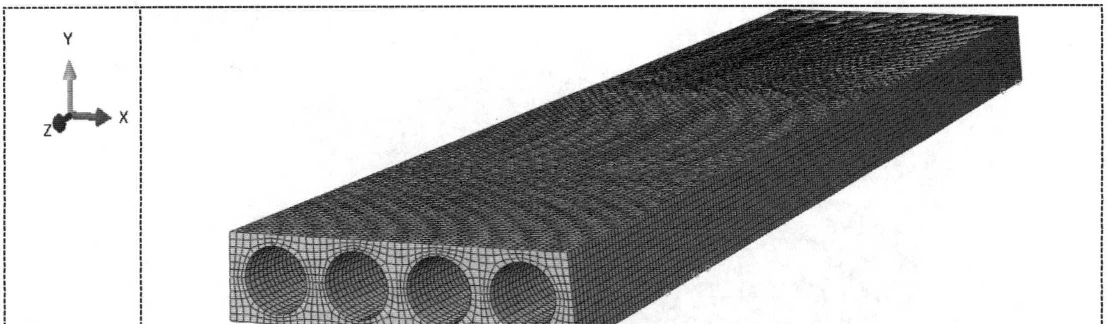

Figure 8. Finite element model geometry.

Numerical analysis is divided into two steps. The first step consists of the application of prestress in the tendons, using the Predefined field option, while in the second step of the analysis, the HCS is subjected to the self-weight as a uniform load on the element and moreover, to a vertical load gradually applied. A displacement control analysis was adopted for the load application. The static geometric nonlinear general method has been chosen to conduct the nonlinear static analysis. A Riks - Wempner analysis has also been performed, but the results obtained were not satisfactory.

The load-deflection relationship represents a significant assessment criterion for a finite element model performance because it depicts the specimens' structural performance during experimental tests. Figure 9 contains the comparison between the data from the finite element modelling and the experimental results. Exp 1 and Exp 2 represent the load-deflection curves obtained from the two experimental tests investigated. It can be concluded that the modelling method adopted estimates the behaviour of HCS in an adequate manner, with reference to the initial stiffness and the failure load.

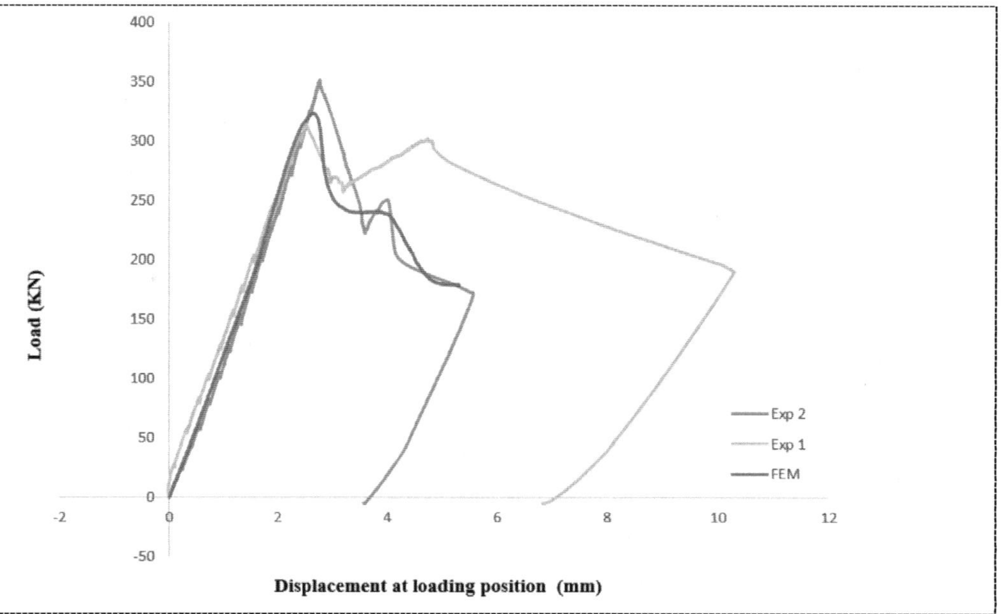

Figure 9. Load-deflection results.

The plastic strain magnitude (PEMAG) is shown in Figure 10, illustrating a classical crack expansion inside the shear-tension region. As noticed from Figures 4 and 5, the results obtained from the numerical analysis match the crack pattern obtained from the laboratory tests.

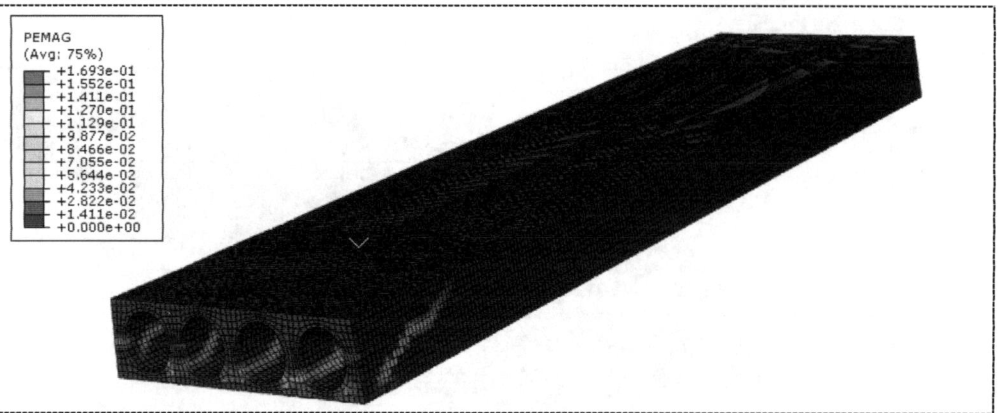

Figure 10. Plastic strain magnitude (PEMAG) results.

4. Conclusions

The experimental program presented in this paper, consisting in testing two hollow core units illustrated the shear behaviour of such structural elements. Finite element simulation undertaken predicts the overall behaviour of the precast prestressed HCS tested. A mesh sensitivity analysis has also been concluded and its results contributed to the overall findings. It is noteworthy that the difference between the experimental data and the numerical data is under 5%, in terms of the load-deflection behaviour. The curves exhibit similar behaviour both in the elastic and in the post-elastic range, which concludes that the characteristics of the numerical model have been thoroughly studied and analysed.

Acknowledgments

The experimental tests were supported by the Asa Cons Romania, a well-known precast manufacturer of high-performance concrete elements. This support is gratefully acknowledged.
The research activities were supported by the project POCU/380/6/13/123927 – ANTREDOC, "Entrepreneurial competencies and excellence research in doctoral and postdoctoral studies programs", project co-funded from the European Social Fund through the Human Capital Operational Program 2014-2020.

5. References

[1] Hawkins N M and Ghosh S K 2015 Shear Strength of Hollow-Core *PCI Journal* pp. 110–114
[2] Walraven J C and Mercx W P M 1983 Bearing Capacity of Prestressed Hollow Core Slabs *Heron* vol **28** no 3
[3] Brunesi E, Bolognini D and Nascimbene R 2012 *Evaluation of the Shear Strength Capacity of Precast- Prestressed Hollow Core Floor Slabs – Part 2 : Code Comparisons and Design Method*
[4] Meng X 2016 *Scholarship at UWindsor Shear Strengthening of Prestressed Hollow Core Slabs Using Externally Bonded Carbon Fiber Reinforced Polymer Sheets*
[5] Dassault Systems 2013 *6.13 Documentation (Abaqus) Abaqus User's Guide* pp 1138
[6] Hafezolghorani M, Hejazi F, Vaghei R, Bin Jaafar M S and Karimzade K 2017 Simplified damage plasticity model for concrete *Struct. Eng. Int. J. Int. Assoc. Bridg.* vol **27** no 1 pp 68–78
[7] Tao Y and Chen J F 2015 Concrete Damage Plasticity Model for Modeling FRP-to-Concrete Bond Behavior *J. Compos. Constr.* vol **19** no 1 p 04014026
[8] Devalapura R K and Tadros M K 1992 Stress-Strain Modeling of 270 ksi Low-Relaxation Prestressing Strands *PCI J.* vol **37** no 2 pp 100–106

FEM analyses on beam-to-column connection for a thin-walled cold-formed steel frame

V M Venghiac[1], I Olteanu-Donțov[1], M S Alexa-Stratulat[1], G Țăranu[1], S A Băetu[1], I O Toma[1]

[1] The "Gheorghe Asachi" Technical University of Iasi, Faculty of Civil Engineering and Building Services, no. 1, Prof.dr.doc. D. Mangeron Street, 700050, Iasi, *Romania*

E-mail: ionut.ovidiu.toma@tuiasi.ro

Abstract. The advantage of fast construction for thin-walled cold-formed steel ground floor structures is somehow lessened by the challenges the engineers encounter when designing the connections. The horizontal forces induced by the seismic motions and not only, represent a serious threat to flexible structures. The paper presents the results obtained from comparative FEM analyses of different types and layouts of beam-to-column connection in case of a thin-walled cold-formed steel frame. The obtained results were analysed and compared from the point of view of maximum vertical displacements of the beam free end, the load-displacement diagrams as well as the principal stresses. The objective was to obtain the most efficient type of connection that ensures the safe transfer of internal forces from the beam to the column.

1. Introduction

Cold formed steel (CFS) structures have seen a tremendous expansion over the past decades being used in a variety of applications ranging from civil engineering to mechanical and even aeronautical engineering [1]. Their first use in civil engineering was somehow limited to secondary structural elements such as roof purlins, side-rails for wall cladding, a.s.o. However, due to their higher strength per unit weight compared to hot rolled steel elements, ease and speed of construction coupled with a higher flexibility in manufacturing, the demand for CFS elements in structural applications has spiked [2, 3]. This resulted in extensive research works being conducted to assess the behaviour to CFS structures.

Such structures are prone to premature local buckling and instability issues due to the small thickness of CFS sections that typically ranges from 0.9 mm to 3.2 mm [4, 5]. This may result in limited applications especially for structures expected to withstand extreme loading scenarios such as moderate to strong earthquakes. The seismic behaviour has been extensively investigated, with the majority of research works being focused on the CFS wall frame panels as one of the major lateral load bearing systems [6]. Only recently, the application of CFS members to moment-resisting frames has started to gain attention and therefore the information available is rather scarce [1, 7].

One of the main advantages of using moment resisting CFS frames is the higher flexibility they offer for space planning and future changes in building destination and intended use. Additionally, moment resisting CFS frames could improve the energy dissipation capacity of a CFS structure during significant seismic events [1, 7].

An important part in investigating the seismic behaviour of any structural system is the correct assessment of the joint behaviour, be it for concrete [8, 9], steel [10-12], timber, CFS [13-15] or aluminium structures.

Numerical simulations have quickly become an indispensable and efficient tool [16, 17] for assessing the behaviour of different structural systems and/or subsystems to a variety of loading scenarios. Once a numerical model was validated, it could be used for a wide range of parameter variations in order to grasp the influencing factors that may govern the response of members or structures to extreme loading events such as moderate to strong earthquakes [15, 18, 19].

Finite Element (FE) modelling is extensively used to assess the behaviour of CFS connections with various configurations [15, 20]. The results of previous studies highlighted the importance of detailed FE models to predict the behaviour of CFS connections under either monotonic or cyclic loading [21, 22].

The paper presents the results obtained by means of linear and non-linear static analyses on the connection joint between the CFS beam and the column of a portal frame. Four different configurations were considered consisting in the geometry of the node, presence of stiffening elements and the number of bolts. The obtained results show significant influence of the node geometry and configuration on the distribution of stresses in all components of the considered joint.

2. Methodology

2.1. Geometry and dimensions

The geometry of the considered model is shown in figure 1a. Both the beam and the column are made of two 150×45×10 mm CFS C-shape lipped profiles, figure 1b, positioned back-to-back, having a thickness of 2 mm.

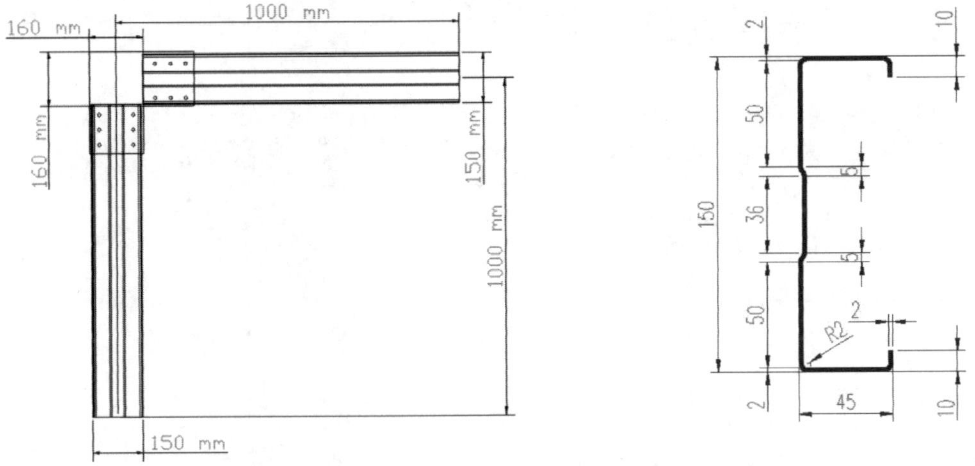

a. Dimensions of the considered model (sizes in mm) **b**. CFS C-shape lipped profile (sizes in mm)

Figure 1. Geometry of the model

2.2. Material

The CFS profiles were made of S235 mild steel with a longitudinal modulus of elasticity, E = 210 GPa, shear modulus, G = 81 GPa, Poisson's ratio 0.3 and a yield strength of f_y = 235 MPa.

2.3. Numerical model

The numerical model was generated following the geometry given in figure 1. ANSYS 2021R1 [23] computer package was used to run the analyses. Four distinct configurations of the connection joint between the beams and the column were considered, as described in table 1.

The maximum mesh size was 20 mm resulting in an average number of 12500 elements for the considered configurations. SHELL181 quadratic plane stress element was used to generate the mesh, as seen in figure 2.

The joint assembly was considered to be fixed at the lower end of the column and the load was applied incrementally at the free end of the beam, pointing downward. Static non-linear finite element analysis (FEA) was run in order to check potential local losses of stability. The obtained results were analysed and compared from the point of view of maximum vertical displacements of the beam free end, the load-displacement diagrams as well as the principal stresses.

The main parameters of the numerical analysis were: the shape of the gusset/connection plate, the influence of stiffeners and the diameter of the bolts.

Table 1. Joint layouts considered in the research.

Designation	Description	Layout
Model 1	Connection plate cut at 90°; thickness of 4mm; M10 bolts – bonded contacts between bolts and steel plate	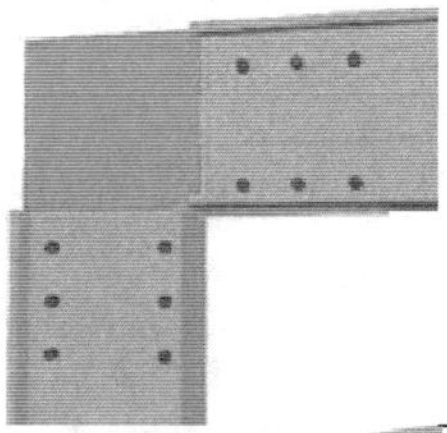
Model 2	Connection plate cut at 90°; thickness of 4mm; M10 bolts (bond contact); exterior stiffening plates – the plate was not connected to the top flange of the beam	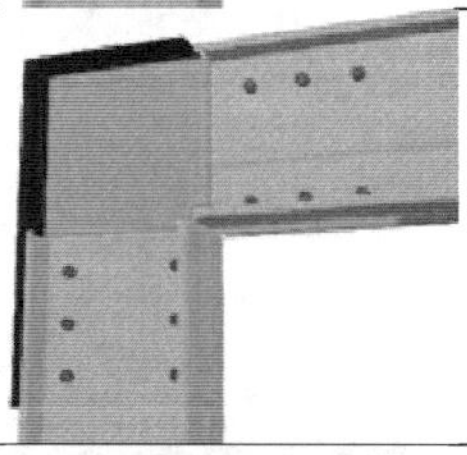
Model 3	Connection plate cut at 45°; thickness of 4mm; M10 bolts (bond contact); exterior and interior stiffening plates	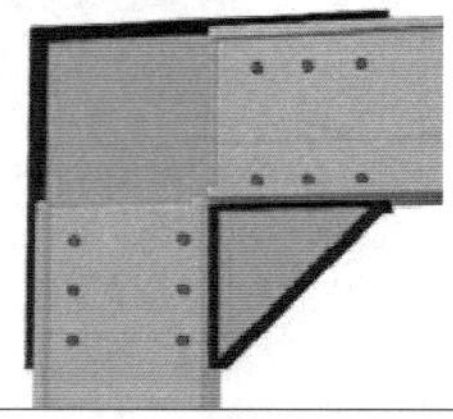
Model 4	Connection plate cut at 45°; thickness of 4mm; M16 bolts (bond contact); exterior and interior stiffening plates	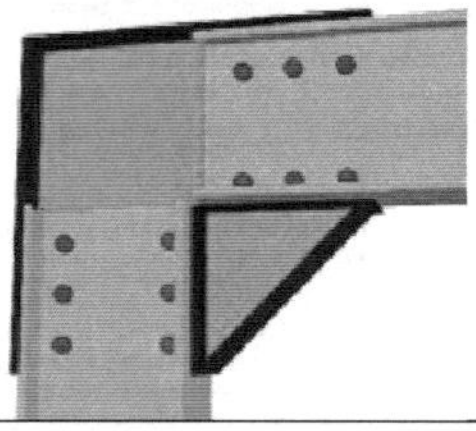

a. Model 1

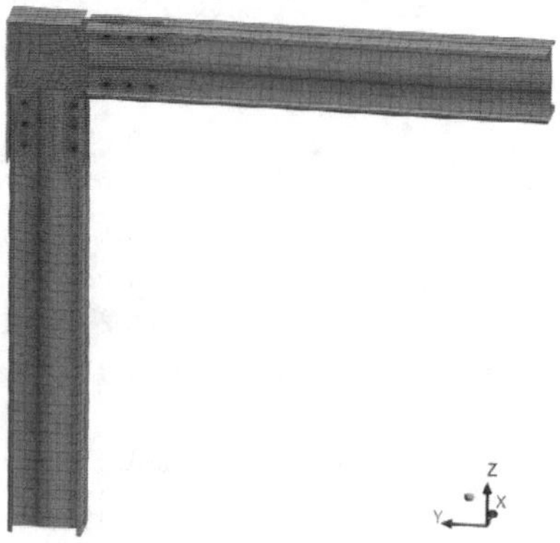

b. Model 2

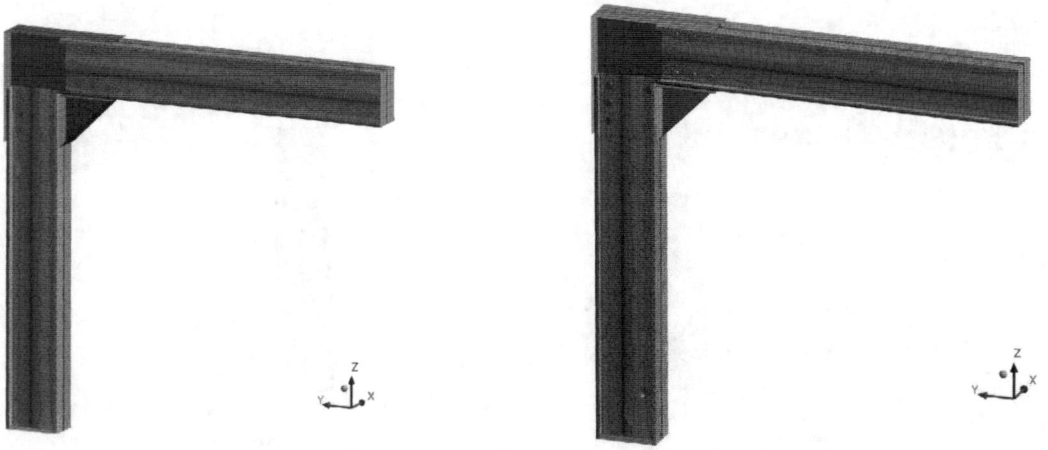

c. Mode 3 **d.** Model 4

Figure 2. Numerical models and meshing

3. Results and discussions

3.1. Displacements

The magnitudes of the vertical displacement at 1000N are shown in figure 3. The highest vertical displacement was obtained for Model 1, 4.68 mm. The presence of the stiffening plate in Model 2 helped reducing of the displacement to 4.12 mm, almost 14% smaller.

A significant reduction in the magnitude of the vertical displacement was observed for Model 3 for which the gusset/connection plate was cut at a 45° angle. The obtained vertical displacement was 2.45 mm, almost half the displacement obtained for Model 1. Increasing the bold diameter from 10 mm to 16 mm helped further reducing of the displacement (increasing the stiffness of the node) by 4% to 2.34 mm.

Increasing the load to 2000 N resulted in failure of Models 1 and 2. They both failed due to lateral buckling of the connection plate, as it can be seen from figure 4a and 4b. The abnormal high values for the displacement of the free end in case of Models 1 and 2 are a clear indication of the failure.

On the other hand, Models 3 and 4, for which a different shape of the connection plate was adopted and additional stiffening plates were used, exhibited displacements of 4.90 mm and 4.69 mm, respectively. Comparing the obtained results with the data for an intensity of the applied load of 1000 N, a linear variation could be observed between the two loading steps.

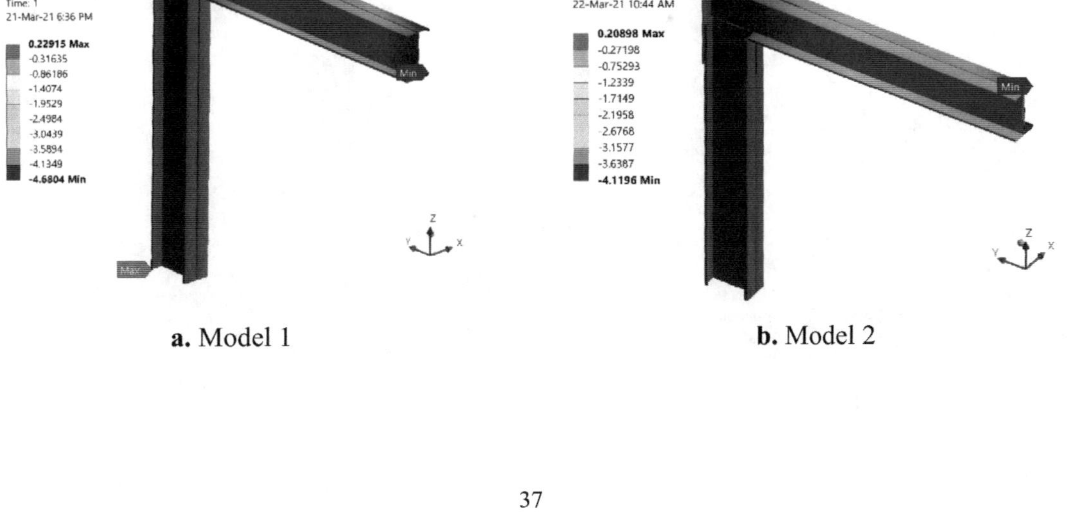

a. Model 1 **b.** Model 2

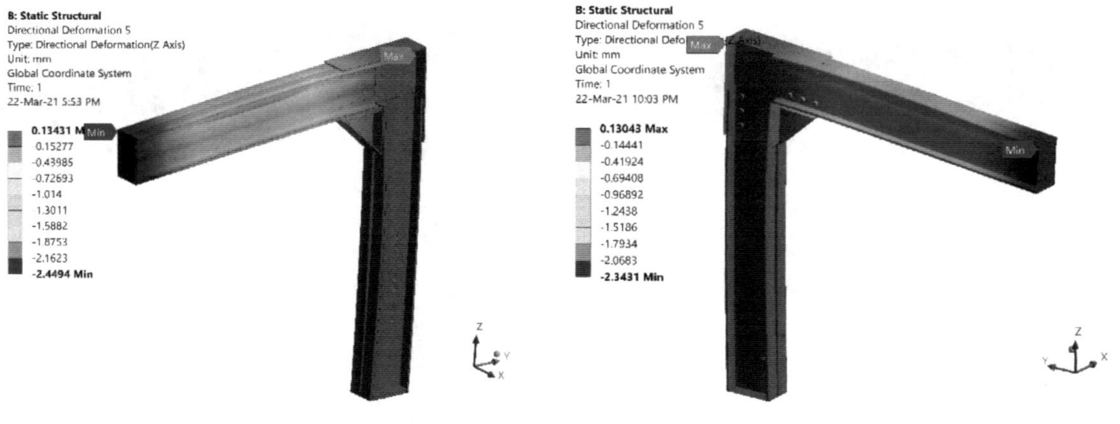

c. Mode 3 **d.** Model 4

Figure 3. Vertical displacements at 1000 N

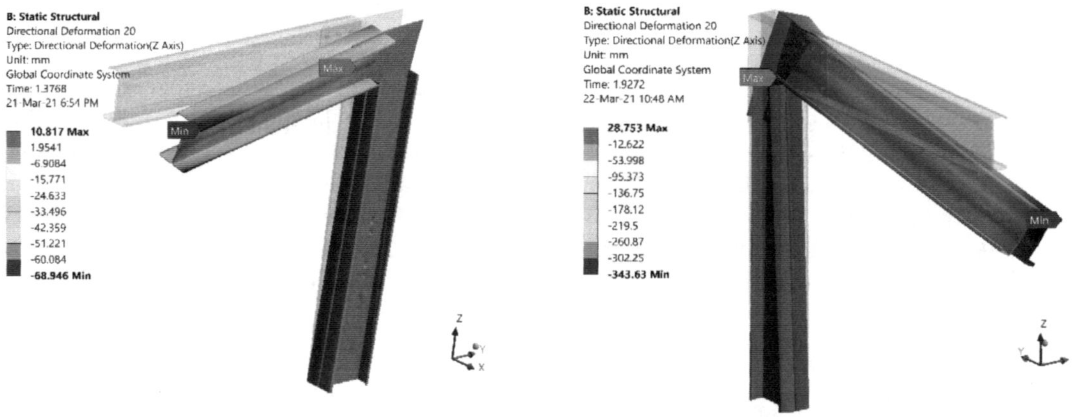

a. Model 1 **b.** Model 2

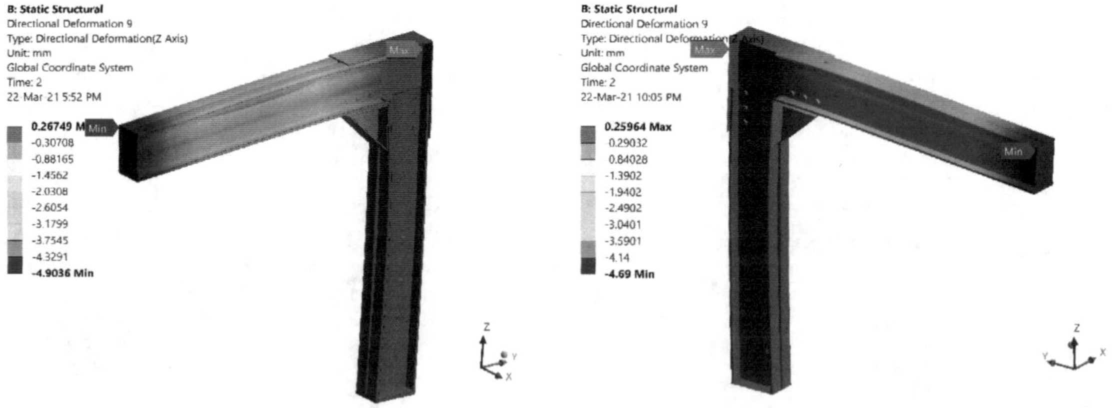

c. Mode 3 **d.** Model 4

Figure 4. Vertical displacements at 2000 N

3.2. Principal stresses

The intensity of the principal stresses was obtained for each calculation step of the numerical analysis. Figure 5 presents the principal stress distribution and the intensity of principal stresses for the magnitude of the applied load of 1000 N. For each of the considered models it can be observed that stress concentrations occurred at the level of the bolts.

The shape of the connection plate plays an important role, as it can be seen from the maximum values of the principal stresses of Model 3 compared to Models 1 and 2, shown in figure 5. There is a 23.3% reduction in the magnitude of stresses for the same intensity of the applied load.

The diameter of the bolt seems to be another significant influencing factor. The stress level drops by 46.15% when increasing the diameter of the bolt from 10 mm to 16 mm, as seen from figure 5d.

The same trends were observed for the ultimate loads obtained for each model. Model 1 failed due to local buckling of the connection plate at a load equal to 1376.8 N. Model 2 failed also due to lateral buckling of the connection plate, at the beam level, but at a higher load, 1889.3 N. This means a 37.22% increase in the load carrying capacity. This behaviour could be explained by the fact that the stiffening plate was not connected to the upper flanges of the beam. Model 3, figure 5c, showed stress concentrations in the upper row of bolts located at the beam end that exceeded the yield strength when the load reached 2400 N. The failure of the assembly changed from loss of stability to local crushing. Model 4 was able to resist the highest load, up to 4226.8 N. The increase of the bolt diameter did not change the failure mode compared to Mode 3 but changed the location where the stress concentrations occurred. According to the numerical investigations, the part of the model that reached the yield strength was located at the level of the top row of bolts in the column, figure 5d. A further increase of the load resulted in general loss of stability of the model.

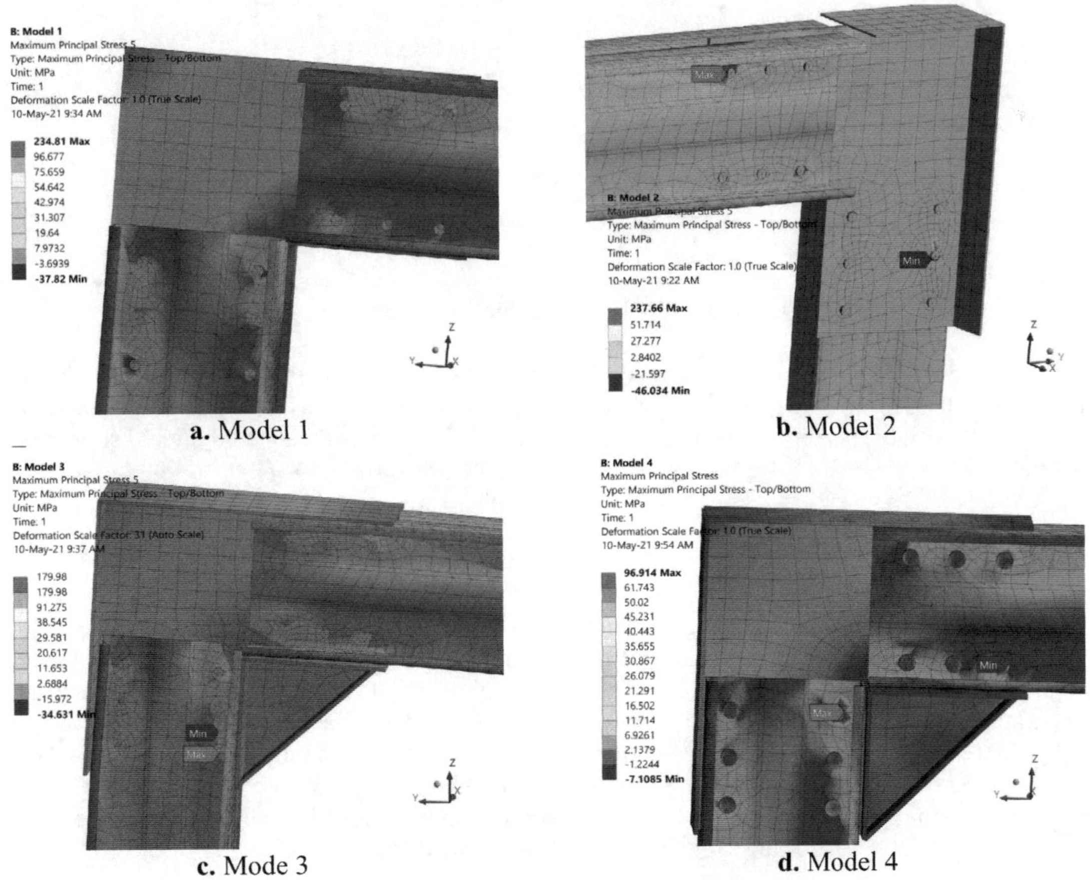

Figure 5. Principal stresses at 1000N

Based on the obtained data, the load carrying capacity of Model 4 increased by 307% compared to Model 1, by 222% compared to Model 2 and by 167% compared to Model 3. The general behaviour of the joint obtained by numerical simulations in the present study is consistent to the results presented in the scientific literature [24].

3.3. Load-displacement curves

The load-displacement curves of the four models are shown in figure 6. It can be observed that Models 1 and 2 have similar initial stiffness, despite Model 2 having an additional stiffening plate. Similarly, Models 3 and 4 have comparative initial stiffness with Model 4 exhibiting a slightly larger value due to the larger diameter bolts used to connect the elements.

The stiffening plate in Model 2 leads to a higher resisting load before local yielding of the material occurs coupled with the loss of stability.

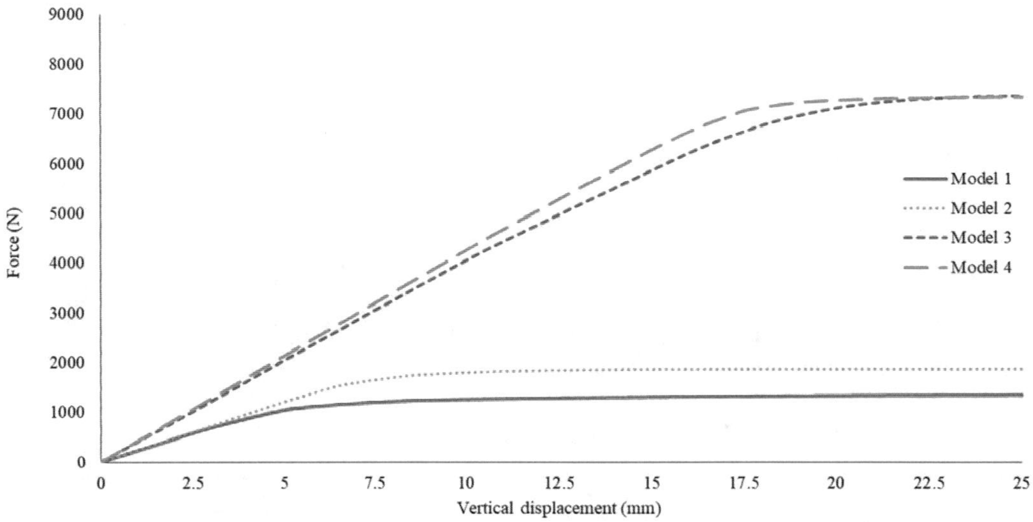

Figure 6. Load-displacement curves

The data is consistent with the observed patterns presented in figure 4 and figure 5. The loss of stability lead to very large displacements recorded for Models 1 and 2, as seen in figure 4a and 4b. For the same intensity of the applied load, the maximum principal stresses reached the yield strength of the material.

4. Conclusions

The paper presents the results obtained by means of non-linear static analysis numerical simulations of a beam-to-column connection joint (eave joint) under four different configurations. Based on the obtained results, the following conclusions can be drawn:

A significant reduction in the magnitude of the vertical displacement was observed for Model 3 for which the gusset/connection plate was cut at a 45° angle. The obtained vertical displacement is 2.45 mm, almost half the displacement obtained for Model 1. Increasing the diameter of the bolt results in a further decrease of the vertical displacement because the joint becomes more stiff.

The shape of the connection plate plays an important role, as it can be seen from the maximum values of the principal stresses of Model 3 compared to Models 1 and 2. There is a 23.3% reduction in the magnitude of stresses for the same intensity of the applied load. The diameter of the bolt is another significant influencing factor. The stress level drops by 46.15% when increasing the diameter of the bolt from 10 mm to 16 mm,

The numerical analyses are conducted in order to shed some light in the failure mode of the connection joint between a beam and a column made of CFS C-shaped lipped profiles. The model needs further calibration and validation based on experimentally obtained data during the next stage of the research. Numerical simulations based on the finite element method are versatile tools that allow for large parametric analyses in order to optimize the design process and to reach a balance between ultimate strength versus material consumption.

Acknowledgments

The authors wish to express their gratitude to the partnership between The "Gheorghe Asachi" Technical University of Iasi, INAS S.A. and ANSYS for making available the full research license for Ansys 2021R1 used in conducting the research.
The authors thank S.C. Panoterm Import-Export S.R.L. for supplying the CFS profiles used in the research as part of the grant no. 4887/03.2019.

5. References

[1] Mojtabaei S M, Kabir M Z, Hajirasouliha I and Kargar M 2018 Analytical and experimental study on the seismic performance of cold-formed steel frames *Journal of Construction Steel Research* **143** 18-31

[2] Rosca O V 2011 Experimental tests on thin-walled steel roof profiles *Bulletin of the Polytechnic Institute of Iasi, Constructions. Architecture Section* **LIV**(4) 85-95

[3] Karabulut B and Soyoz S 2017 Experimental and analytical studies on different configurations of cold-formed steel structures *Journal of Construction Steel Research* **133** 535–46

[4] Lee Y H, Tan C S, Mohammad S, Tahir M M and Shek P N 2014 Review on cold-formed steel connections *The Scientific World Journal* **2014** ID. 951216

[5] Smith B H, Arwade S R, Schafer B W and Moen C D 2016 Design component and system reliability in a low rise cold formed steel framed commercial building *Engineering Structures* **127** 434–46

[6] Schafer B W, Ayhan D, Leng J, Liu P, Padilla-Llano D, Peterman K D, Stehman M, Buonopane S G, M. Eatherton M, Madsen R, Manley B, Moen C D, Nakata N, Rogers C and Yu C 2016 Seismic response and engineering of cold-formed steel framed buildings *Structures* **8** (Part 2) 197–212

[7] Bagheri Sabbagh A, Petkovski M, Pilakoutas K and Mirghaderi R 2012 Experimental work on cold-formed steel elements for earthquake resilient moment frame buildings *Engineering Structures* **42** 371–386

[8] Campian C M, Pop M and Chira N 2017 The influence of the connections between concrete and steel on the joints modelling *ce/papers – Proc. of Eurosteel, September 2017, Copenhagen, Denmark* **1**(2-3) 4680-7.

[9] Sococol I, Mihai P and Olteanu-Dontov I 2019 Ductility-concept for improving the seismic response for structural reinforced concrete frame systems *Bulletin of the Polytechnic Institute of Iasi, Constructions. Architecture Section* **65**(1) 17-30

[10] Budescu M, Ciogradi I P and Rosca OV 2005 Experimental studies of a series of high strength friction grip bolted joints *Bulletin of the Polytechnic Institute of Iasi, Constructions. Architecture Section* **LI**(1-2) 23-9

[11] Melenciuc S C, Stefancu A I and Olteanu I 2009 Seismic design of steel connections *Bulletin of the Polytechnic Institute of Iasi, Constructions. Architecture Section* **LV**(1) 49-56

[12] Mathe A E, Catarig A and Moldovan I 2015 Statics and kinematics of semirigid steel frames under seismic action *Journal of Applied Engineering Sciences* **5**(2) 59-64

[13] Lim J B P, Hancock G J, Clifton G C, Pham C H and Das R 2016 DSM for ultimate strength of bolted moment-connections between cold-formed steel channel members *Journal of Construction Steel Research* **117** 196-203.

[14] Huynh M T, Pham C H and Hancock G J 2020 Design of screwed connections in cold-formed steels in shear *Thin-Walled Structures* **154** ID.106817

[15] Ye J, Mojtabaei S M, Hajirasouliha I and Pilakoutas K 2020 Efficient design of cold-formed steel bolted-moment connections for earthquake resistant frames *Thin-Walled Structures* **150** ID.105926

[16] Pop M, Campian C M and Chira N 2019 Parametric study in composite joints configuration *IOP Conference Series Materials Science and Engineering (Proc. of International Conference Computational Civil Engineering, May 2019, Iasi, Romania)* **586** ID.012014

[17] Toma A M, Taranu G, Nishida T, Mathe A E, Campian C M, Baccay M and Toma I O 2020 Assessing the seismic performance of a R.C. frame structure by numerical simulations – an efficient tool for a sustainable future *Proc. of the International Conference on Building Services and Energy Efficiency, June 2020, Iasi, Romania* 231-240

[18] Mathe A E, Popa A G and Campian C M 2011 The influence of semi-rigid connections upon the performance of steel structures seismically excited *Acta Technica Napocensis: Civil Engineering and Architecture* **54**(3) 268-279

[19] Toma I O, Mihai P, Venghiac V M, Taranu G, Baetu S A, Toma A M and Petrescu C T 2020 Numerical simulations of the seismic behavior of a damaged rc frame retrofitted with composite fabric *Proc. of the World Conference on Earthquake Engineering, September 2020, Sendai, Japan* PaperID 3f-0019

[20] Ye J, Becque J, Hajirasouliha I, Mojtabaei S M and Lim JB P 2018 Development of optimum cold-formed steel sections for maximum energy dissipation in uniaxial bending *Engineering Structures* **161** 55-67

[21] Bagheri Sabbagh A, Petkovski M, Pilakoutas K and Mirghaderi R 2013 Cyclic behaviour of bolted cold-formed steel moment connections: FE modelling including slip *Journal of Construction Steel Researchi* **80** 100-8

[22] Ozturk F and Pul S 2015 Experimental and numerical study on a full scale apex connection of cold-formed steel portal frames *Thin-Walled Structures* **94** 79-88

[23] Ansys 2021 R1, https://www.ansys.com/products/structures/ansys-mechanical

[24] Shahini M, Bagheri Sabbagh A, Davidson P, Mirghaderi R 2019 Development of cold-formed steel moment-resisting connections with bolting friction-slip mechanism for seismic applications *Thin-Walled Structures* **141** 217-231.

Numerical analyses of a base connection for a thin-walled cold-formed profile column

V M Venghiac[1], I Olteanu-Donțov[1], M S Alexa-Stratulat[1], G Țăranu[1], S A Băetu[1], I O Toma[1]

[1] The "Gheorghe Asachi" Technical University of Iasi, Faculty of Civil Engineering and Building Services, Department of Structural Mechanics, no. 1, Prof.dr.doc. D. Mangeron Street, 700050, Iasi, *Romania*

E-mail: ionut.ovidiu.toma@tuiasi.ro

Abstract. Thin-walled cold-formed steel structures tend to gain more attention in the civil engineering field. Ground-floor dwellings, as well as factory warehouses represent the most frequent use of these structural systems. However, due to the inherent flexibility of thin-walled cold-formed steel elements, the designer has to take into consideration the possibility of large deformations occurrence, especially in the case of structures with large openings. Such deformations may lead to significant stresses at the level of the connection joints between elements or between elements and the foundation block. The paper presents a comparative numerical analysis based on the finite element method to assess the efficiency of a connection joint between the columns of a portal frame and the foundation block. The linear and nonlinear analyses were carried out by means of Robot Structural Analysis and Ansys. The column is made of two thin-walled cold-formed C shape profiles positioned back-to-back. The efficiency was assessed both from a stress-state point of view in the column and in the connection parts, as well as from the point of view of the manufacturing process and implementation. Significant differences were observed regarding the stress distribution in the nonlinear analysis.

1. Introduction

In the construction industry there are two types of steel profiles commonly used: hot-rolled and cold-formed steel profiles. The differences between them reside in the used technological process, their shapes and mechanical characteristics [1]. The cold-formed steel profiles (CFSP) were developed as a lightweight alternative for the traditional materials. Nowadays, structures made with this solution cover over 45 percent of the construction market and the number is continuously rising [2,3].

CFS elements have a wide range of application in the construction field from non-structural components (e.g. roof purlins, rails for wall cladding, etc.) to load-bearing systems or subsystems (trusses, rack structures, portal frames, moment-resisting frame, floor decking, etc.) [3,4-6].

Even though CFS members began to be used since the 1850s in the US and Great Britain, it was only after the 1920s and 1930s when it was accepted as a construction material, but with limited usage due to lack of research and official codes. The Virginia Baptist Hospital in Lynchburg, Virginia, constructed in 1925 is considered to be among the first buildings that used CFSP as construction elements. The structural system of the building was load bearing masonry with floors made of double back-to-back cold-formed steel lipped channels. The CFSP usage increased significantly after 1946 when the first prescription for this material appeared.

Despite the advantages of CFS elements, such as higher strength per unit weight compared to hot rolled steel elements, ease and speed of construction coupled with a higher flexibility in manufacturing, there are several draw-backs mainly caused by the very small thickness of the steel section which lies between 0.9 mm and 3.2 mm [7]. The local buckling and general instabilities of CFS elements may prevent their application in structures located in active seismic areas prone to severe earthquakes.

Extensive research was conducted to assess the behaviour of structural systems subjected to seismic excitations. The joints between the structural elements or between the elements and the foundation are sometimes considered to be the weakest link in the whole system and therefore gathered the much-needed attention from researchers worldwide [4,8-12].

The numerical modelling and analysis are one of the most convenient methods for assessing the behaviour of CFS elements and/or structures subjected to a wide range of loading scenarios. The main advantage of numerical modelling over the experimental investigations is the possibility of varying a large number of influencing parameters without the limitations of cost, time and instrumentation of the latter. On the other hand, modelling the behaviour of CFS elements requires careful considerations of geometry, contact and material nonlinearities that may result, if not properly addressed, in convergence issues [13-15].

The paper presents the results obtained from a comparative analysis based on the finite element method (FEM) to assess the efficiency of a connection joint between the columns of a portal frame and the foundation block considering both the linear and nonlinear behaviour of the CFS elements. While linear elastic analysis is the preferred approach in practical design, it lacks the ability to adequately predict the local buckling of CFS elements and should be used with caution. On the other hand, nonlinear FEA, although being much more computationally intensive, manages to render evident such failure modes frequently met in the case of CFS elements.

2. Methodology

2.1. Geometry

The geometry of the considered model is shown in figure 1. The considered assembly consisted of two 150×45×10 mm CFS C-shape lipped profiles positioned back-to-back, having a thickness of 2 mm. The two CFS profiles were connected at the top by means of two M16 bolts via a steel plate. The base of the column was fixed to a gusset plate by means of two rows of 3×M16 bolts located at a distance of 100 mm, as shown in figure 1a.

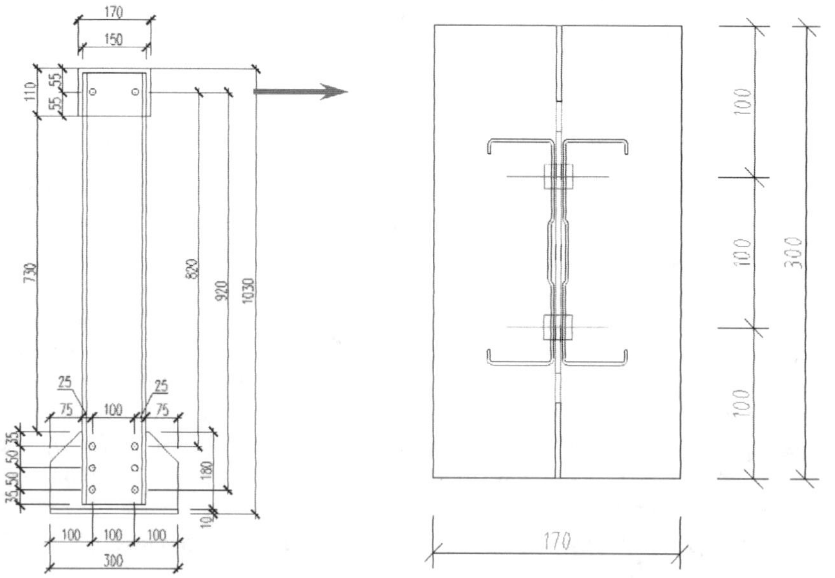

a. Dimensions of the column (sizes in mm) **b.** Column cross-section (sizes in mm)

Figure 1. Geometry of the model

2.2. Materials

The material properties of the CFS profile, as considered in the computer software, are given in table 1. An idealized stress-strain curve was considered for the material, as presented in figure 2.

Table 1. Material characteristics

Material	Young's modulus, E	Tangent modulus	Shear modulus, G	Poisson's ratio	Yield Strength
	[GPa]	[GPa]	[GPa]	-	[MPa]
S235	210	140	81	0.2963	235

2.3. Numerical model

The numerical model was generated following the geometry shown in figure 1. Two computer programs were used and two types of analyses were considered (static linear and non-linear analyses). The two computer programs were chosen as being representatives for their main field of application: design practice (Robot Structural Analysis [16]) and research (Ansys [17]).

A bilinear model for the material behaviour of steel was considered, as shown in figure 2. The material characteristics were the ones presented in table 1.

While Robot Structural Analysis does not allow the user to choose the type of finite element to be used for mesh generation, Ansys gives the user the freedom to choose between a large library of finite element types suitable for the analysis to be conducted. For both linear and non-linear analyses, a SHELL181 quadratic plane stress element type was considered, figure 3.

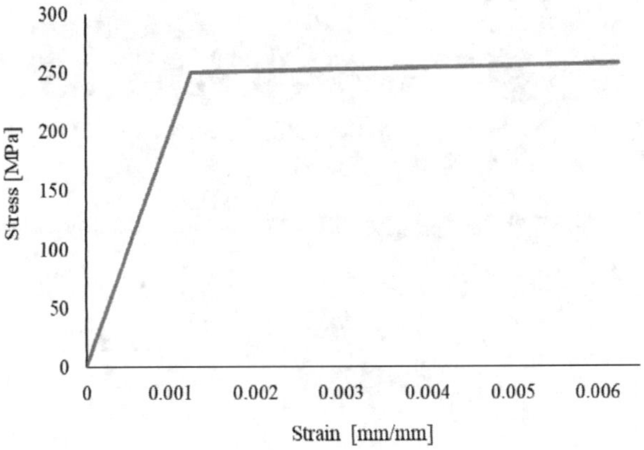

Figure 2. Bilinear material model for CFS

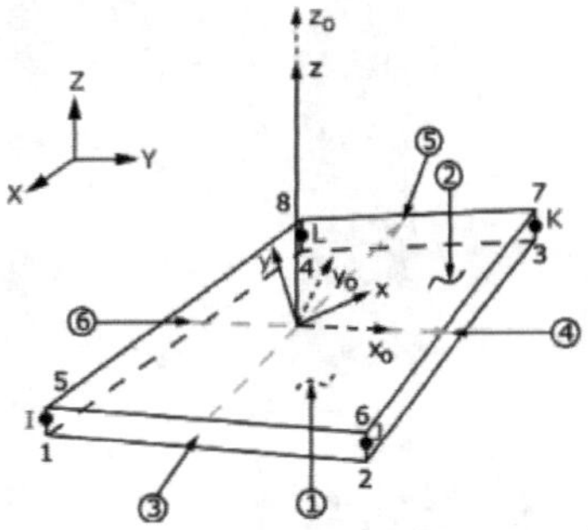

Figure 3. SHELL181 element in Ansys [18]

Three distinct numerical simulation cases were considered: Case 1 – linear static analysis using Robot Structural Analysis software package, Case 2 – linear static analysis using Ansys and Case 3 – non-linear static analysis using Ansys. The non-linear analysis included the residual stress and imperfections, captures the local and overall buckling behaviour, as well as the local distortions of the connections. The maximum mesh size was considered to be 10 mm for all three cases. A brief characterisation of all modelling scenarios is presented in table 2.

The mesh for the models using the two programs is shown in figures 4 and 5.

Table 2. Characteristics of the modelling scenario

<table>
<tr><th>Model</th><th>Analysis type</th><th>FE type</th><th>Maximum element size (mm)</th><th>Number of elements</th><th>Number of nodes</th></tr>
<tr><td>Case 1</td><td>Linear Robot Structural Analysis</td><td>Triangles and squares</td><td>10</td><td>7467</td><td>4096</td></tr>
<tr><td>Case 2</td><td>Linear Ansys</td><td rowspan="2">Squares, quadrilaterals</td><td rowspan="2">10</td><td rowspan="2">10824</td><td rowspan="2">11201</td></tr>
<tr><td>Case 3</td><td>Nonlinear Ansys</td></tr>
</table>

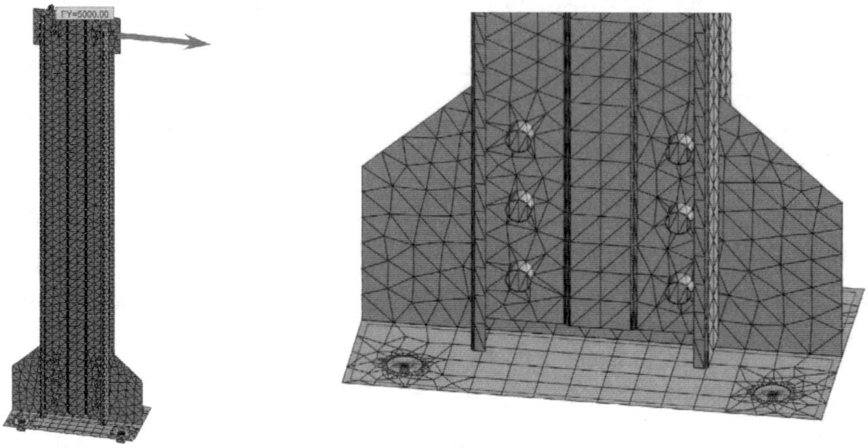

a. General view of the model **b.** Detailed view of the meshing

Figure 4. Generated model using Robot Structural Analysis

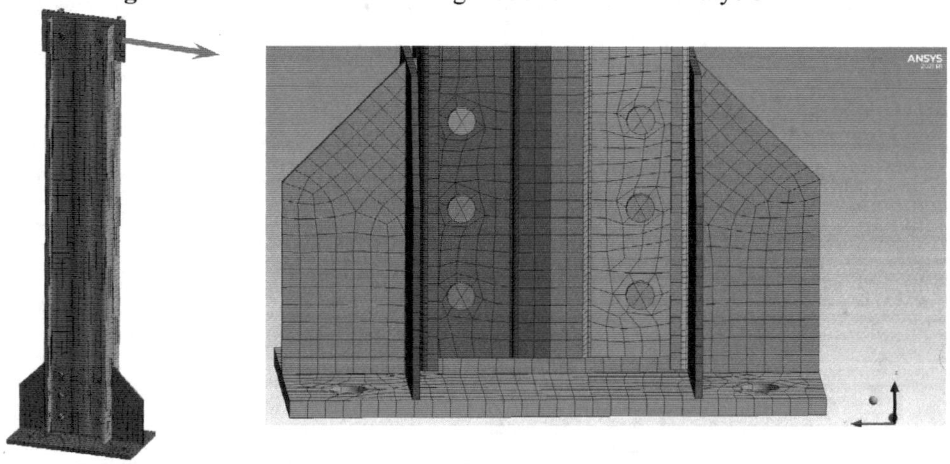

a. General view of the model **b.** Detailed view of the meshing

Figure 5. Generated model using Ansys 2021R1

The load was applied at the top of the column, in Y direction of the coordinate system, at 975 mm from the horizontal steel plate at the base, figure 1a, in incremental steps, until the failure of the model occurred. For the nonlinear analysis case the failure of the element was considered as the onset of the local buckling or stress levels that exceeded the yield strength of the CFS, according to table 1.

The displacements at the base of the steel plate were restrained in all 3 directions, so that to simulate a fixed end. Perfect bond was assumed between the bolts and the connected steel pieces.

The obtained results are analysed from the point of view of stress levels and displacements along the two in-plane directions (X and Y). The analyses will offer some insights on the overall behaviour of the column and the connection at the base and will be further validated and calibrated by means of experimental data obtained from laboratory tests on similar / identical specimens.

3. Results and discussions

3.1. Displacements

The maximum lateral displacements were recorded at the top of the column where the lateral force was applied. As it can be seen from figure 6, both FEM software led to almost the same results considering the linear analysis: 3.99 mm for Robot Structural Analysis (figure 6a) and 4.08 mm for Ansys (figure 6b), with a difference between the obtained results of 2.25%. However, for the same magnitude of the applied lateral force, the non-linear analysis conducted using Ansys 2021 R1 lead to a maximum value for the displacement at the top of the column of 4.34 mm (figure 6c). The 6.11% difference compared to the results obtained by means of linear elastic analysis could be explained by the redistribution of stresses, considering the material behaviour presented in figure 2, during the deformation process.

Figure 7 shows the distribution of the displacements in X direction, perpendicular to the direction of loading, at the same intensity of the lateral force for which figure 6 was plotted.

It can be observed that the maximum values for the displacements in the X direction were obtained for the points located at the mid height of the column while the top part of the column exhibited 0 lateral displacements with respect to the direction of loading [19]. The latter can be explained by the fact that the movement of the top of the column in X direction, outside the loading plane, is restrained due to the applied force. However, the part of the column with unrestrained displacements in X direction, perpendicular to the direction of loading, is the one located between the top of the gusset plate at the base and the load point of application.

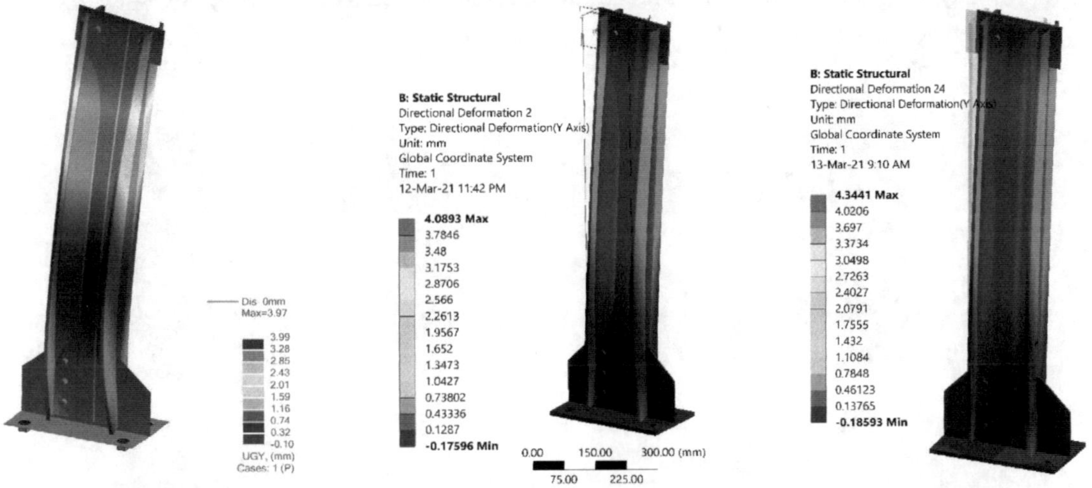

a. Case 1 **b.** Case 2 **c.** Case 3

Figure 6. Horizontal displacements in Y direction (direction of loading)

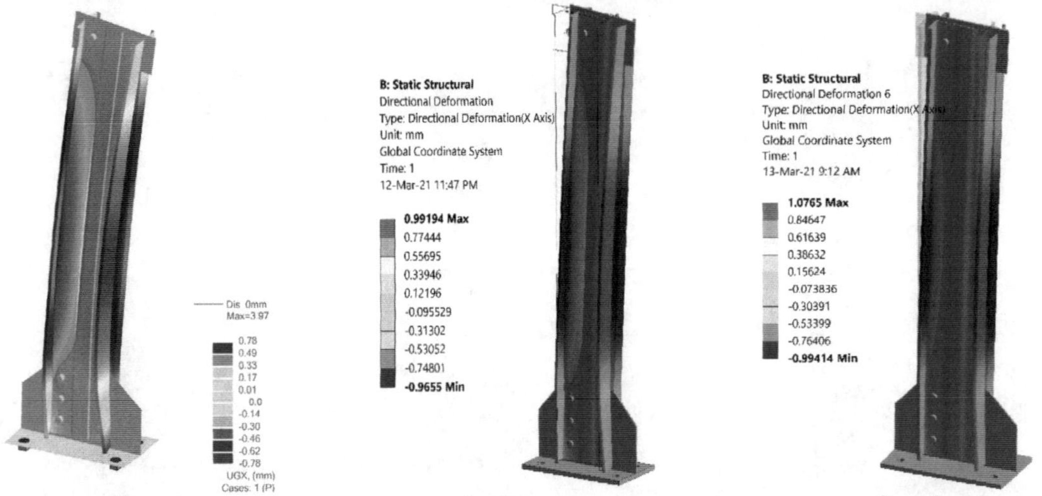

a. Case 1 **b.** Case 2 **c.** Case 3

Figure 7. Horizontal displacements in X direction (perpendicular to the direction of loading)

According to the data presented in figures 6 and 7 the magnitude of the displacement in X direction is almost 25% of the displacement in Y direction due to the fact that the element is very flexible.

This may lead to the conclusion that the column could have failed due to local buckling of the flanges of the CFS C-shaped lipped profile [20].

3.2. Stresses

Figure 8 presents a comparison in terms of the maximum principal tensile stresses obtained in Robot Structural Analysis and Ansys 2021R1. It can be observed that both software packages estimated a value of the maximum principal stress around 500 MPa, meaning the failure of the column. However, these high values were localized at the level of the third bolt on the tensioned side of the column. This would be an indicator of a local failure mechanism. The stress levels for the rest of the element did not exceed the yield limit.

The intensity of the laterally applied load, 5000N, was identical to the one that produced the displacements in X and Y directions presented in figure 7 and figure 6, respectively. According to the data presented in figures 6a, 7a and 8a, the buckling of the compressed flange of the CFS C-shaped lipped profile was observed. The local failure of the specimen at such a small intensity of the lateral load could be explained by the small thickness of the steel profile, 2mm.

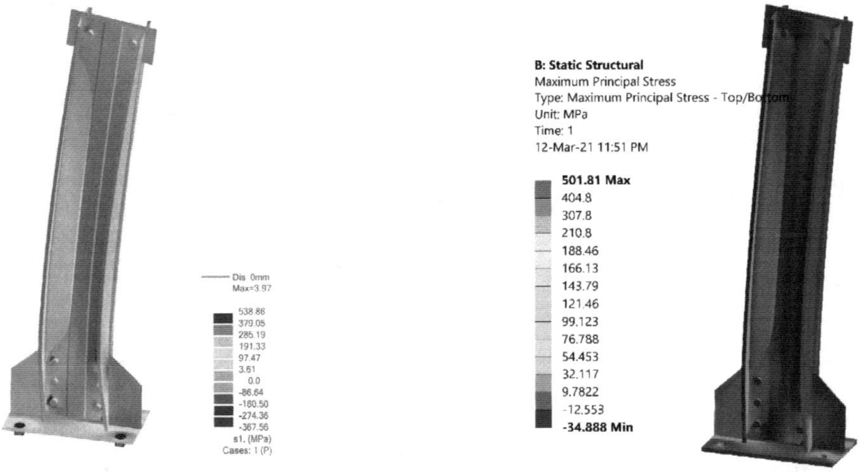

a. Case 1 **b.** Case 2

Figure 8. Principal stress distribution

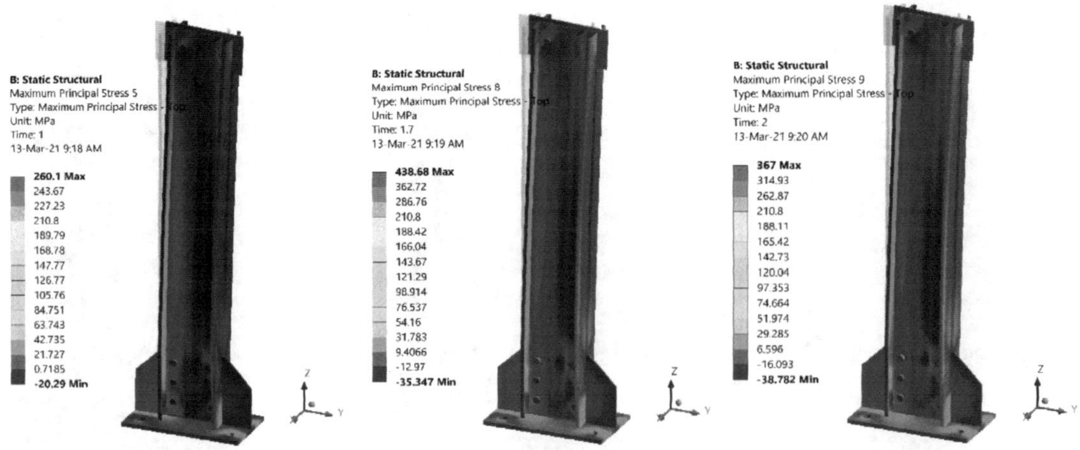

a. 5000N **b.** 8500N **c.** 10000N

Figure 9. Principal stress variation for Case 3 at different loading stages

Figure 9 presents the evolution of the principal stress at different loading stages for Case 3 (table 2).

Compared to figure 8, it can be observed that at the same magnitude of the lateral load, 5000N, the value of the principal tensile stress exceeded the yield limit slightly at the same location. The lower values for the principal tensile stress could be attributed to the possibility of stress distribution in the case of nonlinear static analysis compared to the linear one.

Increasing the load to 8500 N results in a larger area of the CFS column being subjected to tensile stresses that exceed the yield strength. The yielded part tended to concentrate on the tensioned side of the column, right above the gusset plate. As the load increased to 10000 N, 15% increase from 8500 N, the maximum value of the principal tensile stress decreased and an even larger area of the CFS profile yielded.

Localized failure could also be observed at the level of bolts located in the compressed part of the column. This would suggest local tearing of the CFS profile if these stresses are analysed taking into account the out-of-loading-plane behaviour presented in figure 6, showing a separation tendency of the two CFS profiles.

4. Conclusion

The paper presents the results obtained from a comparative analysis based on the finite element method (FEM) to assess the efficiency of a connection joint between the columns of a portal frame and the foundation block considering both the linear and nonlinear behaviour of the CFS elements. The column is made of two 150×45×10 mm CFS C-shape lipped profiles positioned back-to-back, having a thickness of 2 mm.

The values of the lateral displacements in the direction of the applied load are similar between the two employed computer programs as well as between the linear and nonlinear analysis cases. The latter is 6.11% larger compared to the linear analysis case. Important out-of-loading-plane displacements are observed for the points located at the mid height of the column. The magnitude of these displacements is almost 25% of the in-plane lateral ones. This means that the two back-to-back CFS profiles exhibit a separation tendency.

The failure of the column initiates at the level of the connection with the gusset plate. More specifically, local yielding is observed at the contact point between the CFS profile and the top bolt in the tensioned part of the column. The nonlinear static analysis is able to capture the redistribution of stresses after initial yielding. Additionally, local buckling of the compressed CFS profile flange is observed.

The model needs validation and calibration via experimental tests but it manages to offer some insights on the expected failure mode of the column and the magnitude of the laterally applied load.

Acknowledgments

The authors wish to express their gratitude to the partnership between the "Gheorghe Asachi" Technical University of Iasi, INAS S.A. and ANSYS for making available the full research license for Ansys 2021R1 used in conducting the research.

The authors thank S.C. Panoterm Import-Export S.R.L. for supplying the CFS profiles used in the research as part of the grant no. 4887/03.2019.

5. References

[1] Mojtabaei S M, Kabir M Z, Hajirasouliha I and Kargar M 2018 Analytical and experimental study on the seismic performance of cold-formed steel frames *Journal of Construction Steel Research* **143** 18-31

[2] Schafer B W 2011 Cold-formed steel structures around the world *Steel Construction* **4**(3) 141-9

[3] Smith B H, Arwade S R, Schafer B W and Moen C D 2016 Design component and system reliability in a low rise cold formed steel framed commercial building *Engineering Structures* **127** 434–46

[4] Zeynalian M, Shelley A and Ronagh H R 2016 An experimental study into the capacity of cold-formed steel truss connections *Journal of Construction Steel Research* **127** 176-186
[5] Karabulut B and Soyoz S 2017 Experimental and analytical studies on different configurations of cold-formed steel structures *Journal of Construction Steel Research* **133** 535–46
[6] Roure F, Pekoz T, Somalo M R, Bonada J, Pastor M M and Segura A 2019 Cross-aisle stiffness analysis of industrial welded cold-formed steel rack \upright frames *Thin-Walled Structures* **141** 332-44
[7] Lee Y H, Tan C S, Mohammad S, Tahir M M and Shek P N 2014 Review on cold-formed steel connections *The Scientific World Journal* **2014** ID. 951216
[8] Mathe A E, Catarig A and Moldovan I 2015 Statics and kinematics of semirigid steel frames under seismic action *Journal of Applied Engineering Sciences* **5**(2) 59-64
[9] Campian C M, Pop M and Chira N 2017 The influence of the connections between concrete and steel on the joints modelling *ce/papers – Proc. of Eurosteel, September 2017, Copenhagen, Denmark* **1**(2-3) 4680-7
[10] Bondock D H and Salim H A 2017 Failure capacities of cold-formed steel roof trusses end-connections *Thin-Walled Structures* **121** 57-66
[11] Ye J, Becque J, Hajirasouliha I, Mojtabaei S M and Lim JB P 2018 Development of optimum cold-formed steel sections for maximum energy dissipation in uniaxial bending *Engineering Structures* **161** 55-67
[12] Shahini M, Bagheri Sabbagh A, Davidson P and Mirghaderi R 2019 Development of cold-formed steel moment-resisting connections with bolting friction-slip mechanism for seismic applications *Thin-Walled Structures* **141** 217-31
[13] Bagheri Sabbagh A, Petkovski M, Pilakoutas K and Mirghaderi R 2013 Cyclic behaviour of bolted cold-formed steel moment connections: FE modelling including slip *Journal of Construction Steel Researchi* **80** 100-8
[14] Rinchen R and Rasmussen K J R 2019 Behaviour and modelling of connections in cold-formed steel single C section portal frames *Thin-Walled Structures* **143** ID. 106233
[15] Rinchen R and Rasmussen K J R 2019 Numerical modelling of cold-formed steel single C-section portal frames *Journal of Constructional Steel Research* **158** 148-155
[16] Robot Structural Analysis, https://www.autodesk.com/products/robot-structural-analysis/overview?term=1-YEAR
[17] Ansys 2021 R1, https://www.ansys.com/products/structures/ansys-mechanical
[18] https://www.mm.bme.hu/~gyebro/files/ans_help_v182/ans_elem/Hlp_E_SHELL181.html
[19] Arrayago I, Rasmussen K J R and Real E 2020 Statistical analysis of the material, geometrical and imperfection characteristics of structural stainless steels and members *Journal of Constructional Steel Research* **175** ID. 106378
[20] Dhanapal J, Ghaednia H, Das S and Velocci J 2020 Behavior of thin-walled beam-column modular connection subject to bending load *Thin-Walled Structures* 149 ID. 106536

Computer modelling of R.C. shear walls - comparative analyses at different values of the reduction coefficient. Internal forces values.

S Tsvetkov[1]
[1] University of Structural Engineering & Architecture (VSU) "Lyuben Karavelov" Sofia, 175 Suhodolska str., *Bulgaria*

E-mail: st.cvetkov@vsu.bg

Abstract. The report shows a continuation of analyses of reinforced concrete shear walls solved by horizontal forces [14]. In certain places in the walls the stiffnesses are reduced, thus comparing the obtained results between the groups of models. Solutions for wall and mixed structural systems are shown. The decisions were made with specialized Tower software based on work by the Finite Element Method. Interesting conclusions can be seen for the level between the rigid basement and the first floor. Decreases in bending moments and increases in shear forces are observed. Motivation – quick analysis of walls subjected to the action of horizontal forces (hand analysis, without program system); the goals - proving the idea of the influence of the supports from the basement levels; the methodology - solutions with specialized software - only for a certain speed in the decisions. The solutions have characteristic geometry and loads and can be easily parameterized with correction factors.

1. Introduction

Reinforced concrete walls (shear walls) are a key element of the load-bearing structure of buildings with wall and mixed structural systems. Of particular importance is the way of computer modelling (simulations) of the walls, as well as obtaining (interpretation) of the internal forces in them, necessary for their analysis. [14] The structural elements are designed for stress states: eccentric compression and shear. These structures are suitable for the modern level of design of earthquake-resistant structures. The dimensions of the walls section are designed according to the Eurocodes and has a boundary zone with enough thickness. Thicker walls absorb larger shear forces. This is beneficial on the first floors of the walls and especially in the basements. The effect of the slenderness ratio (height to thickness) affects the failure of the compression of the boundary zones of the wall, because with greater slenderness (thinner walls) smaller areas of the load-bearing reinforcement are obtained, compared to the walls with small slenderness. This is because the physico-mechanical reinforcement coefficient is lower. This is a circumstance for not ensuring the local ductility of the wall. The instability outside the plane in the shear walls is expressed precisely by the plastic mechanism of destruction - loss of stability due to its small thickness (from compression) [13]. It is possible that the bearing capacity of shear is not realised for the thin walls. Therefore, engineers must carefully choose the reduction factor α_f [2] by which the effective bending stiffness is determined. This factor reduces the stiffness of reinforced concrete elements to account for the presence of cracks in them (for critical zone). The coefficient does not take into account the real mechanisms of dissipation of seismic energy in various structural elements and systems - does not depend on the type of structural element. In this article, by reducing the stiffness by 50%, the critical zone is realized, and the Poisson's ratio is additionally zeroed. The various directions in which work is done in the worldwide - are shown in the developments [1], [4], [10] - there are been studied problems, describing various theoretical and experimental formulations. Basic information for the development of the models shown in the report was used by [5], [13], where which was illustrated with examples and theoretical postulates. Some of the studies, for example [7] - show a comparison of the deformed and stressed states of shear wall with holes, obtained in different modelling approaches, by comparing two types of computational models, based on finite elements, respectively - beam and plate. Questions about analyses that take into account geometric nonlinearities, i.e. the influence of normal forces on the rigidity of the structure are considered in [8], from where a certain analogy can be made with the frame's computational models. Analysis with ANSYS by modelling the reinforcement for slabs, is shown in [11]. Information on the mechanism of transfer of shear forces in reinforced concrete elements subjected to both shear and compression - with name aggregate interlock, is given in [12]. Reporting the susceptibility of the soil base is a characteristic indicator in the analyses, information can be seen in [6]. The requirements for sustainable development have been the criteria laid down in a number of documents, information can be read in [9].

The idea for the presented analyses was for a "quick" check at the beginning of the design, through a different type of analysis, of the existing scientific base. The report presents a study of vertical load-bearing elements - walls.

2. Information about the loads and groups of computational models

Five groups of computational models have been compiled. Tower 6 software product [15] was used. Comparisons of the values from the obtained results were made (internal forces in characteristic horizontal sections in the height of the R.C. shear walls). For the numerical experiments is accepted concrete class (grade) C25/30 according to BDS EN 206: 2013+A1: 2016 / NA: 2017 [3]. The walls for all models was with lengths 6 m and 25 cm thickness (solid walls) rectangular cross sections in plan, with a storey's height of 3 m. It is customary for the walls to be rigidly clamped (fixed base) in their foundations. Linear supports are modelled between the basement and the floor. The analysis is linear (static), with increasing horizontal forces from E_1 = 50 kN (for the first level) to E_6 = 300 kN (for the last, sixth level) - Figure 1. The dimensions of the finite elements are 10/10 cm. The software product offers an option for automatic "Generation" for discretization.

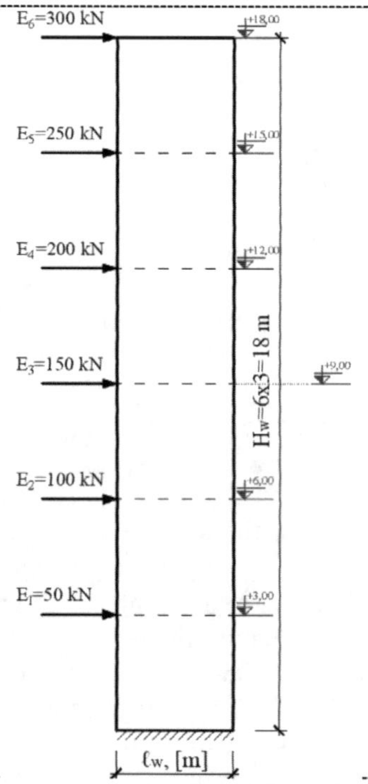

Figure 1. Geometry and applied loads for solid walls (the schemes apply to all considered solutions)

Model groups:

Group I. Models without reduced stiffness - Figure 2.
Group II. Models with 1 critical zone (above the ground – rigid basement) - Figure 3.
Group III. Models with critical zones – the rigid basements and the level above it - Figure 4.
Group IV. Models with critical zones above the ground – the rigid basement - Figure 5.
Group V. Models with critical zones – rigid basements and levels above it - Figure 6.

In the models with (purely) elastic stiffness, without critical zones in height:

$E = 3{,}1x10^7 \text{ kN / m}^2$ and Poisson's ratio ν= 0,2 are used.

Models: 1, 2 and 3. Models with adjusted stiffness only for rigid basements and/or levels above it (for wall system).

Model 4 and Model 5. Models with corrected stiffness (for walls from mixed system).

The stiffness correction is expressed in a 50% reduction of the E - modulus of the material and zeroing of the Poisson's ratio.

Shown are the critical zones by hatching (in red) in Figures 3- 6.

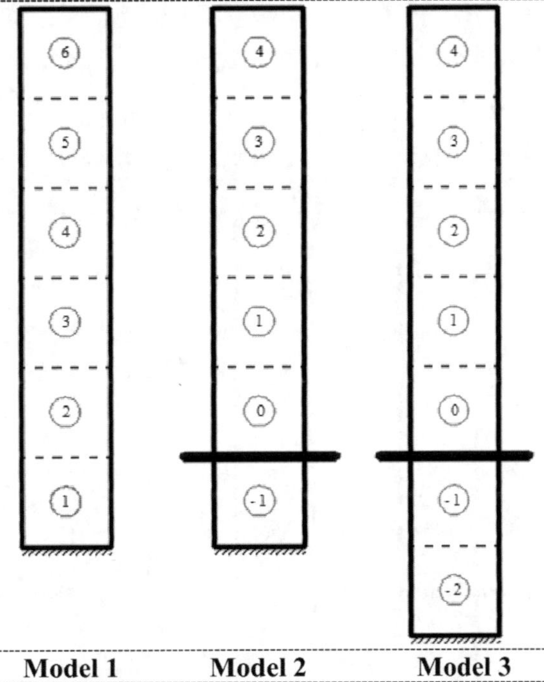

Model 1 **Model 2** **Model 3**

Figure 2. Schemes of Group I

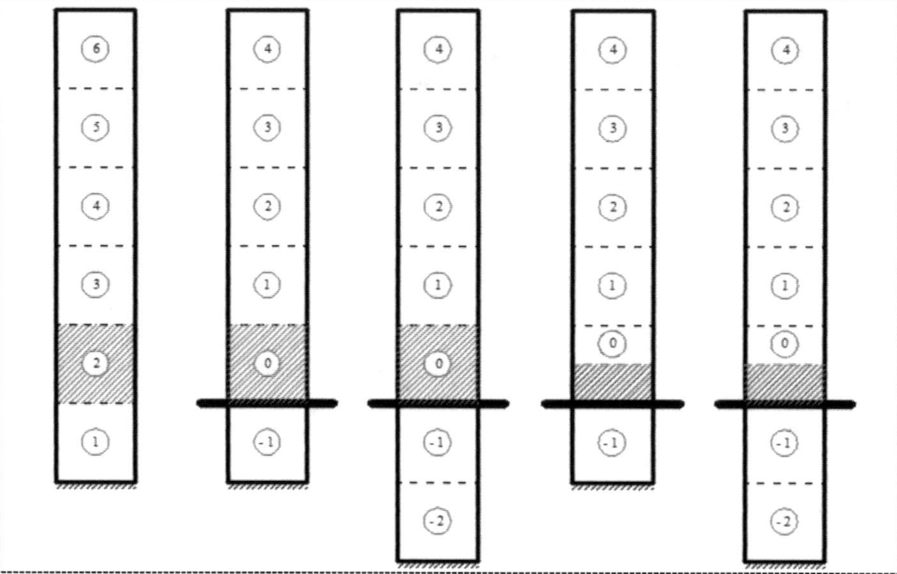

Model 1 Model 2 Model 3 Model 4 Model 5

Figure 3. Schemes of Group II

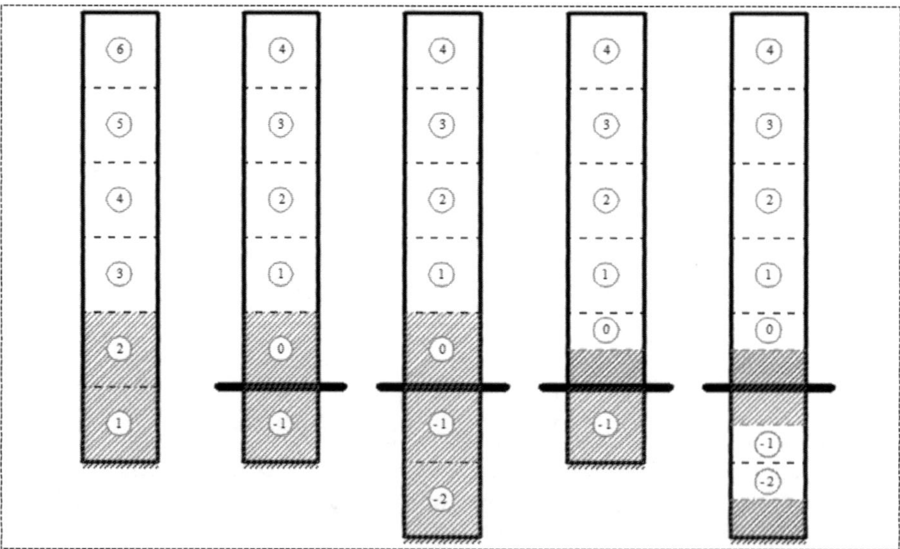

Model 1 Model 2 Model 3 Model 4 Model 5

Figure 4. Schemes of Group III

IOP Conf. Series: Materials Science and Engineering 1141 (2021) 012007 doi:10.1088/1757-899X/1141/1/012007

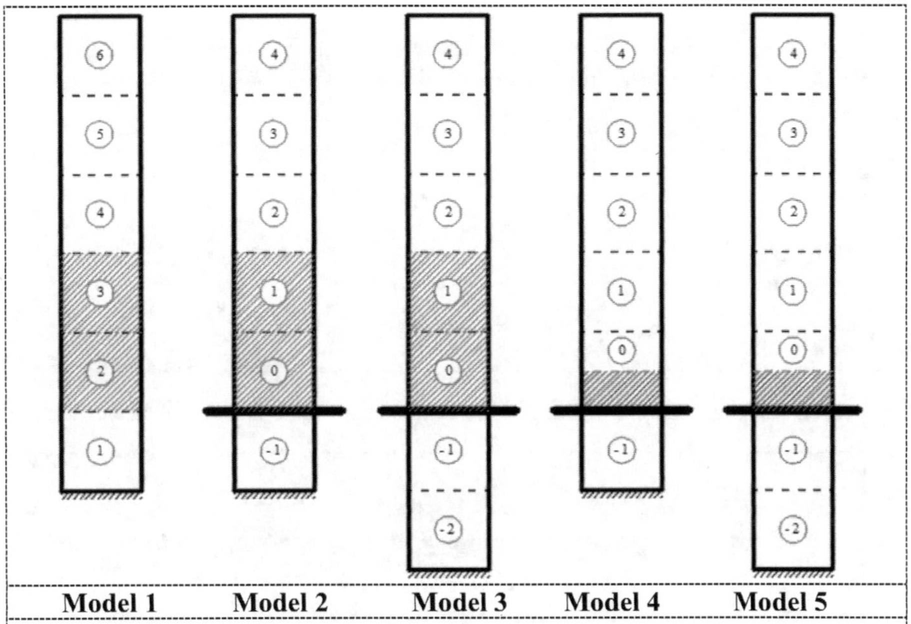

Figure 5. Schemes of Group IV

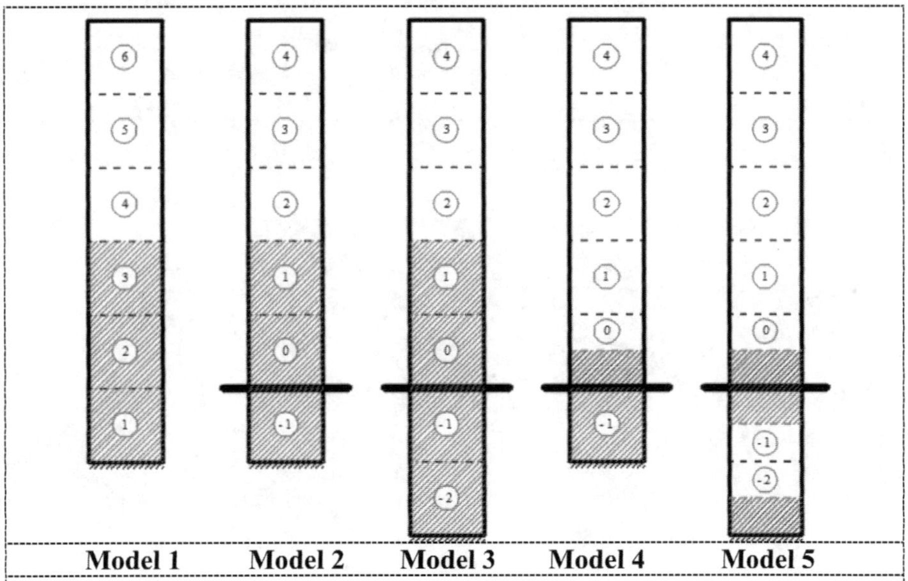

Figure 6. Schemes of Group V

3. Results of numerical experiments

The results for: bending moments M and shear forces V are presented in figures 7 – 16 (directly from the software product). For Models 1 of Figures 2, 3 and 5, the results of [14] are shown.

Note: **Capacitive correction shear force adjustment was required if they were used to analyse the load-bearing reinforcements for the elements!** [14]

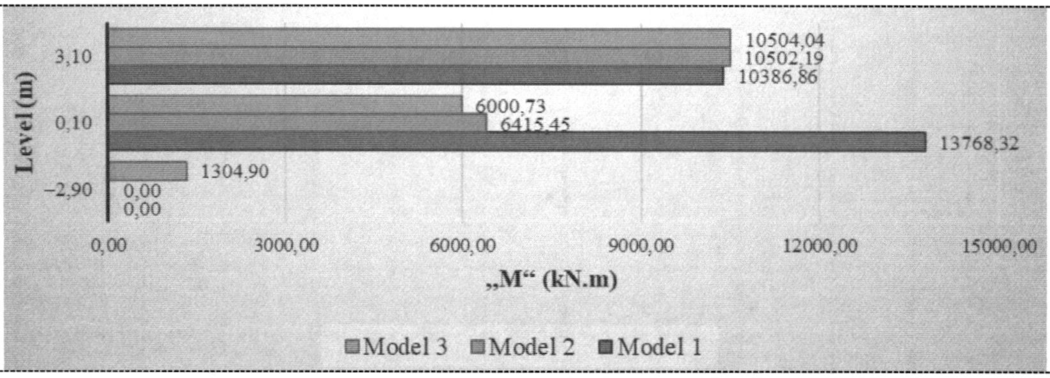

Figure 7. Values of Bending moments in Group I of models in the considered sections.

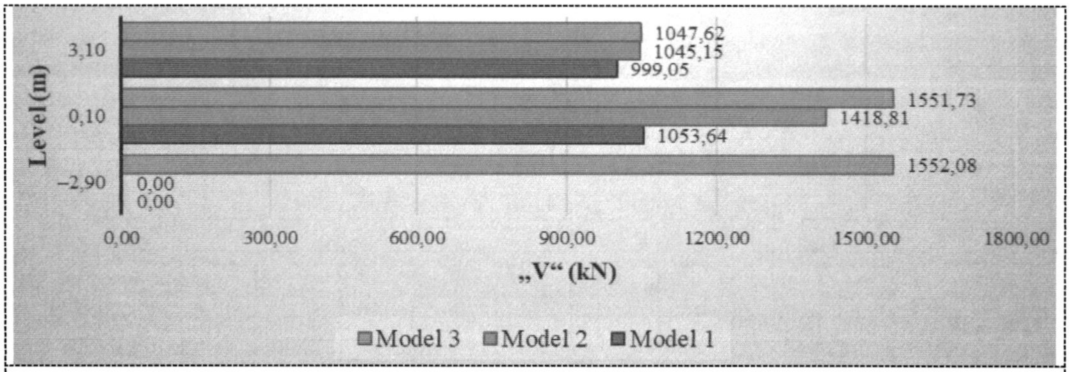

Figure 8. Values of Shear forces in Group I of models in the considered sections.

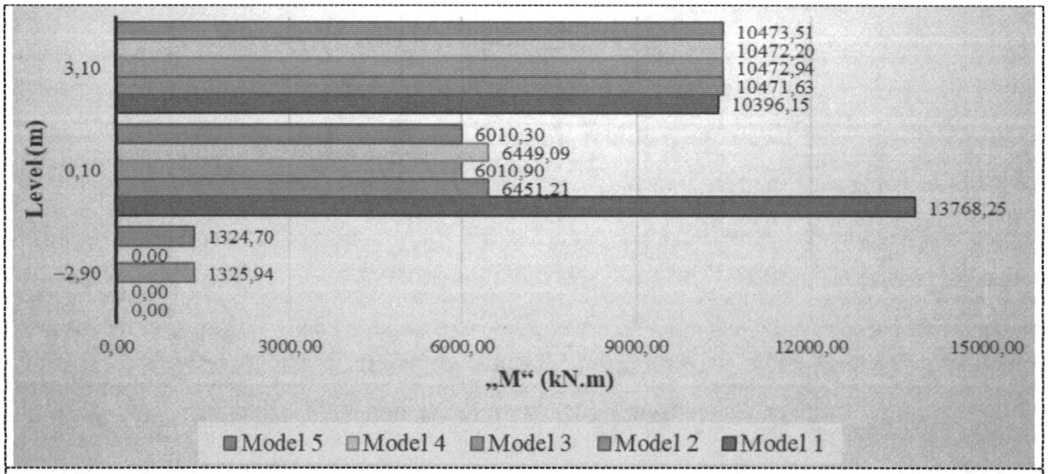

Figure 9. Values of Bending moments in Group II of models in the considered sections.

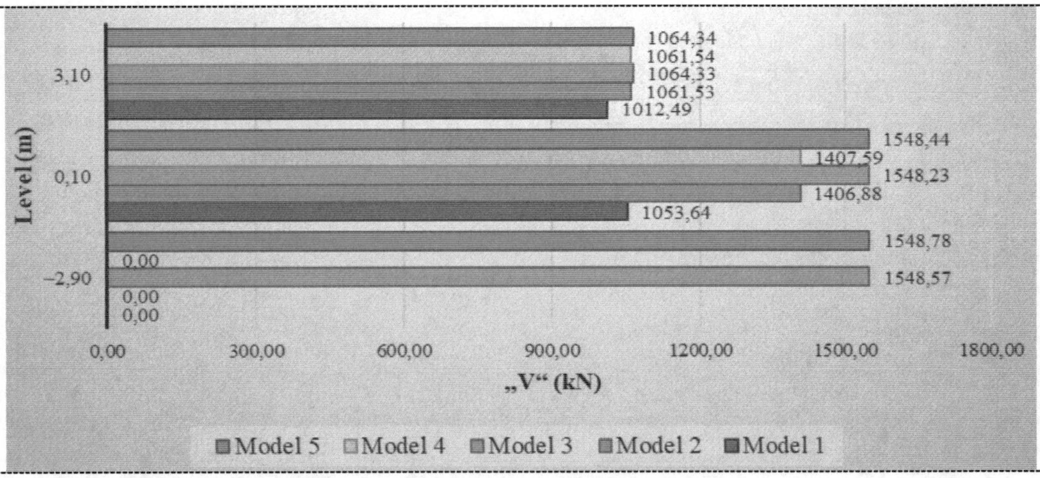

Figure 10. Values of Shear forces in Group II of models in the considered sections.

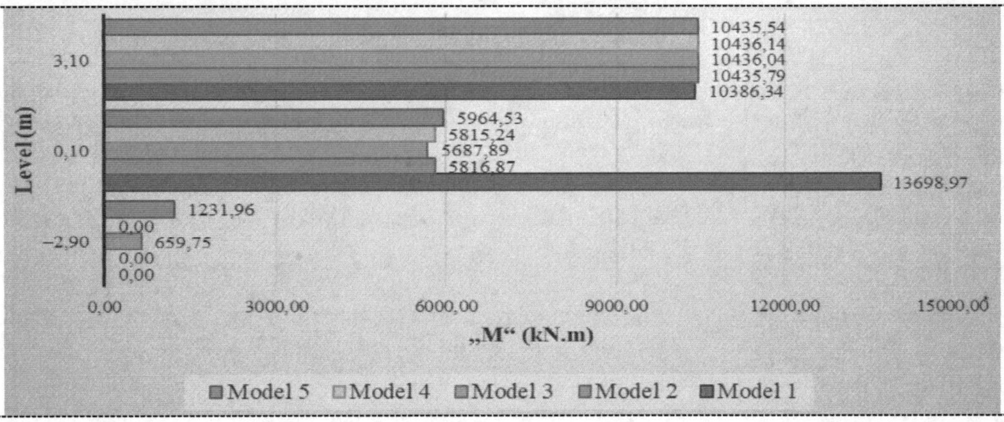

Figure 11. Values of Bending moments in Group III of models in the considered sections.

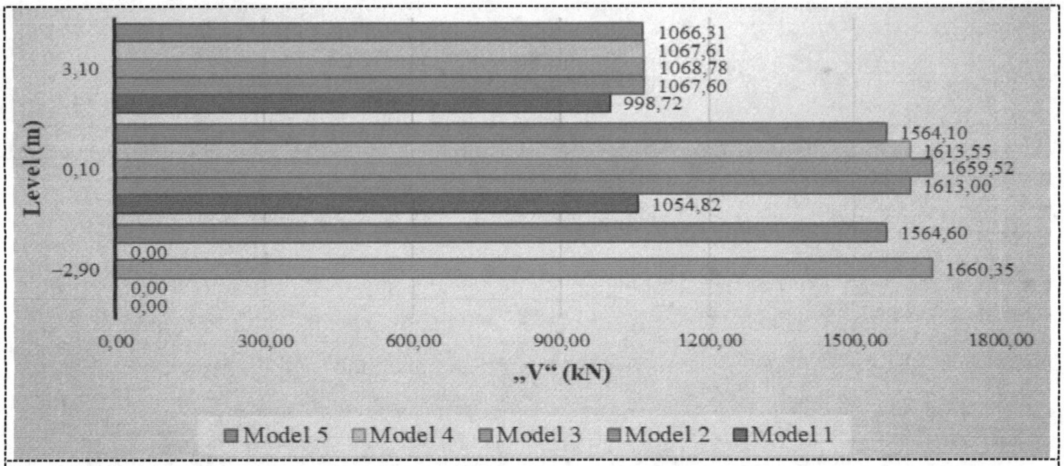

Figure 12. Values of Shear forces in Group III of models in the considered sections.

IOP Conf. Series: Materials Science and Engineering 1141 (2021) 012007 doi:10.1088/1757-899X/1141/1/012007

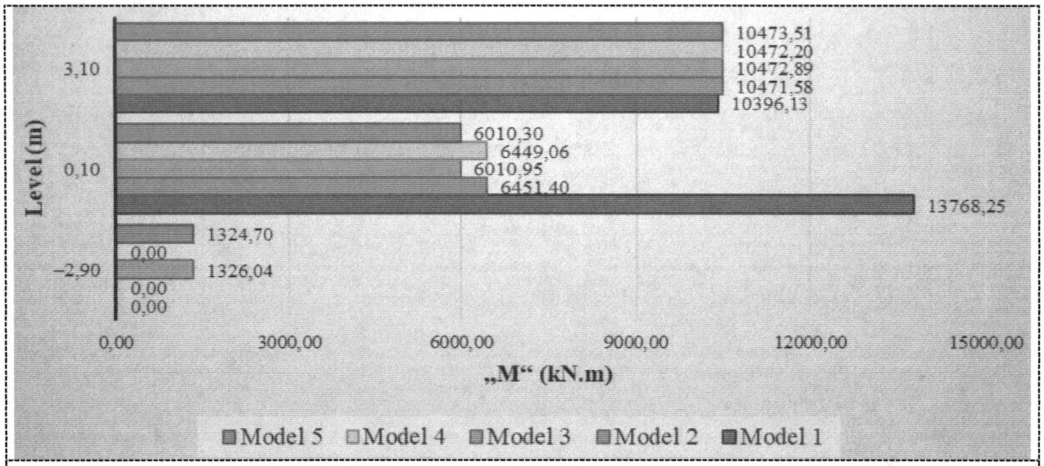

Figure 13. Values of Bending moments in Group IV of models in the considered sections.

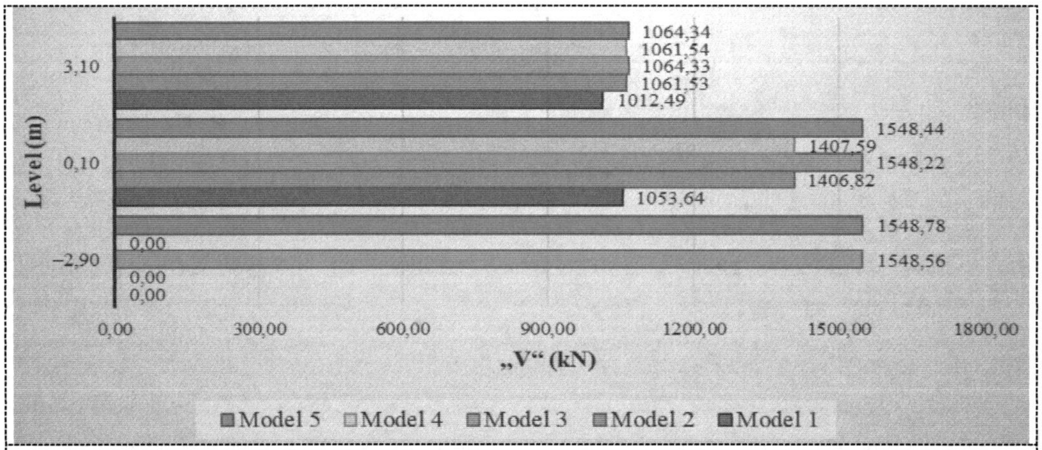

Figure 14. Values of Shear forces in Group IV of models in the considered sections.

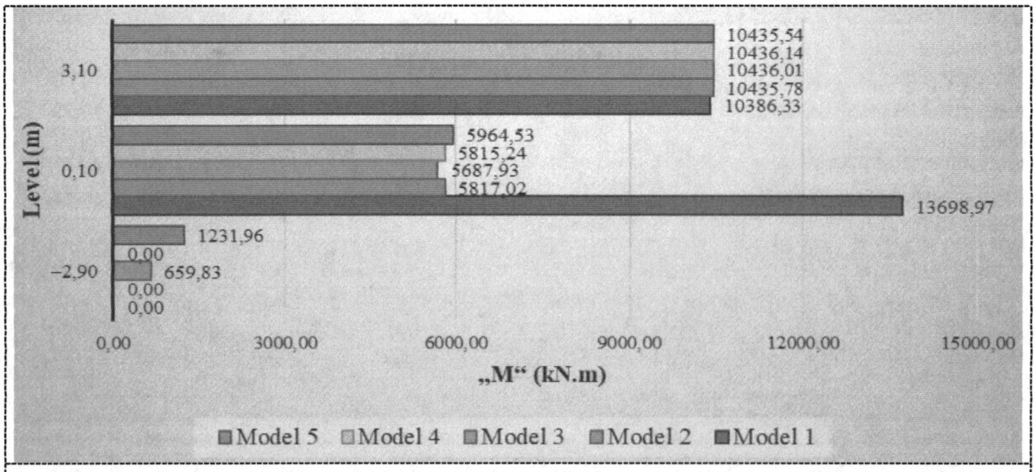

Figure 15. Values of Bending moments in Group V of models in the considered sections.

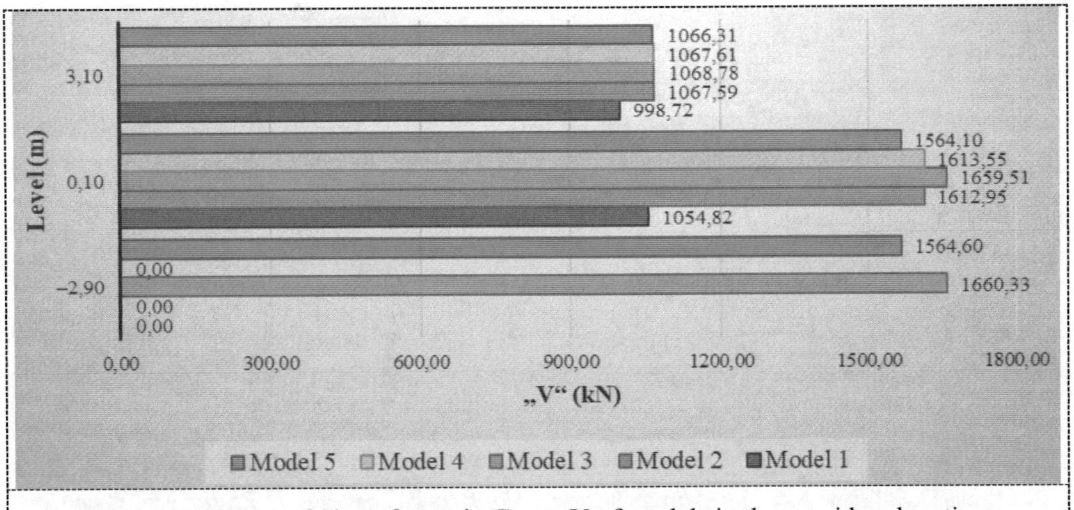

Figure 16. Values of Shear forces in Group V of models in the considered sections.

The diagrams show that:

From the presented internal forces (bending moments and shear forces) for a single section at level +3,10 m, it can be assumed that the considered models of reinforced concrete shear walls behave equally, due to the close values of the obtained results for all groups of models.

At a level +0,10 m there is a sharp limit in the values of the bending moments between the models, first, without including the level of the terrain and second considering the level of the terrain, as in places the values are several times higher.

If we consider models 2 and 3, compared to the respective group of models, it can be seen that with two underground levels for the building, the bending moments decrease and the shear forces increase significantly at +0,10 m level along the height of the shear wall. In models 4 and 5, compared to the respective group of models, the same is observed.

In models 3 and 5 the modelling of a rigid basement at two levels creates reduced bending moments at the models (no rigid basement), at the expense of the shear forces that also increase.

At –2,90 m level, there is no difference in the values of the internal forces of the individual models of the groups, which is analogous at level +3,10 m.

In general, from the presented results for the considered models, it can be concluded that the models presented in groups I, III and V behave in the same mode, which is observed in the types presented in diagrams. The same is observed for the models in groups II and IV.

Adequate reading of the rigid basement leads to a significant difference in the internal forces acting at +0,10 m above ground level, which is expressed as several times lower values of bending moments, and higher values of the shear forces, when the ground level is not calculated.

4. Conclusions

According to the location of the so-called critical zone, it can be clearly seen that when a rigid basement is modelled, the internal forces in the considered sections, i.e. the bending moments respectively decrease, and the shear force increases, regardless of the constructive system for the models.

The adoption of different computational models in the design of reinforced concrete walls leads to different results, which must be analysed and processed with greater care.

There is no significant difference in the values of bending moments and shear forces between the adoption of the two structural systems (wall system and mixed).

5. References

[1] Bertero V 1980 Seismic behavior of R/C wall structural systems. University of California, Berkeley

[2] BDS EN 1998-1: 2006 Eurocode 8: Design of structures for seismic impacts. Part 1: Basic rules, seismic effects and building rules. European Committee for Standardization

[3] BDS EN 206: 2013 + A1: 2016 / NA: 2017 - Bulgarian Institute for Standardization

[4] Chopra A 2011 *Dynamics of Structures: Theory and Applications to Earthquake Engineering*, Forth Edition, Prentice Hall

[5] Georgiev G, Tsvetkov St 2013 *Methodical guide for developing a course project on Reinforced concrete structures according to Eurocode*, Part One, Chapter Two: Seismic design of reinforced concrete wall structures, VSU (USEA), Sofia, Bulgaria p. 207, 238

[6] Hamova M, Frangov G, Zayakova H, Perikliyska M, Mihaylov A 2015 Soil liquefaction in Bulgaria – examples, prognoses and countermeasures. *Proceedings of the XVI ECSMGE Geotechnical Engineering for Infrastructure and Development*, pp. 2243 – 2248

[7] Handruleva An, Kazakov K, Yanakieva A 2010 Comparison of fem computational models of shear walls with holes, *International Conference on Civil Engineering Design and Construction and Application of Eurocodes* (Science and Practice), 9 ÷ 11 September, Varna, Bulgaria, pp. 305-310

[8] Handruleva An 2020 Mechanisms of destruction of spatial frame structure, VSU, Sofia, p.72

[9] Ivanova M, Ivanov Y, V. Stoyanov 2013 Concrete and Requirements for Sustainable Development, In: *Education, Science, Innovations - Proc. III Int. Conf. of the European Polytechnical University*, June 9-10, 2013, Pernik, Bulgaria, pp 109-116

[10] Paulay T and Priestley M 1992 Seismic design of reinforced concrete and masonry buildings. New York: John Wiley & Sons, Inc

[11] Stoynova Il, Hristov Ts 2019 Comparative analysis between by classic solutions methods of flat slab and numerical solutions. Verification of finite elements, *18-th International Symposium MASE 2019*, Ohrid

[12] Stoynova Il, Ivanov R, Kazakov K 2014 On the punching shear arising at column-flat slab connections, *VII International Scientific Conference Design and Construction of Buildings and Facilities, DCB 2014*, 11-13 September, Varna, Bulgaria, pp 507-514

[13] Tsvetkov St 2020 *Reinforced concrete structural shear walls - antiseismic "discs"*, VSU Sofia, pp. 178-186

[14] Tsvetkov St, Yanchev G 2020 Internal forces in R.C. shear walls with different stiffness – Part One (in Bulgarian), presented at *the XI International Scientific Conference „Civil Engineering Design and Construction"*, Sep. 10th – 12th, 2020, Varna, Bulgaria, ISSN 2683-071X (online), pp. 292-300

[15] Tower – http://www.vag.bg/dl/Tower6_user_manual_BUL.pdf (05.2021)

Effect of cooling on power generated by photovoltaic panels

S V Hudişteanu[1,*], N C Chereches[1], C G Popovici[1], M Verdeş[1], V Ciocan[1], M C Balan[1], F E Ţurcanu[1] and I C Scurtu[2]
[1]Gheorghe Asachi Technical University of Iaşi, Faculty of Civil Engineering and Building Services, *Romania*
[2] Naval Academy Mircea cel Batran Constanta, *Romania*

[*] E-mail: sebastian.hudisteanu@tuiasi.ro

Abstract. The paper presents a numerical analysis of the operation of photovoltaic (PV) panels integrated in fixed position on the roofs or facades of the buildings. Knowing that the efficiency of photovoltaic panels is temperature-dependent, and due to fixed PV panel position, the possibility of the improving the conversion is analysed from the point of view of the temperature of the PV cells. The model is simulated using TRNSYS software and the main functioning parameters assessed are the operating temperature of the cells, open circuit voltage, maximum power generated and conversion efficiency. The solution proposed for cooling consists in using water heat exchangers attached to the backside of the photovoltaic panel. The results highlight the direct dependence of the photovoltaic efficiency with the temperature of the panel for different positions in the same geographical location. The energy gain during the cooling interval is about 26.9 Wh/m2 (vertical), 81.9 Wh/m2 (inclined) and 81.7 Wh/m2 (horizontal), which represents an increase of 5.8%, 9.3% and 9.2% respectively, compared to the normal operating conditions.

1. Introduction

Photovoltaic cells are made of several semiconductor materials, but the vast majority of solar cells (95%) are made of silicon (Si). It is one of the most widespread chemical elements in the Earth's crust, representing approximately 25% of it. Therefore, it is available in sufficient quantities, being accessible at a reasonable price [1]. Also, the processing of the material is not aggressive to the environment.

Advantages of using photovoltaic panels:

- direct conversion of solar radiation into electricity;
- no moving elements (no noise);
- long life.

Disadvantages:

- dependence on climatic conditions (clouds, dust etc.);
- diurnal variation of solar radiation;
- dropping of the operating parameters when the temperature of the cells is rising.

When photons from solar radiation strike a photovoltaic cell, they can be reflected, transmitted or absorbed. Only the absorbed photons with a certain level of energy can generate electricity, figure 1. Almost all the solar energy captured by photovoltaic panels not converted into electricity, about 80%, is converted into heat [2].

The performance of photovoltaic panels is known to be temperature-dependent [3]. There are studies regarding the dependence of the conversion efficiency to the cell temperature [4].

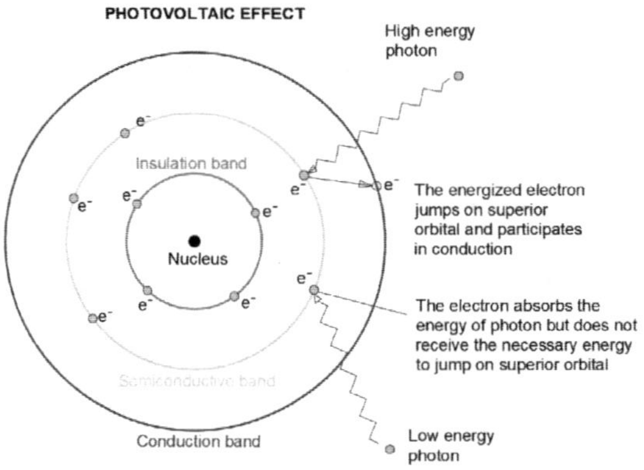

Figure 1. Photons which are determining the photovoltaic effect

A linear variation tendency of the temperature-power dependence, figure 2, is noticed in most studies [5]. Thus, when a temperature increase occurs, the efficiency of converting the energy of solar radiation into electricity decreases. As a guideline value, a reduction of the efficiency of the photovoltaic panels by 0.3 ... 0.5% can be considered [6] (0.45% according to [1]), for each degree of temperature increase. The reduction of the efficiency and of the maximum power produced by the photovoltaic panel occurs due to the decrease of open-circuit voltage, V_{oc}, by 0,30 ... 0,50 %/°C, when the operating temperature of the cells increases.

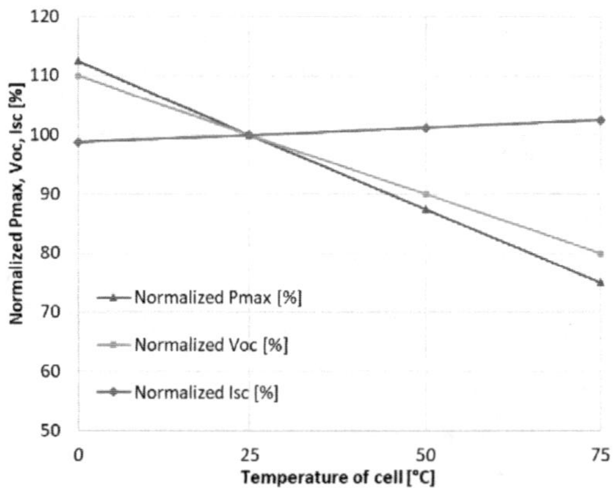

Figure 2. Dependence of P_{max}, V_{oc} and I_{sc} with the operating temperature of the cell

The improvement of the performance of the photovoltaic panels can be achieved by controlling the operating temperature of the cells, while the control of the other parameters implied is more challenging to attain. For example, in the particular case of the placement of photovoltaic panels on the buildings on vertical, inclined or horizontal non-orientable surfaces, the solar radiation becomes an uncontrollable parameter.

Various methods and relationships for determining the dependence between the conversion efficiency of photovoltaic modules and their temperature are presented in the literature [4, 7]. Both the open-circuit voltage (V_{oc}) and the fill factor (FF) are influenced negatively by the temperature, while a small increase of the short-circuit current (I_{sc}) is registered [7]. This linear correlation has the following form:

$$\eta_c = \eta_{tref} \left[1 - \beta_{ref} \left(t_c - t_{ref}\right)\right] \quad (1)$$

where, η_{tref} - electrical efficiency at the reference temperature (t_{ref});

t_{ref} - reference temperature of 25 °C;

β_{ref} - temperature coefficient. It is mainly a material property, having values of about 0.004 K^{-1} for crystalline silicon modules [8].

As the power of a photovoltaic cell varies depending on the temperature and radiation changes, the standard conditions parameters (STC) have been defined (t_{ref} = 25 °C, G = 1000 W/m^2, AM1.5), which produce the so-called Watt-peak [W_p] power [9].

The experimental tests [10, 11] and numerical analysis [12, 13] on monocrystalline or polycrystalline photovoltaic panels are current concerns in the literature [1, 14, 15]. Those researches express the necessity of cooling solutions due to the reduction of the conversion efficiency when the operating temperature of PV panels raises. There are various ways of cooling the photovoltaic panels, but the most wide-spread solutions consist in air cooling [16] and water cooling [17].

The simultaneous conversion of solar energy into electric and thermal energy using photovoltaic-thermal panels is known in the literature as PVT (Photovoltaic-Thermal System) [18, 19]. A particular feature of this solution is the BIPVT (Building Integrated Photovoltaic-Thermal System) concept, which implies the integration of photovoltaic panels into buildings [20, 21]. The generation of two of the most used types of energy is performed under the conditions in which the operation of each of the two systems is beneficial to the other, with priority to the production of electricity and ensuring the efficiency of photovoltaic conversion.

Different technologies to improve the efficiency of photovoltaic panels by passive cooling, using air, are studied in the literature [1, 15, 22-23]. The characteristic values for two photovoltaic cells were compared through experimental tests, under different conditions [24]. This system consists of a network of rectangular fins made of aluminium, being thermally coupled to the photovoltaic cell through a thermal conductive paste. There was an increase in the heat extracted from the photovoltaic cell and a decrease in its temperature, with values between 5 °C and 9 °C. Also, the integration of photovoltaic panels in the building's envelope, as the exterior glazing of the double-skin facades, is considered a viable solution [25-27].

The use of water as a cooling agent for photovoltaic panels is a solution analysed in different studies [28-31]. The hybrid photovoltaic panels studied in literature have attached a water coil with the role of both cooling the PV panel and solar collector. A solution for the cooling of the photovoltaic panels by using water as a heat transfer agent is presented in [28]. This study presents a hybrid photovoltaic panel with an attached functionally graded material backside, with a serpentine inside for water circulation. Besides its primary role of PV panel cooler, this circuit is also used as a solar collector. An example of such a system is presented in [32]. The modelling of heat transfer in the case of photovoltaic-thermal systems under variable atmospheric conditions is studied in [33, 34]. The water-cooling techniques are placed in the category of active cooling solutions and are superior to air cooling, but more expensive.

Also, newer solutions like nanofluids [35], phase change materials [32, 36, 37] or heat pipes [38] are developing fast and emphasizes the real concerns in cooling the photovoltaic panels in order to obtain higher efficiencies.

Most of the studies on the cooling of photovoltaic panels [18, 21, 39-40] are also focused on using the thermal energy extracted in this process, resulting shorter recovery time of the investments comparing to classical stand-alone photovoltaic systems [12]. Since the efficiency of the photovoltaic panels is lower than the solar thermal, it is opportune to use hybrid systems that ensure optimized operation of the whole assembly [24].

The main aim of the proposed work is to determine the quantitative advantages of using a water heat exchanger for cooling the PV panels and to calculate the additional generated power and the required water flow rate [1].

2. Problem description

The main objective of this work is to determine through numerical modelling the operating parameters reported for a 1 m^2 surface of a photovoltaic panel integrated on the roof or in the facade of a building.

IOP Conf. Series: Materials Science and Engineering 1141 (2021) 012008 doi:10.1088/1757-899X/1141/1/012008

The photovoltaic panel consists of 72 solar cells connected in series, being placed in different positions in the same temperate climate area – Constanta, Romania.

The analysis aims to determine the efficiency of PV panel both in normal operation and cooling conditions. The cooling of PV panel is performed by using a heat exchanger, attached to the backside of the panel. In order to achieve high efficiencies, the photovoltaic panel is cooled to a temperature close to the nominal one of t_{ref} = 25 °C.

The heat exchanger must allow the heat to be dissipated from the rear area of the photovoltaic panel. The contact zone between heat exchanger and the photovoltaic panel must be made of a material with a high thermal conductivity (copper or aluminium).

3. Numerical simulation

The external conditions considered are those for a typical day of summer (July) in Constanta, Romania (44.1598° N, 28.6348° E) [41]. The photovoltaic panel is analysed in three main positions: vertical-south, inclined-south and horizontal, while the intensity of the solar radiation is considered to be variable during a day.

The numerical simulation was conducted using TRNSYS software. The model presented in figure 3 has the following blocks for: climatic data, polycrystalline photovoltaic panel, elements of conversion of the measurement units, tracking and output data.

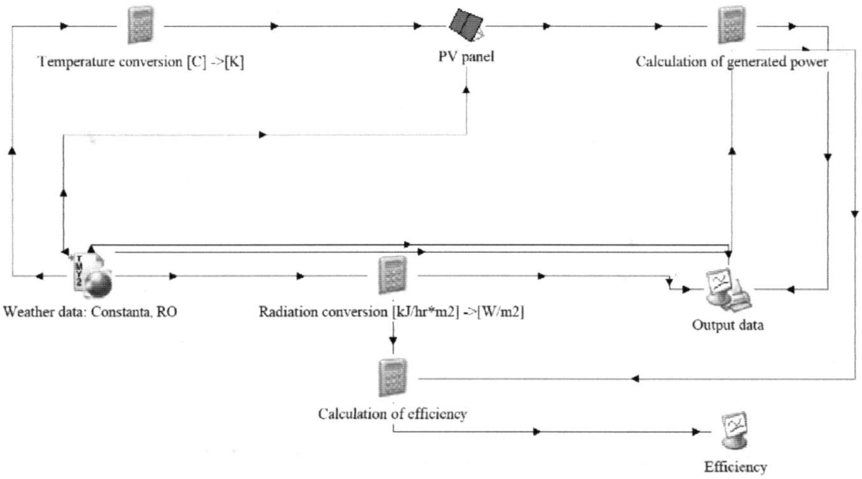

Figure 3. Numerical modelling of the photovoltaic panel with the TRNSYS software.

The recommended unit for the PV panel modelling is Type 94. This block which is used in simulations is appropriate for both monocrystalline or polycrystalline silicon PV modules. The four parameters model [42] used for a more realistic PV panel functioning and evaluation is described by the following equations which are taking into account both series (R_s) and shunt (R_{sh}) resistances [6, 43]. The current, I, generated by PV cell is determined using Eq. 2:

$$I = I_{ph} - I_D\left(exp\frac{q(V+R_sI)}{nKT_{cell}} - 1\right) - \frac{(V+R_sI)}{R_{sh}} \tag{2}$$

The open-circuit voltage (V_{oc}) of the cell is calculated using Eq. 3:

$$V_{oc} = \frac{nKT_{cell}}{q}\ln\left(\frac{I_{ph}}{I_D} + 1\right) \tag{3}$$

The **operating parameters** of the polycrystalline PV panel studied SPR-P17-330-COM [44] are presented in table 1.

Table 1. Operating parameters of the polycrystalline PV panel

short-circuit current	I_{sc} = 8.47 A
open circuit voltage	V_{oc} = 50.9 V
current at the maximum power point	I_{max} = 7.88 A
voltage at maximum power point	V_{max} = 41.9 V
operating temperature of the panel under nominal conditions	t_{NOCT} = 47 °C
net module area	S = 2.06 m^2

The **input** data required for the numerical model are presented in table 2.

Table 2. Input data required for the numerical model

PV panel orientation	South (for vertical and inclined positions)
PV panel position	Vertical (on the façade of the building), inclined and horizontal
climatic conditions	typical day of July according to the Typical Meteorological Year (TMY) of Constanta, Romania
solar radiation intensity	variable

Th **output** data from simulations:

- the operating temperature of the photovoltaic panel;
- current intensity, voltage and electrical power produced by PV panel;
- PV panel efficiency.

4. Results

Figures 4-6 show the values resulted from the simulations, under the conditions described above, for a typical day of production. On the left-hand axis there can be determined the amount of solar radiation [W/m^2] and the power generated by the photovoltaic panel on its maximum operating point [W/m^2]. On the right-hand axis, the values of the outdoor temperature as well as the operating temperature of the photovoltaic panel [°C] are available.

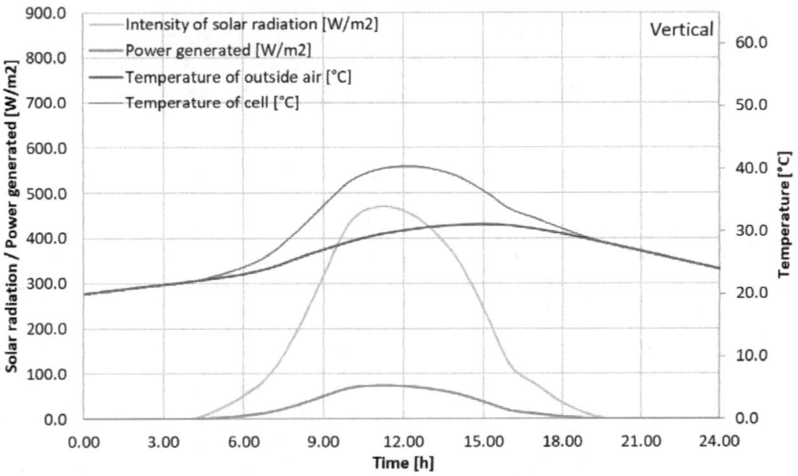

Figure 4. Daily variation of climatic parameters and PV panel output for the vertical position

During the typical day of summer, the temperature of the PV panel has a proportionate variation with the intensity of the solar radiation and with the temperature of the outside air. The power produced by the PV panel is also dependent on the intensity of the solar radiation, but inversely proportional to the temperature of the cells.

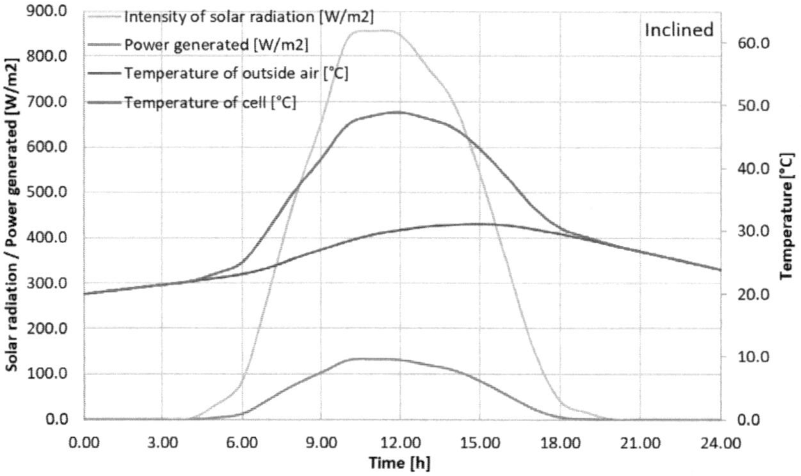

Figure 5. Daily variation of climatic parameters and PV panel output for the inclined position

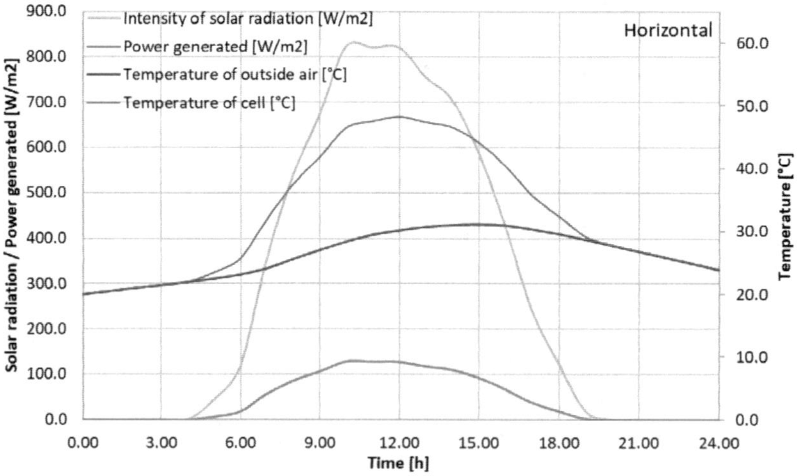

Figure 6. Daily variation of climatic parameters and PV panel output for the horizontal position

Under these conditions the photovoltaic panel operates according to the parameters presented in table 3. The conversion efficiency has values below 16%, with a maximum around 15.96%.

$$\eta = \frac{P_{mp}}{G \cdot S} \quad [\%]$$

Table 3. PV panel parameters under normal operating conditions

Hour [h]	Intensity of solar radiation [W/m²]			Temperature of PV panel [°C]			Efficiency in normal conditions [%]			Electric power generated [W/m²]		
	V /	I /	H	V /	I /	H	V /	I /	H	V /	I /	H
5.00	20.6	31.4	45.0	23.0	23.2	23.5	14.65	14.92	15.15	3.0	4.7	6.8
6.00	51.9	84.0	119.0	24.3	25.0	25.8	15.22	15.49	15.67	7.9	13.0	18.6
7.00	100.7	272.6	357.0	26.4	30.1	32.0	15.57	15.94	15.96	15.7	43.4	57.0
8.00	195.4	481.8	541.0	30.0	36.3	37.6	15.82	15.89	15.86	30.9	76.6	85.8
9.00	318.9	652.5	674.0	34.1	41.4	41.9	15.90	15.73	15.71	50.7	102.6	105.9
10.00	437.5	837.2	822.0	38.0	46.8	46.5	15.86	15.49	15.51	69.4	129.7	127.5
11.00	471.0	855.3	820.0	39.8	48.3	47.5	15.82	15.44	15.49	74.5	132.1	127.0
12.00	461.8	847.4	820.0	40.3	48.8	48.2	15.81	15.44	15.47	73.0	130.8	126.9
13.00	423.5	776.4	755.0	40.1	47.8	47.4	15.82	15.51	15.54	67.0	120.4	117.3
14.00	353.9	698.0	705.0	38.8	46.4	46.5	15.82	15.60	15.59	56.0	108.9	109.9
15.00	244.7	542.4	587.0	36.5	43.1	44.1	15.77	15.74	15.71	38.6	85.4	92.2
16.00	121.3	347.3	424.0	33.6	38.6	40.3	15.52	15.82	15.81	18.8	55.0	67.0
17.00	76.0	156.7	240.0	32.0	33.8	35.6	15.28	15.64	15.79	11.6	24.5	37.9
18.00	36.1	41.5	122.0	30.4	30.5	32.3	14.84	14.93	15.55	5.4	6.2	19.0
19.00	10.8	15.9	18.0	28.9	29.1	29.1	14.04	14.31	14.39	1.5	2.3	2.6

The photovoltaic efficiency is directly influenced by the intensity of the solar radiation but also by the temperature of the panel, which reaches values up to 48.8 °C for solar radiation of 847.4 W/m^2, table 3. In order to decrease the operating temperature and to improve the conversion efficiency of the photovoltaic panel, a cooling down to the operating temperature toward the STC of t_{ref} = 25 °C is analysed. Considering the values of the parameters in table 3, it is recommended that the cooling of the photovoltaic panel must be done between 08:00 and 15:00.

The energy balance of the PV panel is performed in order to determine the necessary heat extraction in order to achieve the standard operating temperature of the photovoltaic panel after cooling [46]. The energy balance is achieved for hourly intervals where the intensity of the solar radiation is high and the conversion efficiency can be improved by reducing the panel temperature. Knowing the specific heat capacity of the components of a PV panel [33], there are determined specific heat of the entire panel and the thermal energy extracted under the same environmental conditions in order to maintain the PV panel at t_{ref}= 25 °C.

Table 4. Properties of the main layer of photovoltaic panels [33]

Layer	Thickness *t*, [m]	Thermal conductivity *k*, [W/m·K]	Density *ρ*, [kg/m³]	Specific heat Capacity *c*, [J/kg·K]
Glass	0.003	1.8	3000	500
ARC	$100 \cdot 10^{-9}$	32	2400	691
PV Cells	$225 \cdot 10^{-6}$	148	2330	677
EVA	$500 \cdot 10^{-6}$	0.35	960	2090
Rear contact	$10 \cdot 10^{-6}$	237	2700	900
Tedlar	0.0001	0.2	1200	1250

Thus, the analysed PV panel has an approximate weight of 20.3 kg and an average specific heat capacity of 594.3 J/kg·K.

In table 5 the values of the parameters of the photovoltaic panel are presented under cooling conditions. The parameters are determined as the average calculated on the hourly intervals.

Because the efficiency of a photovoltaic cell is also influenced by the solar radiation levels, it is noticed that when G is below 1000 W/m^2, even if t_{ref}= 25°C, the efficiency is less than the one for STC.

Table 5. PV panel parameters under cooling conditions

Hourly interval	t_{cell} [°C] V	I	H	t_{ref} [°C]	η_{eff} [%] V	I	H	η_{opt} [%] V	I	H	P_{sup} [W/m²] V	I	H
8.00-9.00	30.0	36.3	37.6		15.82	15.89	15.86	16.18	16.70	16.75	0.7	3.9	4.9
9.00-10.00	34.1	41.4	41.9		15.90	15.73	15.71	16.55	16.89	16.90	2.1	7.6	8.0
10.00-11.00	38.0	46.8	46.5		15.86	15.49	15.51	16.79	17.01	17.01	4.1	12.7	12.3
11.00-12.00	39.8	48.3	47.5	25	15.82	15.44	15.49	16.88	17.06	17.06	5.0	13.9	12.9
12.00-13.00	40.3	48.8	48.2		15.81	15.44	15.47	16.90	17.09	17.09	5.0	14.0	13.3
13.00-14.00	40.1	47.8	47.4		15.82	15.51	15.54	16.89	17.11	17.10	4.5	12.4	11.8
14.00-15.00	38.8	46.4	46.5		15.82	15.60	15.59	16.81	17.10	17.10	3.5	10.5	10.7

Where: t_{cell} – cell temperature in normal operating conditions - without cooling;
η_{eff}, η_{opt} - conversion efficiencies before and after optimization (cooling);
P_{sup} - additional power produced by PV panel in cooling conditions.

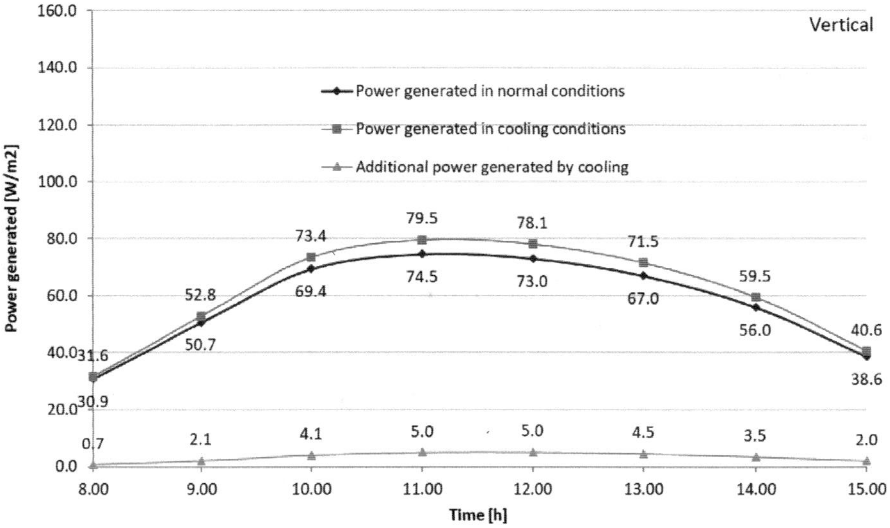

Figure 7. The power generated by PV panel in normal conditions vs cooling conditions – vertical

Considering that the efficiency is decreasing by 0.45% for each degree above the standard temperature of 25 °C [1], the cooling is determining a power increase according to figures 7-9.

According to table 5, when the temperature of the photovoltaic panel is reduced to 25 °C in 08:00 - 15:00 interval, the conversion efficiency reaches up to 17.11% (13.00-14.00).

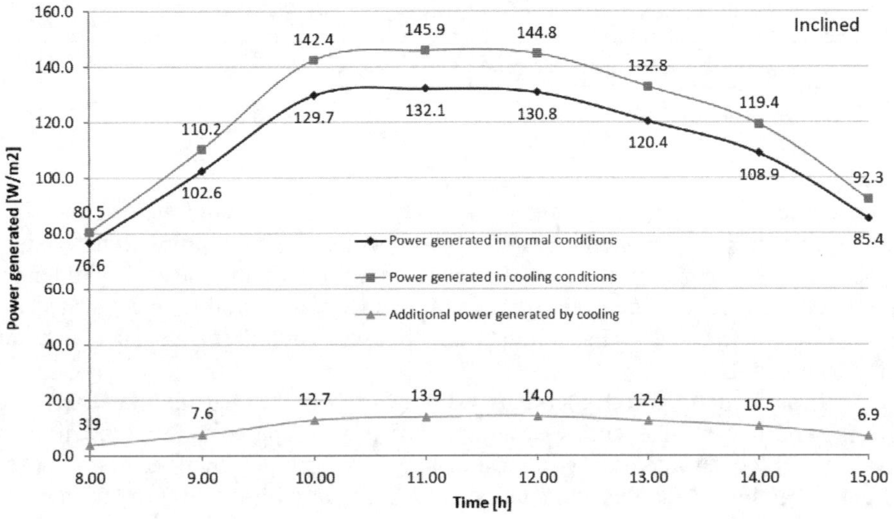

Figure 8. The power generated by PV panel in normal conditions vs cooling conditions – inclined

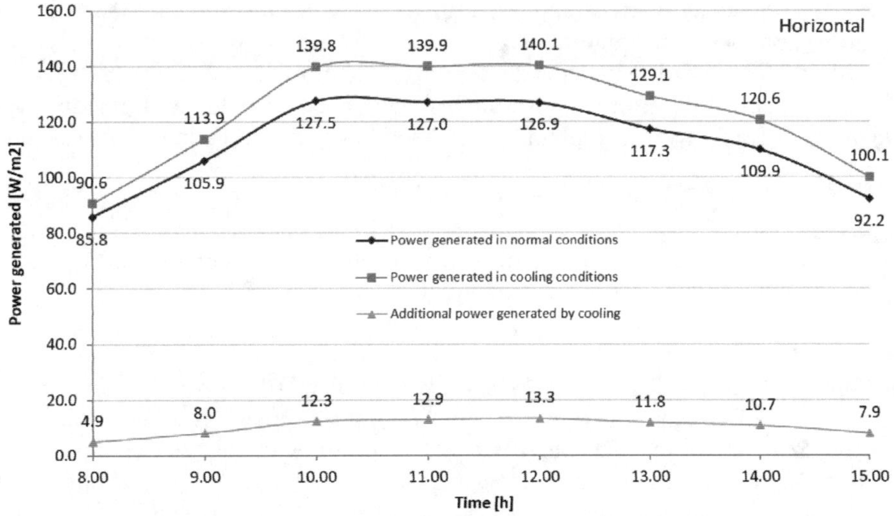

Figure 9. The power generated by PV panel in normal conditions vs cooling conditions – horizontal

Consequently, the electrical power produced by the photovoltaic panel under similar conditions of solar radiation is higher than that obtained under normal operating conditions. The energy gain during the cooling interval, 08:00 - 15:00, is about 26.9 Wh/m^2 (vertical), 81.9 Wh/m^2 (inclined) and 81.7 Wh/m^2 (horizontal), which represents an increase of 5.8%, 9.3% and 9.2% respectively, compared to the normal operating conditions.

Considering that temperature of the supply water is maximum 15 °C for selected location during summer, the following flow rates would be necessary for obtaining the cooling of 1 m^2 of PV panel towards t_{ref}, table 6.

These results show that it can be obtained a large quantity of water at approximate 25 °C, that can be used as a preheated thermal agent in buildings with hot water requirements [46] or guided to be used as conventional clean water for toilets.

Table 6. Average flow rates necessary for cooling 1 m^2 of PV panel for interval 8.00-15.00

	Vertical	Inclined	Horizontal
Flow rate for cooling [l/min]	36.7	56.7	58.3
Flow rate for cooling [m^3/day]	2.2	3.4	3.5

5. Conclusions

Cooling of photovoltaic panels is an advantageous solution for improving their conversion efficiency. The reduction of the operating temperature achieved by using water as a heat transfer agent has many advantages, being a cheap source with optimal parameters for extracting the excess of heat stored in the mass of the photovoltaic panel. The preheating of the cold water available in the water supply system would be useful in further processes in buildings with hot water requirements, such as hotels, laundries and hospitals or industrial.

The use of a heat exchanger and a storage system for the extracted thermal energy determines a superior efficiency of the whole assembly evaluated globally, compared to the stand-alone photovoltaic system. In the case of PV/T systems, the highest efficiency of the photovoltaic panel must always be a priority and the extraction of the thermal energy must be done with the lowest auxiliary consumption possible. In further numerical and experimental studies there will be analyzed other possibilities to improve the performance of these systems, in the conditions of their integration in buildings.

The energy gain during the cooling interval is about 26.9 Wh/m^2 (vertical), 81.9 Wh/m^2 (inclined) and 81.7 Wh/m^2 (horizontal), which represents an increase of 5.8%, 9.3% and 9.2% respectively, compared to the normal operating conditions.

The results underline the direct dependence of the photovoltaic efficiency with the operating temperature of the panel for different positions of PV panel, also highlighting the necessity of a particular analysis for each system implemented, depending on its location, position and time of the year.

Acknowledgments

This work was supported by a grant of the Romanian Ministry of Education and Research, CCCDI - UEFISCDI, project number PN-III-P2-2.1-PED-2019-1294, within PNCDI III.

References

[1] Hudişteanu S V 2016 *Efficiency of the photovoltaic panels integrated in buildings with double skin façade* PhD thesis "Gheorghe Asachi" Technical University of Iasi Romania

[2] Hudişteanu S V, Mateescu T D, Chereches N C and Popovici C G 2015 Numerical study of air-cooling photovoltaic panels using heat sinks *Romanian Journal of Civil Engineering* **6** pp 11-21

[3] Dubey S, Sarvaiya J N and Seshadri B 2013 Temperature dependent photovoltaic (pv) efficiency and its effect on pv production in the world - a review *Energy Procedia* **33** pp 311–321

[4] Skoplaki E and Palyvos J A 2009 On the temperature dependence of photovoltaic module electrical performance: A review of eficiency/power correlations *Solar Energy* **83** pp 614–624

[5] Jafari Fesharaki V, Dehghani M and Jafari Fesharaki J 2011 The Effect of Temperature on Photovoltaic Cell Efficiency *Proceedings of the 1st International Conference on ETEC* **1**

[6] Hudişteanu S V, Mateescu T D and Popovici C G 2015 Five parameter model of photovoltaic panel implemented in Matlab/Simulink *Bulletin of the Polytechnic Institute of Jassy - Construction and Architecture Section* **7** pp 93-102

[7] Zondag H A 2007 Flat-plate PV-thermal collectors and systems – a review *Renew. Sustain. Energy Rev* doi:10.1016/j.rser.2005.12.012.

[8] Notton G, Cristofari C, Mattei M and Poggi P 2005 Modelling of a double-glass photovoltaic module using finite differences *Applied Thermal Engineering* **25** pp 2854–2877

[9] Hudisteanu S V, Mateescu T D, Popovici C G and Chereches C N 2016 The influence of the building integrated photovoltaic panels position on the conversion efficiency *Bulletin of the Polytechnic Institute of Jassy - Construction and Architecture Section* **62** pp 93-103
[10] Benato A and Stoppato A 2019 An experimental investigation of a novel low-cost photovoltaic panel active cooling system *Energies* **12** 1448 doi:10.3390/en12081448
[11] Al-Kouz W and Al-Dahidi S 2020 Assessment of existing photovoltaic system with cooling and cleaning system: Case study at Al-Khobar City *Nader Processes* **8** 9 doi:10.3390/pr8010009
[12] Hudişteanu S V, Popovici C G, MateescuT D and Chereches N C 2017 Efficiency analysis of BIPV systems for different locations in Romania *Energy Procedia* **112** pp 404-411
[13] Kim J and Nam Y 2019 Study on the cooling effect of attached fins on PV using CFD simulation *Energies* **12** 758 doi:10.3390/en12040758
[14] Berthod C, Strandberg R and Odden J O 2015 Temperature coefficients of compensated silicon solar cells influence of ingot position and blend-in-ratio *Energy Procedia* **77** pp 15-20
[15] Hernandez-Perez J G, Carrillo J G, Bassam A, Flota-Banuelos M and Patino-Lopez 2020 A new passive PV heatsink design to reduce efficiency losses: A computational and experimental evaluation *Renewable Energy* **147** pp 1209-1220
[16] Popovici C G, Hudişteanu S V, Mateescu T D and Chereches N C 2016 Efficiency improvement of photovoltaic panels by using air cooled heat sinks *Energy Procedia* **85** pp 425-432
[17] Hudisteanu V S, Mateescu T D and Popovici C G 2015, Comparative study of water film heat exchangers for cooling photovoltaic panels *Bulletin of the Transilvania University of Brasov* **8** pp 365-370
[18] Herrando M, Pantaleo A M, Wang L and Markides C N 2019 Solar combined cooling, heating and power systems based on hybrid PVT, PV or solar-thermal collectors for building applications *Renewable Energy* **143** pp 637-647
[19] Barbu M, Darie G and Siroux M 2019 Analysis of a residential photovoltaic-thermal (PVT) system in two similar climate conditions *Energies* **12** 3595; doi:10.3390/en12193595
[20] Ibrahim A, Fudholi A, Sopian K, Othman M Y and Ruslan M H 2014 Efficiencies and improvement potential of building integrated photovoltaic thermal (BIPVT) system *E Conv and Manag* **77** pp 527-534
[21] Lamnatou C, Notton G, Chemisana D and Cristofari C 2020 Storage systems for (BIPV) and building integrated photovoltaic/thermal (BIPVT) installations: Environmental profile and other aspects *Science of The Total Environment* **699** pp 134-269
[22] Amr A A R, Hassan A A M, Abdel-Salam M and El-Sayed A M 2019 Enhancement of photovoltaic system performance via passive cooling: Theory versus experiment *Renewable Energy* **140** pp 88-103
[23] Al Siyabi I, Khanna S, Sundaram S and Mallick T 2019 Experimental and numerical thermal analysis of multi-layered microchannel heat sink for concentrating photovoltaic application *Energies* **12** 122 doi:10.3390/en12010122
[24] Cuce E, Bali T and Sekucoglu S A 2011 Effects of passive cooling on performance of silicon photovoltaic cells *International J. of Low-Carbon Technologies* **6** pp 299–308
[25] Tripanagnostopoulos Y, Nousia T and Souliotis M 2007 Test results of air cooled modified photovoltaic modules *Proceedings of the 17th European PV solar energy conference* pp 2519-22
[26] Hudişteanu S V and Popovici C G 2019 Experimental investigation of the wind direction influence on the cooling of photovoltaic panels integrated in double skin façades *E3S Web of Conferences* **111** 03045
[27] Zhang T, Wang M and Yang H 2018 A review of the energy performance and life-cycle assessment of building-integrated photovoltaic (BIPV) systems *Energies* **11** 3157 doi:10.3390/en11113157

[28] Yang D J, Yuan Z F, Lee P H and Yin H M 2012 Simulation and experimental validation of heat transfer in a novel hybrid solar panel *International Journal of Heat and Mass Transfer* **55** pp 1076-1082
[29] Sajjad U, Amer M, Ali H M, Dahiya A and Abbas N 2019 Cost effective cooling of photovoltaic modules to improve efficiency. Case Studies *Thermal Engineering* **14** 100420
[30] Hudişteanu S V 2014 Cooling of photovoltaic panels by using water film heat exchanger, *Proc. of the Second International Conference for PhD Students in Civil Engineering and Architecture* pp 445-452
[31] Moharram K A, Abd-Elhady M S, Kandil H A and El-Sherif H 2013 Enhancing the performance of photovoltaic panels by water cooling *Ain Shams Engineering Journal* **4** pp 869-877
[32] Skovajsa J, Kolácek M and Zálešák M 2017 Phase change material based accumulation panels in combination with renewable energy sources and thermoelectric cooling *Energies* **10** 152 doi:10.3390/en10020152
[33] Armstrong S and Hurley W G 2010 A thermal model for photovoltaic panels under varying atmospheric conditions *Applied Thermal Engineering* **30** pp 1488-1495
[34] Adil Khan M, Ko B, Alois Nyari E, Eugene Park S and Hee-Je K 2017 Performance evaluation of photovoltaic solar system with different cooling methods and a bi-reflector PV System (BRPVS): An experimental study and comparative analysis *Energies* **10** 826 doi:10.3390/en10060826
[35] Imtiaz Hussain M, Jin-Hee K and Jun-Tae K 2019 Nanofluid-powered dual-fluid Photovoltaic/Thermal (PV/T) system: Comparative numerical study *Energies* **12** 775 doi:10.3390/en12050775
[36] Liu X, Zhou Y, Li C Q, Lin Y, Yang W and Zhang G 2019 Optimization of a new phase change material integrated Photovoltaic/Thermal panel with the active cooling technique using Taguchi method *Energies* **12** 1022 doi:10.3390/en12061022
[37] Lim J H, Lee Y S and Seong Y B 2017 Diurnal thermal behavior of photovoltaic panel with phase change materials under different weather conditions *Energies* **10** 1983 doi:10.3390/en10121983
[38] Al-Mabsali S A, Chaudhry H N and Gul M S 2019 Numerical investigation on heat pipe spanwise spacing to determine optimum configuration for passive cooling of photovoltaic panels *Energies* **12** 4635 doi:10.3390/en12244635
[39] Turcanu F E, Verdes M, Serbanoiu I, CiocanV and Balan M C 2017 Assessment of Indoor Climate Environment Via Hygrothermal Simulation in Historical Churches *Procedia Engineering* **181** pp 412-417
[40] Turcanu F E, Verdes M, Serbanoiu I 2016 Churches Heating: The Optimum Balance Between Cost Management and Thermal Comfort *Procedia Technology* **22** pp 821-828
[41] Bosneagu R, Scurtu I C, Popov P, Mateescu R, Dumitrache L and Mihailov M E 2019 Hydraulics Numerical Simulation Using Computational Fluid Dynamics (CFD) Method for the Mouth of Sulina Channel *Journal of Environmental Protection and Ecology* **20** pp 2059-2067
[42] Wang Z, Qiu F, Yang W, Zhao X and Mei S 2016 Experimental investigation of the thermal and electrical performance of the heat pipe BIPV/T system with metal wires *Applied Energy* **170** pp 314–323
[43] Aparicio M P, Pelegrí-Sebastiá J, Sogorb T and Llario V 2013 Modeling of Photovoltaic cell using free software application for training and design circuit in photovoltaic solar energy *New Developments in Renewable Energy* pp 121-139
[44] https://us.sunpower.com/sites/default/files/media-library/data-sheets/ds-sunpower-p17-355-commercial-solar-panels.pdf (last accessed on 09.04.2020)
[45] Santbergen R 2008 *Optical Absorption Factor of Solar Cells for PVT Systems* Eindhoven University Press Netherlands pp 35-65
[46] Khordehgah N, Żabnieńska-Góra A and Jouhara H Energy performance analysis of a PV/T system coupled with domestic hot water system *Chem Engineering* **4** 22 doi:10.3390/chemengineering4020022

Seismic response of ½ scaled two storey reinforced concrete moment resisting frame system using nonlinear static analysis

I Sococol[1], I Olteanu-Donțov[2], P Mihai[3] and V-I Iftode[1]
[1] Doctoral School of Faculty of Civil Engineering and Building Services, "Gheorghe Asachi" Technical University of Iasi, Blvd. Mangeron, No. 1, 700050, Iasi, Romania
[2] Department of Structural Mechanics, Faculty of Civil Engineering and Building Services, "Gheorghe Asachi" Technical University of Iasi, Blvd. Mangeron, No. 1, 700050, Iasi, Romania
[3] Department of Concrete, Materials, Technology and Management, Faculty of Civil Engineering and Building Services, "Gheorghe Asachi" Technical University of Iasi, Blvd. Mangeron, No. 1, 700050, Iasi, Romania

E-mail: ion_sococol@yahoo.com

Abstract. Reinforced concrete (RC) frame systems represent one of the structural solutions used in seismic zones by engineers. For this reason, the importance of knowing the real seismic response for these types of structures is essential and presents the research problem studied in this article. Thus, the seismic response for two storey axial RC frame (one bay-one span) system was studied using nonlinear static analysis in ATENA software. The structural degradation of the two storey RC frame system was investigated for several steel reinforcement possibilities and three distinct RC beams cross sections. Particular attention was paid to RC beam-column joints degradation areas and to seismic energy dissipation mechanism in the marginal regions of the RC columns. Thus, it was observed a nonlinear inelastic response in the potential degradation zones with contrary effects with respect to the specified conclusions in the current seismic design norms. Also, it was studied the RC slab bending stiffness influence to horizontal structural elements (RC beams) and to vertical structural elements (RC columns) alongside the cracking mode for horizontal static actions and RC slab local degradation beside RC beam-column joint area. In these conditions, conclusions regarding the seismic response of the moment resisting (MR) RC frame system with low height regime designed according to the current seismic code were specified and an unsatisfactory seismic response was proved.

1. Introduction

The structural systems developed according to the dissipative design concept, begin to be implemented in New Zealand, USA and Japan between 1965-1970 [1-12]. Currently, this design philosophy has become a legal norm through anti-seismic regulations [13-20].

Thus, structural systems, including MR RC frame structures, have become practical solutions for seismic design by structural engineers. The phenomenological specificity regarding this dissipative concept, especially for bare-frame RC systems, consists in sufficient deformation capacity in plastic range with certain specially detailed/ reinforced critical zones which can allow the important lateral displacements of the structure without general collapse [21-25] (see figure 1).

In these conditions, the idealized global seismic response mechanism, represented in figure 2, with plastic hinges at the end of the beams (without considering the height regime of the structures) and dissipative areas developed at the extremities of the columns (in the area of restrain of the columns in the foundations), becomes a basic principle in the seismic design of the RC frame structures [23-26].

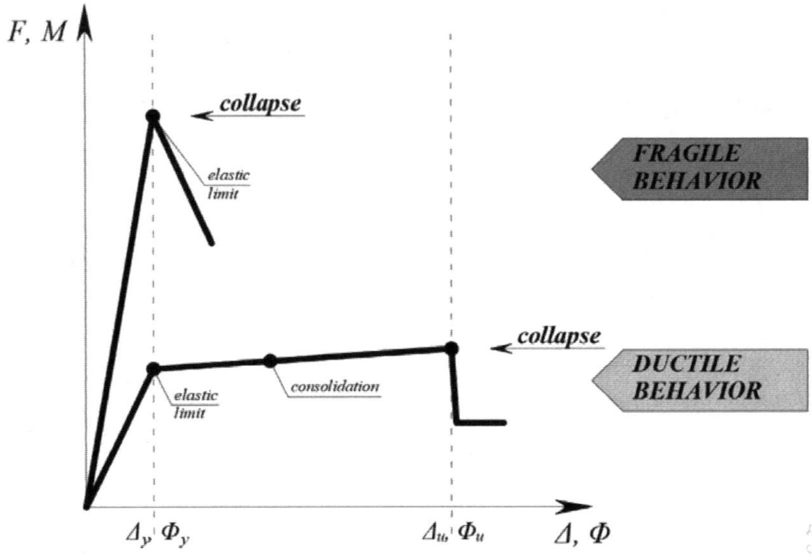

Figure 1. Phenomenological representation of the ductile and fragile behavior of the structure through the F-D relationship [23, 27].

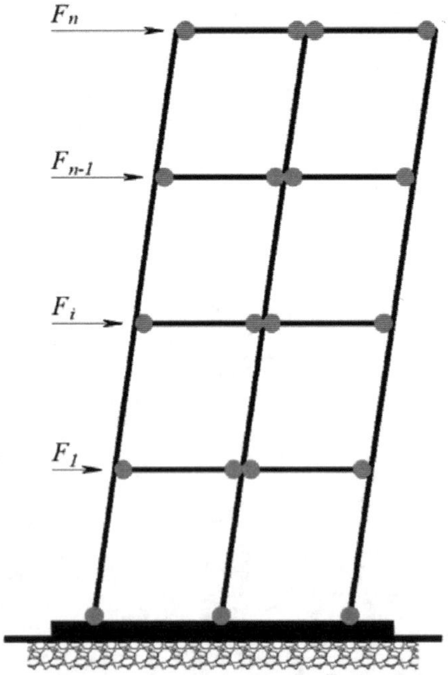

Figure 2. The idealized global plastic mechanism considered in P100-1 code [13] for bare-frame MR RC structures required to severe seismic actions [27-28].

Thus, in the current literature [23-26] some particularities regarding the seismic energy dissipation mechanisms for pure MR RC frame systems through the lateral behaviour of the seismic-structural elements are specified.

In these conditions, it was proposed to check the theoretical specifications regarding plastic hinges development mechanisms and specific zones of local deformations in lateral RC elements for ½ scaled

GF+1F (ground floor + 1 floor) bare MR RC frame system, using nonlinear static analysis performed in ATENA 3D software [29].

In all nonlinear static analyses, the results from the experiments performed on a GF+9F dual reinforced concrete prototype [30-32] with preponderant RC frames, tested on a seismic platform in Japan [33] were taken into account.

2. Particularities regarding the procedure considered for seismic energy dissipation in case of a pure MR RC frame systems using dissipative design concept

2.1. Theoretical perception of the RC columns deformability
Reinforced concrete columns required to severe earthquake action, are considered non-dissipative vertical elements in theoretical perception of structural engineering [13-26].

These lateral elements (RC columns) do not present significant degradations/ important concrete cracks/ major steel reinforcements elongation in nonlinear inelastic domain for any zone, except the bottom of the ground floor and upper end for the top floor [13-26].

Another important aspect regarding RC columns deformability are the capable efforts considered exceeded in the moment of seismic event. Thus, according to P100-1 [13], it is considered a global seismic response of the RC frame structure through local plasticization mechanisms in critical (potential plastic) areas of the RC columns from bending moment effects with structural RC element stiffness/ strength reduction impact.

In these conditions, pure MR RC frame structure present the idealized global plastic mechanism according to figure 2.

2.2. Theoretical perception of the RC beams deformability
Reinforced concrete beams are considered the main dissipative structural elements of the bare MR RC frame system, according to ductile design concept [13-20], [22-27], [34] (see figure 1 and figure 2).
Thus, RC beam end margins represent portions of the beam with plastic deformation/ cracking and plastic rotation according to figure 3 [34], that allow through nonlinear inelastic deformations the lateral displacement of the structure and seismic energy dissipation.

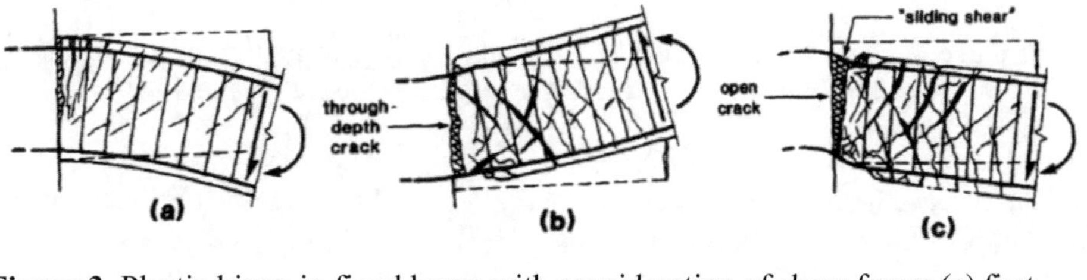

Figure 3. Plastic hinge in fixed beam with consideration of shear force: (a) first loading cycle in one direction; (b) first loading cycle in opposite direction; (c) second loading cycle in one direction [34].

2.3. Theoretical perception of the RC beam-column joints deformability
Reinforced concrete beam-column joints of the pure RC frame systems are considered in the current Romanian seismic standard [13] (standard elaborate according to foreign codes [14-20]), structural rigid components with linear-elastic behaviour/ response in seismic situations.

Lateral stiffness particularity and lateral superior characteristic strength developed in accordance with the current ductile design concept [13-26] represent a consequence of the considered seismic design efforts/ reinforcement method.

Thus, superior stiffness and strength to lateral loading cycles of the RC beam-column joints is similar with the lateral stiffness and strength of the RC columns.

In these conditions, the RC beam-column joints are considered component part of the RC columns [13-26]. The principal explanation of this structural consideration comes from the necessity to plastic hinges development in the marginal areas of the RC beams (practically, in adjacent regions to the RC beam-column joint).

2.4. Theoretical perception of the RC slabs deformability
Reinforced concrete slabs are considered in usual seismic design conditions [13] infinitely rigid structural elements with linear-elastic response to seismic events.

Particular aspect regarding theoretical perception of the RC slabs seismic response is the slab independence in relation to the RC beam-column joint. Partial influence of the RC slab on the lateral stiffness and lateral strength of the RC beams (see figure 4) exists with a minor role on the overall behaviour of the structure (to their nonlinear inelastic deformations).

These theoretical aspects have analytical application in the seismic design phase. Thus, RC beam is calculated with effective width (with T shape – which takes into account only a portion of the real width of the slab) [13], [23-26]. In these conditions, the RC slab idealization considered in the seismic design phase can take another form in the practical seismic response.

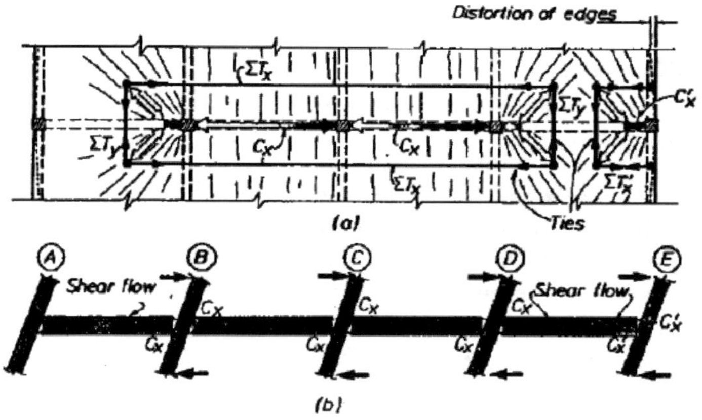

Figure 4. Cracking pattern of the RC slabs with the formation of the plasticizing mechanism at the limitrophe zones of the RC beams [26].

3. Particularities regarding seismic response of a ½ scaled GF+1F bare MR RC frame system designed according to dissipative regulations

3.1. General data
Theoretical perceptions regarding seismic response of the pure MR RC frame systems were proposed to be verified with the numerical simulations for ½ scaled GF+1F pure axial frame MR RC structure represented in figure 5 [35-36]. It was considered necessary to perform numerical simulations on ½ scaled RC frame models because the structural dimensions of the RC frame prototypes are bigger than the dimensions of the seismic platform. Thus, these reductions of the RC frame models are required for a subsequent experimental research study and the obtained results will be compared with the analytical results.

In these conditions, it is desired to recognize the analytical seismic response and seismic energy dissipation mechanisms for this type of structures (pure seismic-resistant RC frame systems with low and medium height regime*) designed in accordance to the dissipative concept [13], [21-27] and described on short in section 2 [37-42].

*Note: In Ishiyama [2] the following classification according to the height regime is presented:

IOP Conf. Series: Materials Science and Engineering 1141 (2021) 012009 doi:10.1088/1757-899X/1141/1/012009

- small scale buildings (H≤12 m, where H=building height);
- medium scale buildings (12 m<H≤20 m, where H=building height);
- large scale buildings (20 m<H≤31 m, where H=building height);
- large scale buildings (31 m<H≤60 m, where H=building height);
- high-rise buildings (H>60 m, where H=building height).

The details regarding the general geometric dimensions of the RC frame are given in figure 5 (a). Similitude relations [30-32], [46] recognized in the engineering literature, were used in the structural scaling stage of the RC frame prototypes.

Seismic design of the RC frame structure was conducted in accordance with current Romanian standards [13], [47-50]. Structural modelling [51] and numerical simulation were performed using ATENA 3D software [29]. Mean values for the material characteristics were considered in the current research, which are presented in table 1 and table 2.

Table 1. Steel reinforcement strength values considered in nonlinear static analyses performed with ATENA 3D software [29].

Material name	Type	Elastic modulus E [MPa]	σ_y [MPa]	σ_t [MPa]	ε_{lim}
Bst 500 S	Bilinear with Hardening	2.000E+05	575.000	632.500	0.060

Table 2. Concrete strength values considered in nonlinear static analyses performed with ATENA 3D software [29].

Material name	Elastic modulus E [MPa]	Poisson's ratio μ	Tensile strength f_t [MPa]	Compressive strength f_c [MPa]
C12/15 concrete strength class	2.700E+04	0.200	1.570E+00	-2.000E+01
C16/20 concrete strength class	2.900E+04	0.200	1.900E+00	-2.400E+01
C20/25 concrete strength class	3.000E+04	0.200	2.200E+00	-2.800E+01

A non-linear static analysis (push-over analyses) [24], [52-54] (figure 5 (b), figure 6) was performed with output data in crack panels configuration for all RC frame models at each lateral loading step. The "Force - Displacement" capacity curves associated to ultimate lateral capable displacements were plotted. General representation of the steel reinforcement and general structure mesh discretization scheme for the RC frame models are attached in figure 5 (c) and figure 5 (d).

IOP Conf. Series: Materials Science and Engineering 1141 (2021) 012009 doi:10.1088/1757-899X/1141/1/012009

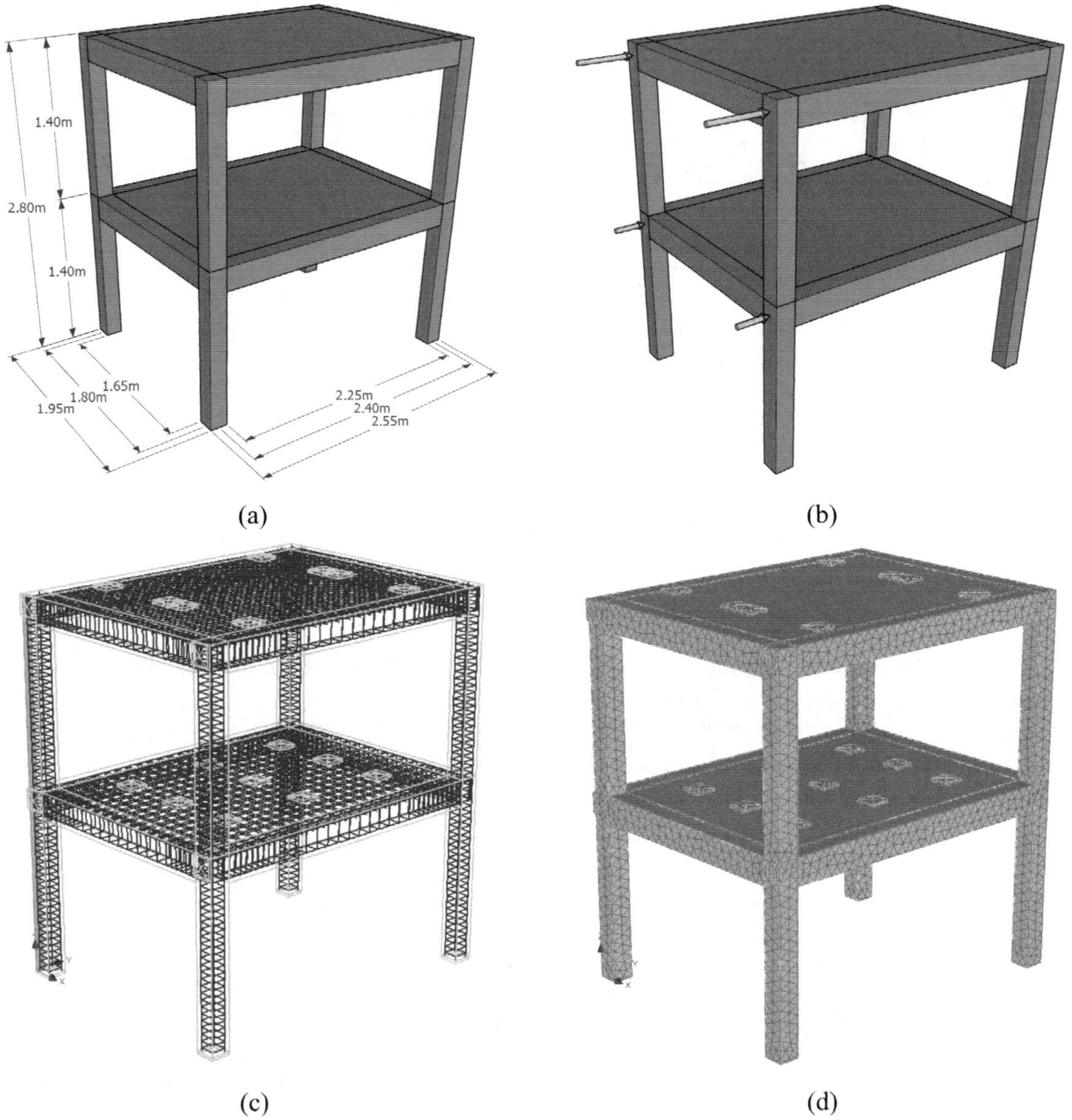

Figure 5. (a) General geometric dimensions of the ½ scaled two storey RC frame models [35-36]; (b) "Pushover loading consideration for RC frame models" [36]; (c) Representation scheme of the steel reinforcement carcase for RC frame models [29], [35-36]; (d) Structural representation of the lateral elements mesh discretization for RC frame models [29], [35-36].

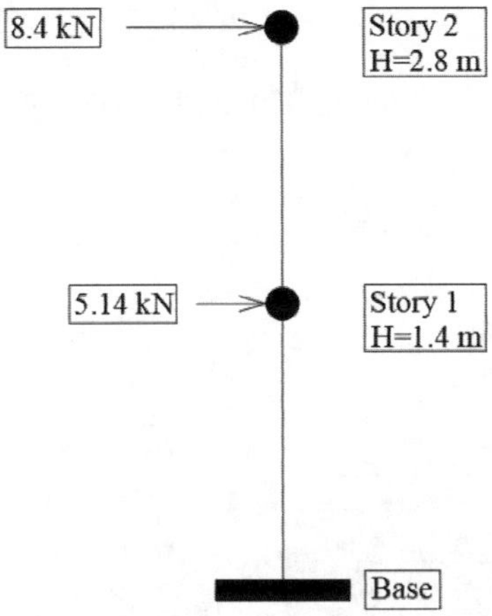

(a)

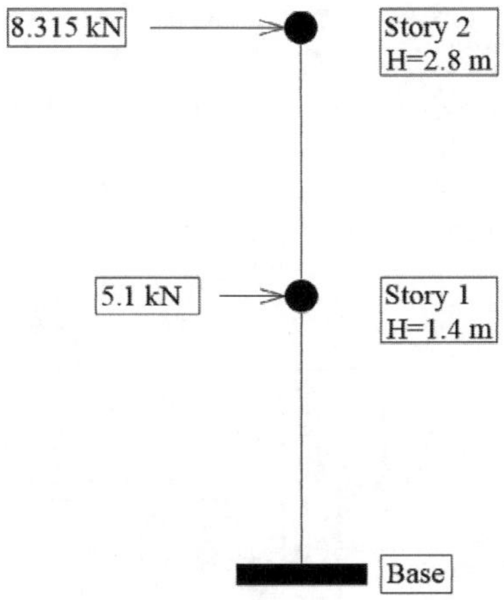

(b)

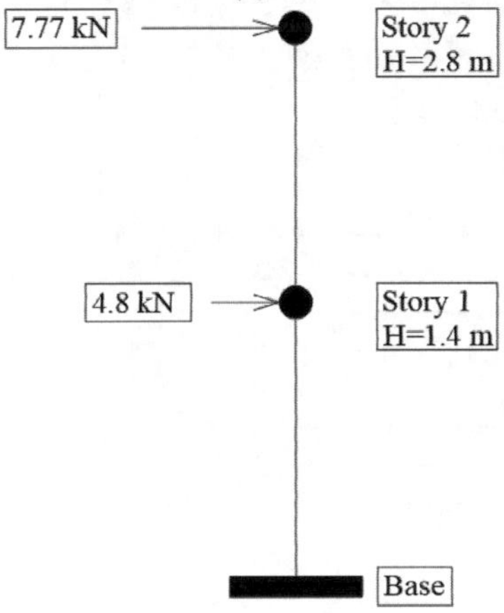

(c)

Figure 6. “Discrete system representation of the RC frame models with lateral loading consideration in the longitudinal direction for: (a) type 1 of the RC frame systems; (b) type 2 of the RC frame systems; (c) type 3 of the RC frame systems” [35] (see figure 5 (b) and figure 7).

Several RC frames models (½ scaled GF+1F bare MR RC frame structures) were considered with the following variables*:

- RC beams cross section [35] (see figure 7);
- RC slabs thickness [36, 43] (see figure 7);
- Concrete strength class [44] (see figure 7);
- RC columns steel reinforcement ratio [45] (see figure 7);
- RC beams steel reinforcement ratio [36] (see figure 7);
- RC slabs steel reinforcement ratio [36] (see figure 7).

*Note: The very large volume of research data led to the specification of general remarks regarding the seismic energy dissipation mechanisms recorded for the analysed RC frame models which were compared with the theoretical mechanisms specified in current seismic design standards.

In several research studies [35], [36], [43], [44], [45] it was implicitly investigated the obtained values through the comparative method.

For this article, the seismic response for three RC frame models was studied. The main variable of this systematization is the cross-section height (h_L) of the RC beams. This variable (h_L) was chosen on the principle that the main dissipative elements of MR RC frame structures are the beams.

Case 1 of RC frame systems has the largest cross section of the beams in the predimensioning condition (h_L=1/8L). Type 2 of RC frame systems was considered with mean dimensions of the RC beams cross sections in the predimensioning condition (h_L=1/12L). Type 3 of RC frame systems has ductile RC beams that have been predimensioned from the h_L=1/16L condition (seismic design condition specified in the new provisions for completing and amending the P100-1/2013 technical regulation).

The detailed dimensions of the RC beams, RC columns and RC slabs cross sections and afferent steel reinforcement ratio with particular concrete strength class consideration are specified in figure 7.

THE REAL SEISMIC RESPONSE OF ONE BAY-ONE SPAN MOMENT RESISTING RC FRAME SYSTEMS

CONCRETE CLASS	NUMERICAL SIMULATION CODE	RC FRAME SYSTEMS - TYPE 1 $h_L=1/8L$ MODEL CHARACTERISTICS					RC FRAME SYSTEMS - TYPE 2 $h_L=1/12L$ MODEL CHARACTERISTICS					RC FRAME SYSTEMS - TYPE 3 $h_L=1/16L$ MODEL CHARACTERISTICS			
		Colum 15x15	GL: 15x27	GT: 10x20	h_p: 7 cm		Colum 15x15	GL: 15x20	GT: 15x20	h_p: 7 cm		Colum 15x15	GL: 10x15	GT: 10x12	h_p: 6 cm
C12/15	V3	4Ø14	4Ø10	4Ø10	Ø6	V20	4Ø14	4Ø8	4Ø8	Ø6	V2	4Ø10+4Ø6	4Ø8	4Ø8	Ø6
	V3_1	4Ø12	4Ø10	4Ø10	Ø6	V20_1	4Ø12	4Ø8	4Ø8	Ø6	V2_1	4Ø10+4Ø8	4Ø8	4Ø8	Ø6
	V3_2	4Ø10	4Ø10	4Ø10	Ø6	V20_2	4Ø10	4Ø8	4Ø8	Ø6	V2_2	4Ø12+4Ø8	4Ø8	4Ø8	Ø6
C16/20	V8	4Ø14	4Ø10	4Ø10	Ø6	V24	4Ø14	4Ø8	4Ø8	Ø6	V7	4Ø10+4Ø6	4Ø8	4Ø8	Ø6
	V8_1	4Ø12	4Ø10	4Ø10	Ø6	V24_1	4Ø12	4Ø8	4Ø8	Ø6	V7_1	4Ø10+4Ø8	4Ø8	4Ø8	Ø6
	V8_2	4Ø10	4Ø10	4Ø10	Ø6	V24_2	4Ø10	4Ø8	4Ø8	Ø6	V7_2	4Ø12+4Ø8	4Ø8	4Ø8	Ø6
C20/25	V13	4Ø14	4Ø10	4Ø10	Ø6	V28	4Ø14	4Ø8	4Ø8	Ø6	V12	4Ø10+4Ø6	4Ø8	4Ø8	Ø6
	V13_1	4Ø12	4Ø10	4Ø10	Ø6	V28_1	4Ø12	4Ø8	4Ø8	Ø6	V12_1	4Ø10+4Ø8	4Ø8	4Ø8	Ø6
	V13_2	4Ø10	4Ø10	4Ø10	Ø6	V28_2	4Ø10	4Ø8	4Ø8	Ø6	V12_2	4Ø12+4Ø8	4Ø8	4Ø8	Ø6
C12/15	V4	4Ø14	4Ø10	4Ø10	Ø5	V21	4Ø14	4Ø8	4Ø8	Ø5					
	V4_1	4Ø12	4Ø10	4Ø10	Ø5	V21_1	4Ø12	4Ø8	4Ø8	Ø5					
	V4_2	4Ø10	4Ø10	4Ø10	Ø5	V21_2	4Ø10	4Ø8	4Ø8	Ø5					
C16/20	V9	4Ø14	4Ø10	4Ø10	Ø5	V25	4Ø14	4Ø8	4Ø8	Ø5					
	V9_1	4Ø12	4Ø10	4Ø10	Ø5	V25_1	4Ø12	4Ø8	4Ø8	Ø5					
	V9_2	4Ø10	4Ø10	4Ø10	Ø5	V25_2	4Ø10	4Ø8	4Ø8	Ø5					
C20/25	V14	4Ø14	4Ø10	4Ø10	Ø5	V29	4Ø14	4Ø8	4Ø8	Ø5					
	V14_1	4Ø12	4Ø10	4Ø10	Ø5	V29_1	4Ø12	4Ø8	4Ø8	Ø5					
	V14_2	4Ø10	4Ø10	4Ø10	Ø5	V29_2	4Ø10	4Ø8	4Ø8	Ø5					
C12/15	V5	4Ø14	4Ø12	4Ø10	Ø6	V22	4Ø14	4Ø10	4Ø10	Ø6					
	V5_1	4Ø12	4Ø12	4Ø10	Ø6	V22_1	4Ø12	4Ø10	4Ø10	Ø6					
	V5_2	4Ø10	4Ø12	4Ø10	Ø6	V22_2	4Ø10	4Ø10	4Ø10	Ø6					
C16/20	V10	4Ø14	4Ø12	4Ø10	Ø6	V26	4Ø14	4Ø10	4Ø10	Ø6					
	V10_1	4Ø12	4Ø12	4Ø10	Ø6	V26_1	4Ø12	4Ø10	4Ø10	Ø6					
	V10_2	4Ø10	4Ø12	4Ø10	Ø6	V26_2	4Ø10	4Ø10	4Ø10	Ø6					
C20/25	V15	4Ø14	4Ø12	4Ø10	Ø6	V30	4Ø14	4Ø10	4Ø10	Ø6					
	V15_1	4Ø12	4Ø12	4Ø10	Ø6	V30_1	4Ø12	4Ø10	4Ø10	Ø6					
	V15_2	4Ø10	4Ø12	4Ø10	Ø6	V30_2	4Ø10	4Ø10	4Ø10	Ø6					
C12/15	V6	4Ø14	4Ø12	4Ø10	Ø5	V23	4Ø14	4Ø10	4Ø10	Ø5					
	V6_1	4Ø12	4Ø12	4Ø10	Ø5	V23_1	4Ø12	4Ø10	4Ø10	Ø5					
	V6_2	4Ø10	4Ø12	4Ø10	Ø5	V23_2	4Ø10	4Ø10	4Ø10	Ø5					
C16/20	V11	4Ø14	4Ø12	4Ø10	Ø5	V27	4Ø14	4Ø10	4Ø10	Ø5					
	V11_1	4Ø12	4Ø12	4Ø10	Ø5	V27_1	4Ø12	4Ø10	4Ø10	Ø5					
	V11_2	4Ø10	4Ø12	4Ø10	Ø5	V27_2	4Ø10	4Ø10	4Ø10	Ø5					
C20/25	V16	4Ø14	4Ø12	4Ø10	Ø5	V31	4Ø14	4Ø10	4Ø10	Ø5					
	V16_1	4Ø12	4Ø12	4Ø10	Ø5	V31_1	4Ø12	4Ø10	4Ø10	Ø5					
	V16_2	4Ø10	4Ø12	4Ø10	Ø5	V31_2	4Ø10	4Ø10	4Ø10	Ø5					

OUTPUT DATA

- Steps
- Force
- Displacement
- Total Strain_Eps_zz
- Principal Fracture Strain_V_max_z
- Cracks panel
- Principal Fracture Strain Max

Figure 7. Principal characteristics of the ½ scaled two storey RC frame models (h_L – height of longitudinal RC beams, L – opening, GL – longitudinal RC beam, GT – transversal RC beam, h_p – slab thickness) [35-36], [44-45], [56]. (Note: in numerical simulation code column, **V** notation comes from the „Version" word).

3.2. Results

3.2.1. General results

Type 1 of the RC frame models (figure 7, figure 8) laterally loaded with static forces [52-54], present the following conclusions:

- Superior longitudinal steel reinforcement ratio in RC columns, leads to superior lateral strength/ stiffness for the structure [45].
- RC columns present a ductile behaviour with incursions in the nonlinear inelastic domain [45].
- RC frame models are loaded with a higher lateral force for superior concrete strength classes [44].
- Superior longitudinal steel reinforcement ratio in longitudinal RC beams leads to superior lateral strength for the structure [36].
- Large cross sections of the longitudinal RC beams determine the appearance of the principal stress/ strain concentration effects at the top/ inferior end margins of the RC columns with the nonlinear inelastic deformations of the RC beam-column joints [35].
- RC beams are partially deformed/ cracked and they are not the principal dissipative structural elements [35].
- Steel reinforcement ratio in RC slabs has insignificant influence on the global seismic response of the structures [36], [55].

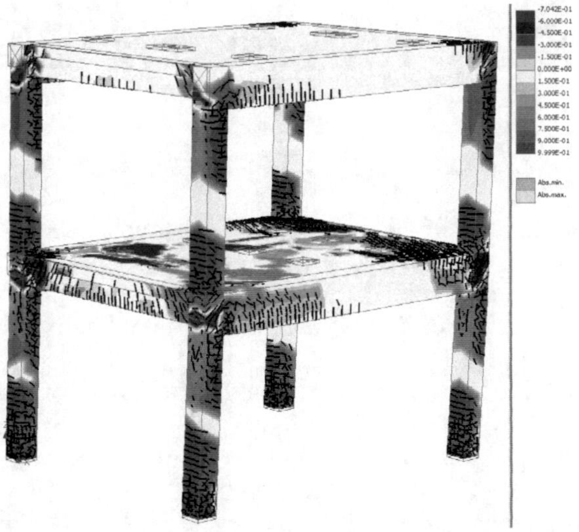

Figure 8. Representation of the fracture specific strain (in the global direction z) concentration [29] in limitrophe zones of the RC columns for V16_2 RC frame model (type 1 RC frame system – see figure 7) with concrete cracks panel in ultimate lateral loading step. (Note: V16_2 is the numerical simulation code – see figure 7).

Type 2 of the RC frame models (figure 7, figure 9) longitudinally loaded with equivalent static forces [52-54], present the following conclusions:

- Superior concrete strength classes in combination with higher steel reinforcement ratios in RC slabs lead to superior ultimate lateral displacements [44].
- Superior concrete strength classes in combination with higher longitudinal steel reinforcement ratios in RC columns and minimum longitudinal steel reinforcement ratios in RC beams lead to maximum global (horizontal) displacements of the RC frame structures. In these cases, maximum principal deformations in tensioned regions of the RC beams with ample stress/ strain concentrations in the marginal areas of the RC columns is registered [44].

- Superior concrete strength classes in combination with maximum longitudinal steel reinforcement ratios in RC columns, lead to higher lateral load forces of the RC frame structures (RC frame models) [44].
- RC columns present significant nonlinear inelastic degradations in the marginal areas [45].
- RC beam-column joints are influenced by stress-strain panel at the end regions of the RC columns and RC beams [35-36], [44-45].
- RC beam-column joints present important nonlinear inelastic degradations [35-36], [44-45].

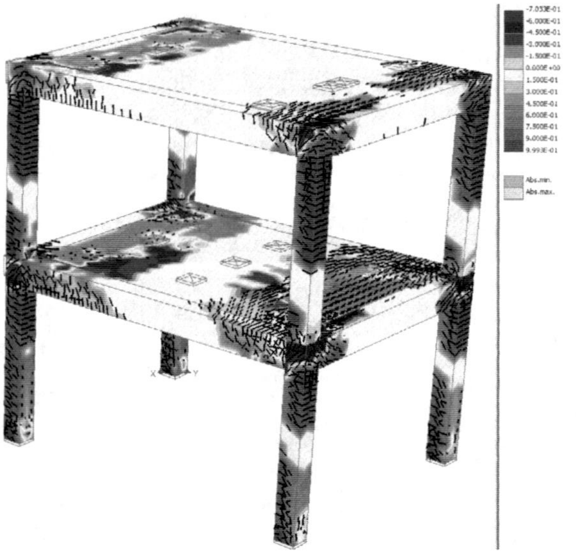

Figure 9. Representation of the fracture specific strain (in the global direction z) concentration [29] in marginal areas of the RC columns with the possibility of plastic hinges formation and general collapse of the V30 structural system (type 2 RC frame model – see figure 7). (Note: V30 is the numerical simulation code – see figure 7).

Type 3 of the RC frame models (figure 7, figure 10) horizontally loaded with equivalent static forces [52-54], present the following conclusions:

- Superior concrete strength class (C20/25) [44] in combination with superior longitudinal steel reinforcement ratio in RC columns (2.90%) [45] present the peak point of the seismic response in terms of lateral forces (V12_2 RC frame model– see figure 7).
- The ultimate specific strain in the global direction z of the RC columns registering the maximum values for C20/25 concrete strength class in combination with minimum longitudinal reinforcement ratio in RC columns (1.89%). Continuous deformation effect of the structural vertical elements (RC columns) occurs due to the:
  - transverse steel reinforcement mode;
  - longitudinal steel reinforcement yielding (elongation);
  - superior concrete strength class.
- Steel reinforcement ratio in RC slabs has insignificant influence on the global seismic response of the RC frame structure [36].
- RC beams present nonlinear inelastic deformations and do not represent the essential dissipative elements [35-36], [43-45].
- RC columns suffer significant degradations/ rotations/ cracks throughout their height [45].
- RC beam-column joints describe important cracks and do not remain in the linear-elastic domain [35-36], [43-45].

Figure 10. Representation of the fracture specific strain (in the global direction z) concentration [29] on the total RC columns height/ on the entire RC beams length for ultimate lateral loading step of the V12 structural system (type 3 RC frame model – see figure 7). (Note: V12 is the numerical simulation code – see figure 7).

In addition to the general remarks of the current analytical study regarding seismic response of the RC frame models, the "F-D" capacity curves for each RC frame model type was plotted (figure 11, figure 12 and figure 13), taking into consideration: lateral deformation mode for each lateral loading step, structural degradation mode and other deformation indicators etc.

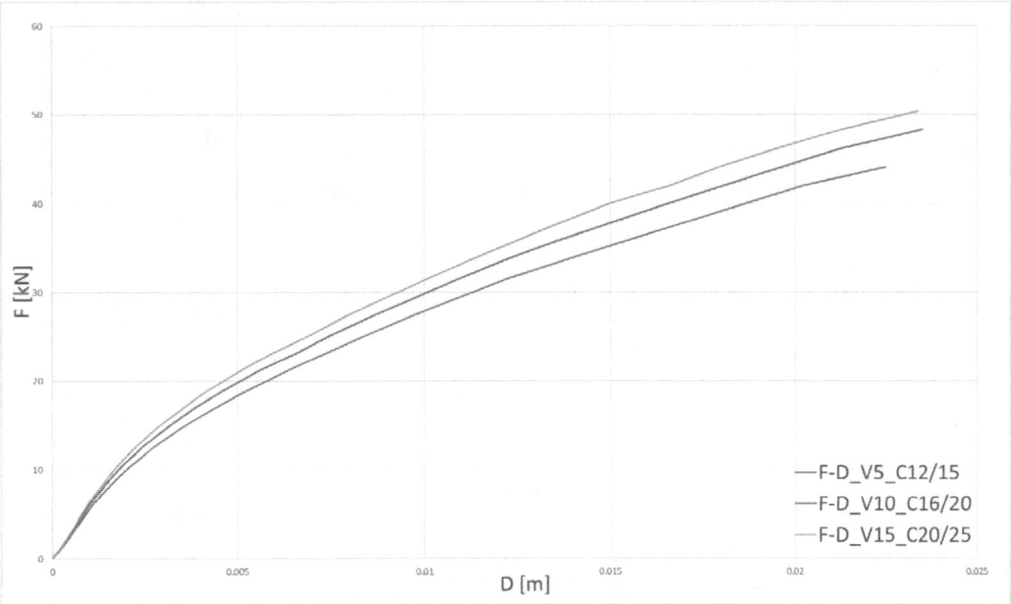

Figure 11. "F-D" capacity curves for some type 1 RC frame models (V5, V10 and V15 RC frame models), (Note: V5, V10 and V15 are the numerical simulation codes – see figure 7).

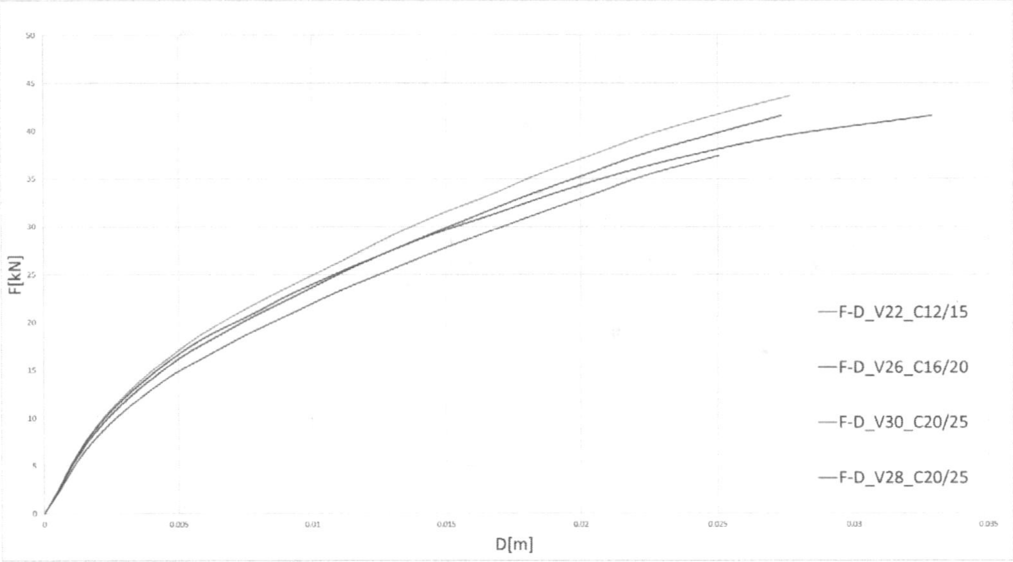

Figure 12. "F-D" capacity curves for some type 2 RC frame models (V22, V26, V28 and V30 RC frame models), (Note: V22, V26, V28 and V30 are the numerical simulation codes – see figure 7).

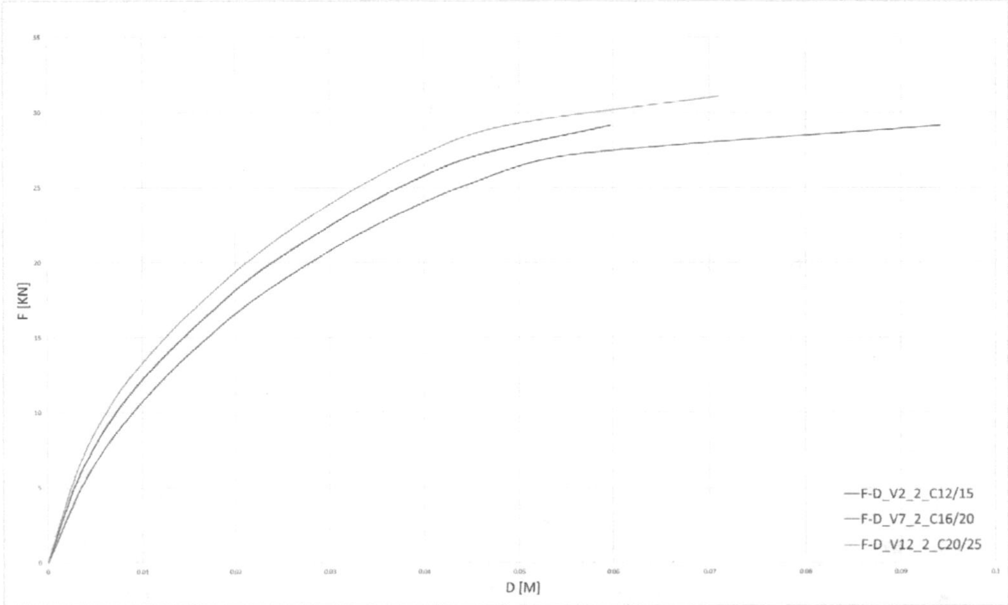

Figure 13. "F-D" capacity curves for some type 3 RC frame models (V2_2, V7_2 and V12_2 RC frame models), (Note: V2_2, V7_2 and V12_2 are the numerical simulation codes – see figure 7).

It is observed that type 3 RC frame models present superior seismic response with appropriate "F-D" curve shapes represented in figure 13. However, the global seismic energy dissipation mechanism for this RC frame structures type, does not correspond to the dissipative concept considered in structural engineering design [13], [22-26], [42].

3.2.2. Analytical seismic response of the RC columns

Numerical analyses of the three types scaled bare MR RC frame models, present RC columns as important dissipative structural elements.

In terms of specific strain, RC columns reach 35% of the total specific deformations of the structure (see figure 14 and figure 15) for type 1 and type 2 models. In case of type 3 models, RC columns reach 20% of the total specific deformations.

In these conditions, RC columns develop nonlinear inelastic seismic response [39, 41], [57-58] for all three types of the RC frame models (see figure 8, figure 9 and figure 10) with complex structural degradation mechanism.

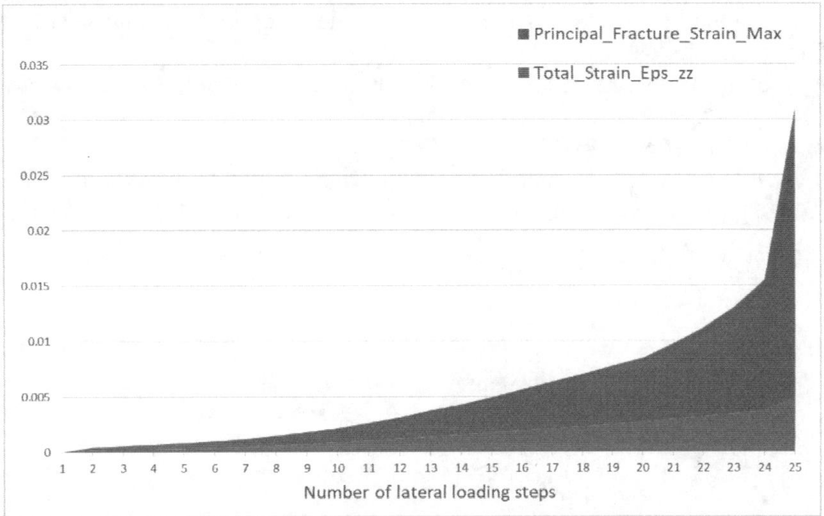

Figure 14. Value representation of the all RC columns specific strain in the global direction z and fracture specific strain for V15 RC frame model (type 1 RC frame model), (Note: V15 is the numerical simulation code – see figure 7).

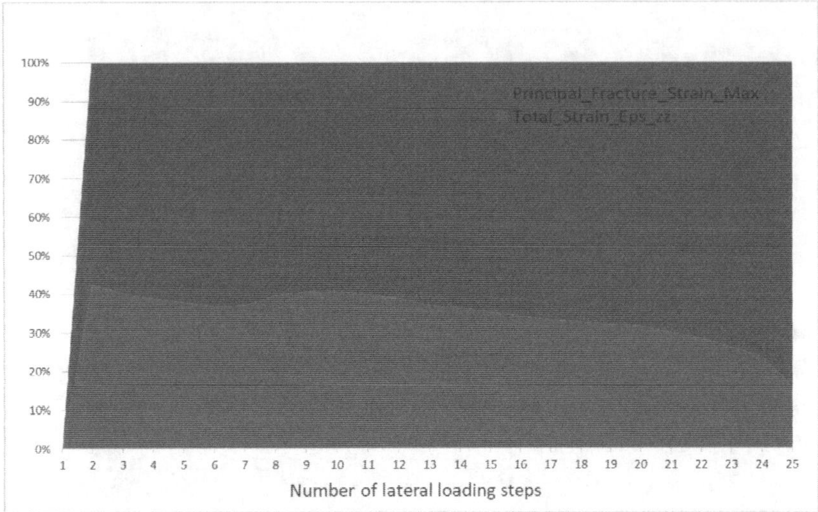

Figure 15. Percentage (ratio) representation of the all RC columns specific strain in the global direction z and fracture specific strain for V15 RC frame model (type 1 RC frame model), (Note: V15 is the numerical simulation code – see figure 7).

Concrete cracking panel representation of the RC columns for V30 RC frame model (where V30 is explained in figure 7) in each lateral loading step, can be studied in figure 16. Thus, it can be observed the destructive process by which the RC columns are the first required structural elements with a multitude of concrete cracks in large areas (especially for ultimate lateral loading steps).

In each horizontal loading step of the V30 RC frame structure, the plastic zones formation in marginal areas of the RC columns is emphasized. These areas are the limitrophe tensioned/ compressive regions from the RC columns with contrary seismic dissipative mechanism declared in P100-1 standard [13].

In conclusion, a different seismic response (nonlinear inelastic response) of the RC columns is observed compared to the linear elastic seismic response declared in the actual seismic design standard [13] for all RC frame model types.

RC columns present important concrete cracks, significant steel reinforcement buckling and steel reinforcement elongation in the tensioned marginal areas with large inter-storey displacements and high global displacements (figure 16).

The major risk of the structural system is collapse through potential hinges, located in the column's ends [35-36], [39-41], [43-45], [57-58].

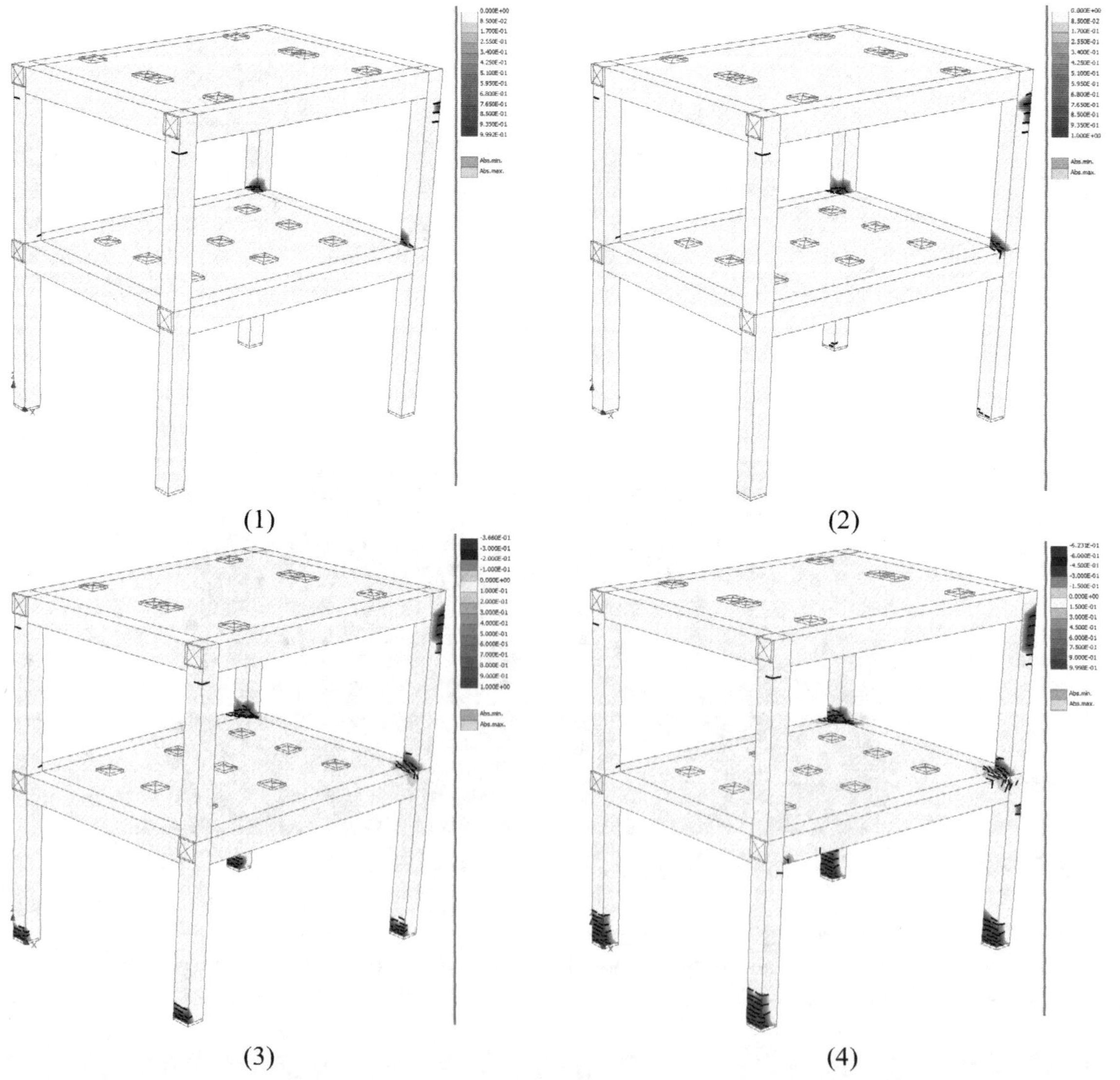

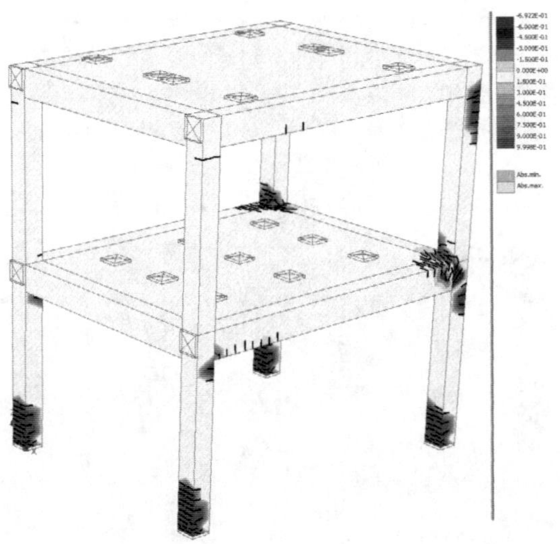
(5)

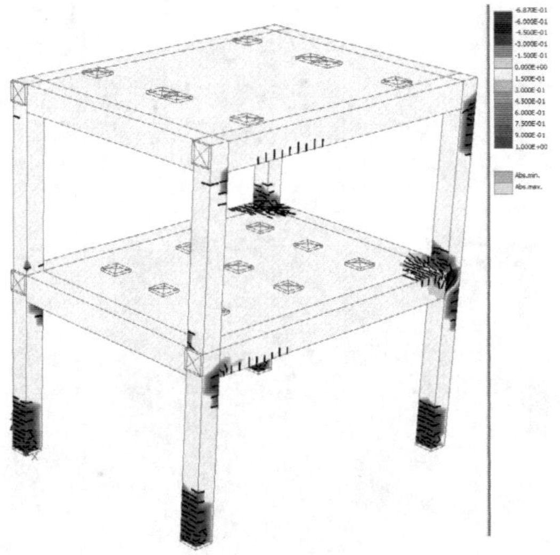
(6)

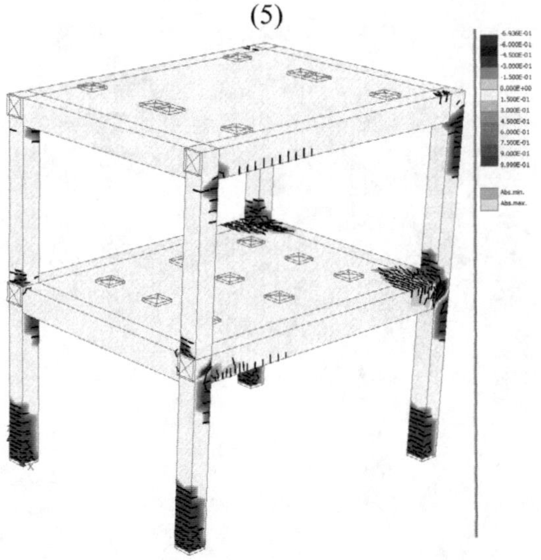
(7)

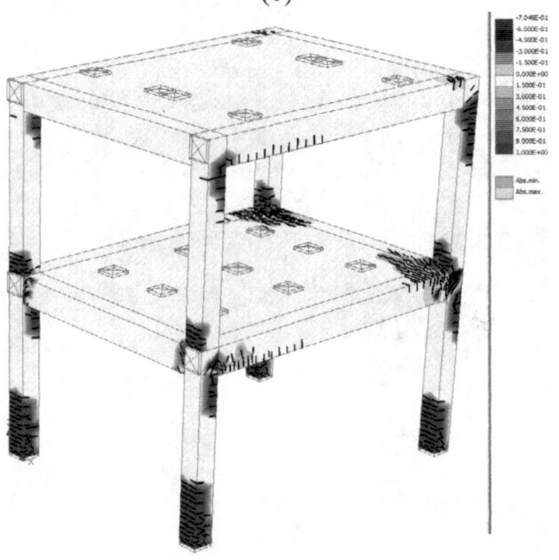
(8)

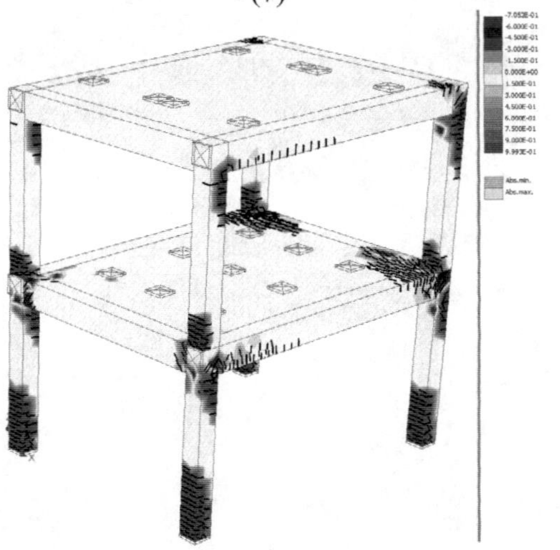
(9)

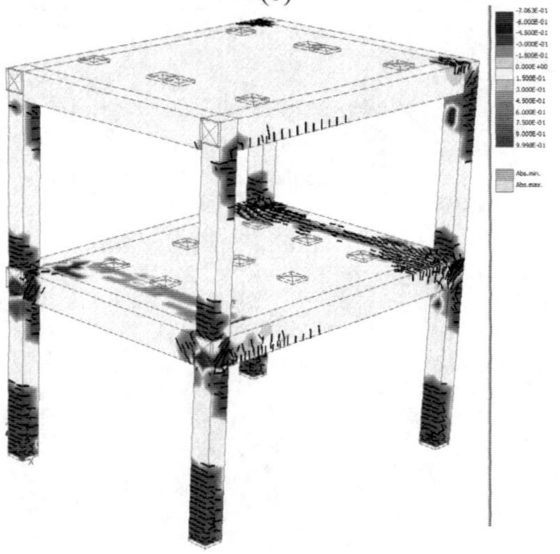
(10)

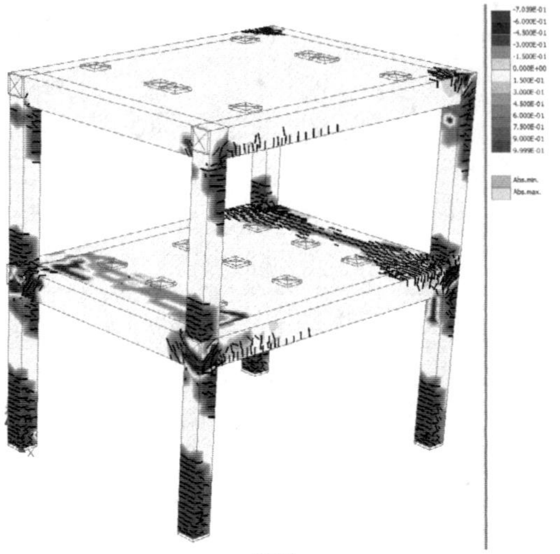

(11)

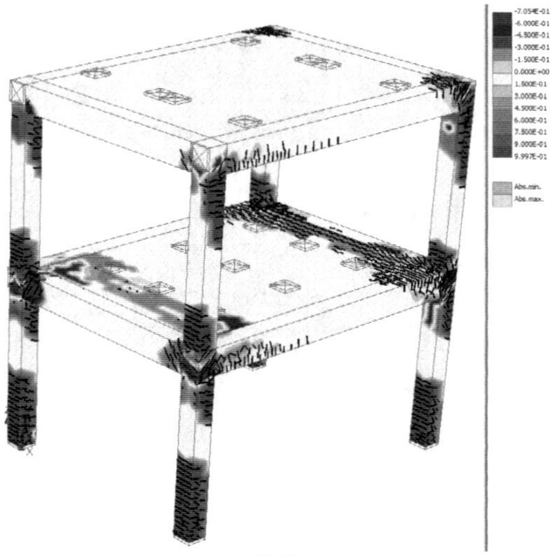

(12)

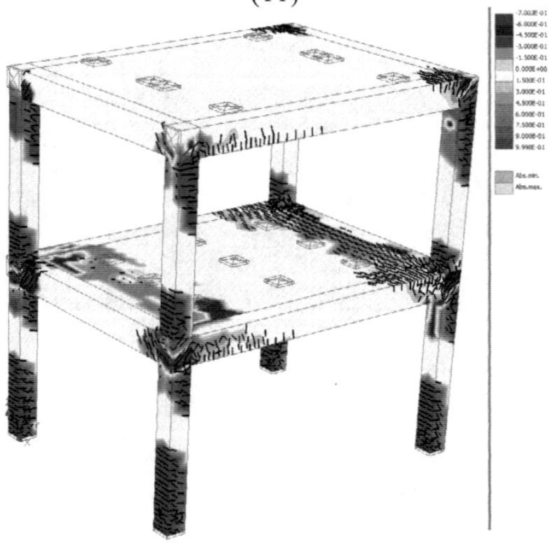

(13)

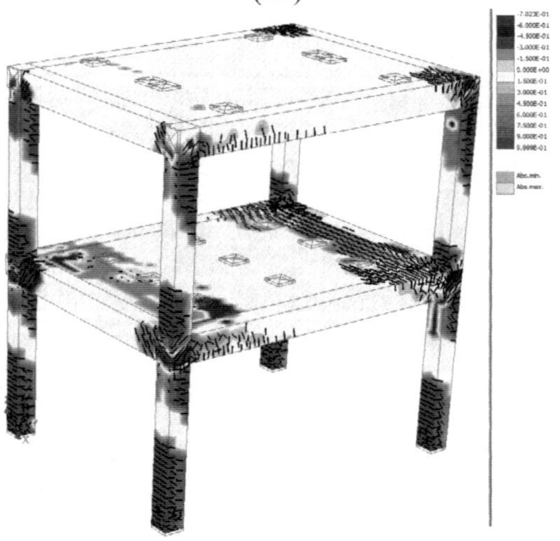

(14)

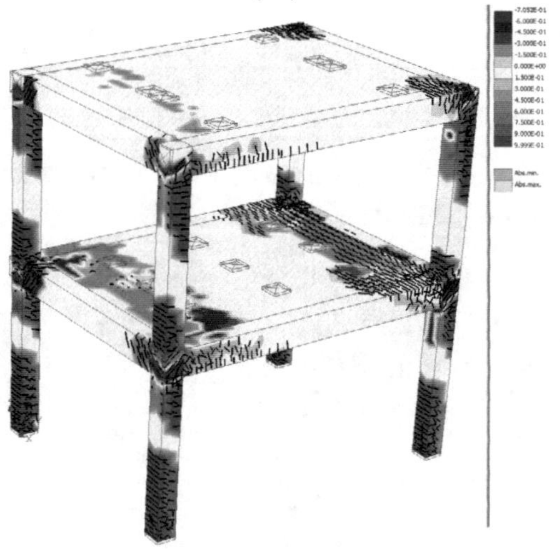

(15)

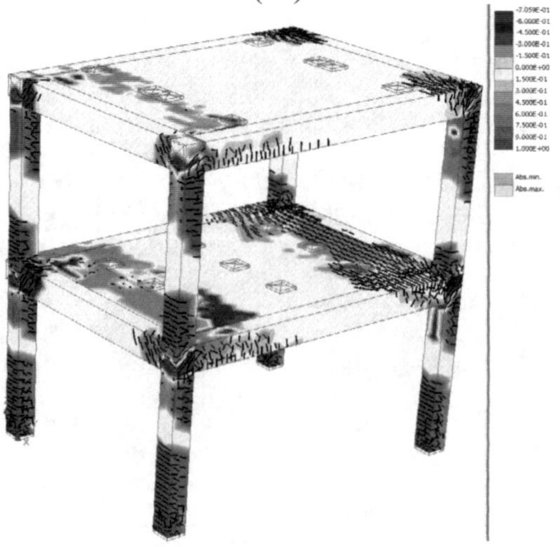

(16)

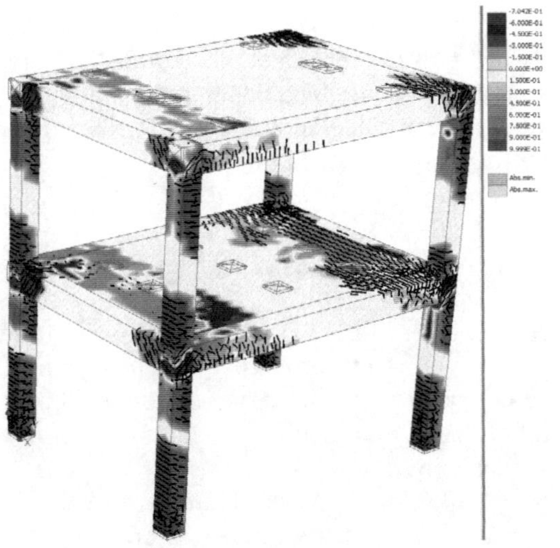

(17)

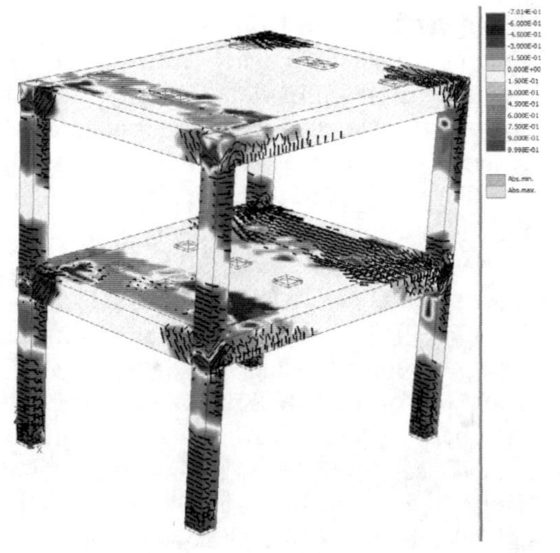

(18)

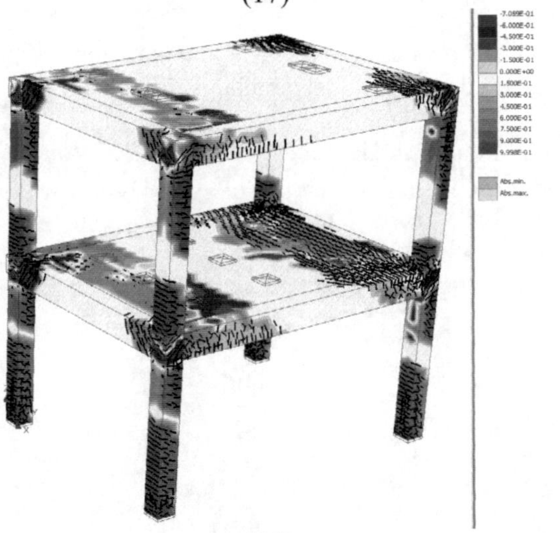

(19)

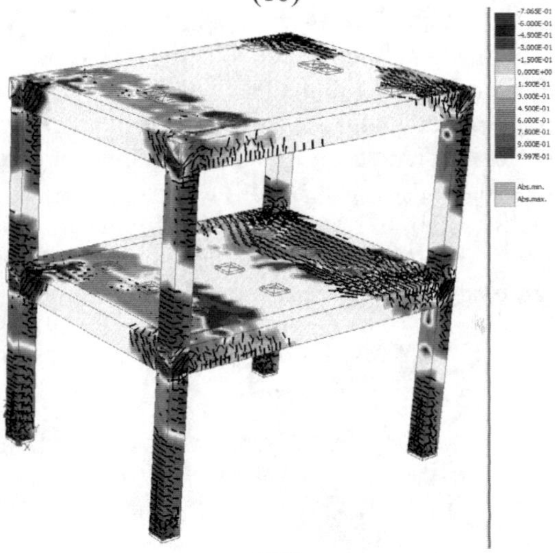

(20)

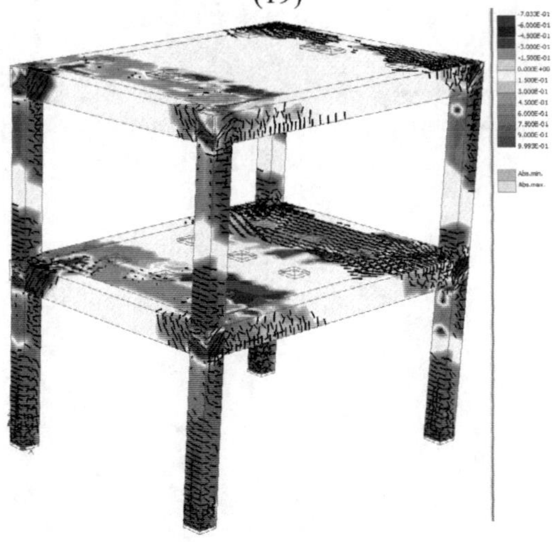

(21)

Figure 16. Concrete cracking representation and structural deformation mode (fracture specific strain (in the global direction z) concentration at the marginal regions of the RC columns) of the V30 type 2 RC frame model – for each lateral loading step [29], (Note: V30 is the numerical simulation code – see figure 7).

3.2.3. Analytical seismic response of the RC beams

RC beams designed in accordance with the ductile concept [13-26] represent structural dissipative elements for all three models. These horizontal dissipative elements provide a concrete cracking panel (figure 16) that develop in limited potential plastic tensioned areas [38-39]. However, RC beams are sufficiently stiff and they can rotate only together with the RC slab and RC beam-column joints.

In other words, the RC beam develop plastic deformations in terms of cracked concrete/ buckled steel reinforcements in combination with other structural material (see figure 16).

In these conditions, the potential plastic hinge length of the RC beams is determined by the nonlinear degradation length of the RC slab.

Basically, the idealized global plastic mechanism represented in figure 2 for pure MR RC frame structures (with marginal plastic hinges of the RC beams) is not provided by the nonlinear deformations of the RC beams in extremities. Reinforced concrete beams form a common rigid block (with superior bending stiffness and slightly deformable) with RC slab and RC beam-column joints. This aspect can be observed through the graphical representation in figure 16.

Thus, it does not ensure the formation of the **S**trong **C**olumns – **W**eak **B**eams (SCWB) [27] mechanism for all RC frame models types (see figure 7), presented in subsection 2.2 (figure 2).

3.2.4. Analytical seismic response of the RC beam-column joints

Numerical analyses of the RC frame models present important nonlinear inelastic deformations in the RC beam-column joints (see figure 17). It can be observed that the RC frame nodes do not develop a common body with the RC columns (see subsection 2.3).

Consequently, the RC beam-column joints of all RC frame type do not „borrow" the RC columns strength and stiffness. Reinforced concrete frame nodes develop a common rigid body with RC beams and RC slab and extremely deform in all existing situations (see figure 16 and figure 17). It was observed the possibility of developing potentially plastic rupture regions in RC frame nodes (see figure 18).

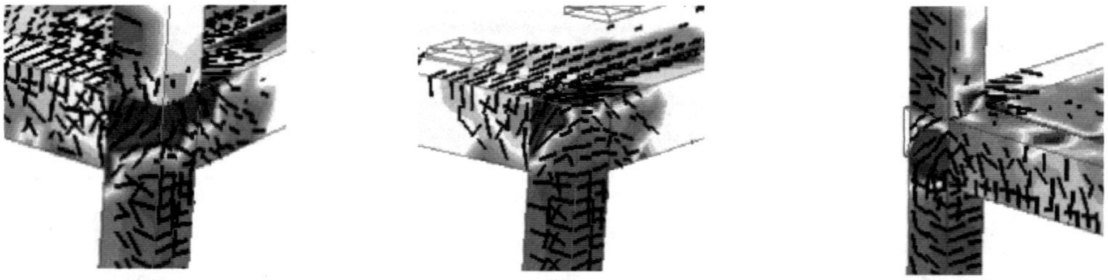

(a) (b) (c)

Figure 17. Nonlinear inelastic fracture specific strain (in the global direction z) for the ultimate horizontal loading step of the V30 RC frame model (type 2 RC frame model) [29], (where V30 is the numerical simulation code – see figure 7): (a) intermediate RC beam-column joint between the first storey and upper storey; (b) superior RC beam-column joint – upper storey; (c) intermediate RC beam-column joint between the first storey and upper storey.

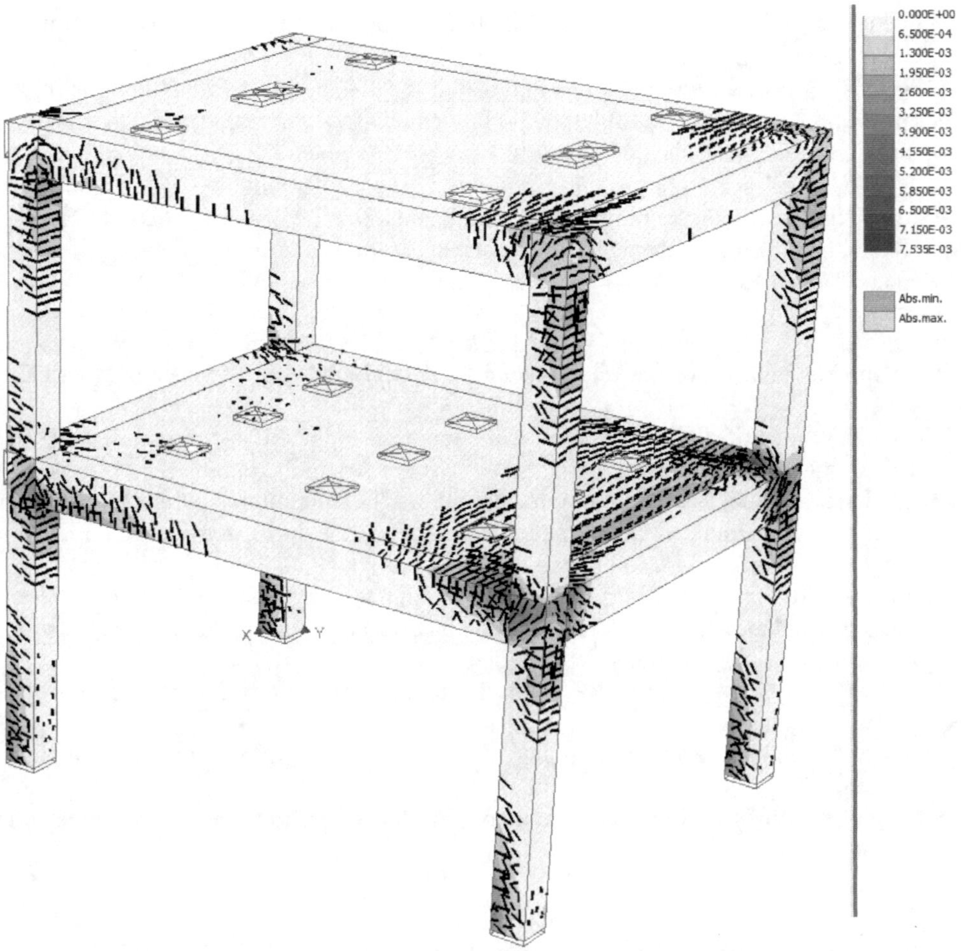

Figure 18. Maximum fracture specific strain of the V30 RC frame model (type 2 RC frame model) (where V30 is the numerical simulation code – see figure 7) for step 25 lateral loading [29].

3.2.5. Analytical seismic response of the RC slabs

Reinforced concrete slabs of all RC frame model types ensure the horizontal forces transfer through the RC beams to vertical elements (RC columns) [55] in a different form compared to the effect specified in the literature [26], [13-26] (see subsection 2.4).

The RC slabs develop a common rigid body in combination with the RC beams and RC beam-column joints [35-36], [43-45], determining nonlinear deformations/ cracking of the marginal length (considered plastic joint) for the RC beam and synchronizing RC slabs and RC beams displacements with the RC frame nodes (see figure 16).

In these conditions, reinforced concrete beams achieve important bending strength and significant bending stiffness, „borrowing" these characteristics to adjacent lateral elements (such as RC beam-column joints).

Thus, the formed common rigid „RC slab-beams-frame nodes" block [35-36], [43-45], negatively influences the global seismic response of the RC frame models designed according to the ductile concept [13-26].

In conclusion, the following aspects regarding the reinforced concrete slabs consideration in design stage can be specified:

- The RC slab makes a common rigid body with the RC beams and RC frame nodes, naturally deforms and dictates the possible plastic hinges length in the RC beams. In this case, RC slab cannot present a linear-elastic behaviour.
- The RC slab makes a common rigid body with the RC beams and RC frame nodes, rotates simultaneously with the reinforced concrete beams. Practically, the efficient plastic rotation of the RC beams, entirely depends on the RC slab plastic rotation.

4. Conclusions

Nonlinear static analyses demonstrate significant structural degradation in every structural element for all RC frame models and active participation of the RC columns in the seismic energy dissipation mechanism.

The RC slab bending stiffness contributes to the increase of the RC beams bending stiffness. Thus, RC beams are not the seismic fundamental dissipative members in the pure MR RC frame studied.

Nonlinear static analyses revealed the formation of the common rigid „RC slab-beams-frame nodes" block for all RC frame models types, which causes the local nonlinear deformations in RC columns (in adjacent regions of the RC beam-column joints).

RC columns and RC frame nodes do not form a common rigid body specified in the engineering literature. Also, RC columns and RC frame nodes have a nonlinear inelastic seismic response with important plastic material degradations/ cracks.

RC beams with large cross-section led to plastic hinges concentration mechanism at the marginal regions of the RC columns.

RC beams with minor cross-section led to minimal negative effects on the nonlinear inelastic seismic response of the RC columns.

Superior concrete strength classes have positive influences on the deformation mechanism of the pure MR RC frame models.

5. References

[1] Arnold C *et al* 2006 *Designing for Earthquakes. A Manual for Architects. FEMA 454* (California: NEHRP)
[2] Ishiyama Y 2011 *Introduction to Earthquake Engineering and Seismic Codes in the World* (Japan: Hokkaido University)
[3] Park R and Paulay T 1975 *Reinforced concrete structures* (USA: John Wiley & Sons)
[4] Fenwick R and MacRae G 2009 *Bulletin of the New Zealand Society for Earthquake Engineering* **42** No **3**
[5] BOCA 1975 *National Building Code. The BOCA Basic Building Code* (Chicago: Building Officials & Code Administrators)
[6] BSLJ 1981 *Building Standard Law in Japan* (Tokyo)
[7] ICBO 1970 Uniform Building Code vol 1 *Int. Conf. of Building Officials* (Pasadena: California)
[8] MOW 1970 *Ministry of Works. Code of Practice. Design of Public Buildings* (USA: Office of Chief Structural Engineer)
[9] NZS 1900 1965 *New Zealand Standards Institute* (New Zealand: Standards Association New Zealand)
[10] NZS 4203 1976 *Code of Practice for General Structural Design and Design Loadings for Buildings* (New Zealand: Standards Association New Zealand)
[11] UBC 1970 *Uniform Building Code* (USA)
[12] Sococol I 2019 *Creatii Universitare* **13** 137-47
[13] Popa V *et al* 2013 *Cod de proiectare seismică, Partea-I, Prevederi de proiectare pentru clădiri, Indicativ P100-1* (București: UTCB)
[14] Seismic ASCE 7 2006 *Minimum Design Loads for Buildings and Other Structures* (USA:

American Society of Civil Engineers)
[15] NRC 2005 *National Building Code of Canada* (Canada: National Research Council of Canada)
[16] EC 8 2006 *Eurocode 8: Design of Structures for Earthquake Resistance. Part 1: General Rules, Seismic Actions and Rules for Buildings* (UE: ASRO)
[17] NZS 1170.5 2004 *Structural Design Actions – Part 5: Earthquake Design Actions* (New Zealand: New Zealand Standards Executive)
[18] AIJ 1990 *Design Guidelines for Earthquake Resistant Reinforced Concrete Buildings Based on Ultimate Strength Concept* (Japan)
[19] Nch433Of.96 1996 *Earthquake Resistant Design of Buildings* (Chile)
[20] MCBC 2004 *Mexico City Building Code* (Mexic)
[21] Sococol I, Mihai P and Donțov I 2019 *The Bulletin of the Polytechnic Institute of Jassy, Construction. Architecture Section* **65** No **1** 17-30
[22] Akiyama H 2002 *Annals of Geophysics* **45** No **6** 791-98
[23] Budescu M and Ciongradi I 2014 *Inginerie Seismică* (Iași: Politehnium)
[24] Postelnicu T 2012 *Proiectarea Structurilor de Beton Armat în Zone Seismice* (București: MarLink)
[25] Stratan A 2014 *Dinamica Structurilor și Inginerie Seismică. Note de Curs* (Timișoara: UPT)
[26] Paulay T and Priestley M J N 1992 *Seismic Design of Reinforced Concrete and Masonry Buildings* (New York: John Wiley & Sons)
[27] Sococol I, Toma I-O, Mihai P, Țăranu N and Budescu M 2021 *Proceedings of 1st Croatian Conference on Earthquake Engineering* (Zagreb: University of Zagreb Faculty of Civil Engineering) (in course of publication)
[28] Murty C V R, Goswami R, Vijayanarayanan A R and Mehta V V 2012 *Some Concepts in Earthquake Behaviour of Buildings* (Gandhinagar: Gujarat State Disaster Management Authority)
[29] ATENA software *http://www.cervenka.cz*
[30] El-Attar A G, White R N and Gergely P 1991 *Shake Table Test of a 1/6 Scale Two-Story Lightly Reinforced Concrete Building* (New York: Cornell University)
[31] Harris H G and Sabnis G M 1999 *Structural Modeling and Experimental Techniques*, Second Edition (Boca Raton: CRC Press)
[32] Moncarz P D and Krawinkler H 1981 *Theory and Application of Experimental Model Analysis in Earthquake Engineering* (San Jose: Stanford University)
[33] https://www.ibtimes.co.uk/japan-researchers-test-10-storey-concrete-building-resilience-against-new-kobe-earthquake-1532970
[34] Farzad N 2001 *The Seismic Design Handbook. Second Edition* (New York: Springer Science+Business Media)
[35] Sococol I, Mihai P, Toma I-O, Olteanu-Donțov I and Venghiac V-M 2020 *The Bulletin of the Polytechnic Institute of Jassy, Construction. Architecture Section* **66** No **4** xx-xx (in course of publication)
[36] Sococol I, Mihai P, Budescu M and Olteanu-Donțov I 2021 *Acta Technica Napocensis: Civil Engineering & Architecture* **XX** No **X** xx-xx (in course of publication)
[37] Cui Y, Lu X and Jiang C 2017 *J. Engineering Structures* **132** 684-94
[38] Liu J, Zhang J, Li X and Cao W 2021 *J. Engineering Structures* **232** 111810
[39] Hu B, Lv H-L and Kundu T 2019 *J. Construction and Building Materials* **217** 473-86
[40] Arruda M R T and Castro L M S 2021 *J. Advances in Engineering Software* **153** 102965
[41] Rizwan M, Ahmad N and Khan A N xxxx *J. Structures* **xxx** xxx-xxx
[42] Dalal S P and Dalal P 2021 *J. Structures* **29** 1154-64
[43] Sococol I, Mihai P, Iftode V I and Olteanu-Donțov I 2019 *Computational Civil Engineering Conf. (Romania)* (Iasi: Materials Science and Engineering)
[44] Sococol I, Mihai P, Toma I-O, Venghiac V-M and Olteanu-Donțov I 2020 *The Bulletin of the Polytechnic Institute of Jassy, Construction. Architecture Section* **66** No **2** 95-108

[45] Sococol I, Mihai P, Toma I-O, Venghiac V-M and Olteanu-Donţov I 2020 *The Bulletin of the Polytechnic Institute of Jassy, Construction. Architecture Section* **66** No **3** xx-xx (in course of publication)
[46] Lu X, Fu G, Shi W and Lu W 2008 *J. The Structural Design of Tall and Special Buildings* **17** 181-201
[47] CEN 2006 *Eurocod 2: Proiectarea structurilor de beton, Partea 1-1: Reguli generale şi reguli pentru clădiri, indicativ SR EN 1992-1-1* (Bucureşti: ASRO)
[48] GR 2012 *Cod de proiectare. Evaluarea acţiunii zăpezii asupra construcţiilor, indicativ CR 1-1-3/2012* (Bucureşti: Ministerul Dezvoltării Regionale şi Turismului)
[49] GR 2012 *Cod de proiectare. Evaluarea acţiunii vântului asupra construcţiilor, indicativ CR 1-1-4/2012* (Bucureşti: Ministerul Dezvoltării Regionale şi Turismului)
[50] GR 2012 *Cod de proiectare. Bazele proiectării construcţiilor, indicativ CR 0/2012* (Bucureşti: Ministerul Dezvoltării Regionale şi Turismului)
[51] Bitencourt Jr. L A G, Manzoli O L, Trindade Y T, Rodrigues E A and Dias-da-Costa D 2018 *J. Finite Elements in Analysis and Design* **149** 32-44
[52] Ghayoumian G and Emami A R 2020 *J. Structures* **28** 1077-1107
[53] Jalilkhani M, Ghasemi S H and Danesh M 2020 *J. Engineering Structures* **213** 110528
[54] Zou X-K and Chan C-M 2005 *J. Engineering Structures* **27** 1289-1302
[55] Men J, Xiong L, Wang J and Fan G 2021 *J. Engineering Structures* **229** 111639
[56] Sococol I, Mihai P, Toma I-O, Olteanu-Donţov I and Venghiac V-M 2020 *The Bulletin of the Polytechnic Institute of Jassy, Construction. Architecture Section* **66** No **4** xx-xx (in course of publication)
[57] Murugan K and Sengupta A K 2020 *J. Structures* **27** 487-505
[58] Taheri A, Tasnimi A A and Moghadam A S 2020 *J. Structures* **27** 163-173

IOP Conf. Series: Materials Science and Engineering 1141 (2021) 012010 doi:10.1088/1757-899X/1141/1/012010

Numerical flexural strengthening investigation of timber-CFRP composite beams

D Harrach[1], M Movahedi Rad[1]
[1]Department of Structural and Geotechnical Engineering
Faculty of Architecture, Civil Engineering and Transport Sciences
Széchenyi István University, Egyetem tér 1, Győr 9026, Hungary.

E-mail: harrach.daniel@sze.hu

Abstract. Timber is among the most important construction materials with unique characteristics in comparison with other materials for instance steel and concrete. It is one of the first sustainable materials and continues to be a favoured selection in the current infrastructure. In the last decade, to refine the mechanical properties of timber, structural elements reinforcing fiber polymers have developed. In this study, to improve the flexural properties of timber beams, carbon fiber-reinforced polymer (CFRP) laminates are attached externally and internally between timber layers. For numerical simulations, first finite element model was validated according to the data obtained from experimental tests, then a series of simulations with different parameters were executed to investigate the flexural behaviour. The considered parameters include CFRP laminates' existence and position. The linear elastic orthotropic material model was used to represent linear behaviour while the 3D anisotropic Hill material model was assumed to define the nonlinear behaviour of the material. Finite element simulations are executed by commercial software ABAQUS.

1. Introduction

The use of timber as a building material has greatly expanded in recent centuries, with wood being widely used in architecture, even for the construction of bridges and buildings. Mainly, it is characterized by its easily formed and its good mechanical properties. In addition to its aesthetic characteristics, its low density contributes to the construction of lightweight structures. Wood will continue to be an important building material due to the spread of environmentally conscious thinking. Moreover, it is considered a truly renewable and biodegradable material with lower production and processing energy requirements than other structural materials.

Protecting the forests and reducing the disadvantages compared to reinforced concrete or steel requires the efficient use of wood in structural applications. In order to increase the load-bearing capacity of wood, several attempts have been made to reinforce glue laminated timber (GLT) beams [1], which can be done with steel, polymer fabric, fiberglass, or carbon fiber [2]. Placing the reinforcing layer in the case of a timber support on the pulled part is most effective. According to practical experiments, up to a two-fold increase in load capacity and a stiffness increase of approximately 20% can be achieved. [3] The expected failure of a laminated-glued beam reinforced in this way begins with the rupture of the cover lamella. This phenomenon can be explained by the high modulus of elasticity of the resin layer, as the tensile stresses that the resin, which is about 20 times stiffer than wood, can still withstand and easily exceed the tensile strength of the wood. A characteristic feature of the failure process is that the rupture of the cover lamella does not mean complete failure. The wood system can withstand even more load increases after a smaller load capacity drop, as shown by researchers Blaß and Romani. [4].

De Jesus et al. [5] investigated experimentally and numerically the effect of different lengths of carbon fiber-reinforced polymer (CFRP) laminates on the strengthening of GLT beams in bending. M. Khelifa et al. [6] and [7], Khennane et al. [8] and Kawecki and Podgórski [9] investigated flexural strengthening of timber beams with CFRP, using Abaqus software to numerically predict the performance of GLT beams in flexion after strengthening by use of CFRP sheets of distinct lengths, externally attached to the soffit of timber beams. Researchers are using the method of gluing strips to the bottom of a girder on the entire cross-section, because fiber-reinforced polymer (FRP) is most efficient under tension. Fiorelli [3] carried out studies on beams reinforced with Glass Fiber Reinforced Polymer (GFRP) and CFRP while Kossakowski [10] used laminates made of glass, aramid and carbon.

This paper presents the results of an experimental program of 3-point bending tests of GLT beams reinforced with unidirectional CFRP laminates. In the experimental program, two different reinforcement vertical positions are investigated and obtained results are compared with a non-reinforced beam result.

Furthermore, a constitutive model based on timber plastic damage was developed and used to investigate the bending strength of GLT beams. In this study, a constitutive model based on timber plastic damage developed and used to investigate the bending strength of GLT beams. The novelty of the article is that it contains not only the usual external plane reinforcements, but also examines the effects of CFRP lamellas incorporated during the manufacture of wooden beams. Moreover, CFRP strips were applied to enhance the bending strength of GLT beams. All these parameters were considered and their effects were studied.

The present model considered the Hill mechanical model for orthotropic timber, isotropic adhesive and isotropic CFRP behaviours. Nowak et al. [11] used the same Hill model for the plasticity of wood defined by using the Hill yield criterion and the generalized Hill yield criterion, which take into account the differing yield stresses during compression and tension.

2. Behaviour of materials and constitutive modelling

2.1. Timber

The behaviour of GLT in compression parallel to grain shows some softening in general; it can be regarded as elastic perfectly plastic. The basic anisotropic yield criterion which can be applied in timber is the Hill yield criterion [10]. Generalization of this criterion for anisotropic materials with different plasticity limits during compression and tension, so defining the generalized Hill yield criterion.

The Hill yield criterion [12] is one of the criteria used in the numerical analysis of wooden elements. The theory results from a generalization of the Huber-Mises-Hencky hypothesis for anisotropic materials. It allows a connection between material strengths and anisotropic directions.

When this criterion is used with the isotropic hardening option, the yield function is given by:

$$f(\sigma) = \sqrt{\frac{3}{2}(\sigma)^T \cdot [M] \cdot (\sigma)} - \sigma_0{}^{(\bar{\varepsilon}^p)} \tag{1}$$

where:

σ_0 reference yield stress;
ε^P equivalent plastic strain;
$[M]$ Mass matrix.

When it is used with the kinematic hardening option, the yield function takes the following form:

$$f(\sigma) = \sqrt{\frac{3}{2}\big((\sigma) - (\alpha)\big)^T \cdot [M] \cdot \big((\sigma) - (\alpha)\big)} - \sigma_0 \tag{2}$$

where:

(α) yield surface translation vector.

In a coordinate system that is aligned with the anisotropy coordinate system, the Hill yield stress potential can be expressed as follows:

$$\begin{aligned} f(\sigma, \sigma_y) = {} & F(\sigma_{22} - \sigma_{33})^2 + G(\sigma_{33} - \sigma_{11})^2 + H(\sigma_{11} - \sigma_{22})^2 + \\ & +2L\sigma_{23}{}^2 + 2M\sigma_{31}{}^2 + 2N\sigma_{12}{}^2 - \sigma_y{}^2 = 0 \end{aligned} \tag{3}$$

where:

F, G, H, L, M, N are constants which are determined experimentally of the material in different orientations.

$$F = \frac{1}{2} \cdot \left(\frac{1}{R_{22}^2} + \frac{1}{R_{33}^2} - \frac{1}{R_{11}^2}\right) \quad (4)$$

$$G = \frac{1}{2} \cdot \left(\frac{1}{R_{33}^2} + \frac{1}{R_{11}^2} - \frac{1}{R_{22}^2}\right) \quad (5)$$

$$H = \frac{1}{2} \cdot \left(\frac{1}{R_{11}^2} + \frac{1}{R_{22}^2} - \frac{1}{R_{33}^2}\right) \quad (6)$$

$$L = \frac{3}{2 \cdot R_{23}^2} \quad (7)$$

$$M = \frac{3}{2 \cdot R_{13}^2} \quad (8)$$

$$N = \frac{3}{2 \cdot R_{12}^2} \quad (9)$$

where:

$R_{i;j}$ are anisotropic yield stress ratios.

This criterion can be applied in modelling wood and processed wood products, titanium alloys, zirconium alloys and fiber matrix composites.

2.2. CFRP

This material may be considered anisotropic with transverse isotropy. The CFRP material was considered as linear elastic isotropic until failure. The elastic modulus in the fiber direction of the unidirectional CFRP material used in the present study was specified by the manufacturer as E_{CFRP} = 170 GPa while ν_{CFRP} = 0.3 was used for the isotropic model.

3. Materials and methods

3.1. Finite element modelling

Finite element analysis was carried out to model the nonlinear behavior of GLT beams. For modelling the laminated timber beams 8-node solid (C3D8) element was used. The 4-node (doubly curved shell) element (S4) was selected to stand for the CFRP. The contact between CFRP and timber is considered as surface to surface contact, and the glue contact is defined by using friction coefficient equals 0.1 which is available in the ABAQUS manual [13]. For gluing the timber lamellas together cohesive contact was used too. Steel bearing plates (length = 200 mm, width = 120 mm and thickness = 30 mm) were installed at points of loading to prevent local failure caused by crushing.

Figure 1 shows beam geometry and loading and supporting conditions where all beams were tested under monotonic loading up to failure, with one concentrated load on the center of the beams.

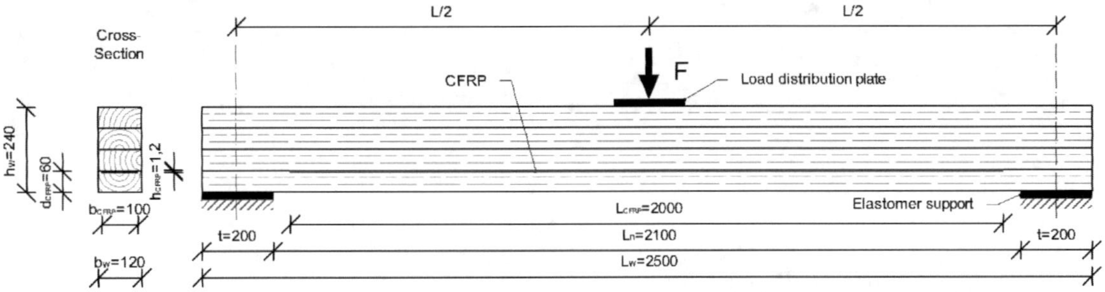

Figure 1. Beams geometry and loading condition.

Symbols used on Figure 1:

b_w width of GLT beam in mm dimension;
h_w height of GLT beam in mm dimension;
L_w length of GLT beam in mm dimension;
b_{CFRP} width of CFPR lamella in mm dimension;

h_{CFRP} height of CFPR lamella in mm dimension;
L_{CFRP} length of CFPR lamella in mm dimension;
d_{CFRP} distance of CFRP lamella from the bottom of GLT beam in mm dimension.

To model the load distribution plates, the (C3D8) 8-node solid element is used. The boundary condition at the first support is a roller to produce rotation and horizontal movement and a pin set at the other support to produce rotation. The loading conditions were the same for all the models, a vertical concentrated load was considered at each plate which placed at the top of the beam and this load was distributed by the coupling effect. Moreover, a fine mesh was applied to obtain results of sufficient accuracy where the total elements number of the beams was approximately equal to 60000 elements. Figure 2 shows the numerical model used, while Table 1 shows the numerical specimens 'properties.

Table 1. Material properties for numerical modelling of timber.

Elasticity properties			Plasticity properties		
E_1= 11.5 GPa	E_2= 1.2 GPa	E_3= 1.2 GPa	σ_{yield}=43.1 MPa	$Q=130$ MPa	$b=21$
G_{12}=0.98 GPa	G_{13}=0.34 GPa	G_{23}=0.34 GPa	$F=0,5$	$G=0.5$	$H=0.5$
v_{12}= 0.29	v_{13}=0.29	v_{23} = 0.29	$L=1.5$	$M=1.5$	$N=1.5$

It can also be clearly seen in Figure 2 that an important aspect in the creation of the model was that the GLT beam was not taken as a homogeneous timber material, but the timber lamellas were modelled separately and then connected to each other by a cohesive connection.

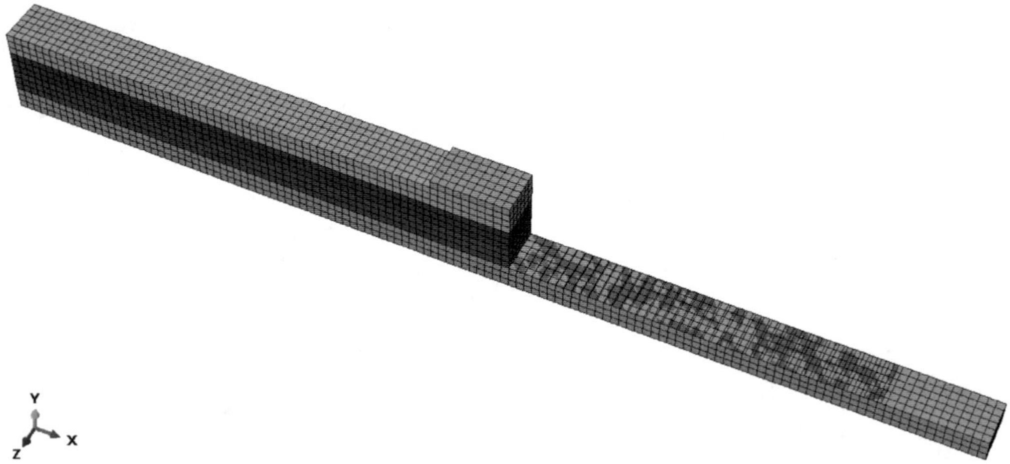

Figure 2. Numerical model.

3.2. Experimental program

In this research, we performed three-point bending tests following de Jesus [13]. According to this experimental work, nine beams of 2500 mm long GLT beams - with a cross-sectional area of 120x240 mm - were tested. Besides, reinforcement Sika CarboDur S-1012 CFRP laminates were used with a width of 100 mm, a thickness of 1.2 mm, and a length of 2000 mm which was glued using adhesives SIKA products. Adhesion tests of timber and CFRP were performed before the start of the tests. Based on this, it can be stated that the adhesive strength between the two materials is higher than the tensile strength of the timber perpendicular to grain, which was a condition of the buildability. Commercially available GLT beams were used, the properties of which were specified by the manufacturer. The experimental properties of the used materials are presented in Table 2.

Table 2. Material properties.

Material	Type	Flexural strength [N/mm^2]	Compression strength [N/mm^2]	Tensile strength [N/mm^2]	Shear strength [N/mm^2]	Elastic modulus [N/mm^2]
Timber	GL24h	$f_{m,k}$= 24.0	$f_{c,0,k}$ = 21.5 $f_{c,90,k}$= 2.5	$f_{t,0,k}$= 19.2 $f_{t90,k}$= 0.5	$f_{v,k}$= 3.5	$E_{0,mean}$ = *11.5* $E_{90,mean}$ = 0.30
CFRP	Carbodur	-	-	3.100	-	170.000
Epoxy glue	Sikadur-30	-	85 - 95	26 - 31	16 - 19	9.600

The first three test specimens were the normal GLT beam without strengthening (TYPE-0). The CFRP laminates were glued on the lower plane of the other three beams (TYPE – A while for the rest three beams, CFRP laminates were embedded between timber layers (TYPE-B), as shown in Figure 3.

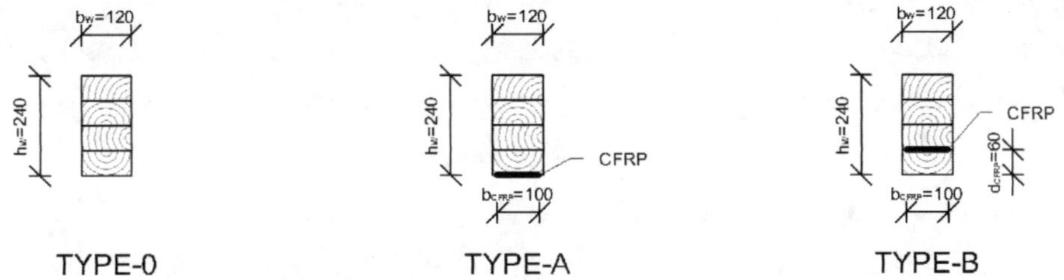

Figure 3. Cross sections of the specimens.

Displacement transducers were used to measure the deflection of the beams along the girder, as shown in Figure 4. These measuring points can be used to define the deflection line of the beams under the loading. The beams were tested during the project GINOP-2.2.1-15-2016-00030.

Figure 4. Experimental test layout.

4. Results and discussion

At the beginning of the simulation work, the aim was to validate the models (without reinforcement - TYPE-0, CFRP used at the bottom - TYPE-A and the strengthening built into the structure - TYPE-B). Then the parameters of the timber material model and the adhesive properties of the GLT beams were determined, and then the parameters of the CFRP and epoxy were set. By comparing the numerical results with the experimental test results, it was realized that the ultimate loads and stiffness properties for the three specimens' types are almost similar.

The main measurement results are shown in Table 3.

Table 3. The numerical and experimental test results.

Specimen	Experimental results			Numerical results		
	Deflection at 70 kN Δ [mm]	Deflection at failure Δ_u [mm]	Ultimate force P_u [kN]	Deflection at 70 kN Δ [mm]	Deflection at failure Δ_u [mm]	Ultimate force P_u [kN]
TYPE-0	12.85	25.79	121.08	12.99	27.43	116.14
TYPE-A	12.30	35.01	133.00	11.01	36.06	136.54
TYPE-B	12.36	30.03	119.50	12.47	33.12	124.89

Figures 5-7, show the deflections of the middle cross section of the validated models compared to the experimental tests. 0-1 to 0-3 means the three TYPE-0 beams, A-1 to A-3 means the three TYPE-A and B-1 to B-3 means the three TYPE-B beams that is used in the experimental tests.

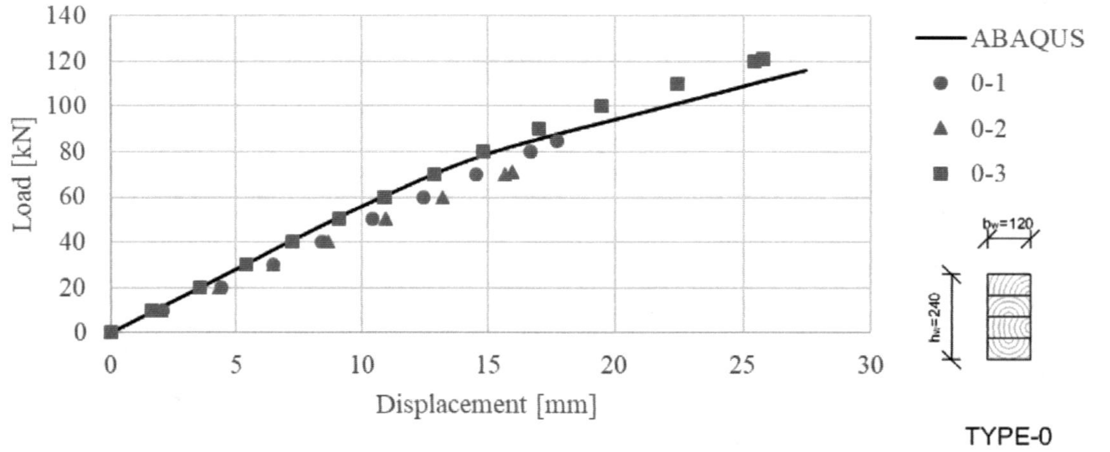

Figure 5. Comparison of experimental and numerical results on TYPE-0.

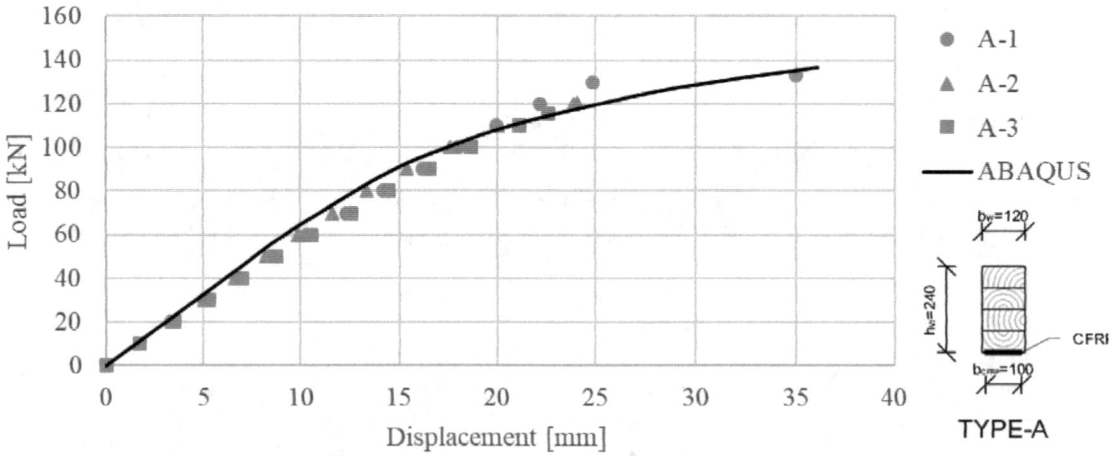

Figure 6. Comparison of experimental and numerical results on TYPE-A.

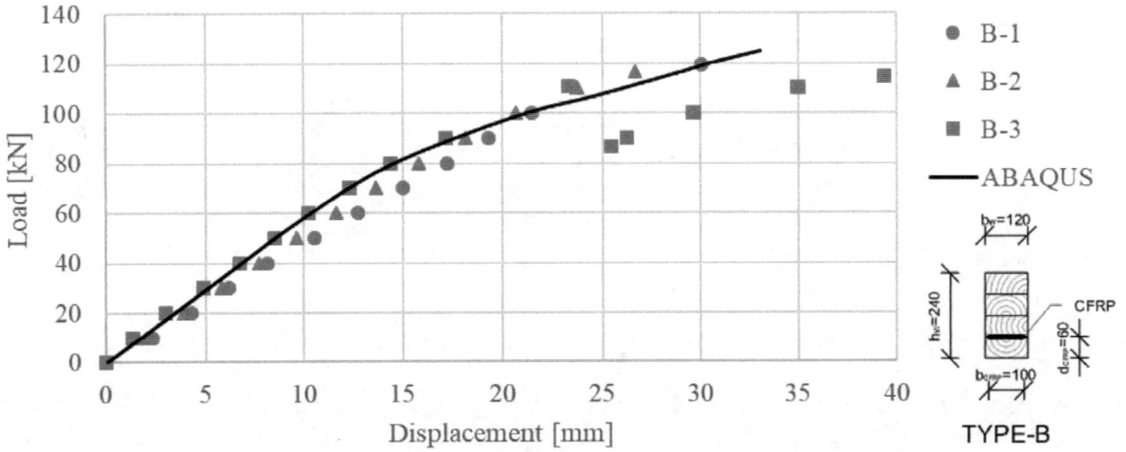

Figure 7: Comparison of experimental and numerical results on TYPE-B.

Using the parameters of the fabricated models, additional models were constructed by changing the vertical position of the CFRP laminates along with the height of the beam in 10 mm steps. The process was repeated until the neutral axis of the strengthened beam was reached because by then the CFRP laminates would have worked in the compression zone. The measurement results show that the stiffness (resistance to deflection) of the beam increased continuously as the CFRP laminates were placed towards the bottom plane of the beam. Figure 8 shows the results of the numerical experiments, the different curves indicating the location of the CFRP from the bottom of the beam.

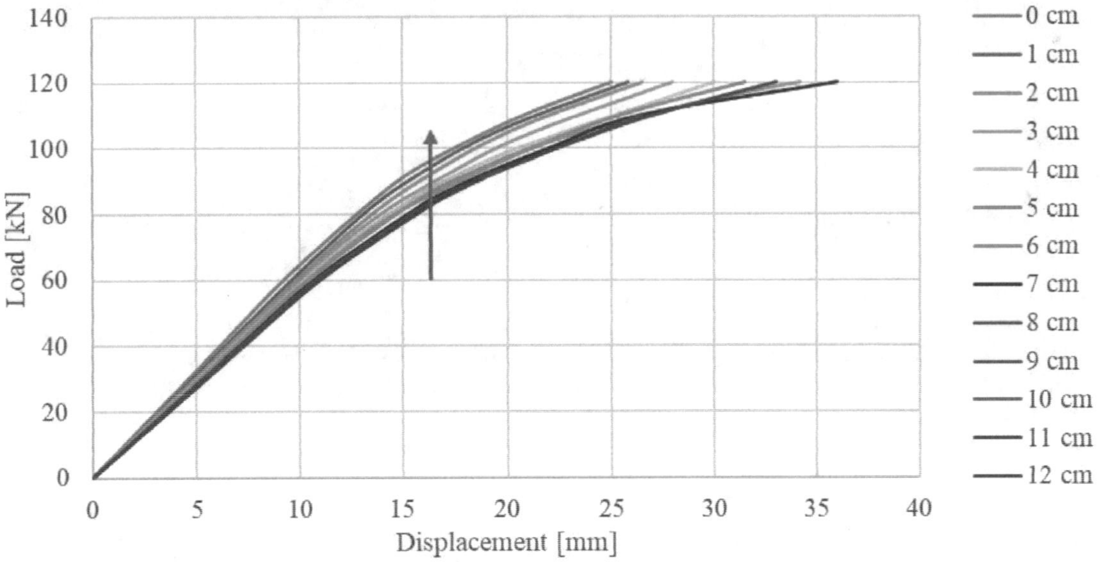

Figure 8: The load-displacement curves of the strengthened GLT beams.

As can be seen, where the CFRP laminates in the lower position were applied, results in the greater strength of the GLT beam. The load capacity of the beams was not given in this case, the concentrated load was maximized at 120 kN, because it was only intended to study the extent to which the behaviour of the girder changes.

IOP Conf. Series: Materials Science and Engineering 1141 (2021) 012010 doi:10.1088/1757-899X/1141/1/012010

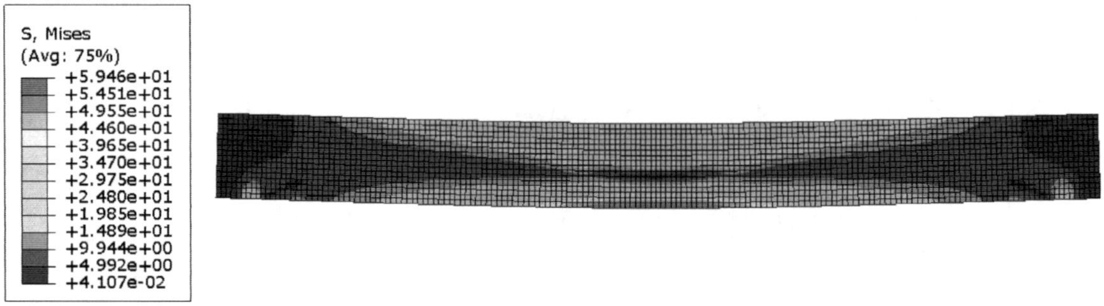

Figure 9: The stresses of the beam (2cm) under 110 kN load.

The stresses arising in the beam are also as expected, i.e. the highest tensile stress values occurred in the most bent zone of the girder under the loading. Figure 9 shows the beam, where the CFRP laminate was placed 2 cm from the bottom under 110 kN concentrated load.

We also investigated the effect of different CFRP positions on the neutral axis of bent beams where zero stress arises in the elastic range. Figure 10 shows how the position of the neutral axis moved due the different CFRP positions.

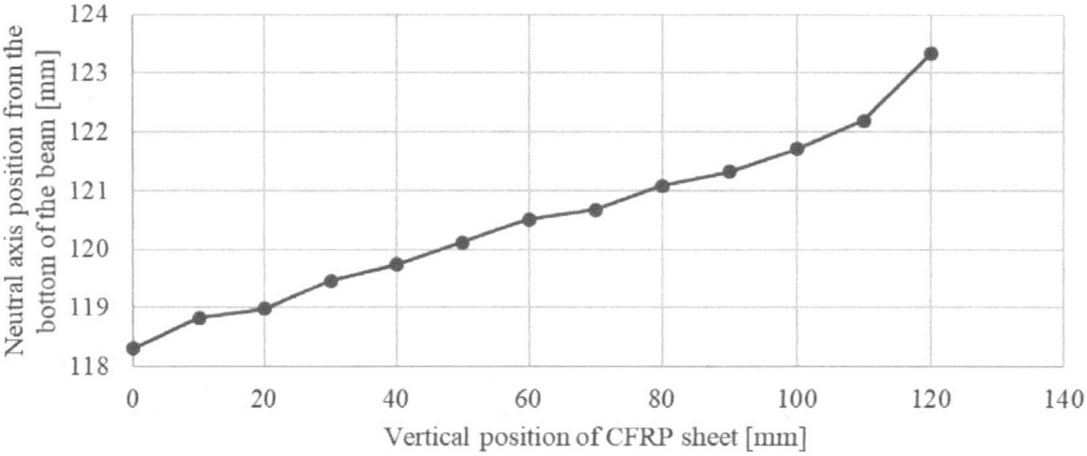

Figure 10: The neutral axis position displacement due to the use of CFRP.

5. Conclusion

Comparing between numerical and experimental results it is noted that the existence of carbon fiber reinforced polymer laminates causes an increment in the load capacity significantly and the rigidity of the glue laminated timber beams.

The experimental and the numerical results show that the use of CFRP laminates for bent GLT beams is most effective when applied in the lower, tensioned zone. However, there is a limit to this eccentricity beyond which the application of the CFRP no longer brings such benefits.

By using CFRP reinforcement, up to a 10% increase in load capacity and stiffness can be achieved for glue laminated timber beams during practical application.

Further testing of CFRP laminates used - as non-reinforcing after installation - at the production of wooden beams is necessary, because a more accurate understanding of the behaviour of these types of structures is needed.

6. References

[1] Kim Y J, Harries K A 2010 Modeling of timber beams strengthened with various CFRP composites *Eng. Struct.* **32** pp 3225–3234

[2] Lee Y, Park J, Hong S, Kim S 2015 A Study of Bond of Structural Timber and Carbon Fiber Reinforced Polymer Plate *Mater. Sci.* **21** pp 563–567

[3] Fiorelli J, Dias A A 2003 Analysis of the strength and stiffness of timber beams reinforced with carbon fiber and glass fiber *Mater. Res.* **6**, pp 193–202

[4] Blaß H J, Romani M 2001 CIB Working Commission W18: Timber Structures Design Model for FRP Reinforced Glulam Beams.

[5] De Jesus AMP, Pinto José MT, Morais José JL 2012 Analysis of solid wood beams strengthened with CFRP laminates of distinct lengths *Constr Build Mater.* **35** pp 817–828

[6] Khelifa M, Auchet S, Méausoone P J, Celzard A 2015 Finite element analysis of flexural strengthening of timber beams with Carbon Fibre-Reinforced Polymers *Eng. Struct.* **101** pp 364–375

[7] Khelifa M, Celzard A 2014 Numerical analysis of flexural strengthening of timber beams reinforced with CFRP strips *Composite Structures* **111** pp 393–400

[8] Khennane A, Khelifa M, Bleron L, Viguier J 2014 Numerical modelling of ductile damage evolution in tensile and bending tests of timber structures *Mechanics of Materials* **68** pp 228–236

[9] Kawecki B, Podgórski J 2020 The Effect of Glue Cohesive Stiffness on the Elastic Performance of Bent Wood–CFRP Beams *Materials* **13** (22) 5075

[10] Kossakowski P G 2011 Load—Bearing Capacity of Wooden Beams Reinforced with Composite Sheets, *Structure* **3** pp 1–9

[11] Nowak T, Patalas F, Brol J 2018 The use of Hill anisotropic yield criterion timber elements reinforced with CFRP strips *Annals of Warsaw University of Life Sciences - SGGW, Forestry and Wood Technology* **104**, pp.219-227.

[12] Hill R 1983 *The Mathematical Theory of Plasticity.* New York, Oxford University Press.

[13] Simulia D S 2018 ABAQUS User's manual Dassault Systems, analysis user's guide volume IV: Elements.

IOP Conf. Series: Materials Science and Engineering 1141 (2021) 012011 doi:10.1088/1757-899X/1141/1/012011

The socio-economic efficiency of computer-integrated transport modelling

S Rakytska[1]**, O Kambur**[1]**, O Zhus**[1]
[1] Odessa State Academy of Civil Engineering and Architecture – Odessa, 4 Didrihsona st., *Ukraine*,

E-mail: rakitska@ogasa.org.ua

Abstract. In the article has been analysed the possibilities of using modern digital technologies to optimize the transport system of a big city. The Urbanization's processes lead to high rates of urban development, which provides for the creation of assorted transport supplying. However, there are situations when the geographic, landscape, natural and climatic area features, architectural solutions (characteristics) or blunders in the design of the city do not allow to create the road infrastructure of the necessary level. This leads to congestion of the main transport networks and generate traffic jams, increasing the degree of environmental pollution. The authors consider the horizons of creating a transport model of the Odessa city as an effective method of traffic management. It has been made out a case that the digitalization of the urban highways management reduces the trip time, alleviates the level of the road network congestion, increases the reliability and safety of transport, minimizes the number of non-standard traffic situations, and increases the economic efficiency of the city's transport system. The using of powerful content solution allows you to build a mental (virtual) model, taking into account the actual traffic situation at this time and predict its changing when you have to make a certain management decision. This article is dedicated to the economic issue of the building a city transport model, the contemplation for its use to improve the efficiency of the city economy operation.

1. Introduction

The transport system of the city is a significant part of supporting its life-activity, it combines varied districts of the city, allows you to get a work, shops and holiday destination. The city's transport system includes a street and road network, transportation items, cases of transferring (public transport, private motor transport, trucks, special vehicles, etc.), likewise the relevant infrastructure - bicycle and foot walks, parking places, gas stations, public transport stops. The structure and length of the road network, its branching and severity level, the quality of road surface directly affect the safety and reliability of transport networks, passenger and freight traffic scope, the situation of the environment, therefor it is one of the important life quality indicators in the city living. The growing number of residents increases the load on the city's transport system. The discrepancy of the city's transport system and modern requirements, deficiency of proper infrastructure significantly affects to socio-economic situation: it leads to appreciable wastes of working time, increases traffic accidents, interferes efficiency, reduces social mood, increases conflicts in various areas. The optimization of traffic streams brings appreciable, social and commercial effects, which may not always be obvious, but at the same time can be identified and calculated by using modern methods of analysis. An important part of this optimization is the development of the city's transport model.

2. Methodology

The transport model of the city is an implement that is used to reason some management decisions in strategic planning of city development. It allows to estimate the distribution of passenger traffic on the existing or projected transport network on the basis of functional and spatial characteristics of the city, items of demand for transport, supply of transport. This will determine the transport time costs and wastes in the way of comparing different options of the city development and carry out their economic evaluation.

The growth of computer power counting, software development, availability of information of the transport system condition, its load and population movements, geographic information systems allow to develop and use detailed transport models of cities and take into account many different factors influencing road user behaviour.

It is used the PTV VISUM software package [1], data on the city's transport system (OpenStreetMap [2], 2GIS map data [3], and Google Earth Pro with the historical and photographic information [4].

PTV VISUM is a modern information and analytical decision support system, which allows to do a strategic and operational transport planning of cities and regions, forecasting traffic intensity, optimization of passenger transport and furthermore justification of the volume and economic feasibility of investment in transport infrastructure modernization. Many researchers use this system to optimize traffic streams [5, 6].

The PTV VISUM software complex integrates all road users (car and freight transport, passenger transport, cyclists, ganger etc.) into a single mathematical transport model. PTV VISUM combines data from geographic information systems, data on the population, ideas and directions of its movements in space, data on transport support- into a single database with several levels.

Strategic macroscopic modelling and construction of the demand model provides: Strategic planning, Operational analysis, Real time traffic management. To build a model of transport demand, the source data is used: behavioural data; network model; land use data.

Output data: data of the passenger traffic distribution in different ways, identification of "bottlenecks", forecasting changes in passenger traffic during the commissioning of new facilities, meaning losses of time during the trip. The new transport infrastructure facilities typically reduce the load on the transport system by redistributing flows. And new facilities that are the source of trips or the destination of travel can both increase and decrease the load on the transport system. To determine the demand for transport services, a 4-step demand determination program is used (Figure 1).

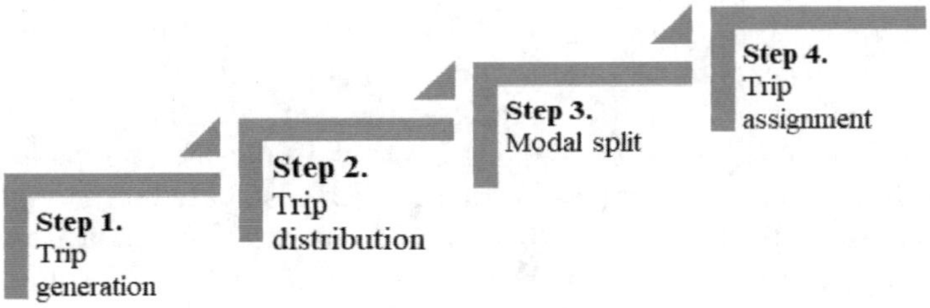

Figure 1. Classic 4-step transport model.

At every step, the model provides for its verification and calibration, which involves returning to the previous steps if necessary. The model can be used both to analyse the current situation and to forecast the development of demand for future periods.

- In the first stage, the potential value of the flow is estimated. On the socio-economic and demographic basis factors the purposes of trips and places of passenger flows attraction are defined.
- In the second stage, the distribution of traffic flows in different directions is analysed and a matrix of inter-district correspondence constructs.
- In the third stage, the correspondence is distributed by different modes of transport. In the estimation way, it takes into a usefulness of a particular kind of transport for different groups of people and the purpose of trips.
- In the fourth stage, the intensity of traffic flows on different alternative routes is calculated.

IOP Conf. Series: Materials Science and Engineering 1141 (2021) 012011 doi:10.1088/1757-899X/1141/1/012011

The amount of transport demand in the area depends on socio-economic and demographic characteristics. Residents of the district can be segmented by the availability of transport, comings, age, etc.

The choice of a certain district depends on a goal of the trip is influenced on the one hand by its attractiveness, on the other - the resistance by the moving from the start place to destination, which consists of travel time, fare, travel comfort, etc....

The volume of traffic flow Q from district i is calculated by the formula [1]:

$$Q_i = \sum_g agSGg(i) \tag{1}$$

where *SGg* is summed up by all structural properties;
SGg(i) designates the value of *SGg* in zone *i,* (the number of people);
The coefficient *ag* is an efficiency which describes how many trips per structural property unit accomplished.

For a similar purpose, the volume of traffic flow to the target *Zj* is calculated.

The total volume of traffic flows from start place for this layer of demand to all districts is usually equal to the total volume of traffic to the target:

$$\sum_i Q_i = \sum_j Z_j \tag{2}$$

where Q_i –efficiency of zone i (the number of people moving out zone i);
Z_j – attraction of zone j i (the number of people moving in zone i).

To appraise the feasibility of measures to unload road traffic, it is necessary to determine the effectiveness of investment.

The potential efficiency of road infrastructure investments is reflected in transport accessibility improving. The consequence is a reduction in trip time from one point to another, i.e. the individual social effect. The total result will be the time savings for all transport correspondence. And although the social effect is obvious here, it is quite challenging to determine the commercial effect, which is based on benefits and costs. One of the using approaches is to estimate the cost of working hours in the work market in this city on the basis of statistics on the average salary. This allows you to recalculate in monetary terms the total amount of time saved and compare it with the amount of invested finance [7]. Evaluation of the investments effectiveness in transport projects should be based on the general principles of project analysis, ie. taking into account the time factor and the use of discounted estimates. The World Bank's recommendations in this area are also based on this point [8]. These notes contain recommendations on the use of a number of summary economic indicators giving an overall measure of the project's performance in cost-benefit terms. These include the Internal Rate of Return (IRR) and the Net Present Value (NPV). NPV is calculated by the formula [9]:

$$NPV = \sum_{k=1}^{n} \frac{C_k}{(1+r)^k} - \sum_{j=1}^{m} \frac{IC_j}{(1+r)^j} \tag{3}$$

Where C_k is the currency flow in year number k (pecuniary units);
IC_j is the investment in year number j (pecuniary units);
n is the number of years during which the currency flow takes place;
m is the number of years during which the investment takes place;
r is the discount rate (decimal).

Internal Rate of Return is a discount rate at which the NPV of the project is 0, i.e. the condition is fulfilled:

$$\sum_{n=1}^{n} \frac{C_k}{(1+r)^k} = \sum_{j=1}^{m} \frac{IC_j}{(1+r)^j} \tag{4}$$

In practice, the formula of linear interpolation is used to determine this indicator:

$$IRR \approx r_+ + \frac{NPV(r_+)}{NPV(r_+) + |NPV(r_-)|} \cdot (r_- - r_+) \tag{5}$$

Where r_+ is discount rate at which NPV is positive (decimal);
r_- is discount rate at which NPV is negative (decimal);
$NPV(r_+)$ is NPV value at r_+ (pecuniary units);
$NPV(r_-)$ is NPV value at r_- (pecuniary units).

In our opinion, you should also use the PI indicator, which can be used in choosing between project alternatives.

$$PI = \frac{\sum_{k=1}^{n} \frac{C_k}{(1+r)^k}}{\sum_{j=0}^{m} \frac{IC_j}{(1+r)^j}} \tag{6}$$

Also, it is possible to offer the formula for definition of the project break-even point level. For example, to find the break-even point of a project to optimize the transport system, you need to use the equation:

$$PI = \frac{\sum_{k=1}^{n} \frac{Ts \cdot AS}{(1+r)^k}}{\sum_{j=0}^{m} \frac{IC_j}{(1+r)^j}} \tag{7}$$

Where *Ts* is amount of time saved (man-hours);
AS is the average salary per hour (pecuniary units).

3. Case study

The city of Odessa is a prominent, historical, cultural and resort-recreational center of Ukraine. The formation of the architectural and planning structure of the city can be divided into the following stages:

- 1st stage (1794 - 1819) - the period of the city's foundation and construction according to the plan drawn up by engineer-colonel F. Devolan. The plan is based on a planning module - rectangular blocks. The pace of development of the city in the XIX century had no analogues in Europe due to the port complex, through which passed more than 70% of all grain exports of Russia and imported goods from Europe, the Middle East and America.
- 2nd stage (1819 - 1858) - The central part of the city is being built and the main architectural ensembles are being formed. A significant role in the development of the city was played by the implementation in 1819 of the "Porto Franco" regime. As a result of accelerated economic growth, the city's population increased rapidly and the settlement area expanded. Around the historic center, behind the line of defense of the barracks, built in the early XIX century, there were belts of suburbs: Moldavanka, Peresyp, Slobodka, Near and Far Mills. After De Volan's plan, in the planning development of Odessa for a long time there were inconsistencies. The city was considered as a sum of state, municipal and private lands, and not as a whole organism.
- 3d stage (1859 - 1917) - After the liquidation of "Porto Franco" in 1859, a period of unplanned territorial growth of the city began. Urban growth is characterized by an increase in industrial and transport production, which as a result led to the formation of a kind of ring around the central part of the city and to some extent blocked its further territorial development. There were arrays of spontaneous residential and industrial development in the area of "Odessa-Malaya", there were areas of residential development in the Soldier's settlement near Zhevahova mountain.

- 4th stage (1918 - 1941) - Restoration of the city after the Civil War of 1917-1922. From 1920, plans for the development of the city began to be developed, the old districts were reconstructed, new construction was carried out according to the general plan of 1930, the settlement of the city was expanded, and large industrial enterprises were built by forming new industrial districts.

- 5th stage (1945 -1965) - Restoration of the city after the Second World War. A significant part of the buildings was destroyed, the unity of the architectural style was not observed during the restoration of the city, and monumental houses of the "Stalinist" style and panel buildings without architectural refinements (so-called "Khrushchevs") were built next to the most beautiful buildings of the 19th century.

- 6th stage (1965-1990). The construction was carried out according to the general plan of 1966 with the development of agricultural lands adjacent to the city both in the north and in the south. (residential blocks of the village named Kotovsky, Yuzhny and Yugo-Zapadny). The number of the population of the place increased significantly, the place-supporters grew up - Illichivsk and Pivdenne. Multi-apartment high-rise buildings were carried out, which were mainly of a focal nature. Public service establishments were located, as a rule, in places of public transport stops. The low level of car mobilization of the population in the Soviet Union determined the adoption of planning decisions without the allocation of parking and garage areas. In the 1989 general plan, the city of Odessa, according to the directives, was assigned to the category of resort cities of all-Union significance.

- 7th stage (1991 - present) After Ukraine's secession from the Soviet Union and the transition to a market economy, the processes of housing construction in the city of Odessa are focused primarily on obtaining a commercial effect and are predominantly elemental way. The existence and area of lots that can be used by building influencing on the city development. The mass development of coastal areas and beaches with high-rise residential buildings begins. In the general plan, adopted in 2015, in the development of a functional planning model for the development of the city, one of the ways of the rational use of intracity territories is the reorganization of industrial territories. Lack of free areas for living forget in the central part of the city due to the creation of new wake-up calls according to the point principle.

Chaotic development revealed imbalances in the planning structure of the city of Odessa, significantly increased the load on the city's transport infrastructure. These problems were exacerbated by natural geographic factors. Historically, Odessa was formed as a large seaport and resort town, therefore, geographically, it is strongly stretched along the coast. Features of the terrain, the proximity of groundwater, the threat of landslides and the presence of catacombs are the reasons for the refusal to build the metro. The street and road network of Odessa turned out to be unable to perceive the increase in population density and the sharply increased level of motorization. At the same time, there is a reserve of extensive growth through the expansion of the road to the maximum extent.

The construction of the "Arcadia" district of the Odessa city, which has been a favourite vacation part of Odessa for guests in the town. Proximity sea with a comfortable beach, lots of greenery, coziness and comfort of living formed an increased demand for housing in this neighbourhood. Now there is a significant amount high-rise building, landscaping has decreased significantly, and the available transport, engineering, communal and social infrastructure does not meet the needs that are caused significant increase in population density (Figures 2, 3).

Figure 2. "Arcadia" district, 2004.

Figure 3. "Arcadia" district, 2020.

One of the microdistricts of Arcadia, the Gagarin table land, has undergone active development since the beginning of the 90s. The number of the population increased during this period from 0.5 thousand to 18.85 thousand people. At the present time, residential buildings are still under construction, which are planned to be commissioned within 5 years. Thus, by 2025, it is expected that the population of the Gagarin table land will be more than 33 thousand people, what is shown in figure 4.

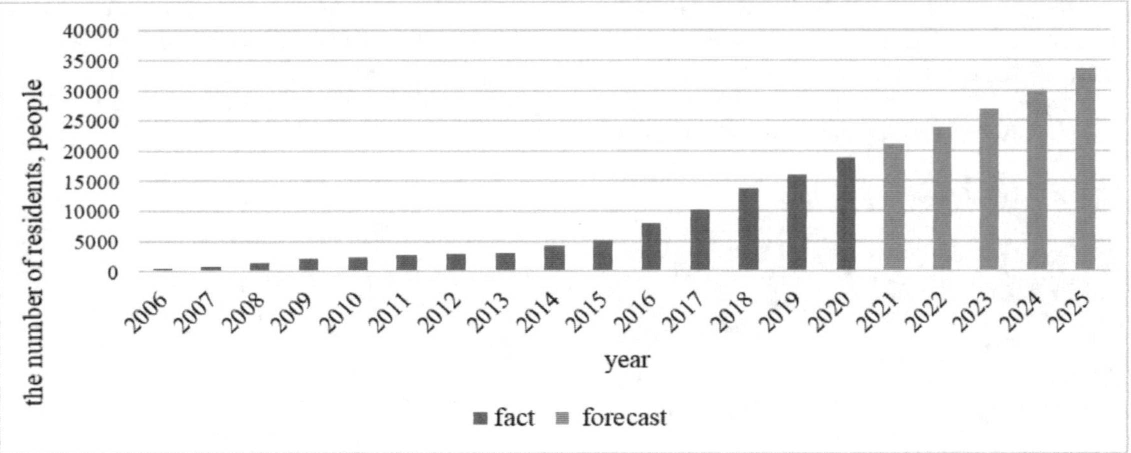

Figure 4. Dynamics of the population of the microdistrict "Gagarin table land "

The population density on the Gagarin table land is 16830 people / km^2, and when the houses under construction are commissioned and populated it will reach 30,000 people / km^2.

Currently, you can get to the Gagarin table land either along the 4-lane Genuezskaya street, 1240 m long, or along the 2-lane Posmitnogo street, 1180 m long.Trolleybus, fixed-route taxis in the general flow of cars and the tram travel along the dedicated lane along Genuezskaya street. Private cars and fixed-route taxis drive along Posmitnogo Street. In addition, you can get to the area by car along Kurortny Lane with a partly one-way traffic.

Traffic intensity during peak hours along Genuezskaya Street is up to 1300 vehicles / hour in one direction, along Posmitnogo Street up to 500 cars, along Kurortny Lane up to 200 cars. The roads adjacent to the microdistrict were not designed for such traffic intensity, and now there are congestions on them most of the time. what is shown in figure 5.

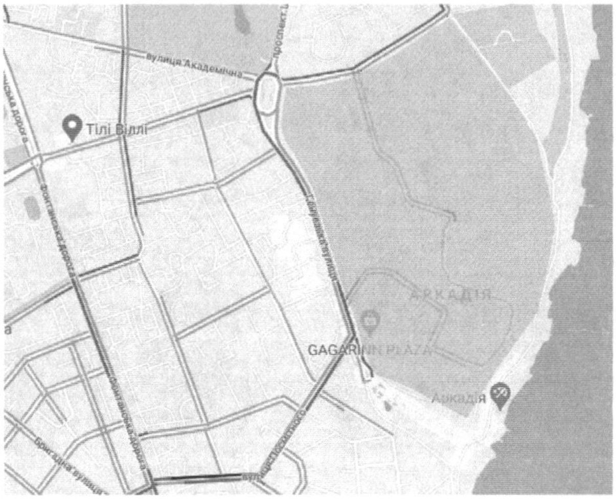

Figure 5. Traffic congestion in the Arcadia area

According to research [10] there are 300 cars per 1000 residents of Odessa. Thus, the actual presence of cars among residents of the Gagarin table land is about 5700 cars. Assuming that about 70% of cars leave the Gagarin table land every day, the total traffic flow from private cars is 3990 in the morning and evening, which, taking into account the capacity of the roads, corresponds to the picture of traffic jams in the area. The average speed in this section is 7 kilometres per hour.

The graphs below show the average congestion level, for several main purposes of travel for residents of the Gagarin table land. Congestion level shows the percentage of travel lengthening in congestion conditions relative to travel on an empty transport network.

At the trip to and from the Pryvoz market area (figure 6), morning traffic jams are practically combined with evening ones and the congestion level is 40-92%, what is shown in figure 7.

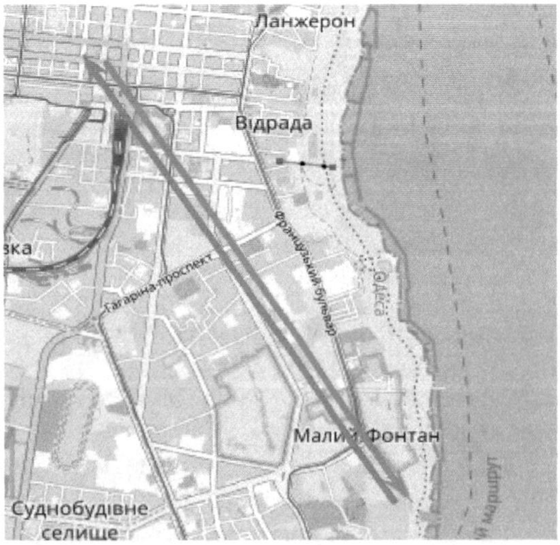

Figure 6. Location of districts Arcadia and Privoz

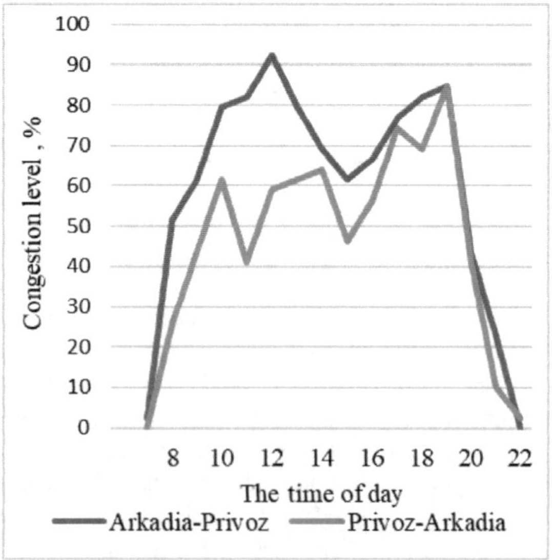

Figure 7. Congestion level between the districts Arcadia and Privoz

A similar point is observed by trip to the city center, what is shown in figures 8 and 9. This is connected with the trips of the residents to and from work and with the trips of tourists.

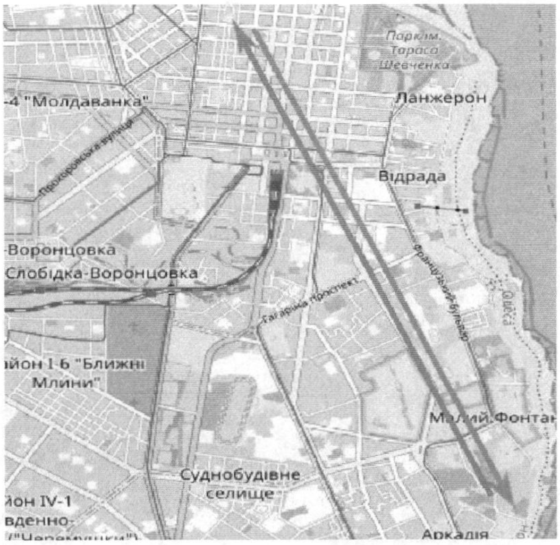

Figure 8. Location of districts Arcadia and City Center

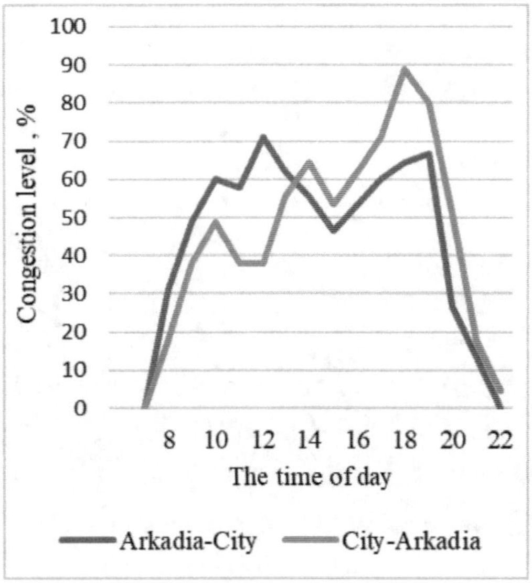

Figure 9. Congestion level between the districts Arcadia and City Center

For trips to the area of the Polytechnic University (figure 10) the peak of morning and evening traffic jams are more pronounced and the congestion level reaches 130%, what is shown in figure 11.

Figure 10. Location of districts Arcadia and University

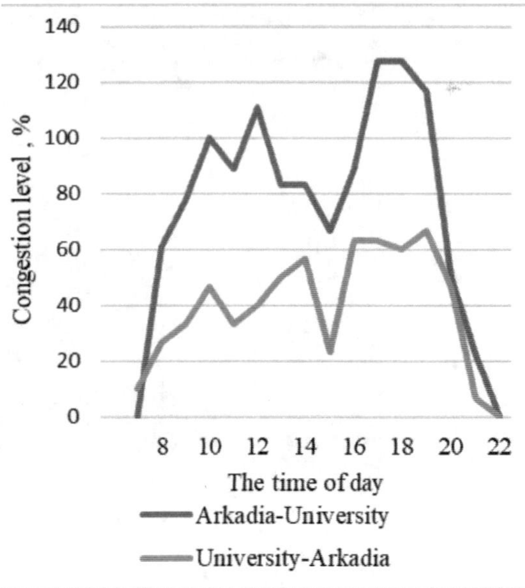

Figure 11. Congestion level between the districts Arcadia and University

For trips from the district to the shopping center in the Gagarin table land area (figure 12), there is also a high load throughout the day, what is shown in figure 13. The morning peak may be due to the general high congestion of alternative roads towards the city center.

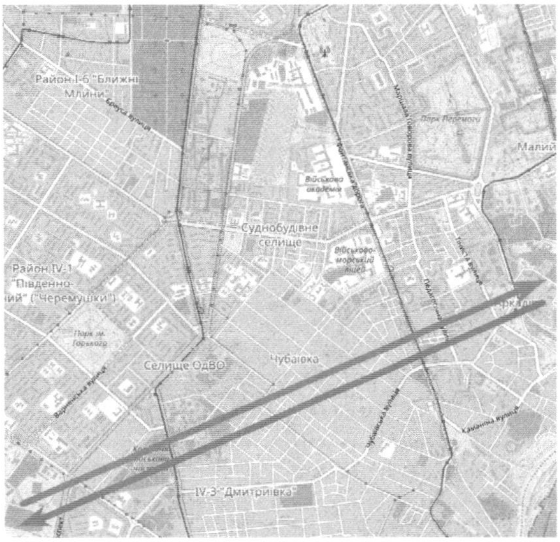

Figure 12. Location of districts Arcadia and shopping center

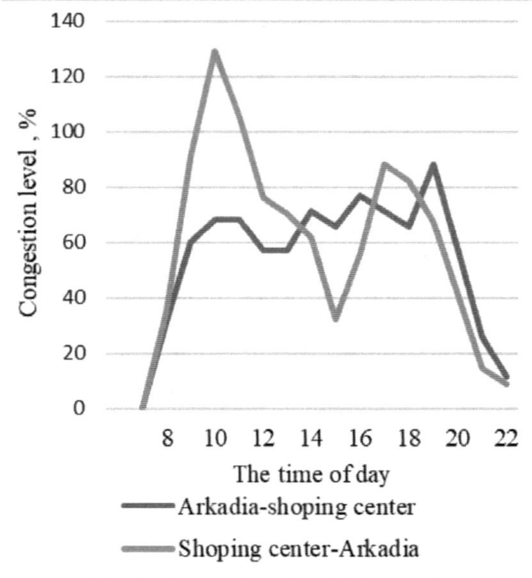

Figure 13. Congestion level between the districts Arcadia and shopping center

PTV Visum was used to simulate the traffic load on this transport network section (figure 14).

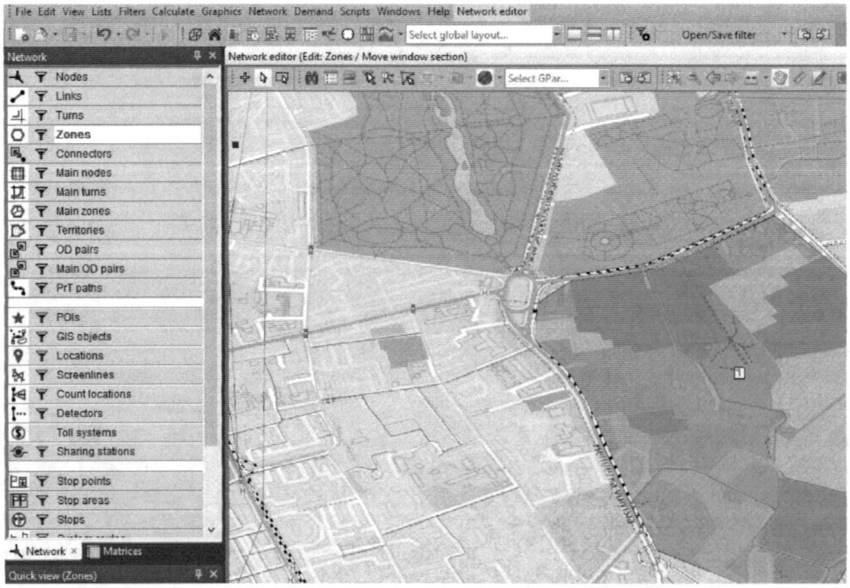

Figure 14. Network editor of PTV Visum

The Demand matrix in figure 15 shows the movement of the population between the zones, in this case it shows isolating movements from and to Arcadia.

5x5			1	2	3	4	5
	Name		Arkadia	University	Privoz	City Center	Shopping Center
		Sum	3990.00	690.00	1300.00	1350.00	650.00
1	Arkadia	3990.00	0.00	690.00	1300.00	1350.00	650.00
2	University	690.00	690.00	0.00	0.00	0.00	0.00
3	Privoz	1300.00	1300.00	0.00	0.00	0.00	0.00
4	City Center	1350.00	1350.00	0.00	0.00	0.00	0.00
5	Shopping Center	650.00	650.00	0.00	0.00	0.00	0.00

Figure 15. Demand matrix for the existing population of the Gagarin table land (the number of trips per day)

Can calculate skims such as travel times between the individual zones, what is shown in figure 16.

5x5			1	2	3	4	5
	Name		Arkadia	University	Privoz	City Center	Shopping Center
		Sum	77.12	13.30	24.05	24.75	19.95
1	Arkadia	82.05	0.00	13.30	24.05	24.75	19.95
2	University	13.10	13.10	0.00	0.00	0.00	0.00
3	Privoz	21.07	21.07	0.00	0.00	0.00	0.00
4	City Center	23.52	23.52	0.00	0.00	0.00	0.00
5	Shopping Center	19.43	19.43	0.00	0.00	0.00	0.00

Figure 16. Skim matrix for the existing population of the Gagarin table land (average duration of the trip in minutes)

With the commissioning and occupancy of houses under construction, the number of cars in the area will increase to 10,080, the number of leaving cars, while maintaining proportions, to 7056 (figure 17), which, taking into account the capacity of roads, will increase the duration and intensity of morning and evening traffic jams.

5x5			1	2	3	4	5
	Name		Arkadia	University	Privoz	City Center	Shopping Center
		Sum	7182.00	1242.00	2340.00	2430.00	1170.00
1	Arkadia	7182.00	0.00	1242.00	2340.00	2430.00	1170.00
2	University	1242.00	1242.00	0.00	0.00	0.00	0.00
3	Privoz	2340.00	2340.00	0.00	0.00	0.00	0.00
4	City Center	2430.00	2430.00	0.00	0.00	0.00	0.00
5	Shopping Center	1170.00	1170.00	0.00	0.00	0.00	0.00

Figure 17. Demand matrix for the projected population of the Gagarin table land (the number of trips per day)

The results of congestion modelling with an increase in the population to 33 thousand people, that is, 1.8 times are shown in the matrix, what is shown in figure 18.

5x5			1	2	3	4	5
	Name		Arkadia	University	Privoz	City Center	Shopping Center
		Sum	138.24	29.14	40.05	43.60	30.50
1	Arkadia	143.29	0.00	29.14	40.05	43.60	30.50
2	University	28.11	28.11	0.00	0.00	0.00	0.00
3	Privoz	38.68	38.68	0.00	0.00	0.00	0.00
4	City Center	41.76	41.76	0.00	0.00	0.00	0.00
5	Shopping Center	29.69	29.69	0.00	0.00	0.00	0.00

Figure 18. Skim matrix for the projected population of the Gagarin table land (average duration of the trip in minutes)

To calculate the difference between matrix indices, we find the increase in time spent on one trip for each direction. By multiplying the number of trips, you can get the overall increase in time spent and calculate the economic loss.

The following matrix (figure 19) shows the loss of time if, with an increase in the number of residents, measures are taken to increase the capacity of roads.

5x5			1	2	3	4	5
	Name		Arkadia	University	Privoz	City Center	Shopping Center
		Sum	132.24	28.70	39.20	42.10	30.10
1	Arkadia	140.10	0.00	28.70	39.20	42.10	30.10
2	University	27.90	27.90	0.00	0.00	0.00	0.00
3	Privoz	36.12	36.12	0.00	0.00	0.00	0.00
4	City Center	40.07	40.07	0.00	0.00	0.00	0.00
5	Shopping Center	28.15	28.15	0.00	0.00	0.00	0.00

Figure 19 Skim matrix with the projected population of the Gagarin table land and an increase in road throughput (average duration of the trip in minutes)

Time savings will amount to 96360 hours per year.

We will calculate the efficiency of investments in increasing road capacity.

Average salary in Odessa 11646 UAH. Hourly salary for a 168 working hours per month - 69 UAH / hour [11].

According to the Odessa City Administration, the average rate on loans attracted in 2020 was 18% per year. This rate determines the minimum level of return on investment and therefore was taken in the calculations as the discount rate for cash flows.

We can calculate at what savings hours per million investments, the PI efficiency index will be greater than or equal to 1, that is, the investment will pay off.

Based on the intervals between major road infrastructure repairs, we will take a planning horizon of 10 years. By formula (7):

$$PI = \frac{\sum_{k=1}^{n} \frac{x \cdot AS}{(1+r)^k}}{IC} \geq 1 \tag{8}$$

$$PI = \frac{\sum_{k=1}^{10} \frac{x \cdot 69\,UAH}{(1+0{,}18)^k}}{1000\,000\,UAH} \geq 1$$

We have solved the equation, to determine that x> = 3210 man-hours.

Now, at the exit from the Gagarin table land, up to 199,500 hours a year are spent.

The city authorities are planning some investments in the reconstruction and extension of the cycle path, regulation of traffic lights, and repair of the road surface in the amount of UAH 20 million. If the time of each train is reduced by at least 5 minutes, then the total time savings per year for residents of the Gagarin land table will amount to 83,125 hours. In this case, the condition PI> = 1 will be fulfilled.

It should be borne in mind that the transport infrastructure of a city is a complex system with many different interconnections, and changes in its efficiency are best evaluated in a comprehensive manner. Therefore, for more accurate forecasts, it is necessary to develop a transport model of the city as a whole, simulate the complete road situation and evaluate the efficiency of investments in its optimization.

4. Conclusions

The modern traffic situation in overcrowded microdistrict is complicated by many subjective and objective actualities. Modelling the traffic situation in the city by the PTV Visum software package makes it possible to predict the traffic intensity, traffic jams, therefor it is possible to estimate the economic wastes or benefits from the new transport infrastructure constructions. The consumption of transport modelling at the period of urban development planning will ensure a balance between transport infrastructure and the load on it and reduce economic wastes. It is simulating the transport situation in the article for the one of an extensively developing Odessa city microdistricts, predicts the load on the transport system at the end of its development, proved the correlation between the economically justified level of transport infrastructure investing and reducing the waste of time in traffic jams. In the course of depletion of the unloading transport system extensive possibilities (expansion of roads, creation of detour opportunities, additional cycle paths, installation of "smart" traffic lights, etc.), it is necessary to consider the alternative possibilities for optimizing road traffic. The promoting of online services, spacing of social infrastructure items within walking distance, the popularization of a healthy lifestyle and an increase in the share of cycling will contribute to reducing the level of traffic congestion.

It is future-oriented but cost-intensive the further development of public transport, the start-up of a high-speed tram, the organization of a ferry crossing the Black Sea.

5. References

[1] PTV Visum. *The software for multimodal transport planning & macroscoping traffic simulations*, https://www.ptvgroup.com/en/solutions/products/ptv-visum/
[2] OpenStreetMap, https://www.openstreetmap.org/copyright
[3] Mapping data, https://2gis.ua/odessa/
[4] Historical cartographic data from Google Earth Pro, https://earth.google.com/
[5] Jacyna M, Wasiak M, Kłodawski M and Gołębiowski P 2017 Modelling of Bicycle Traffic in the Cities Using VISUM *Procedia Engineering* **187** pp. 435-441.
[6] Heyken Soares P, Ahmed L and Mao Y 2021 Public transport network optimisation in PTV Visum using selection hyper-heuristics *Public Transport* **13** pp. 163–196.
[7] Osetrin M, Bespalov D, Hahen A and Kutsyna I 2018 Research of demand and supply of mobility of the urban population on the example of Uzhgorod *Urban planning and spatial planning* (Kiev National Academy of Civil Engineering and Architecture) **68** pp 409- 419.
[8] Mackie P, Nellthorp J and Laird J 2005 Notes on the Economic Evaluation of Transport Projects: Fiscal Impacts *Transport Notes Series* (Washington, DC: World Bank) https://openknowledge.worldbank.org/handle/10986/11794
[9] Carlin W and Mayer C 2003 Finance, investment, and growth *Journal of financial Economics* **69** pp. 191-226.
[10] General plan of Odessa, http://old.omr.gov.ua/images/File/DODATKI_2015/GORSOVET/6489.doc
[11] Main Department of Statistics in Odessa region: labor market statistics, http://od.ukrstat.gov.ua/stat_info/doxod/doxod18.html

Project management software in Ukraine

I Azhaman[1],*, N Petryshchenko[2], M Hronska[1], A Pandas[2] and N Pushchina[3]

[1]Odessa State Academy of Civil Engineering and Architecture, Educational and Scientific Institute of Business and Information Technology, Department of Management and Marketing, 4, Didrihsona Str., Odessa 65029, Ukraine
[2]Odessa State Academy of Civil Engineering and Architecture, Educational and Scientific Institute of Business and Information Technology, Department of Economics and Entrepreneurship, 4, Didrihsona Str., Odessa 65029, Ukraine
[3]Odessa State Academy of Civil Engineering and Architecture, Educational and Scientific Institute of Business and Information Technology, Department of Information Technologies, 4, Didrihsona Str., Odessa 65029, Ukraine

*E-mail: ag_irisya@ukr.net

Abstract. Modern consumers are interested in receiving goods within a specified time frame, of the required quality and quantity. Such requirements apply to all goods and services, including those in the construction industry. Features of the construction industry are due to a variety of technologies, materials, equipment used for the construction of facilities. This justifies the need to implement automation tools for planning processes, determining the volume and delivery schedule of production stocks, using labour resources, forming a financing schedule, identifying risks affecting processes and approaches to minimizing them. To solve such problems, special attention in modern conditions is paid to project management in construction and its software. A project is understood as a process which is a set of sequential actions, the implementation of which contributes to the achievement of the set goal. The project is characterized by limitations in terms of time, cost and resources used. In project management, software products such as Primavera Systems, Inc (Oracle), Spider Project, Microsoft Office Project, YouControl. It is their use, advantages and disadvantages that are discussed in the paper. Particular attention is paid to the Microsoft Office Project software product, which has the following advantages: user-friendly interface and ease of use, flexibility of adaptation to user needs, clear and convenient analytical charts and the ability to manage them, availability of support that allows you to quickly master and use the software product, availability of reference materials.

1. Introduction

Digitalization is relevant for any type of activity in the modern period. Any manufacturer or distributor pays more and more attention to the improvement of their activities. Improvements should be comprehensive in the following areas: marketing, calculation of financial flows, calculation of production processes, economic security, logistics and others. The construction industry is no exception. Companies in the construction industry must also speed up their processes in order to maintain their competitiveness and position in the market and to develop.

The purpose of the construction industry in the state economy is to create new funds for housing, social facilities, infrastructure and production assets [2,3]. The need to speed up data processing has long been a necessity. Construction companies, using computer technology or partnerships with digital enterprises, are more competitive [1].

Much attention is paid to logistics in construction, digitalization of information in the construction supply chain [10]. Revealed a significant backlog of enterprises in this industry in the implementation and use of digital technologies [7]. Also, studies have been carried out on digitalization in the environment of the construction industry [13]. However, the topic of introducing digital technologies directly into construction production in Ukraine remains poorly understood.

2. Basic software in construction

Projects for the construction of residential buildings and commercial real estate are developed systematically, taking into account the needs of the market, the characteristics of the settlement, industry. It also takes into account the requirements of environmental protection, legal regulations and restrictions.

The main design task is to develop the most optimal project for both stakeholders and the contractor. The most relevant criterion for the implementation of projects in construction should be considered the term of work without loss of quality.

It is customary to understand a project as a process that consists of a set of coordinated, managed activities. The project must have a start and end date. The project always has a goal, restrictions on funds and resources [6]. The implementation of projects in construction is based on many factors: the supply of materials, the logistics component of the enterprise, reliable suppliers, the project management system, the flow of financial resources, and more. The digitalization of these flows will increase both the terms of work execution and the quality of products, the quality of management, the competitiveness of the enterprise in the market.

In Ukraine not all construction companies use project management software. These programs provide an opportunity to use them in modelling the construction of residential buildings and other functions. The article provides an overview of the possibilities of using such programs as: Primavera Systems, Inc (Oracle); Spider Project; Microsoft Office Project; YouControl.

2.1. Primavera Systems, Inc.

Software Primavera Systems, Inc. – this is the specialized software of the company Primavera Systems, Inc. for project management. The software entered the US market in 1983. It was originally developed for strategic project management and has evolved continuously over the years.

Oracle offers to improve construction management software with the help of its product. Primavera will provide transparency and control, optimize supply chain tracking and data security [4]. Also, when using this software, productivity will increase, risks in project processes will decrease. The software is widely used in many countries and industries. In 2005, before the release of a new version, the company announced the unification of the names of all numerous software modules. Now professional project management software is simply called - Primavera.

The Primavera Software allows the company to record and manage all data sets on projects and financial capabilities of the organization. The software is aimed at project managers operating in various industries. A variety of software modules will help to optimize all the processes occurring during the implementation of a particular project.

The company continues to develop and improve specialized project management software. Primavera is a powerful and reliable way to plan, execute and manage projects. The product allows to manage construction projects of any size and complexity.

Today Primavera is practically the standard in the field of project management software in many countries of the world. The capabilities of the software package certainly make it possible to simulate the processes of erection of residential buildings.

However, the disadvantages of this product include a rather high cost and difficulty in mastering the program. Often an organization is forced to bring in additional project scheduling specialists.

2.2. Spider Project

In recent years a Russian development - the Spider Project software complex - has become quite widespread in the project management software market. The first version of the system was released in 1992 and already in 1993 was demonstrated at exhibitions in Russia, as well as in Germany.

Among the users of the system are representatives of many sectors of the economy - construction, space industry, oil refining, tourism, energy and many others [8]. Among the users of the system are representatives of all sectors of the economy - Transstroy, Russian Engineering Company, Rocket Center im. Makeeva, Lukoil, Petrobras, Hill International, Odebrecht, Hidroconstructia, Mostogradnja, Thyssen Shachtbau, Embrayer, Van Oord, Romtelecom, FUJIFILM Diosynth Biotechnologies, Brasilia Electric Company and many others [8].

In the process of expanding the functionality of the program, the peculiarities and needs of the market of the post-Soviet countries are taken into account. The methodology behind the Spider Project has a number of features compared to traditional approaches to project management. In this regard, the capabilities of the Spider Project product differ in many respects from what is offered by Western software vendors in the field of project management.

Spider Project software is designed for different user groups. It includes the professional system Spider Project Professional, as well as cheaper versions of Desktop (single-user version of the professional system) and Lite (version with limited functionality).

Some of the advantages of Spider Project as stated on their website are listed below:

1. The ability to schedule the timing of work based on their physical volume This function of the program is especially relevant when simulating construction processes. In the Primavera software package considered above, the concept of "scope of work" is absent.
2. The ability to use in projects any additional characteristics of work, resources and purposes, it is also very convenient when modelling construction processes.
3. Extensive opportunities for cost and resource analysis of projects. In one project you can conduct a parallel analysis of costs in different units and under different regulatory frameworks.
4. The built-in accounting system allows not only to adjust the remaining duration and volume of work but also to receive reports on the implementation of the project in any section and for any period of time.
5. Built-in project management guide fully covers international standards and takes into account the specifics of project management in post-Soviet countries.

This software package can be used and will be convenient for modelling the construction of residential buildings. The product has not yet received such widespread distribution as its Western counterparts (Primavera and Microsoft developments). But of course, it is a promising tool for building models of construction processes [5].

2.3. Microsoft Office Project.

The first commercial version of the Microsoft Office Project product went on sale in 1985 and since then for many years the program by the number of copies sold has been the most widely used project management tool in the world. The advantages of this program are a wide range of applications, advanced innovations that are implemented in each new version of the product.

Traditionally the product is available in two versions. These are Standard (single-user version) and Professional (corporate version).

Using Microsoft Project allows the following [9]:

- draw up a work production plan, which includes the timing of the work, the need for resources (people, mechanisms, materials, finances, etc.);
- calculate the project budget and distribute the planned time expenditures;
- calculate in time the project's resource requirements;
- develop an optimal financing scheme for work, supply of materials and equipment;
- analyse risks;
- determine reserves of time;
- effectively monitor the implementation of the plan;
- analyse deviations of the actual progress of work from the plan, timely and reasonably adjust the planned indicators;
- analyse deviations of the actual progress of work from the plan, timely and reasonably adjust the planned indicators.

This software has a number of undeniable advantages that have made it so popular in the project management environment. The main advantages are:

1. Intuitive interface and easy to learn.
2. Flexibility in customizing the program for specific needs.
3. Easily and efficiently analyse and manage charts.
4. Interactive step-by-step planning assistant that allows you to quickly master project management processes.
5. Ability to track the progress of a project at all stages of its life cycle.
6. A rich set of built-in reference materials designed for both newbies and project management professionals.

Microsoft Project is easy to use, convenient and in our opinion allows the user to solve almost any problem related to modelling the processes of building residential buildings.

2.4. YouControl.

The work of a construction company is based not only on project management. It is equally important to obtain reliable information when contacting subjects of the external environment. The YouControl software package is one of the software options for the Ukrainian market, which allows to identify unscrupulous counterparties.

YouControl is an analytical system for business intelligence and verification of counterparties in Ukraine. In this program, on the basis of open data, a dossier is formed for each company that operates within the country and changes in state registers are tracked. The capabilities of this program allow to visualize the connections between affiliated persons [11]. The technology used by this program allows, in a fairly short time, to obtain information on organizations or individual entrepreneurs from more than fifty state registers that is relevant at the time of the request.

One of the advantages of this software is that it has an "end-to-end search" feature that provides detailed searches on a single attribute. For example, a query by the name of a director will show to get information about all companies where he is a director, a founder. A complete dossier provides data on the status of the enterprise, its managers and beneficiaries, the financial and industrial group that the enterprise belongs to.

This function allows to instantly receive information about the reliability of future partners when working with external counterparties, enterprises or individual entrepreneurs. The automatic monitoring function is carried out every day and is based on the possibility of promptly receiving alerts by e-mail about new data on the bankruptcy of the counterparty, criminal cases, the occurrence of tax debt and other events.

The promptness of obtaining such information makes it possible to make timely appropriate decisions on interaction with counterparties that are losing their reliability. Checking the search for related persons makes it possible to preliminary determine the feasibility of cooperation with potential partners. Market analysis reveals for the sales department potential customers for goods and services by type of activity, the commercial department to compile a rating of potential competitors by type of activity and region, and the analytical department to identify industry leaders in terms of revenue and personnel.

The use of all these factors will significantly increase the company's competitiveness in the market and the efficiency of its activities. The study of the capabilities of the YouControl software package showed that the information obtained in advance about the financial condition, reliability and financial stability of potential partners allows us to assess the prospects for cooperation with them already at the stage of developing a strategic plan [12]. Further use of such software makes it possible to track and analyse the production and financial activities of counterparties in order to timely respond to negative changes.

3. Conclusion

In the modern world, quick decision-making is required at any level of project implementation. This is especially true for construction companies where large financial resources are involved in production and the coherence of the project stages affects the timing of the result, its quality and competitiveness.

Using project management software will provide many benefits to a construction organization. This will reduce the construction time, production costs, improve intra-industrial communications and logistics flows, and the quality of management.

It should also be noted that the use of modern information technologies allows not only to effectively build communication channels for internal affairs, but can also improve the establishment of relations with subjects of the external environment during joint activities.

If compare the capabilities of the three programs (Spider Project, Primavera and Microsoft Project) then Primavera has the widest functionality. However, due to the peculiarities of the development of the Ukrainian market today for enterprises that are engaged in the construction of buildings, it will be more relevant to use the Spider Project which covers international standards and takes into account the specifics of project management in post-Soviet countries. For enterprises that specialize in the production of construction products will can recommend Primavera or Microsoft Project. In this case it should be noted that Primavera is quite an expensive product and will not be available for small businesses. YouControl is a companion program the work of which is not directly related to the calculation and modelling of projects. However, it complements the project management capabilities of any enterprise and helps ensure economic security.

4. References

[1] Aghimien D, Aigbavboa C, Oke A, Wellington T, Moripe P 2020 Digitalization of construction organisations – a case for digital partnering, *J. International Journal of Construction Management* Available at: https://doi.org/10.1080/15623599.2020.1745134

[2] Ayodele E and Alabi O 2011 Abandonment of construction projects in Nigeria: causes and effects. *J. Emerg Trends Econ Manag Sci.* **2 (2)** pp 142–145

[3] Bidzhieva F 2015 Features of construction as a branch of the economy *J. Economics* **4(5)** Available at: https://scienceproblems.ru/osobennosti-stroitelstva-kak-otrasli-ekonomiki/1.html

[4] *Drive Performance Across the Project and Asset Lifecycle.* Available at: https://www.oracle.com/industries/construction-engineering/

[5] Goncharenko D, Menejlyuk I 2019 Determination of indicators of civil engineering in Odessa at variant organizational and technological designing *J. Nauchnyj vestnik stroitelstva.* **2(96)** pp 197-203

[6] Gutnov A, Glazychev V 2000 *The world of architecture* (Moskow, Molodaya gvardiya) p 351

[7] Hautala K, Järvenpää M, Pulkkinen P 2017 Digitalization transforms the construction sector throughout asset's life-cycle from design to operation and maintenance *J. Stahlbau* **86** pp 340-345

[8] *Integrated project management system Spider Project.* Available at: http://www.spiderproject.com/ru/

[9] *Introducing Project - a simple, powerful and innovative solution for everyone.* Available at: https://www.microsoft.com/uk-ua/microsoft-365/project/project-management-software

[10] Liu R and Chua V 2016 Theoretical Digitalization of Information Flow in the Construction Supply Chain. *J. International Journal of Management Research and Business Strategy.* **5** Available at: http://citeseerx.ist.psu.edu/viewdoc/download?doi=10.1.1.741.8448&rep=rep1&type=pdf

[11] Official site *YouControl.* Available at: https://youcontrol.com.ua/.

[12] Petrishenko N, Serogina N 2018 Modern technologies in the activity of hotel and restaurant business enterprises *J. Naukovo-virobnichij zhurnal «Biznes-navigator».* **6 (49)** Available at: http://business-navigator.ks.ua/journals/2018/49_2018/27.pdf

[13] Tereshko E, Romanovich M, Rudskaya I 2021 Readiness of Regions for Digitalization of the Construction Complex. *J. Open Innov. Technol. Mark. Complex.* **7** 2 Available at: https://doi.org/ 10.3390/joitmc7010002

An analytical design approach for post-tensioned concrete slabs with unguided tendons

T-N Toader[1], Z Kiss[1,2], S-G Țere[2] and A Puskás[1,3]
[1] Department of Structures, Faculty of Civil Engineering, Technical University of Cluj-Napoca, Romania
[2] Plan 31 Ro, Cluj-Napoca, Romania
[3] KÉSZ Romania, Cluj-Napoca, Romania

E-mail: traian.toader@dst.utcluj.ro

Abstract. This article contains practical information on the design and dimensioning of prestressed concrete solid flat slabs using unbonded post-tensioned tendons, in accordance with the European standard EN 1992-1-1 and German standard DIN 1045-1. The calculation method and the mathematical equations necessary for the dimensioning of the non-adherent unguided tendons are given. The paper presents a case study where a slab floor is dimensioned for deflection using prestressing only in the column strips. The deflection of the slab is calculated using equivalent loads (unguided tendon), a mixt tendon profile with 4th degree curve segments (unguided tendon) and parabolic tendon profile with 2nd degree curve segments (guided tendon – classical solution).

1. Introduction

Today`s practice shows that architects and builders prefer the use of slender slab floors in the design and construction of office buildings, apartment buildings, administrative buildings, as well as underground garages [1]. Slab floors allow the optimization of the story height, clear ceiling surfaces for installations that also determine the acceleration of the construction process. Therefore, structural engineers are increasingly faced with the task of designing the thinnest possible slabs that are to meet with the requirements of EN 1992-1-1 [1], both in the ultimate limit state (ULS) and service limit state (SLS).

The design limitation of large spans reinforced concrete slabs is, particularly the maximum allowed camber. An extension of these limits is possible by reducing the slab thickness (self-weight) and prestressing the slab, respecting the maximum allowed values for deflections given in the standards and those imposed by the design theme. The minimum thickness of a solid flat slab is 150 mm [2] [3] and 200 mm when shear reinforcement is used [3] [4]. The appropriate limiting values of deflection should not exceed L/250 (for quasi-permanent loads) and L/500 (deflection after construction) [5] [6], where L is the distance between two adjacent supports. As a recommendation, the values given in table 1 can be used to pre-dimension the thickness of the solid flat slab.

In the first solid flat slabs built in the USA, most of the time strands (figure 1, left) were evenly distributed in both directions both on the field strips as well as on the column strips (figure 2, left).

IOP Conf. Series: Materials Science and Engineering 1141 (2021) 012013 doi:10.1088/1757-899X/1141/1/012013

Table 1. Recommended span/depth ratios for a solid flat slab with and without drop panel when a L/250 deflection is imposed [1] [2] [3] [4] [5] [7] [8] [9]

	Total imposed load [kN/m²]	Span/depth ratios (6 m < span < 13 m)	
		Post-tensioned	Reinforced concrete (with steel S500)
Solid flat slab 	2,5	40	30
	5,0	36	24
	10,0	30	17
Solid flat slab with drop panel 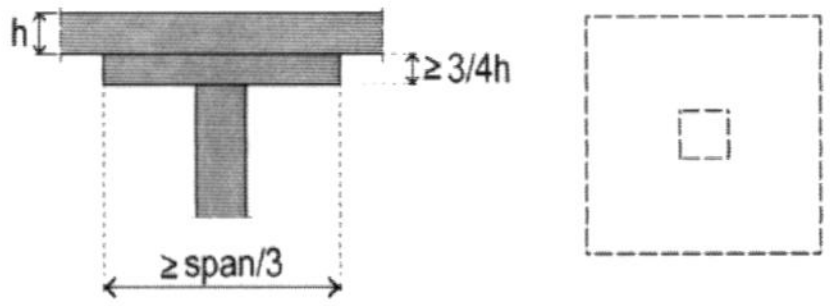	2,5	44	34
	5,0	40	28
	10,0	36	21

Notes:
1) The total imposed load represents the sum of all vertical loads with their characteristic value carried by the slab, without slab`s self-weight;
2) L = the maximum span of the considered flat slab panel;
3) The table is applicable for multi-span flat slabs;
4) The span/thickness ratio is applicable for all internal panels and for external panels when external beams are used;
5) When no external beams are used, the thickness of the external slabs should be increased with approx. 15%;
6) The ratio between the distance of adjacent columns of the same flat slab panel should be $0{,}66 \leq L_x / L_y \leq 1{,}50$, while the maximum distance between columns in the two orthogonal directions should be limited to $0{,}8 \leq L_{x,max} / L_{y,max} \leq 1{,}2$.

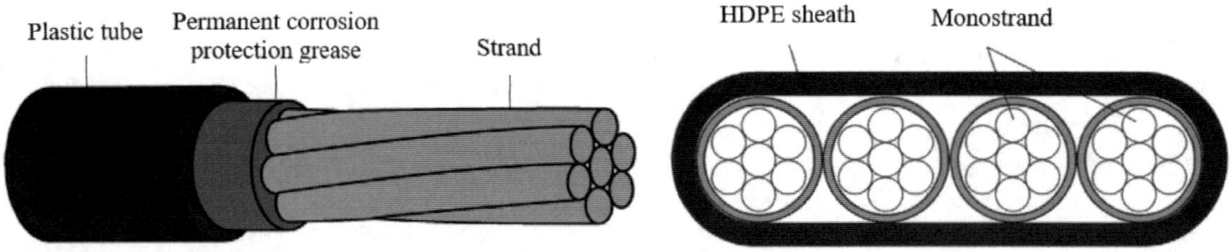

Figure 1. Description of an unbonded monostrand and of a band with 4 strands. [1] [10]

Research in the early 1970s carried out in Switzerland has indicated that the arrangement of tendons in the form of compact groups crossing the columns (figure 2, right) is advantageous in terms of statics of the flat slabs [11]:

- due to the posttensioning of the tendons on the slab act upward-facing deflection forces, while downward-facing deflection forces are introduced directly into the columns;
- by concentrating the tendons in the support strips, part of the punching force is taken up by the tendons on the inclined segment, and thus the shear stress to be taken up by the concrete is reduced.

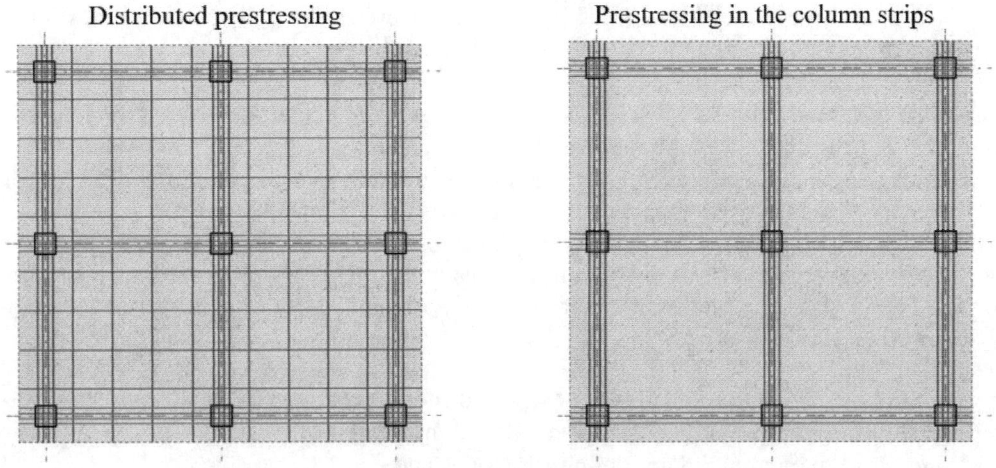

Figure 2. Types of prestressing for solid flat slabs.

In the solid flat slabs, the bending moments given by the applied loads are concentrated on the column strips, especially in the vicinity of the columns. Reinforcement in two rows above the columns in both directions is sometimes necessary. Consequently, prestressing measures must mainly contribute to a „normalization" of the layout of rebars. This can be achieved by prestressing only the column strips (figure 3) [11].

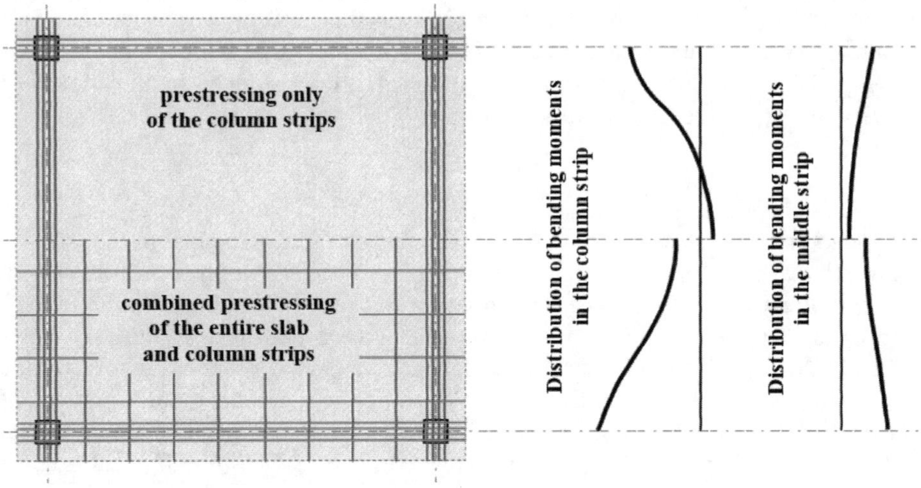

Figure 3. The effect of prestressing only the column strips vs. combined precompression of the slab [11].

The comparison between combined prestressing (entire slab + column strips prestressing) and prestressing only the column strips gives the following results:

- the appearance of the bending moment diagrams produced by prestressing exclusively the column strips is no longer like the appearance of the bending moment diagrams from the gravitational loads (figure 3, right);
- in the case of prestressing exclusively the column strips, bending moments appear in the column strip whose value is much greater than 50% of the value of the total bending moments (determined with one of the methods: the coefficients method or the equivalent frame method [2] [3]) which would be for the same slab. If a combined prestressing were to be used, a double total prestressing force would be required [11];

- the total axial force induced in the slab by exclusively prestressing the column strips is only half of the value of the induced axial force necessary for a combined prestressing.

It is recommended [11] that:

- combined prestressing should be chosen if it is necessary to check for decompression limit state or to reduce camber, thus using a high degree of prestressing;
- prestressing exclusively of the column strips to be used, especially when it is desired to minimize the number of tendons used, thus resulting in a low degree of prestressing.

In the case of slab panels with combined prestressing that do not have a square shape in plane, an additional prestressing is necessary by adding tendons in the longer (with larger spans) column strip, to better adjust and balance the necessary reinforcement amount in both directions of the slab, amount mainly placed in the column strips [11].

2. Prestressing by unbonded and freely positioned tendons (unguided tendons)

The purpose of using prestressing with unbonded tendons consist of:

- fulfilling of the maximum deflection limit for slabs and,
- increasing the punching shear capacity of the flat slab floors.

The role of passive, bonded steel reinforcement is:

- ensuring the load-bearing capacity and
- limiting the crack openings.

The consequences of how the tendons are placed are the followings [11]:

- the state of stresses resulted from the prestressing has opposite sign and is almost similar, but not identical to the one resulted from external loads, and the position of the tendons is not exactly corresponding to cancel the effects of the application of external loads;
- in the case of slabs is an option to install the tendons freely positioned in elevation, influenced by the relatively high cost due of deviators (chairs) and their lack of series production, usually made from stirrups.

2.1. Tendon profile in elevation and equivalent loads

In general, the tendon profile is composed of parabolic and straight segments. In the vertical plane, the grouped tendons (banded tendons) are often constructed from partly parabolic profiles (Figure 4) [9]. In practice, the parabolic span profile of the tendon is lowered from the support to the field; it is almost impossible to obtain straight profiles. The tendon will have a path of the altitude which gradually decreases, often following a 2nd order curve for guided tendons and 4th order curve for unguided tendons, respectively. In execution, making a profile only from parabolic segments involves many chairs, which must be placed and fixed to the tendon, therefore the tendon retains its parabolic profile before concreting, during concreting and vibration of the concrete. During construction, instead of using many chairs to force the tendon to follow a parabolic path along its entire length, it is easier to fix it only in the lifting areas (on supports), and in the field areas (directly on the bonded reinforcement at the bottom), this way between the upper and bottom fixation points the tendon will hang under its self-weight, thus describing a parabola. The latter solution was justified by several [1] [9] [11] [12]. The difference between the performance of the two types of tendon profiles in elevation is insignificant, but the use of freely positioned tendons contributes to savings in terms of time, effort, and money.

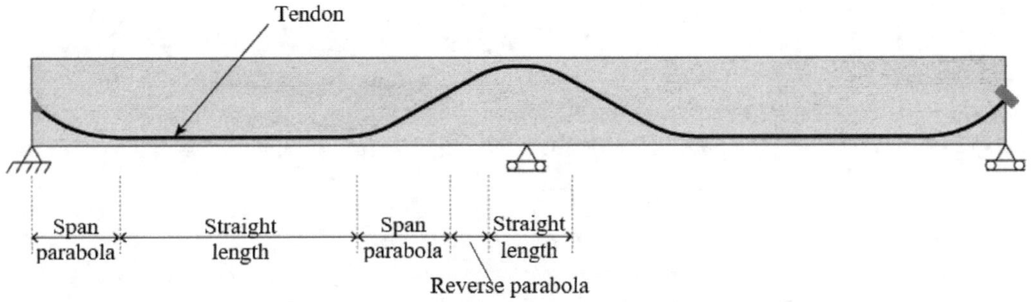

Figure 4. Mixt vertical layout of a tendon.

The tendon will hang under its self-weight and will be supported by the bonded mesh reinforcement installed at the bottom at a clearly established distance (figures 5 and 6). The equations describing the path of the tendons, defined by Maier and Wicke in [12] are:

Lifting of the tendon at the edge support of the slab

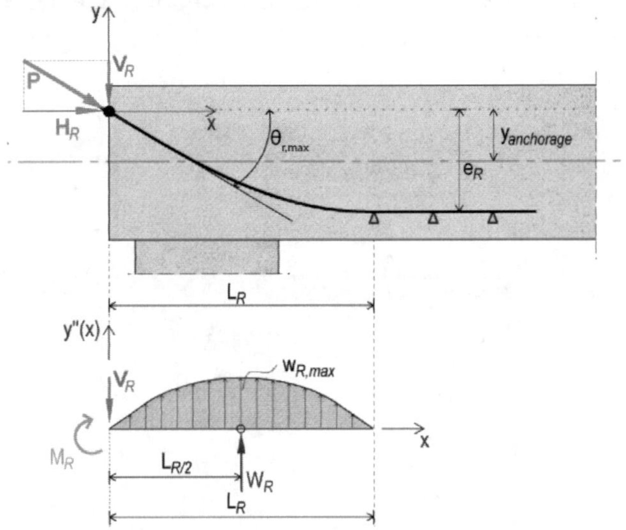

Figure 5. Lifting of the unguided tendon at the edge support of the slab.

The path of the tendon:

$$y(x) = -e_R \cdot \left[\frac{x^4}{L_R^4} - 2 \cdot \frac{x^3}{L_R^3} + 2 \cdot \frac{x}{L_R}\right] \tag{1}$$

Tendon inclination:

$$\tan[\theta_R(x)] = y'(x) = -e_R \cdot \left[4 \cdot \frac{x^3}{L_R^4} - 6 \cdot \frac{x^2}{L_R^3} + 2 \cdot \frac{1}{L_R}\right] \tag{2}$$

Curvature of the tendon:

$$y''(x) = -e_R \cdot \left[12 \cdot \frac{x^2}{L_R^4} - 12 \cdot \frac{x}{L_R^3}\right] \tag{3}$$

Considering the curve of bending in linear bar theory, the length on which the tendon hangs under its own weight (length of free bending) in horizontal projection L_R is calculated using the equation:

$$L_R = \sqrt[4]{\frac{24 \cdot E \cdot I \cdot e_R}{g}} \tag{4}$$

The equivalent load from the post-tensioned tendon overturned on the concrete:

$$w_R(x) = y''(x) \cdot P = -12 \cdot e_R \cdot P \cdot \left[\frac{x^2}{L_R^4} - \frac{x}{L_R^3}\right] \tag{5}$$

$$w_{R,max} = \frac{3 \cdot e_R}{{L_R}^2} \cdot P \tag{6}$$

The resultant of the distributed equivalent load:

$$W_R = V_R = \int_0^{L_R} w_R(x)\, dx = \frac{2 \cdot e_R}{L_R} \cdot P \tag{7}$$

When the tendon is not anchored in the axis of the flat slab, an additional bending moment appears, which acts on the edge of the slab:

$$M_R = P \cdot y_{anchor} \tag{8}$$

Lifting the tendon on the intermediate support

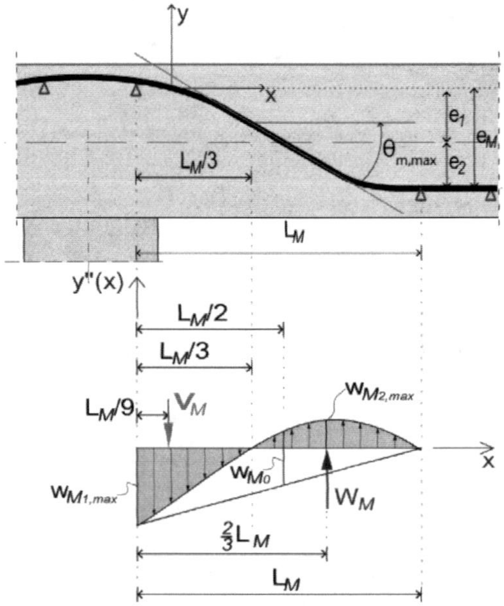

Figure 6. Lifting the tendon on the intermediate support.

Tendon path:

$$y(x) = e_M \cdot \left[-3 \cdot \frac{x^4}{L_M^4} + 8 \cdot \frac{x^3}{L_M^3} - 6 \cdot \frac{x^2}{L_M^2} + 1\right] - e_1 \tag{9}$$

IOP Conf. Series: Materials Science and Engineering 1141 (2021) 012013 doi:10.1088/1757-899X/1141/1/012013

Tendon inclination:

$$\tan[\theta_M(x)] = y'(x) = e_M \cdot \left[-12 \cdot \frac{x^3}{L_M^4} + 24 \cdot \frac{x^2}{L_M^3} - 12 \cdot \frac{x}{L_M^2}\right] \tag{10}$$

Tendon curvature:

$$y''(x) = e_M \cdot \left[-36 \cdot \frac{x^2}{L_M^4} + 48 \cdot \frac{x}{L_M^3} - 12 \cdot \frac{1}{L_M^2}\right] \tag{11}$$

Considering the curve of bending in the linear bar theory, the length on which the tendon hangs under its own weight (free bending length) in horizontal projection L_M is calculated using the equation:

$$L_M = \sqrt[4]{\frac{72 \cdot E \cdot I \cdot e_M}{g}} \tag{12}$$

The equivalent load from the post-tensioning tendon transmitted on the concrete:

$$w_M(x) = y''(x) \cdot P = -12 \cdot e_M \cdot P \cdot \left[3 \cdot \frac{x^2}{L_M^4} - 4 \cdot \frac{x}{L_M^3} + \frac{1}{L_M^2}\right] \tag{13}$$

$$w_{M1,max} = \frac{12 \cdot e_M}{L_M{}^2} \cdot P \tag{14}$$

$$w_{M2,max} = \frac{4 \cdot e_M}{L_M{}^2} \cdot P \tag{15}$$

The resultant of the distributed equivalent load:

$$W_M = V_M = \int_0^{L_M/3} w_M(x)\,dx = \int_{L_M/3}^{L_M} w_M(x)\,dx = \frac{16}{9} \cdot \frac{e_M}{L_M} \cdot P \tag{16}$$

2.2. Prestress force

2.2.1. Overview

The mean value of the initial prestress force P_{m0} is determined considering the following factors ([5] 5.10.5 and [13] 8.7.2)

- elastic deformations of the element (elastic shortening of concrete);
- short-term relaxation of prestressing steel;
- stress losses due to friction;
- slip of the tendons in the anchorage.

The mean value of the prestressing force P_{mt} at time $t > t_0$ is determined considering also the following factors ([5] 5.10.6 și [13] 8.7.2):

- losses due to creep and shrinkage of the concrete;
- long-term relaxation of prestressing steel.

The concrete in the post-tensioned elements should already have minimum resistance during the tensioning of the reinforcement. The German standard DIN 1045-1/2001, [14]8.7.2, requires observing an average compressive strength during the partial prestressing of the reinforcement, respectively double of this value at the end of the prestressing (table 2). In the Romanian norm NE 013/2002, [15] 7.10, it is required that the transfer of the prestressing force to be achieved when the compressive strength of the concrete foreseen in the project is reached. In case it is not specified it is necessary to respect the strengths values given in the norm (see table 2).

Table 2. Minimum concrete compressive strength during prestressing with post-tensioned or unbonded members according to [14] and [15]

Concrete strength class	Minimum concrete cylinder compressive strength (mean value) a time t_j, $f_{cmj,min}$ in N/mm² [14]		Minimum concrete cylinder/cube compressive strength (calculated according to Annex 7.3) in N/mm² [15]
	Partial prestressing	Final prestressing	
C25/30	13	26	23,5/28
C30/37	15	30	27,5/33,5
C35/45	17	34	30/38,5
C40/50	19	38	34/42
C45/55	21	42	37,5/45,5
C50/60	23	46	41,5/50
C55/67	25	50	45/54,5
C60/75	27	54	49/61
C70/85	31	62	57/69
C80/95	35	70	65/77
C90/105	39	78	73,5/85,5
C100/115	43	86	85,5/93,5

2.2.2. Prestress force at time $t = t_0$

The mean value of the prestress force P_{m0}, for the position x on the abscissa, at the time immediately after the tendons were tensioned [4] [5] [13] [16], can be written as:

$$P_{m0}(x) = P_0 - \Delta P_c - \Delta P_r(x) - \Delta P_\mu(x) - \Delta P_{sl}(x) + \Delta P_{g+q}(x) \tag{17}$$

P_0 = maximum force at tensioning, measured at the line of the anchorage
ΔP_c = loss during transfer due to elastic shortening of the concrete
$\Delta P_r(x)$ = losses due to short-term relaxation of prestressed steel (applies only to adherent tendons)
$\Delta P_\mu(x)$ = losses due to friction between the tendon and the sheath
$\Delta P_{sl}(x)$ = loss of prestressing force due to anchorage slip
$\Delta P_{g+q}(x)$ = additional elongation of the tendon at time $t = t_0$ from another action (e.g. self-weight), but which was not included in P_0

According to [5] 5.10.2.1 and [13] 8.7.2, the maximum force applied to a tendon, $P_{0,max}$, and the mean initial prestress force immediately after the moment of prestressing, $P_{m0,max}$, in the construction elements will be limited as follows:

$$P_{0,max} \leq \min(A_p \cdot 0{,}8 \cdot f_{pk}, A_p \cdot 0{,}9 \cdot f_{p0,1k}) \tag{18}$$

$$P_{m0,max} \leq \min(A_p \cdot 0{,}75 \cdot f_{pk}, A_p \cdot 0{,}85 \cdot f_{p0,1k}) \tag{19}$$

At the beginning of the prestressing process, the tendons are individually prestressed and the anchoring is done one by one in the hardened concrete. During the tensioning of each tendon occurs an elastic shortening of the construction element, $\Delta\varepsilon_{cp,p}$ (figure 7). To avoid the decrease of the prestressing force in the first tendons below the minimum value, these tendons must be overstretched, while the last tendon can be tensioned without overstretching (figure 7) [11].

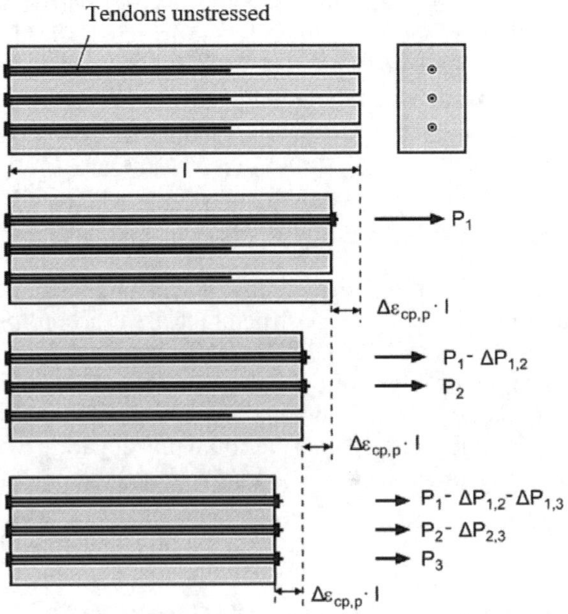

Figure 7. Loss during transfer due to elastic shortening of the concrete [16].

The loss of prestressing force for all n tendons is calculated in [16] as:

$$\Delta P_c = A_p \cdot \sigma_{cp,p} \cdot \frac{E_p}{E_{cm}} \cdot \frac{n(n-1)}{2n} \tag{20}$$

$$\sigma_{cp,p} \approx \Delta\sigma_{cp,p} \cdot n = E_{cm} \cdot \Delta\varepsilon_{cp,p} \cdot n \tag{21}$$

In general, for pre-dimensioning $\Delta P_{\mu}(x)$ can be considered maximum 20% of P_0 for adherent post-tensioned tendons, and maximum 10% of P_0 for non-adherent post-tensioned fittings [16].

2.2.3. *Prestress force at time t→∞*

At any time, t, the average value of the pretensioning force P_{mt} it is supplementary reduced with time-dependent losses [5] [13] [4] [6] [10] [16], which for the position x on the abscissa are written:

$$P_{mt}(x) = P_{m0}(x) - \Delta P_{c+s+r}(x) + \Delta P_{g+q}(x) \tag{22}$$

$\Delta P_{c+s+r}(x)$= time dependent losses of prestress force due to creep and shrinkage of the concrete, and due to reduction of stress in the steel tendon caused by the relaxation under tension, at the x abscissa

$\Delta P_{g+q}(x)$ = additional elongation of the tendon at the time t→∞ from another action (e.g. self-weight), at the x abscissa, but which was not included in P_{m0}

In general, in pre-dimensioning, $\Delta P_{c+s+r}(x)$ can be considered between 10% and 20% of P_{m0} for post-tensioned tendons [10] [16]. In common practice, for flat and horizontal flat slabs, the losses due to the additional elongation of the tendon from an action other than the pretensioning itself, is considered zero.

2.3. *Determination of the prestress degree*

Thick floor slabs and those heaving small openings are not very suitable for partial prestressing. In the case of slabs with great slenderness (L / h ≥ 30) and large live loads ($q_{k,N}$ > 10 kPa) partial prestressing leads to an optimum reinforcement consumption from the total reinforcement required. In the case of less slender slabs (L / h < 30) and reduced live loads ($q_{k,N}$ < 5 kPa), especially for distances between column below 7,0 m, the solution can be considerably improved by partial prestressing compared to the reinforced concrete slab solution, sometimes with only a little extra effort [11].

The highest possible degree of prestressing is determined by constructive conditions, namely the maximum number of tendons positioned in the column strip. In the case of partial prestressing, in the support areas at the top, a single row reinforcement is arranged in both directions to obtain an effective depth of a cross section of the slab as high as possible. For average live loads ($q_{k,N} \leq 10$ kPa) it is economical that approx. 30% from the bending moments on the support from external loads to be taken up by partial prestressing. In the case of such a low degree of prestressing, it is appropriate to prestress only the column strips [11].

Decompression checking is not required for buildings and ordinary structures [11], having a designed service life of 50 years (buildings classified in the structural class D according to DIN 1045-1, equivalent to the structural class S4, according to EN 1992-1-1). To determine the required number of tendons, n, in the column strips, the maximum camber of the column strips can be limited. In the present study, it is assumed that the camber of the prestressed floor slab is accordingly limited when the deflection of the support strips due to its own weight is zero. Thus, the following equation mathematically condenses the condition that the camber from its own weight (f_{gk1}) be equal to the pre-camber due to prestressing n tendons ($f_{p,\infty}$):

$$f_{gk,1} + n \cdot f_{p,\infty} = 0 \tag{23}$$

Another approach for determining an economic degree of prestressing is a combination of passive reinforcement and active reinforcement so chosen as to ensure both the limitation of crack opening and the bending bearing capacity [11].

2.3.1. *Deflection due to pretensioning $f_{p,\infty}$*

Cambers can be calculated using a unit tension force P = 1 kN on an equivalent system for the support strips, f_s. Based on these values, the camber $f_{p,\infty}$ is determined at time t = ∞ using equation:

$$f_{p,\infty} = f_s \cdot P_{m,\infty} \tag{24}$$

2.3.2. *Deformation due to slab`s self-weight $f_{gk,1}$*

Deformations of the support strips under the action of the slab`s self-weight, $g_{k,1}$, is determined using a replacement system like the system used with prestressing. For this purpose, the transfer of the uniformly distributed load from slab to the column strips was determined by applying equivalent linear loads distributed triangularly and trapezoidally along the strip, according to [2] [3]. The column strip is considered undeformed in relation to the support line.

2.3.3. *Determining the number of tendons and their arrangement*

The required number of tendons is obtained by dividing the value of the camber from its self-weight to the value of the pre-camber from the prestressing at time t = ∞, for the considered column strip:

$$n = \frac{f_{gk,1}}{f_{p,\infty}} \tag{25}$$

3. Case study

A flat slab is designed as a post-tensioned concrete slab with free tendon layout, as shown in figure 8. A partial prestress is used. By partially prestressing the concrete slab, the goal is to limit appropriately the values of deflection by using unbonded tendons. Besides tendons, to provide enough load-bearing capacity and to limit the crack widths, steel reinforcement is required. In this example is calculated only the necessary number of tendons and their distribution. It is considered a single-story structure with columns fully fixed in the foundations. The column length is L_{col} = 3,50 m.

The slab is prestressed using unbonded monostrands which are banded in both directions and placed exclusively in the column strips. Due to the free tendon layout, it is enough to have the tendons fixed on chairs only over the columns.

In this example, we aim the camber of the column strips due to slab`s self-weight to be approximately compensated by the pre-camber gained through stressing the tendons. This means the concrete in the column strips will remain uncracked and the ideal slab thickness can be easily predimensioned using table 1.

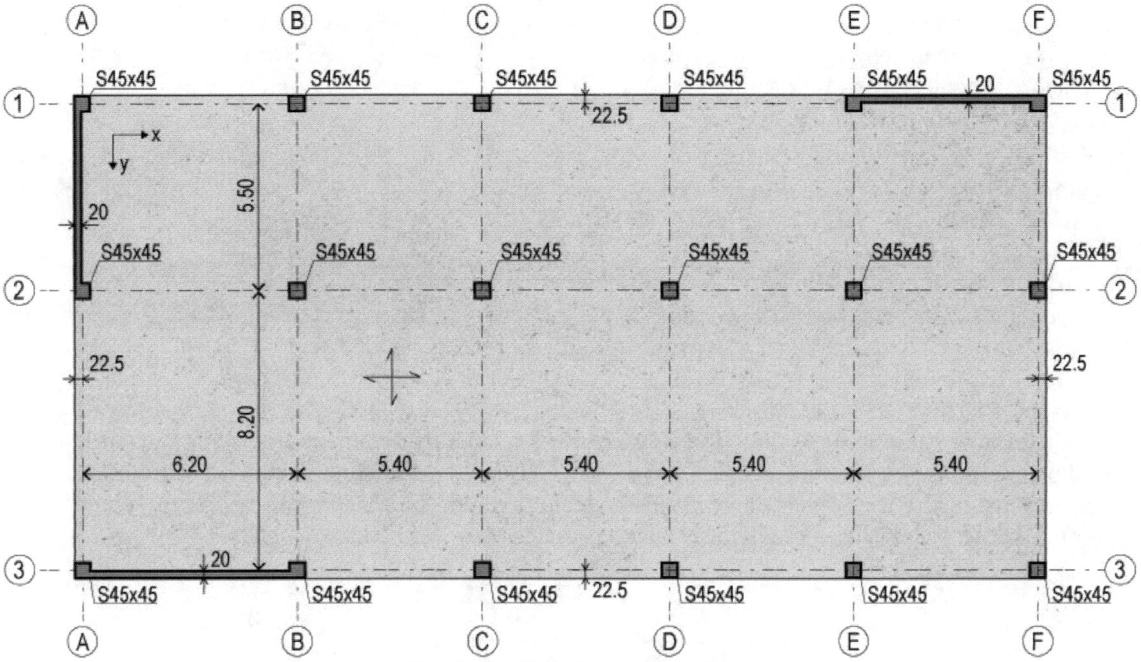

Figure 8. Slab formwork plan.

3.1. Materials

For the slab, a C30/37 concrete strength class is chosen with f_{ck} = 30N/mm^2, steel reinforcement S500 B and unbonded steel tendons (table 3). The tendons used are made of steel Y1860S7 with HDPE-sheathing and corrosion protected with grease (trade named CONA CMF BT n05-100) and they can be supplied as individual monostrand or as bands.

Table 3. Technical characteristics for the tendons used [17] [18].

Manufacturer: BBR VT International Ltd, Switzerland (www.bbrnetwork.com) Tendon type: Unbonded steel monostrands (single or grouped in bands) Product name: BBR VT CONA CMM Single S1 and S2 – monostrand unbonded or bonded post-tensioning system BBR VT CONA CMF BT – Internal Post-tensioning System with Flat Anchorages and 02 to 06 Strands European technical approvals: European Technical Assessment ETA-06/0165 European Technical Assessment ETA-12/0076 Delivery: individual monostrands or grouped in bands of 2, 3, 4, 5 or 6 strands	
Nominal diameter for the 7-wire prestressing steel strand	Ø 12,9 mm
Nominal cross-sectional area	100 mm^2
Characteristic tensile strength	f_{pk} = 1860 N/mm^2

External diameter of HDPE-sheathing	$\geq$ 17 mm
Mass of prestressing steel	0,78 kg/m
Nominal mass per tendon (strand + protection)	$\geq$ 0,90 kg/m
Characteristic value of maximum force	F_{pk} = 186 kN
Maximum value of maximum force	$F_{m,max}$ = 214 kN
Characteristic value of 0,1% proof force	$F_{p,01}$ = 164 kN
Minimum elongation at maximum force, $L_0 \geq 500$ mm	A_{gt} = 3,5%
Modulus of elasticity	E_p = 195000 MPa
Minimum concrete strength at time of stressing	f_{ck} = 20 N/mm^2 (class C20/25)
Slip at stressing anchor and fixed anchor	Δl_{sl} = 7 mm
Friction coefficient (for monostrand)	μ = 0,06 rad^{-1}
Wobble coefficient (for monostrand)	k = 0,009 rad/m
Friction coefficient (for band with smooth steel duct)	μ = 0,18 rad^{-1}
Wobble coefficient (for band with smooth steel duct)	k = 0,005 rad/m
Minimum radius of curvature	2,00 m

3.2. Loads, predimensioning and concrete cover

For the floor plan shown in figure 9, a solid flat slab is chosen, loaded with 10 kN/m^2 (characteristic value) and free spanning over maximum L = 8,20 m. Therefore, according to table 1, a necessary slab thickness of 280 mm is necessary.

The tendon position in the y direction (see figure 9) results from the condition that the tendon to be placed on top of the bottom reinforcement layer with an overall minimum cover. The minimum concrete cover for unbonded tendons is mentioned in the corresponding European technical approval. A minimum concrete cover c_{nom} = 20 mm is necessary.

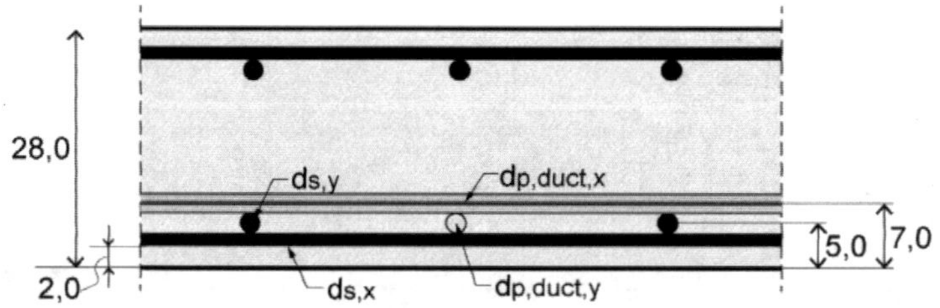

Figure 9. Sketch with reinforcement and tendon positioning in a middle field slab cross-section (axis B), dimensions are in cm.

Slab`s self-weight:	$g_{k,1} = 25\ kN/m^3 \cdot 0,28\ m = 7,00\ kN/m^2$
Other permanent loads:	$g_{k,2} = 2,50\ kN/m^2$
Live load:	$q_k = 7,50\ kN/m^2$

3.3. Tendon layout

For a Ø 12,9 mm monostrand, unbonded, greased and protected in a sheath, using the data given in the technical assessment [18] and the equations from (1) to (16) for the slab in the *y-direction*, one obtains the results condensed in table 4. In addition, in figure 10 is represented the tendon profile for *axis B*.

Table 4. The vertical profile of the tendon in the y-direction and the equivalent loads for a prestressing force in the monostrand of P=1kN.

Diameter of the steel strand	Ø 12,9 mm
Nominal cross-sectional area	100 mm^2
Number of wires	7
Diameter of the wire	Ø 4,26 mm
Moment of inertia of the wire	16 mm^4
Moment of inertia of the strand	113 mm^4
Tendon`s weight (strand + sheath + grease)	9 N/m
e_R	90 mm
$y_{anchorage}$	0 mm
L_R	1517 mm
$W_R = V_R$	$118{,}65 \cdot 10^{-3}$ kN
M_R	0 kN·m
e_M	140 mm
L_M	2230 mm
$W_M = V_M$	$111{,}63 \cdot 10^{-3}$ kN

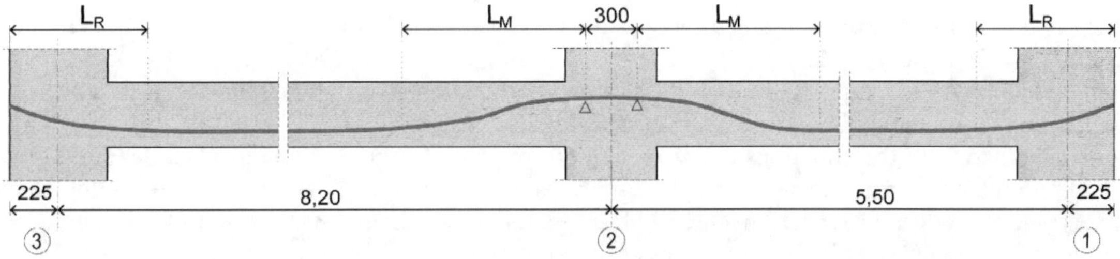

Figure 10. Vertical profile of the tendon in axis B, distances expressed in [mm] and [m].

3.4. Equivalent loads and equivalent frame system

For the column strip in axis B, the equivalent system is a frame loaded with the prestressing force applied as an equivalent load (figure 11). The point loads near support 2-B are offset with 150 mm on each side due to the fixation points of the tendon on the top non-prestressed rebar layer.

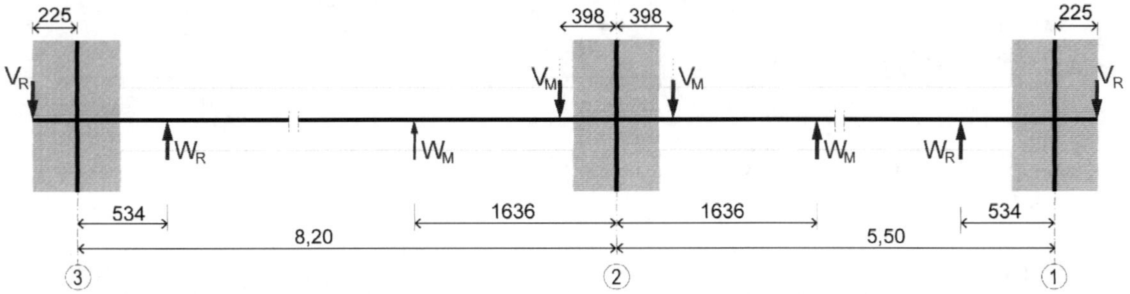

Figure 11. Equivalent frame system for the strip column in axis B, distances in [mm].

3.5. Determination of the required number of tendons

To determine the required number of tendons, the force losses from the prestressing of the concrete must be calculated with equations from (17) to (22), using the characteristics of the tendon (table 3) and its profile. In a simplified way, the following calculations use the average value of the prestressing force per tendon at time $t = t_0$, $P_{m0} = 125$ kN , respectively at time $t = \infty$, $P_{m,\infty} = 105$ kN.

IOP Conf. Series: Materials Science and Engineering 1141 (2021) 012013 doi:10.1088/1757-899X/1141/1/012013

In the case of unguided tendons, the tendon exerts a parabolic pressure on the concrete. It should be noted that the equivalent loads due to the tendon action in the inner support area do not start from the support's centre, but they are offset with 150 mm because of the two chairs (fixation points) at the top reinforcement layer level. According to [17] [18], in the inner support area the unguided tendon is enough if it is fixed in two points, spaced between 300 mm and 1 m. In this example the distance is set to 300 mm (figure 10).

The static calculation for a flat slab with uneven spans is usually performed using software based on the finite element method. Since the degree of prestressing for the column strip has been chosen so that the strip deflection loaded with slab`s self-weight is zero, the slab can be calculated as a continuous beam (the equivalent frame method), the prestressing being done in two directions of the floor. In the present calculation example, the prestressing of the slab is modelled in two distinct ways, the first by equivalent loads and the second by modelling the tendon with its actual vertical profile. This means that in both models the secondary effects of prestressing an undetermined static system are implicitly included. For the calculation with real tendon layout, two models were used for verification. First, a model with mixed vertical layout, including segments with 4th degree curves (real geometry of unguided tendon in the support zone), respectively a model with parabolic-only vertical layout, the tendon path being modelled with 2nd degree curves (guided tendon – ordinary solution).

The deflection of the column strip measured at mid span is calculated using an equivalent system modelled with beam elements, while the prestress effect is introduced as an equivalent load. Then, the same column strip was calculated using equivalent frames, with beams having the width equal to half the sum of the widths of the adjacent slab panel according to [3], the tendon being integrated inside the beam element with its real path. In the model with unguided tendons, the tendon layout is a mixed one - with straight and parabolic segments (span and reverse parabola as 4th degree curves), as in figure 4 and determined with the equations (1) to (9). In the model with guided tendons, the tendon layout is parabolic only, consisting of span parabolas and reverse parabolas as 2nd degree curves. Static analysis was carried out with ETABS v19 from CSI. The results of the calculations are shown in table 5 for the column strip in axis B, the bending stiffness of the concrete cross-section being reduced to 0,3·Ecm·Ic [2].

Table 5. Deflections and number of tendons chosen.

Column strip in axis B	Upward deflection due to a single monostrand at time t = t$_0$ and t = t$_\infty$						Deflection due to slab`s own weight	Number of tendons	
	with unguided tendon (figure 11)		with equivalent loads (figure 12)		with guided tendon (2nd degree curve)			necessary	chosen
	$f_{p,0}$ [mm]	$f_{p,\infty}$ [mm]	$f_{p,0}$ [mm]	$f_{p,\infty}$ [mm]	$f_{p,0}$ [mm]	$f_{p,\infty}$ [mm]	$f_{gk,1}$ [mm]	n_{nec} [buc.]	n_{ef} [buc.]
B/3-2	0,345	0,290	0,322	0,274	0,380	0,319	8,927	33	6x6=36
B/2-1	0,169	0,142	0,160	0,136	0,178	0,150	0,738	6	2x6=12

The distribution of the tendons in the column strips is summarized in figure 13, these being distributed over a maximum width of 0,1 times the width of the slab panel on either side of the column [11] [19]. In the same area where it is recommended that a minimum of 50% of the upper total reinforcement amount in the column strip is concentrated (reinforcement in tension for negative bending moment), according to [4]. In addition, in order for the whole number of tendons to contribute to the punching shear resistance of the slab, tendons must be distributed over the width of the control perimeter [5] [11] [13]. Table 6 summarizes the values of the maximum deflection of the column strip on the axis B, with and without precompression of concrete, choosing as tendons 6 bands of 6 monostrands for span

2-3, respectively 2 bands of 6 monostrands for span 1-2. Figure 12 shows the distribution of the tendons in the studied slab for deflection control.

Table 6. Pre-cambers (upward deflections) and final cambers (downward final deflections).

Column strip in axis B	Camber due to quasi-permanent loads [20] $1,1\cdot(g_{k,1}+g_{k,2})+0,6\cdot q_k$	Pre-camber with the final number of chosen tendons, due to equivalent loads	The total final camber	Limit value for deflection [5] L/250	Camber after construction [5] $1,1\cdot g_{k,2}+0,6\cdot q_k$	Limit value for deflection after construction [5] L/500
	f_{qp} [mm]	$f_{p,\infty}$ [mm]	$f_{qp}+f_{p,\infty}$ [mm]	f_{adm} [mm]	$f_{qp,exec}$ [mm]	$f_{adm,exec}$ [mm]
B/3-2	19,1	9,9	9,2	32,8	9,3	16,4
B/2-1	1,6	1,6	1,0	22,0	0,8	11,0

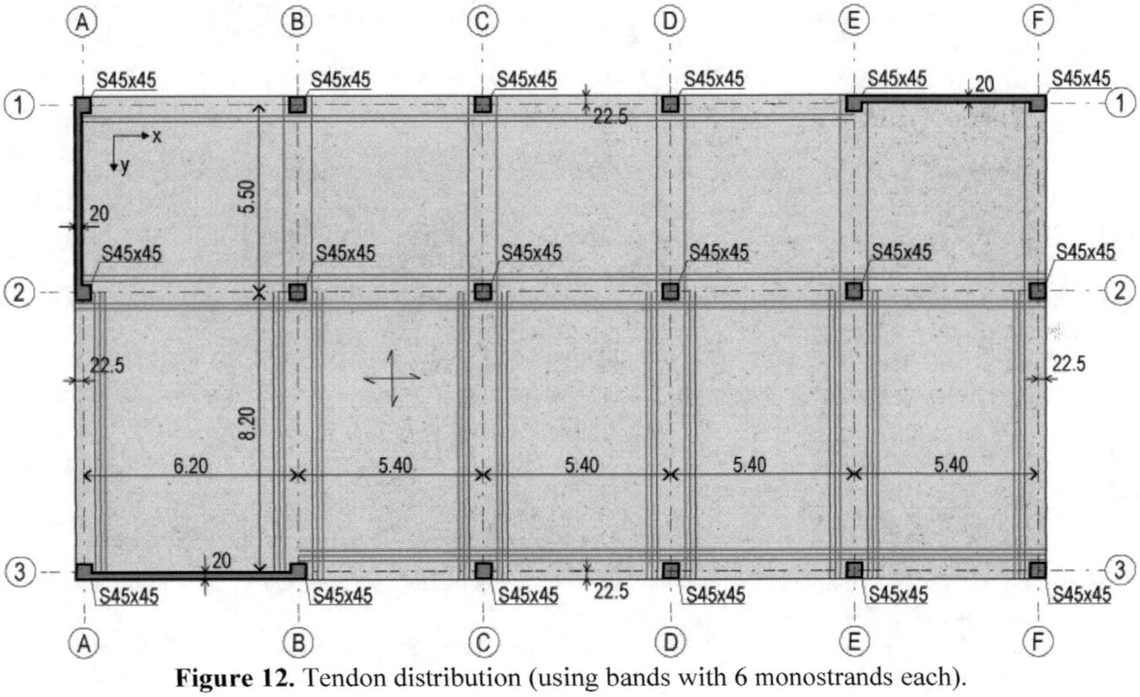

Figure 12. Tendon distribution (using bands with 6 monostrands each).

4. Conclusions

The analytical design approach for post-tensioned concrete slabs with unguided tendons, presented above, proved to be easy to use and offered fast results. Using an equivalent load method for determining the deflection and inner forces on a flat slab, when compared with a finite element analysis, showed differences of maximum 7%, which can be considered acceptable in terms of precision. Besides, the performance of the prestress slab when unguided tendons are used instead of the classical guided ones, decreased only by 12%. Based on the exposed study case, one can find the advantages for using unguided tendons for prestressed concrete slabs and the possibility of designing them in the absence of a dedicated software for prestressed concrete.

Acknowledgements
The authors are thankful to Ovidiu Prodan, representing CSI America in Romania, who made it possible to use the ETABS software for the statical analysis in the present research, as free of charge.

References

[1] Morgen K, Nölting D and Wollrab E 2004 Flachdecken mit Vorspannung ohne Verbund nach DIN 1045-1 *Beton- und Stahlbetonbau* **99** Heft 11 pp 914-928
[2] Universitatea Tehnică de Construcţii Bucureşti 2012 *Ghid pentru proiectarea planşeelor dală în zone seismice. GP 118-2012 (proiect)* MLPDA Bucureşti România
[3] Universitatea Tehnică de Construcţii Bucureşti 1998 *CR 2 Proiectarea structurilor din beton. Subprogram CR 2-01-A. Planşee dale şi planşee ciuperci (redactarea a 2a - finală)* M.L.P.A.T. Bucureşti România
[4] Kiss Z and Oneţ T 2010 *Proiectarea structurilor de beton după SR EN 1992-1 (ediţia a 2a revizuită)* Cluj-Napoca Editura Abel
[5] CEN European Committee for Standardization 2004 *Eurocode 2: Design of concrete structures - Part 1-1: General rules and rules for buildings* Bruxelles CEN/TC 250
[6] Pascu R and Zybaczinski A 2012 *Beton precomprimat: calcul după Eurocode 2* Bucureşti Conspress
[7] Fédération Internationale du Béton (fib) 2010 *Model Code - First complete draft, Volume 2* Germany DCC Document Competence Center Siegmar Kästl e.K.
[8] The Concrete Society 2005 *Post-tensioned concrete floors. Design Handbook - Concrete Society Technical Report No. 43 (second edition)* Camberley UK The Concrete Society
[9] Aalami B O 2014 *Post-tensioned buildings: Design and Construction* China International Edition
[10] Geßner S, Niedermeier R, Ahrens M A, Hegger J, Fischer O and Markt P *2017 BetonKalender 2017: Spannbeton, Spezialbetone* Ernst & Sohn GmbH & Co. KG chapter 1 pp 1-100
[11] Grünberg J and Hansen M 2004 Vorgespannte Deckenplatten *FRILO-Magazin Fachthema DIN 1045-1* pp 8-20
[12] Maier K and Wicke M 2000 Die Freie Spanngliedlage *Beton- und Stahlbetonbau* **2** 95 pp 62-71
[13] DIN Deutsches Institut für Normung 2008 *DIN 1045-1:2008-08 Tragwerke aus Beton, Stahlbeton und Spannbeton - Teil 1: Bemessung und Konstruktion* Berlin Beuth Verlag GmbH
[14] DIN Deutsches Institut für Normung 2001 *DIN 1045-1:2001-07 Plain, reinforced and prestressed concrete structures* Berlin Beuth Verlag GmbH
[15] M.D.R.A.P. 2002 *NE 013-2002: Cod de practică pentru execuţia elementelor prefabricatelor din beton, beton armat şi beton precomprimat* Bucureşti Matrix Rom
[16] Zilch K and Zehetmaier G 2010 *Bemessung im konstruktiven Betonbau (2.,neu bearbeitete und erweiterte Auflage)* Leipzig Springer
[17] BBR VT International Ltd 2018 *BBR VT CONA CMM Single S1 and S2, in Unbonded Post-tensioning System with 01, 02 and 04 Strands* Schwerzenbach (Switzerland) www.bbrnetwork.com pp 1-60
[18] BBR VT International Ltd 2019 *BBR VT CONA CMF (ETA-12/0076), in Internal Post-tensioning System with Flat Anchorages and 02, 03 and 04 Strands*, Schwerzenbach (Switzerland) www.bbrnetwork.com pp 1-76
[19] Grasser E and Thielen G 1991 Hilfsmittel zur Berechnung der Schnittgrößen und Formänderungen von Stahlbetontragwerken *Deutscher Ausschuss für Stahlbeton*, Heft 240, 1991.
[20] CEN European Committee for Standardization 2002 *Eurocode 0: Basis of structural design* Bruxelles CEN/TC250

Experimental statistical models for the study of cement-concrete for road pavements

L Chintea[1], A Mishutin[2], I Grynyova[3], M Zavoloka[4]

[1] Postgraduate student, Research associate; Odesa State Academy of Civil Engineering and Architecture, Didrihson, 4, 65029, Odesa, *Ukraine*

[2] Department of Roads and Airfields, Odesa State Academy of Civil Engineering and Architecture; Didrihson, 4, 65029, Odesa, *Ukraine*

[3] Department of Architectural structures; Odesa State Academy of Civil Engineering and Architecture; Didrihson, 4, 65029, Odesa, *Ukraine*

[4] Department of production of building products and constructions; Odesa State Academy of Civil Engineering and Architecture, Didrihson, 4, 65029, Odesa, *Ukraine*

E-mail: grynyova@ogasa.org.ua

Abstract. The article presents the calculation and statistical analysis of the experimental statistical (ES) model of the several samples of concrete and reinforced concrete for road pavements. For the analysis, the Program (dialogue) Complex COMPEX was used, which was developed at the Odessa State Academy of Civil Engineering and Architecture by scientists of the scientific school of prof. V.A. Voznesensky. In the research of the structure and properties of rigid road pavements concrete with an optimal plan, a 4-factor experiment was carried out, in which the following factors of concrete composition were varied: X1 - the amount of Portland cement PC II / A-Sh-500, from 400 to 500 kg / m^3; X2 - the amount of polypropylene fiber from 0 to 2 kg / m^3; X3 - the amount of Metakaolin, from 0 to 30 kg / m^3; X4 - the amount of complex action additive (polycarboxylate type) Coral ExpertSuid-5, from 0.6 to 1% of the cement mass. An increase in the amount of Coral ExpertSuid-5 additive from 0.6 to 0.8% of the cement mass is noticeable, by 10...12%, and decreases the W/C ratio of mixture. An increase in the amount of this additive of complex action to 1% causes a further, but less noticeable, decrease in W/C ratio.

1. Introduction

Rigid cement concrete pavements are widely used in the European Union, the United States and other developed countries. These pavements have shown their high-performance properties and durability, which confirmed the economic feasibility of their use. In recent years, rigid cement concrete pavements are increasingly used in domestic practice of road construction in Ukraine and other countries with temperate climates. Frost resistance is one of the main indicators of the quality of concrete pavements, ensuring the durability of materials. Nowadays, one of the most effective materials for cement concrete pavements are modified fiber concrete. These materials are able to provide both high strength and durability of the pavement. Accordingly, the task of developing modified fiber concrete rigid pavements based on domestic binders, aggregates and modifiers is actual.

2. Analysis of recent research and publications

In the conditions of constant influence of transport and weather factors, the service life of rigid road pavements is two and more times longer in comparison with asphalt concrete pavements [1]. Another important advantage of cement concrete pavement is that bitumen is not used for their production. Oil reserves for bitumen production are more limited than raw material reserves for cement production. It is also important that most of the bitumen in Ukraine is imported [2].

During the manufacture of rigid concrete pavements, the achievement of the required level of efficiency simultaneously with high durability and strength of the material is possible only when using modern additives modifiers and fiber [3, 4]. Today polycarboxylate type superplasticizers are one of the most effective concrete modifiers. For concrete pavements also an effective method of increasing durability is the use of active mineral additives of pozzolanic type [5], in particular metakaolin [6, 7]. Metakaolin significantly affects the process of structure formation of concrete. These changes in the structure increase the durability of concrete by reducing permeability and shrinkage, as well as increasing frost resistance and corrosion resistance [4-7].

Disperse reinforcement is also a known method of improving the mechanical properties of concrete [4, 5, 8, 9]. The use of fiber can increase crack resistance, frost resistance, tensile strength and wear resistance of concrete, which is important for paving materials [4, 9]. Accordingly, disperse-reinforced concrete modified with a polycarboxylate-type superplasticizer and an active pozzolanic-type mineral additive are promising materials for rigid pavements that are operated in severe conditions.

3. The purpose of this research

The purpose of this research is to determine the optimal concentrations of microsilica and superplasticizer to improve the rheological and physical-mechanical properties of plaster solutions, as well as to assess the sensitivity of the system to changes in the mineralogical and chemical compositions of microsilica.

4. Objects and methods of research

Numerical modelling based on a modern software and computational complex is an effective approach, which makes it possible to determine the composition of cement-concrete under the influence of a transport load. For the calculation and statistical analysis of experimental-statistical (ES) models of the investigated concretes and mixtures composition influence on their physical and mechanical properties, the Program (dialogue) Complex COMPEX [10, 11] was used, which was developed at the Odessa State Academy of Civil Engineering and Architecture by Head of the scientific school by prof. V.A. Voznesensky. Calculations of ES models were carried out taking into account the accepted experimental error at a two-sided risk of 10%, which means α=0,1. For a given risk level, the hypothesis was tested about the difference in the estimates of the coefficients of the ES models from zero. The significance of the coefficients was checked. After a sequential analysis, the ES model with all significant estimates of the coefficients was tested for adequacy according to the Fisher's criterion. To calculate the ES models, a transition was made from full-scale variables to coded ones, which should be in the range from - 1 to +1. For this, the usual formula (1) was used:

$$x_i = \frac{X_i - \dfrac{X_{i.\max} + X_{i.\min}}{2}}{\dfrac{X_{i.\max} - X_{i.\min}}{2}} \qquad (1)$$

During research of the structure and properties of rigid road pavements concrete with an optimal plan, a 4-factor experiment was carried out, in which the following factors of concrete composition were varied:

X1 - the amount of Portland cement PC II / A-Sh-500, from 400 to 500 kg / m^3;

X2 - the amount of polypropylene fiber from 0 to 2 kg / m^3;

X3 - the amount of Metakaolin, from 0 to 30 kg / m^3;

X4 - the amount of the complex action additive (polycarboxylate type) Coral ExpertSuid-5, from 0.6 to 1% of the cement mass.

Crushed stone of fraction 5-10 mm and quartz sand with a modulus of 2.7 were used as aggregates for fiber concretes. The compressive strength of concrete was determined according to DSTU B B.2.7-224:2009. The tensile strength in bending was determined on samples of beams with measurements 4×4×16 cm. Frost resistance of rigid pavements concretes was determined by the accelerated method in salt water when frozen to –50°C [12].

The amount of aggregates, that is, crushed stone and quartz sand in concrete mixtures, during the experiments, was adjusted in accordance with the known methods for selecting the composition of concrete to ensure the required calculated volume of material per 1 m^3, depending on the amount of cement, water, fiber and additives.

All mixtures of concrete and fiber-reinforced concrete investigated in both series of experiments had equal mobility M2 (cone slump (CS) from 6 to 8 cm), which was achieved by selecting the amount of water. This mobility was chosen to meet the most common requirements for concrete mixtures of rigid road pavements.

An extended plan of the experiment with coded samples and real values of the factors of variation of concrete and fiber-reinforced concrete of rigid road pavements are presented in Table 1.

Table 1. The plan of the experiment and the compositions of the concrete and fiber-reinforced concrete of rigid road pavements.

№	Levels of factors				Concrete mixtures						
	X1 (cement)	X2 (fiber)	X3 (metakaolin)	X4 (add-on)	Cement (kg / m^3)	Crushed stone (kg / m^3)	Sand (kg / m^3)	Metacaolin (kg / m^3)	Fiber (kg / m^3)	Additive (kg / m^3)	Water (*l* / m^3)
1	0	0	0	0	450	1115	655	15	1	3.6	183
2	-1	-1	1	-1	400	1130	668	30	0	2.4	191
3	-1	1	-1	1	400	1135	692	0	2	4.0	193
4	1	-1	-1	1	500	1105	635	0	0	5.0	168
5	1	1	1	1	500	1100	603	30	2	5.0	181
6	1	1	-1	-1	500	1105	621	0	2	3.0	192
7	0	1	1	-1	450	1115	617	30	2	2.7	206
8	0	-1	-1	-1	450	1115	684	0	0	2.7	161
9	0	-1	1	1	450	1115	651	30	0	4.5	164
10	1	0	1	-1	500	1105	574	30	1	3.0	203
11	-1	0	-1	-1	400	1130	691	0	1	2.4	192
12	-1	0	1	1	400	1130	672	30	1	4.0	182
13	1	-1	0	-1	500	1105	601	15	0	3.0	199
14	-1	1	0	-1	400	1130	676	15	2	2.4	199
15	-1	-1	0	1	400	1135	702	15	0	4.0	158
16	1	-1	1	0	500	1105	590	30	0	4.0	183
17	-1	1	1	0	400	1130	662	30	2	3.2	189
18	-1	-1	-1	0	400	1135	710	0	0	3.2	163

Based on the experimental data, the ES model of the influence of composition factors on the W/C of concrete mixtures of rigid road pavements of equal mobility (2) was built:

$$\begin{aligned} W/C = 0{,}396 &- 0{,}043x_1 + 0{,}015x_1^2 - 0{,}010x_1x_2 \pm 0x_1x_3 \pm 0x_1x_4 \\ &+ 0{,}021x_2 - 0{,}013x_2^2 - 0{,}008x_2x_3 + 0{,}004x_2x_4 \\ &+ 0{,}009x_3 \pm 0x_3^2 - 0{,}010x_3x_4 \\ &- 0{,}018x_4 + 0{,}009x_4^2 \end{aligned} \quad (2)$$

5. The results of research

According to the ES model (2), one-factor diagrams were constructed and shown on Figure 1, reflecting the influence of variable factors on the W/C ratio of mixtures in the extremum zones, through which the maximum (red lines) and minimum (blue lines) values pass. During constructing these graphs, the level of three factors is not reflected in each of them and were fixed at the values that provide, respectively, the maximum and minimum values of the W/C ratio of mixtures. During constructing these graphs, the level of three factors not reflected in each of them were fixed at the values that provide, respectively, the maximum and minimum values of the W/C ratio of mixtures.

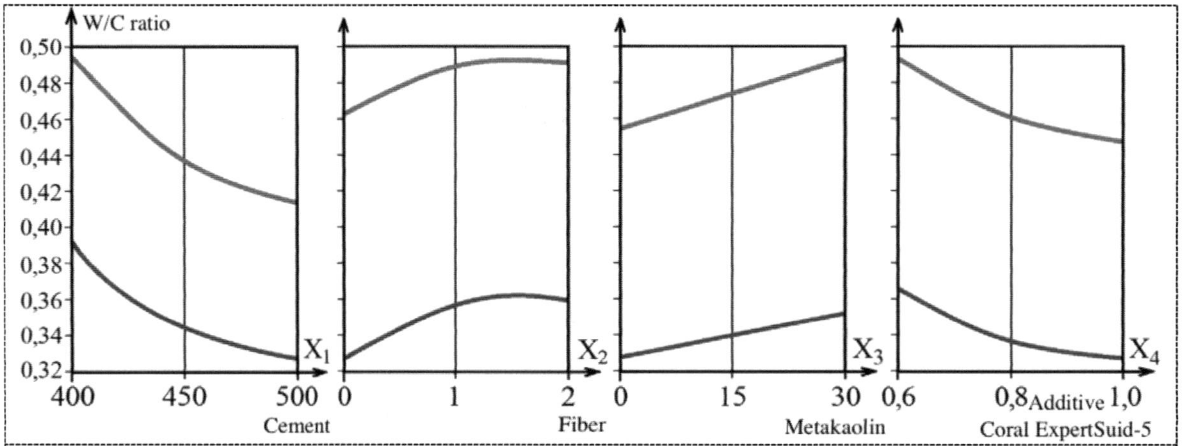

Figure 1. One-factor diagrams with influence of X1, X2, X3, X4 factors

An analysis of the diagrams shown in Fig. 1 shows that with an increase in the amount of Portland cement W / C, the ratio of mixtures is expected to decrease. The introduction of polypropylene fiber necessitates an increase in the W / C ratio in order to maintain the necessary mobility of the mixture. Due to the additional water requirement of Metakaolin, with the introduction of this active mineral additive, the W / C ratio of the mixture also increases. Thus, due to the introduction of the maximum amount of Metakaolin the W / C ratio increases by 8...9%.

The values of the compressive strength of concrete and fiber concrete of rigid road pavements, which were determined during a full-scale experiment, investigated at 18 points of the planned experiment on the 3rd, 7th, 28th and 180th days after hardening are given in Table 2.

Table 2. Compressive strength of the researched concretes and fibro-concretes rigid pavements at different ages

№	Compressive strength (MPa)			
	3 days, $f_{ck.cube.3}$	7 days, $f_{ck.cube.7}$	28 days, $f_{ck.cube.28}$	180 days, $f_{ck.cube.180}$
1	41.2	45.3	59.4	69.4
2	31.6	36.6	49.3	56.3
3	30.7	33.2	45.2	51.9
4	44.1	49.8	66.1	76.9
5	49.8	53.9	65.0	77.6
6	41.1	43.0	54.4	64.3
7	36.5	39.6	51.1	61.8
8	36.5	41.2	54.8	62.3
9	47.7	51.2	65.4	75.5
10	42.6	45.4	55.5	63.8
11	27.4	31.2	41.8	47.9
12	36.3	38.6	49.1	56.0
13	39.8	43.6	56.6	63.4
14	29.6	34.3	46.3	52.8
15	39.1	45.2	60.9	67.8

16	46.6	50.5	65.2	74.5
17	36.1	39.3	51.1	57.5
18	31.2	36.4	49.4	58.3

ES models (3,4) which are constructed according to the data given in Table 2, are presented below:

$$\begin{aligned} f_{ck.cube.3}(MPa) = 40{,}9 &+ 5{,}8x_1 - 1{,}7x_1^2 + 0{,}8x_1x_2 \pm 0x_1x_3 \pm 0x_1x_4 \\ &- 0{,}4x_2 + 0{,}9x_2^2 \pm 0x_2x_3 - 0{,}6x_2x_4 \\ &+ 2{,}1x_3 - 0{,}8x_3^2 + 1{,}0x_3x_4 \\ &+ 3{,}0x_4 - 1{,}1x_4^2 \end{aligned} \quad (3)$$

$$\begin{aligned} f_{ck.cube.7}(MPa) = 45{,}4 &+ 5{,}7x_1 - 1{,}5x_1^2 + 1{,}1x_1x_2 \pm 0x_1x_3 + 0{,}8x_1x_4 \\ &- 0{,}9x_2 + 1{,}2x_2^2 \pm 0x_2x_3 - 0{,}6x_2x_4 \\ &+ 2{,}0x_3 - 1{,}7x_3^2 + 0{,}7x_3x_4 \\ &+ 3{,}2x_4 - 1{,}2x_4^2 \end{aligned} \quad (4)$$

Based on the ES models (3,4) in the PC COMPEX, the influence of variable composition factors on the compressive strength of the researched samples of concrete and fiber concrete of rigid road pavements at an early age: 3 days, $f_{ck.cube.3}$ (Figure 2) and 7 days, $f_{ck.cube.7}$ (Figure 3), was modelled and presented as a three-factor volumetric diagram.

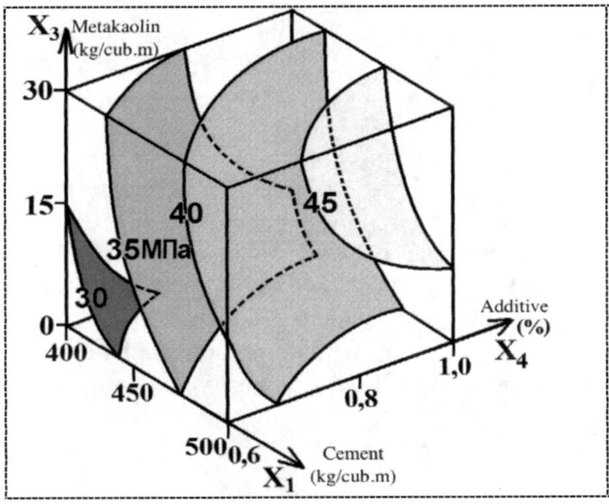

Figure 2. Influence of the amount of Portland cement, Metakaolin and Coral ExpertSuid-5 complex action additive on the compressive strength of fiber-reinforced concrete on the 3rd day ($x_2 = 0$)

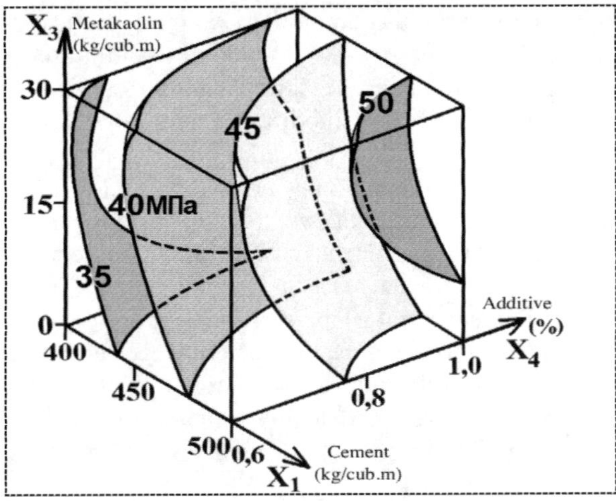

Figure 3. Influence of the amount of Portland cement, Metakaolin and Coral ExpertSuid-5 complex action additive on the compressive strength of fiber-reinforced concrete on the 7th day ($x_2 = 0$)

The maximum compressive strength on the 3rd day was $f_{ck.cube.3.max}$ = 50.5 MPa at the point with coordinates $x_1 = x_3 = x_4 = 1$, $x_2 = -1$. The maximum compressive strength on the 7th day was $f_{ck.cube.7.max}$ = 55.2 MPa at a point also with close coordinates $x_1 = x_4 = 1$, $x_2 = -1$, $x_3 = 0.69$. That is, on the 3rd and 7th days, concrete has the greatest compressive strength with the maximum amount of cement and Coral additives and without fiber. The maximum strength on the 3rd day with an amount of Metakaolin of 30 kg / m^3 and on the 7th day - 24...25 kg / m^3 was observed.

Analysis of the diagrams shown on Figure 1, Figure 2 shows that with an increase in the amount of cement in the composition of concrete, its strength is expected to increase, but this effect is somewhat non-linear both on the 3rd day and on the 7th day after hardening. That is, with an increase in the amount of binder from 400 to 450 kg / m^3, the compressive strength of composites increased more intensively than with an increase in the amount of binder from 450 to 500 kg / m^3. Metakaolin, as an active mineral additive, affects the processes of structure formation, which increases strength of concrete both in the early age and in the branded age.

The strength of concrete on the 28th day can be called strength at the "standard" age, since it is this strength that determines the class of concrete or its strength grade. The minimum class of concrete for road pavements is set by the relevant regulatory documents. Therefore, it is the compressive strength on the 28th day that is largely decisive from the point of view of the possibility of using concrete in a particular structure, including rigid road pavements.

The influence of varying composition factors on the compressive strength of the investigated concretes and fiber concretes of rigid road pavements at the “standard” age of 28 days is described by the ES model (5) given below:

$$\begin{aligned} f_{ck.cube}\,(\text{MPa}) = 59{,}4 &+ 5{,}9x_1 - 2{,}4x_1^2 + 1{,}1x_1x^2 \pm 0x_1x_3 + 1{,}1x_1x_4 \\ &- 2{,}0x_2 + 2{,}4x_2^2 \pm 0x_2x_3 - 1{,}0x_2x_4 \\ &+ 1{,}6x_3 - 3{,}1x_3^2 \pm 0x_3x_4 \\ &+ 1{,}5x_4 - 1{,}9x_4^2 \end{aligned} \quad (5)$$

The field of properties of this ES model shows the minimum strength $f_{ck.cube.min}$ = 41.3 MPa at the point with coordinates $x_1 = x_3 = x_4 = -1$, $x_2 = 0.43$, which corresponds to compositions with a minimum amount of cement and additive Coral ExpertSuid- 5, without Metakaolin and with a fiber amount of approximately 1.3 kg / m^3. The maximum strength $f_{ck.cube.max}$=70.9 MPa was observed at the point with coordinates $x_1 = x_4 = 1$, $x_2 = -1$, $x_3 = 0.24$, which corresponds to compositions with the maximum amount of cement and Coral additive, without fiber and with the amount Metakaolin 18...19 kg / m^3.

6. Conclusions

The results of the experiment and recommendations for the manufacture and use of modified concrete for hard road surfaces are presented. An increase in the amount of Coral ExpertSuid-5 additive from 0.6 to 0.8% of the cement mass by 10…12% reduces the W / C ratio of the mixture. An increase in the amount of this additive of complex action to 1% causes a further, but less noticeable, decrease in the W / C ratio. Due to the use of a rational number of modifiers, the strength of the investigated concretes and fiber-reinforced concretes on the 3rd day, depending on the amount of cement in their composition, reaches from 40 to 50 MPa, and on the 7th day - from 45 to 55 MPa. That is, due to the use of modifiers, concrete has an increased early strength, which, as mentioned above, makes it possible to accelerate the start of operation of the road surface. With an increase in the amount of the additive of complex action Coral, both due to a decrease in the W / C ratio of the mixture, and due to the action of the additive as a hardening accelerator, the early strength of concretes increases, while the positive effect of the additive will be more noticeable for concretes and fiber-reinforced concretes with the addition of Metakaolin. In general, mixtures were researched using an increased amount of Coral ExpertSuid-5 additive (superplasticizer) and with a limitation of up to 20 kg / m^3 the amount of Metakaolin have a W / C ratio of no higher than 0.40, and a strength of at least 40 MPa, for 28 days, which allows use them for rigid road pavements.

The economic effect is achieved by increasing the service life and saving materials and repair costs. In the future, it is planned to continue research with greater variability.

7. References

[1] Afinogenov O P 2004 Design of non-rigid road surfaces *Kemerovo: Kuzbassvuzizdat.*
[2] Voznyy S P 2017 Prospects for the development of cement concrete roads in Ukraine *Modern technologies and methods of calculation in construction* no 6 pp 47-55
[3] Sobko Y M, Shevchuk H Y, Topilko N I and Novytskyy Y L 2018 Road cements based on modified additives of the new generation *Bulletin of the National University" Lviv Polytechnic". Series: Theory and Practice of Construction* (Lviv: Lviv Politics Publishing House) no 888 pp 133-137
[4] Chen Y, Cen GP and Cui YH 2018 Comparative study on the effect of synthetic fiber on the preparation and durability of airport pavement concrete *Construction and Building Materials* vol 184 pp 34-44 https://doi: 10.1016/j.conbuildmat.2018.06.223.
[5] Gupta S, Rao V V L K and Sengupta J 2008 Evaluation of polyester fiber reinforced concrete for use in cement concrete pavement works *Road materials and pavement design* vol 9 Issue 3 pp 441-461 https://doi: 10.3166/RMPD.9.441-461.
[6] Busari A, Akinmusuru J and Dahunsi B 2019 Mechanical properties of dehydroxylated kaolinitic clay in self-compacting concrete for pavement construction *Silicon* vol 11 Issue 5 pp 2429-2437 https://doi: 10.1007/s12633-017-9654-6.
[7] Salimi J, Ramezanianpour A M and Moradi M J 2020 Studying the effect of low reactivity metakaolin on free and restrained shrinkage of high-performance concrete *Journal of Building Engineering* vol. 28 101053 https://doi: 10.1016/j.jobe.2019.101053.
[8] Rabinovich F N 2004 Composites based on dispersed reinforced concrete. Questions of theory and design, technology, construction *Moscow:ASV.*
[9] Shen H R, She Y X and Gao P W 2012 The influence of polypropylene fiber on the performance of concrete pavement *Applied Mechanics and Materials* vol 178-181 pp 1099-1103 https://doi: 10.4028/www.scientific.net/amm.178-181.1099.
[10] Voznesensky V and Lyashenko T 1998 Experimental-statistical modelling in computational materials science *Odessa: Astroprint.*
[11] Voznesensky VA, Lyashenko TV and Ogarkov BL 1989 Numerical methods for solving construction and technological problems on a computer *Kyiv: Vysha shkola.*
[12] Mishutin A, Kos Z, Grynyova I and Chintea L 2021 Durability of Modified Fiber Concrete for Rigid Pavements *Croatian Regional Development Journal* Vol 1 No 2 pp 35-46. https://www.crdj.eu/wp-content/uploads/2021/03/03_CRDJ_vol2_no1_2021_pp_35-46.pdf

IOP Conf. Series: Materials Science and Engineering 1141 (2021) 012015 doi:10.1088/1757-899X/1141/1/012015

Seismic response of setback buildings with multiple underground levels

C V Gafincu[1], P Mihai[2], I Lungu[3]
[1] Ph.D student at the Faculty of Civil Engineering and Building Services, "Gheorghe Asachi" Technical University of Iaşi, 700050, Romania
[2] Department of Concrete Structures, Building Materials, Technology and Management, Faculty of Civil Engineering and Building Services, "Gheorghe Asachi" Technical University of Iaşi, 700050, Romania
[3] Department of Transportation Infrastructure and Foundations, Faculty of Civil Engineering and Building Services, "Gheorghe Asachi" Technical University of Iaşi, 700050, Romania

E-mail: cosminav.gafincu@tuiasi.ro

Abstract. The topic of this paper exceeds the provisions of the present regulations concerning the structural design issues related to the connection or the behaviour of the setback buildings on slopped ground surface. Setback buildings means that the spans progress in a step-like succession from one side of the building along the entire length. Due to the increased number of buildings positioned on sloping ground this paper aims to highlight aspects related to the influence of a variable number of underground levels. This research follows the differences between the loading effects acting on three types of setback buildings that differ by the number of the underground levels (one, two, respectively three), on the length of two spans. The conclusions of the paper refer to the influence of the number of underground levels on the seismic response of reinforced concrete structural elements: columns, beams and walls. The obtained results are compared in the form of internal forces: axial force, shear force and bending moment. The structural modelling and analysis of the buildings have been performed by using ETABS software. In the end these results will show the influence of the number of underground levels in the alteration of the structural elements above ground in terms of material consumption.

1. Introduction

The construction sites considered favourable for the stability of the constructions, with horizontal ground surfaces are less and less, resulting an increased number of buildings positioned on natural ground slopes.

To ensure the general stability of the construction site, the ideal situation is a ground surface with 0° slope that is rarely the real case, especially in the area of large urban agglomerations, which includes also the city of Iasi.

In many situations, due to the architecture imposed by asymmetrical shapes, the evaluation of the soil-structure interaction becomes complex, influenced by the variation of the mass and stiffness of the structure. In addition, for buildings located in a seismic zone, the values of internal forces: axial force, shear force and bending moment become significantly higher, regardless of the structural solution chosen.

The structural solutions specific for buildings on a slopped ground surface belong to one of the following categories:

- Setback buildings mean that the aboveground spans progress in a step-like succession from one side of the building along the entire length; in this case the footing is referred to a local 0° slope, upstream being proposed either a retaining structure or an embankment slope work (figure 1a);

- Stepback buildings mean that the underground spans progress in a step-like succession from one side of the building along the entire length (figure 1b);
- Stepback - setback buildings mean that the aboveground spans and the underground spans progress in a step-like succession from one side of the building along the entire length (figure 1c).

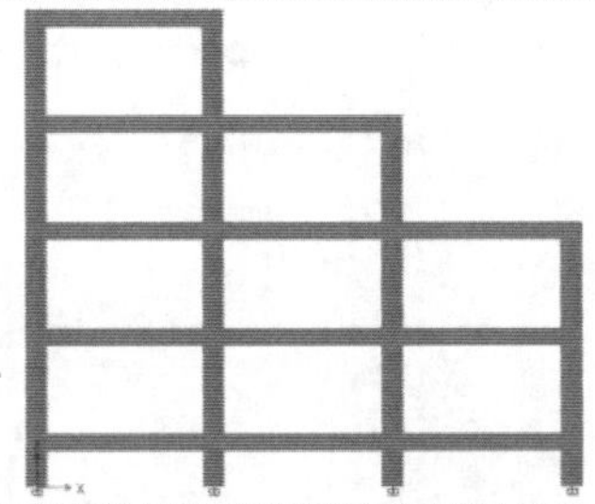

Figure 1a. Setback buildings.

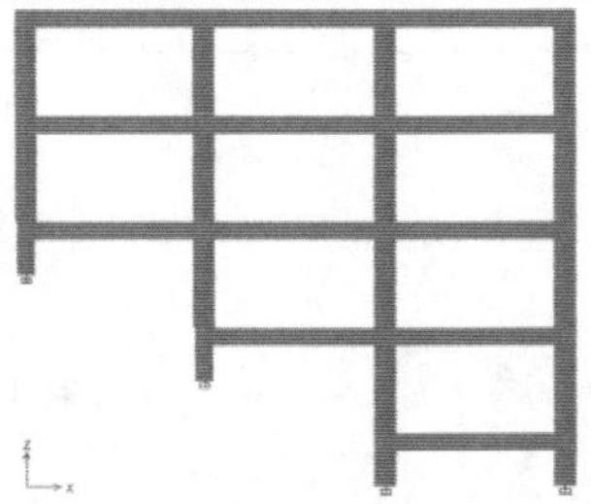

Figure 1b. Stepback buildings.

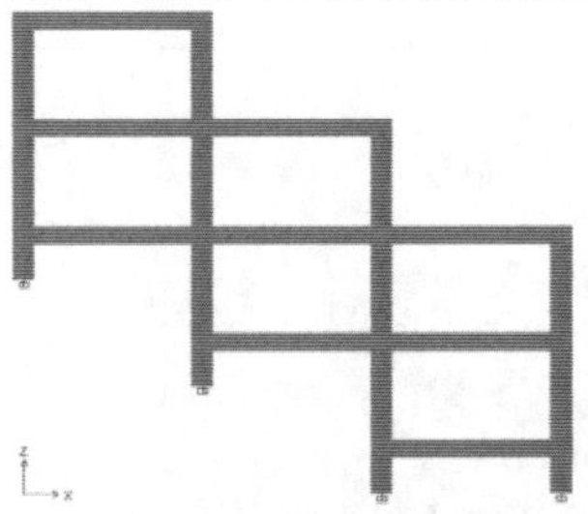

Figure 1c. Stepback - setback buildings.

The topic of this paper exceeds the provisions of the present regulations concerning the structural design issues related to the behaviour of the **setback buildings** on slopped ground surface.

2. General data

In order to demonstrate the efficiency of the **setback buildings**, a case study located in Iaşi was carried out on a building which has six spans of 4.00 m (SBI), 5.00 m (SBII) and 6.00 m (SBIII), presented schematically in figure 2.

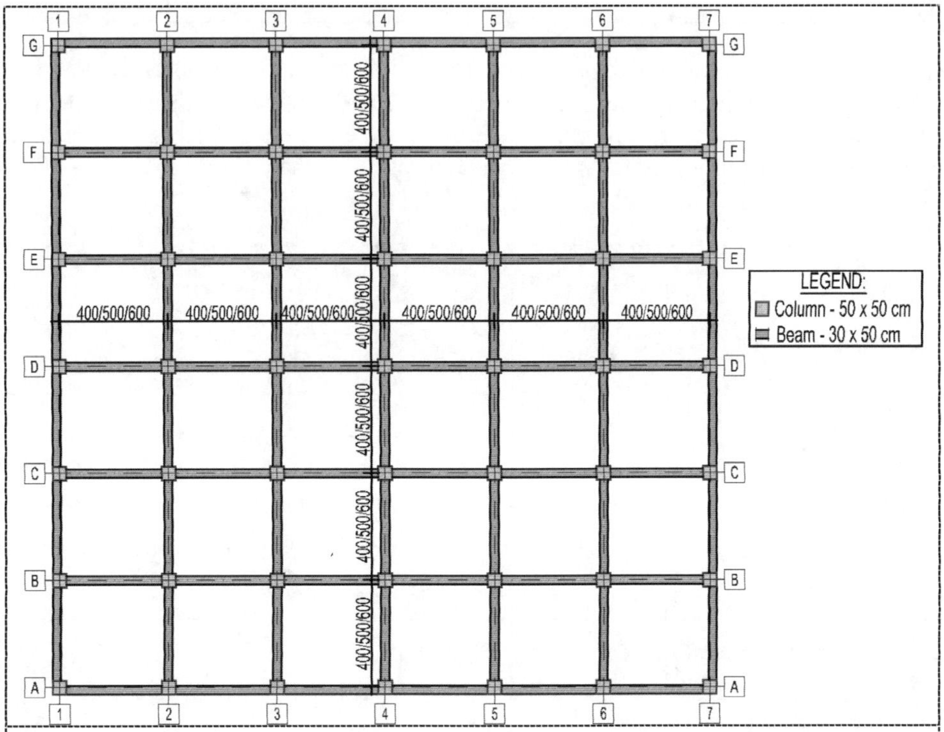

Figure 2. Floor plan – spans of 4.00 m (SBI), 5.00 m (SBII) and 6.00 m (SBIII).

The structure is made of class C20/25 reinforced concrete frame building, reinforced with independent concrete-steel bars BST500C. The columns have constant section along their entire height (50 x 50 cm) and are connected at the floor levels by reinforced concrete beams (30 x 50 cm).

Table 1. Data for buildings configuration

Title			Description		
Type of the building			SB I	SB II	SB III
Occupancy				residential buildings	
Floor height				3.00 m	
Spacing in X direction			4.00 m	5.00 m	6.00 m
Spacing in Y direction			4.00 m	5.00 m	6.00 m
Beam cross-section				30 x 50 cm	
Column cross-section				50 x 50 cm	
Slab thickness				15 cm	
Concrete wall				25 cm	
Raft foundation				1.00 m	
Floor loads	Floor	Permanent load		2.00 kN/m^2	
		Live load		2.00 kN/m^2	
	Roof	Permanent load		2.00 kN/m^2	
		Live load		2.00 kN/m^2	
		Snow load		2.00 kN/m^2	
Beam loads		Dividing walls		7 kN/m	
Strength class				C20/25	
Grade of steel				BST500C	

According to Romanian Design Code P100-1 / 2013 *"Seismic design code. Part 1. Design provisions for buildings"* (amended and supplemented in 2019), the site is situated in the seismic zone with the value of the peak design ground acceleration, determined for average recurrence interval IMR=225 years for ultimate limit state, $a_g = 0.25g$ (figure 3) and the corner period $T_c = 0.7$ sec (figure 4).

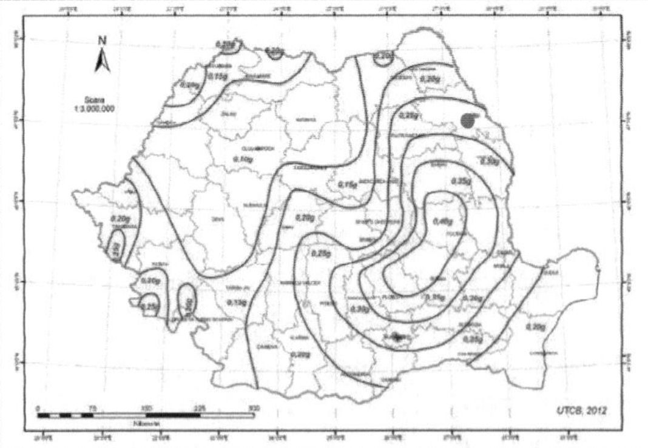

Figure 3. Peak ground acceleration according to P100/-2013.

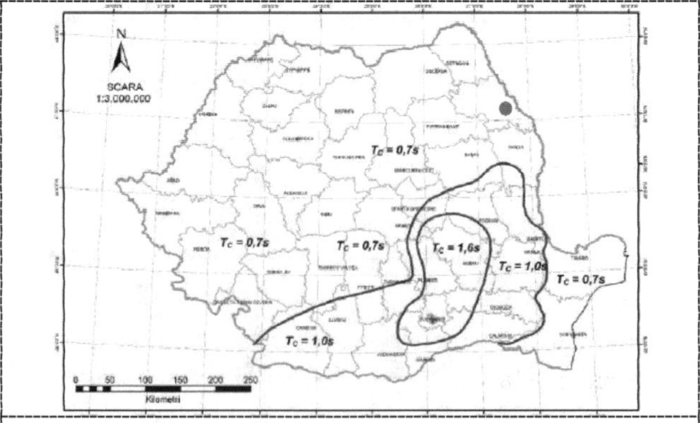

Figure 4. Corner period according to P100/-2013.

The structural modelling and analysis of the buildings have been performed by using ETABS software, using a spatial model defined by help of surface finite elements of shell type and linear elements of frame type (beam and column). According to P100-1 / 2013 amended and supplemented in 2019, for the analysed reinforced concrete structures, the behaviour factor is designed for high ductility class (table 2.). Table 3 presents the relevant data required to consider the soil-structure interaction for this case study.

Table 2. Behaviour factors

Type of the structure	q		
	DCH	*DCM*	*DCL*
frame structures, shear wall structures, dual structures	$5\ \alpha_u/\alpha_1$	$3{,}5\ \alpha_u/\alpha_1$	2,0

For common cases, the following approximate values of the α_u/α_1 ratio can be adopted:
(a) For frames or dual structures with predominant frames:

- buildings with one level: $\alpha_u/\alpha_1 = 1.15$;
- buildings with several levels and with a single span: $\alpha_u/\alpha_1 = 1.25$;
- *buildings with several levels and several spans:* $\alpha_u/\alpha_1 = 1.35$.

Table 3. Data for soil-structure interaction in the local seismic zone for Iasi city

Title	Description
Peak ground acceleration, a_g	0.25g
Corner period, T_c	0.70 sec
Importance factor, $\gamma_{I,e}$	1.00
Behaviour factor, q	6.75
Soil Type	sandy soil
Angle of internal friction of soil, φ	25°
Unit weight of soil, γ	20 kN/m³

The lateral influence of the soil at the basement level is addressed by at-rest earth pressure, which is an assumed state where the soil and the structure develop no displacement (figure 5, table 4). The soil type used for the analyses is non-cohesive with internal friction angle of 25° (medium compacted sand).

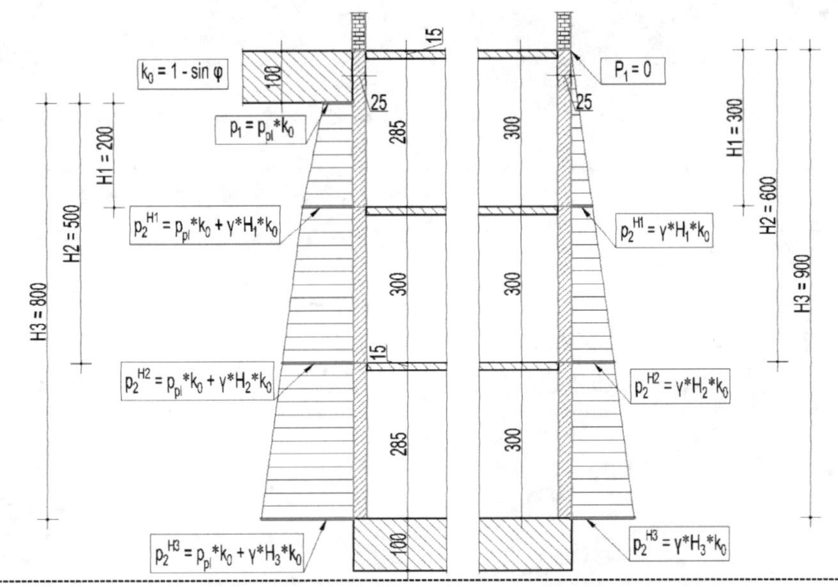

Figure 5. Earth pressure diagrams on the inner and outer basement walls.

Table 4. Calculation of lateral earth pressure

Lateral earth pressure on the interior wall: $p = p_{max} * k_0$ - at the top where ($p_{max} = p_{pl}$) $p = p_{max} * k_0 + \gamma * H * k_0$ - at the bottom			Lateral earth pressure on the exterior walls: $p = 0$ - at the top $p = \gamma * H * k_0$ - at the bottom		
Soil internal friction angle φ	25.00	°	Soil internal friction angle φ	25.00	°
Coefficient of earth pressure at rest: $k_0 = 1 - \sin \varphi$	**0.58**		Coefficient of earth pressure at rest: $k_0 = 1 - \sin \varphi$	**0.58**	
$p_1 = p_{pl} * k_0$	**47.78**	*kN/m²*	$p_1 = 0$	**0.00**	*kN/m²*
Height: H_1	2.00	*m*	Height: H_1	3.00	*m*
Height: H_2	5.00	*m*	Height: H_2	6.00	*m*
Height: H_3	8.00	*m*	Height: H_3	9.00	*m*
$p_2^{H1} = p_{pl} * k_0 + \gamma * H_1 * k_o$	**70.88**	*kN/m²*	$p_2^{H1} = \gamma * H_1 * k_o$	**34.65**	*kN/m²*
$p_2^{H2} = p_{pl} * k_0 + \gamma * H_2 * k_o$	**105.53**	*kN/m²*	$p_2^{H2} = \gamma * H_2 * k_o$	**69.31**	*kN/m²*
$p_2^{H3} = p_{pl} * k_0 + \gamma * H_3 * k_o$	**140.19**	*kN/m²*	$p_2^{H3} = \gamma * H_3 * k_o$	**103.96**	*kN/m²*

3. Numerical analysis

A number of nine alternative buildings are analysed in ETABS software, six spans of 4.00 m (SBI), 5.00 m (SBII) and 6.00 m (SBIII), that differ by the number of the underground levels (one, two, respectively three), on the length of two spans (figures 6, 7 and 8). The characteristics of the structure configurations are summarized in table 1.

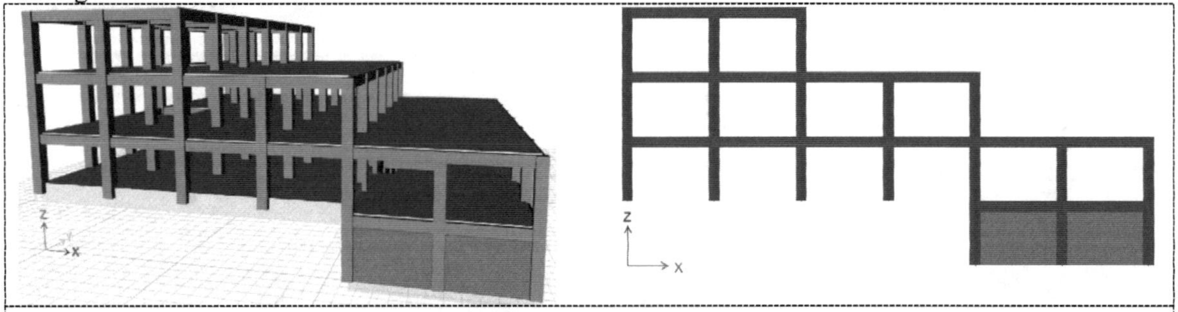

Figure 6. Building with one underground level.

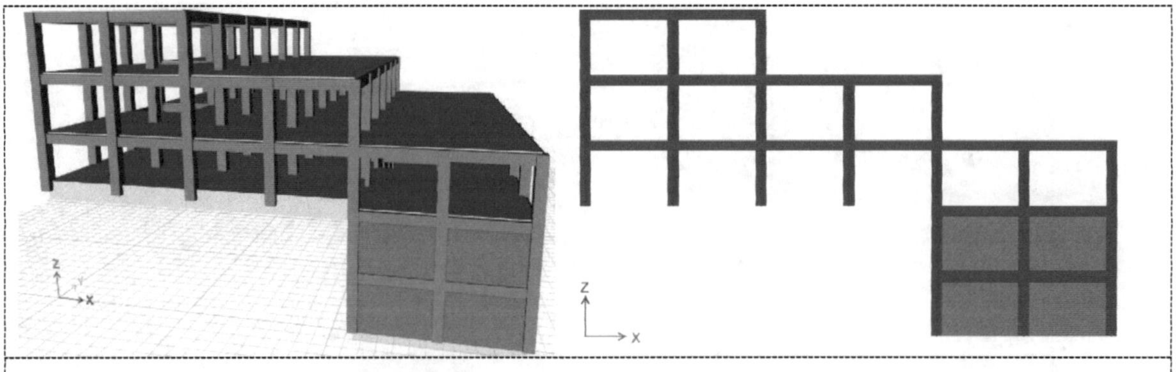

Figure 7. Building with two underground levels.

IOP Conf. Series: Materials Science and Engineering 1141 (2021) 012015 doi:10.1088/1757-899X/1141/1/012015

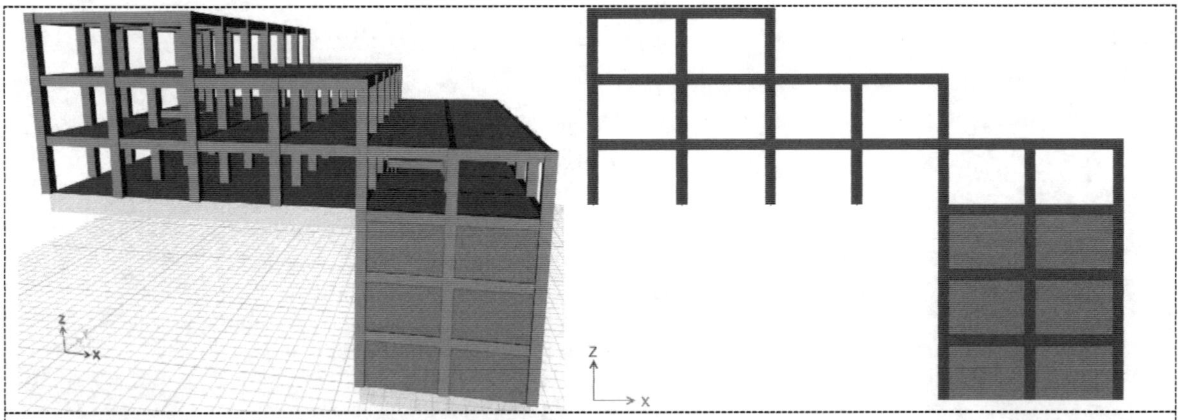

Figure 8. Building with three underground levels.

4. Results and discussions

The obtained results are compared in the form of internal forces: axial force, shear force and bending moment on relevant structural elements.

The first step is to identify the maximum values of the internal forces. These are extracted from the envelope combination type, which is overlapping values resulting from ten load combinations (8 seismic combinations, one fundamental combination and geotechnical loads combination = ENVE). The maximum values of the internal forces were identified for the column positioned at the intersection of the 5 – F axes (ground floor), for the beams situated between axes 5 – 6 / F (ground floor), 4 – 5 / F (first floor), 2 / G – F (second floor) and for the wall situated between axes 5 / C – D (basement 1), in all nine structural models (figures 9, 10, 11, 12).

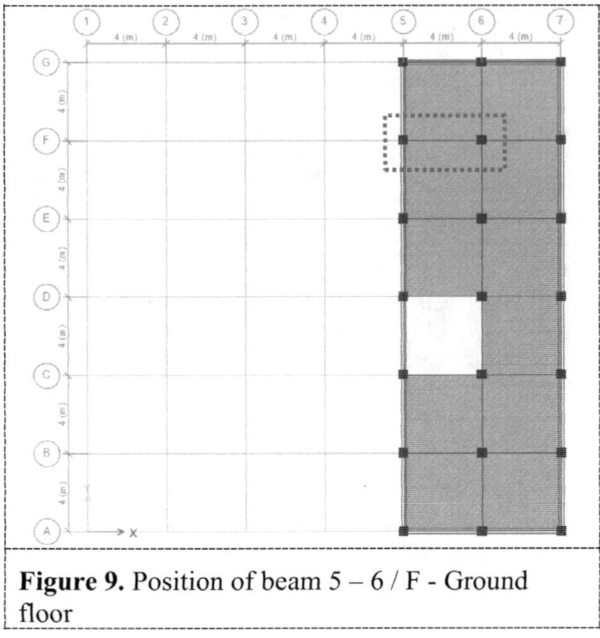

Figure 9. Position of beam 5 – 6 / F - Ground floor

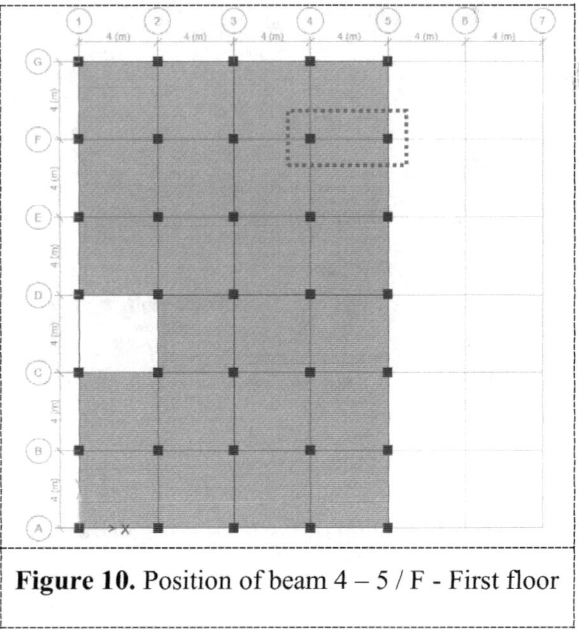

Figure 10. Position of beam 4 – 5 / F - First floor

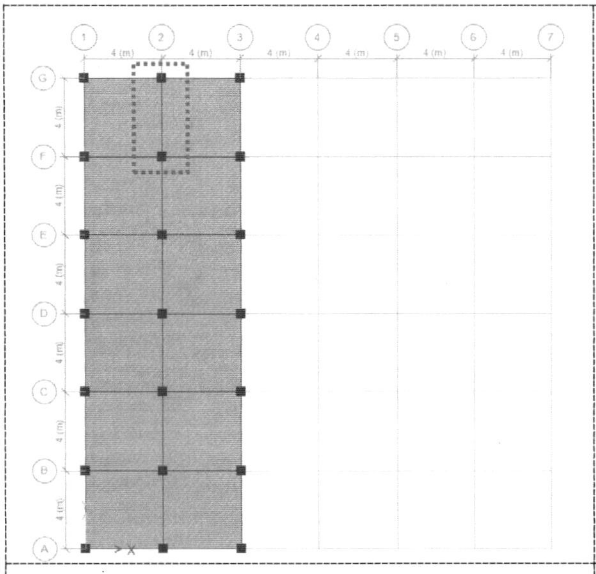

Figure 11. Position of beam 2 / G – F - Second floor

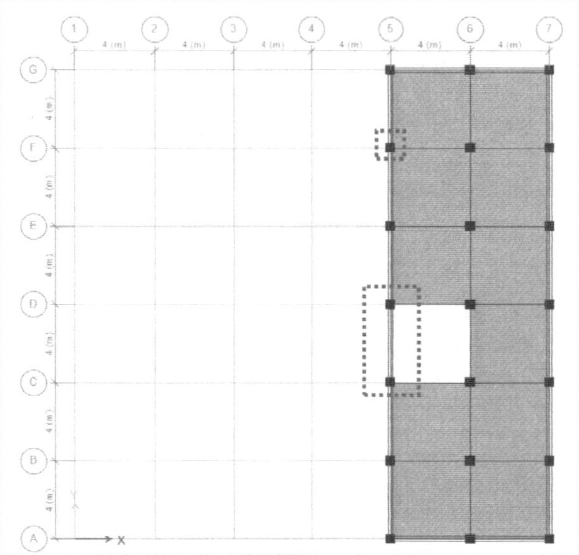

Figure 12. Position of column 5 – F - Ground floor and position of wall 5 / C – D - Basement 1

4.1. Internal forces in the beams

Figures 13, 14 and 15 show that for each analysed situation, the values of internal forces do not present relevant variations, in sense of increase or decrease.

It is noted that the percentage variation (from three underground levels building to two underground level building and from two underground levels building to one underground level building) for shear forces and bending moments, is between 0.01 % and 7.53 %.

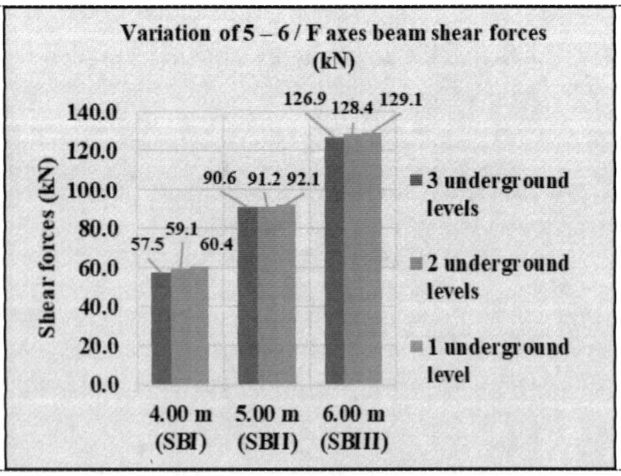

Figure 13a. Variation of 5 – 6 / F axes beam shear forces.

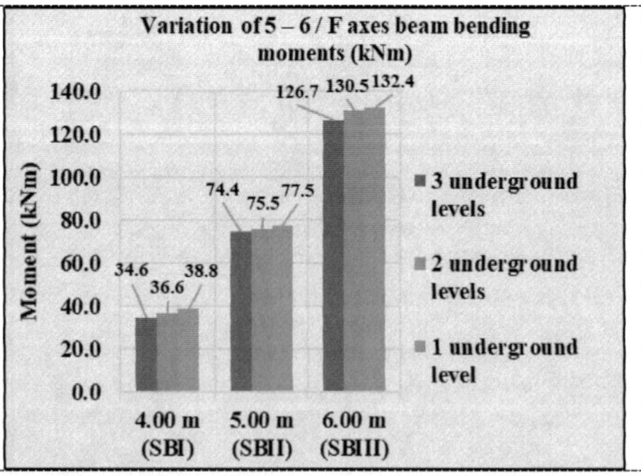

Figure 13b. Variation of 5 – 6 / F axes beam bending moments.

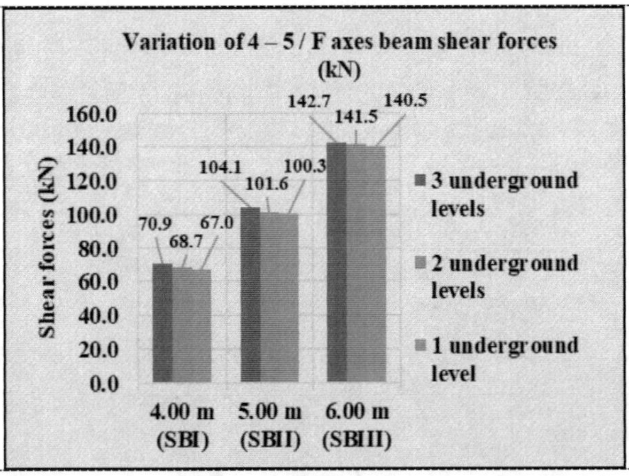

Figure 14a. Variation of 4 – 5 / F axes beam shear forces.

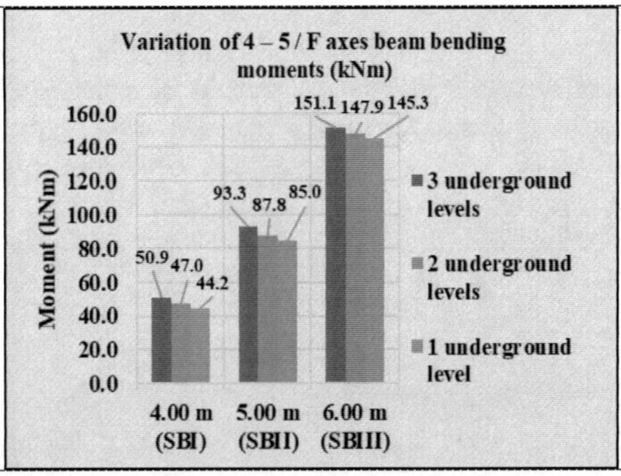

Figure 14b. Variation of 4 – 5 / F axes beam bending moments.

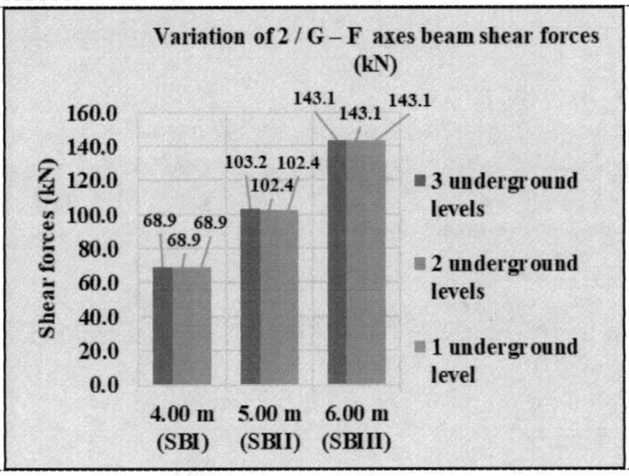

Figure 15a. Variation of 2 / G – F axes beam shear forces.

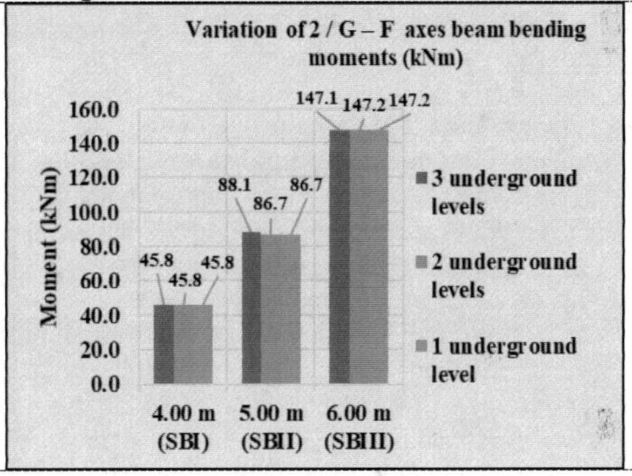

Figure 15b. Variation of 2 / G – F axes beam bending moments.

4.2. Internal forces in the column

For the analysed cases (SBI, SBII, SBIII), from three underground levels building to two underground level building and from two underground levels building to one underground level building, it is showing the following conclusions:

- the value of axial forces increases once the number of underground levels decreases and the percentage variation for axial forces is less than 1.00 % (figure 16a);
- the value of bending moments decreases once the number of underground levels decreases and the percentage variation for bending moments is between 4.17 % and 14.19 % (figure 16b).

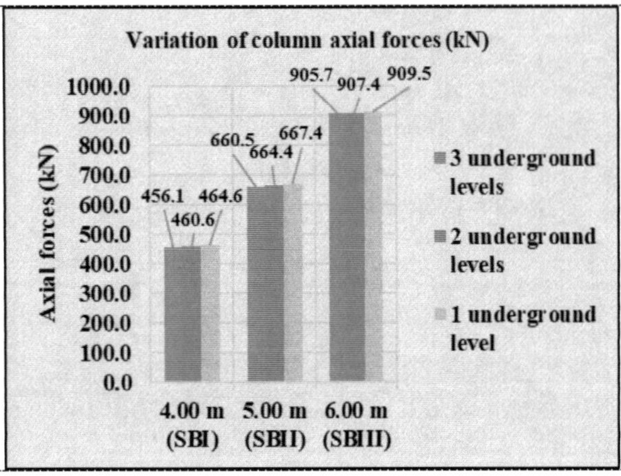

Figure 16a. Variation of 5 – F axes column axial forces

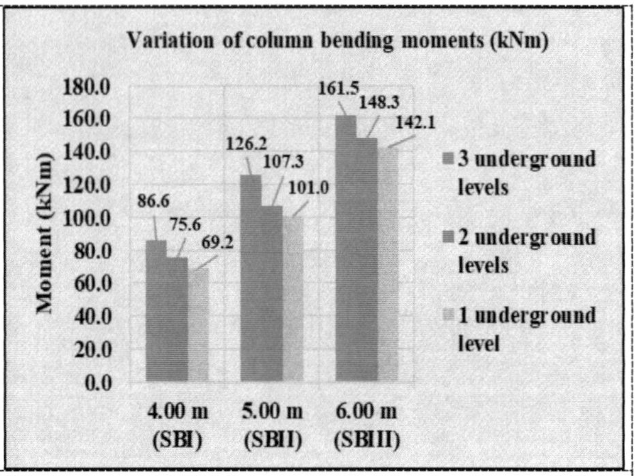

Figure 16b. Variation of 5 – F axes column bending moments

4.3. Internal forces in the underground wall

For the analysed cases (SBI, SBII, SBIII) the following conclusions are relevant:

(a) the values of M11 moments (figure 17a):

- from three underground levels building to two underground level building, the values of M11 moments decrease by approximately 22.22 % – 45.71 %;
- from two underground levels building to one underground level building, the values of M11 moments decrease by approximately 78.57 % – 84.21 %.

(b) the values of M22 moments (figure 17b):

- from three underground levels building to two underground level building, the values of M11 moments decrease by approximately 16.08 % – 29.29 %;
- from two underground levels building to one underground level building, the values of M11 moments decrease by approximately 43.75 % – 65.66 %.

Figure 17a. and figure 17b. show that for all nine analysed situations, the values of M11 and M22 moments decreases with the decrease of the number of underground levels.

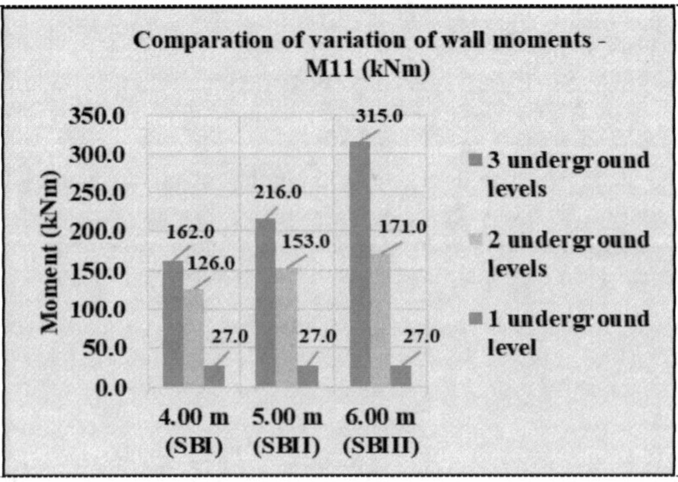

Figure 17a. Variation of M11 moments for 5 / C – D axes wall

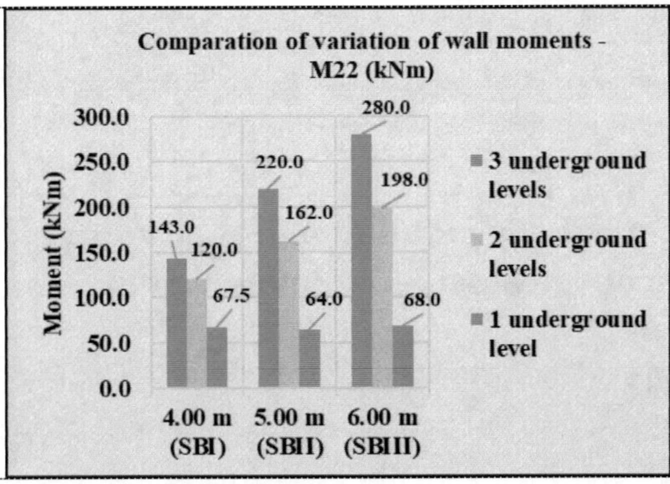

Figure 17b. Variation of M22 moments for 5 / C – D axes wall

5. Conclusions

The following conclusions can be drawn from the results obtained for the buildings presented above:

- the increase or the decrease of internal forces directly influences the material consumption;
- the number of underground levels is affecting in a small manner the amount of materials required for the superstructure of the building (beams and columns);
- the values of internal forces for underground walls present a decrease which is directly proportional with the number of underground levels; thus, for buildings with the same number of underground levels, the internal forces increase when the values of spans increase; also, for the building with the same spans values (SBI, SBII, SBIII), the bending moments decrease when the number of underground levels decrease;
- the number of underground levels should be optimized based on multiple: required functionality, the expected structural response, availability of the materials required, available budget in terms of cost and time, etc.;
- for the building with one underground level, the maximum values of the internal forces are obtained from a seismic combination, while for the building with two and three underground levels the maximum values of the internal forces are obtained from the load combination including the lateral earth pressures.

References

[1] Stratan A 2014 *Structural dynamics and seismic engineering.* Course notes, Timişoara
[2] Norm NP 112 / 2014: *Norm for the design of surface foundations*
[3] Norm NP 124 / 2010: *Norm for geotechnical design of abutment constructions*
[4] SR EN 1997-1, Eurocode 7: *Geotechnical Design. Part 1: General rules*
[5] Norm P100-1 / 2013: *Seismic design code. Part 1. Design provisions for buildings (amended and supplemented by order 2956/2019)*
[6] Indicative CR 0 - 2012, *Design code. Basics of construction design*

IOP Conf. Series: Materials Science and Engineering 1141 (2021) 012016 doi:10.1088/1757-899X/1141/1/012016

Effects of chemical degradation on longitudinal fracture of non-homogeneous beams: an analytical study

V Rizov[1]

[1]Department of Technical Mechanics, University of Architecture, Civil Engineering and Geodesy, 1046 - Sofia, Bulgaria

E-mail: v_rizov_fhe@uacg.bg

Abstract. An analytical study of the chemical degradation of the material and its effects of on longitudinal fracture behaviour of continuously non-homogeneous cantilever beam is carried-out. The chemical degradation is described by decreasing of the elasticity modulus of the non-homogeneous material with time. The fracture is studied by analyzing of the strain energy release rate. For this purpose, the strain energy release rate is derived as a function of time with taking into account the chemical degradation and the non-homogeneity of the material. The time-dependent fracture is analyzed also by applying the *J*-integral with considering the chemical degradation for verification. The variation of the strain energy release rate with the time due to the chemical degradation is studied and discussed in detail.

1. Introduction

Continuously non-homogeneous materials are used very often in various load-bearing structural applications in modern engineering. These materials exhibit continuous variation of their properties in one or more coordinates in the structural member. Hence, the material properties (the modulus of elasticity in this paper) are smooth functions of the coordinates. The scientific literature shows that among the various kinds of non-homogeneous materials, functionally graded materials have attracted significant attention in the last decades [1, 2, 3, 4, 5, 6, 7, 8, 9, 10, 11]. In fact, the interest in continuously non-homogeneous materials is due in a high degree to the extensive utilization of functionally graded materials in advanced engineering technologies. The functionally graded materials are continuously non-homogeneous composites made usually of two or more phases. Gradual change of microstructure and material properties of functionally graded materials is formed by gradual variation of volume fractions of the constituent materials during manufacturing.

The analysis of fracture behaviour of continuously non-homogeneous (functionally graded) materials plays an important role in safety design of engineering structures. The objective of this paper is to analyze the effect of chemical degradation of material on longitudinal fracture in continuously non-homogeneous beam structures since this problem has not been studied in the previous publications [12, 13, 14]. One of the motives for carrying-out this analysis is that during their lifetime, continuously non-homogeneous beams are often exposed to the influence of aggressive chemical agents such as acid rain attack, carbonation, sulphate attack.

2. Non-homogeneous beam with longitudinal crack under chemical degradation

The continuously non-homogeneous cantilever beam subjected to chemical degradation is under consideration (figure 1). The cross-section of the beam is a rectangle with dimensions, $b \times h$. The beam has length, l. The static scheme of the beam is cantilever with clamping in section, D. A longitudinal vertical crack of length , a, is located arbitrary in the beam as displayed in figure 1. The cross-sections of the right-hand and left-hand crack arms are rectangles with different widths designated by b_1 and b_2, respectively. Besides the chemical degradation, the beam structures is loaded also by external loading that includes two bending moments, M_1 and M_2, applied at the free ends of the two crack arms, as displayed in figure 1.

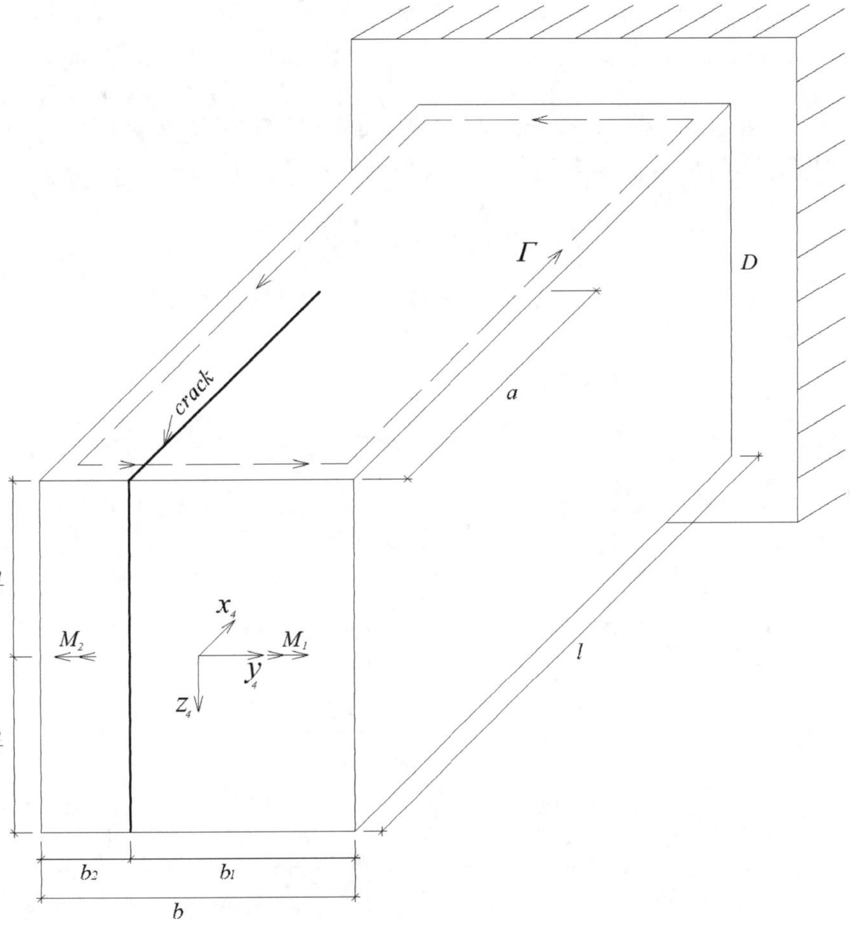

Figure 1. Geometry and loading of the non-homogeneous cantilever beam.

The beam is made of material that is continuously non-homogeneous in both thickness and length directions. The distribution of the elasticity modulus, E, along the beam thickness is written as

$$E = E_U + \frac{E_L - E_U}{h^n}\left(\frac{h}{2} + z_4\right)^n \tag{1}$$

where $-h/2 \leq z_4 \leq h/2$, E_U and E_L are, respectively, the values of the modulus of elasticity at the upper and lower surfaces of the beam, n is a material property that controls the variation of the modulus of elasticity along the thickness. The distributions of E_U and E_L along the length of the beam are expressed as

$$E_U = E_{UF} + \frac{E_{UC} - E_{UF}}{l^m} x_4^m \tag{2}$$

$$E_L = E_{LF} + \frac{E_{LC} - E_{LF}}{l^f} x_4^f \tag{3}$$

where $0 \leq x_4 \leq l$, E_{UF} and E_{UC} are, respectively, the values of E_U at the free and clamped ends of the beam, E_{LF} and E_{LC} are, respectively, the values of E_L at the free and clamped ends of the beam, m and f control the variations of E_U and E_L along the beam length, respectively.

The upper and lower surfaces of the structure are exposed to influence of aggressive chemical agent. The chemical degradation of the material leads to decrease of the elasticity modulus with time. The decrease of E_{UF}, E_{UC}, E_{LF} and E_{LC} are expressed as [15]

$$E_{UF} = E_{UF0} e^{-\beta_{UF} t - \psi_{UF} \sqrt{t}} \tag{4}$$

$$E_{LF} = E_{LF0} e^{-\beta_{LF} t - \psi_{LF} \sqrt{t}} \tag{5}$$

$$E_{UC} = E_{UC0} e^{-\beta_{UC} t - \psi_{UC} \sqrt{t}} \tag{6}$$

$$E_{LC} = E_{LC0} e^{-\beta_{LC} t - \psi_{UC} \sqrt{t}} \tag{7}$$

where t is time, β_{UF}, ψ_{UF}, β_{LF}, ψ_{LF}, β_{UC}, ψ_{UC}, β_{LC} and ψ_{LC} are material properties which control the change of the modulii of elasticity with time, E_{UF0}, E_{UC0}, E_{LF0} and E_{LC0} are, respectively, the values of E_{UF}, E_{UC}, E_{LF} and E_{LC} at $t = 0$.

The beam under chemical degradation exhibits non-linear elastic behaviour that is described by the following non-linear stress-strain relationship [15]:

$$\sigma = \frac{E\varepsilon}{\sqrt{1 + \varepsilon^2}} \tag{8}$$

where σ is the stress, ε is the strain. In fact, (8) is a stress-strain-time relationship since the elasticity modulus varies with time as a result of the chemical degradation of the material.

The longitudinal fracture in the beam under chemical degradation is time-dependent and is studied by analysing of the strain energy release rate. In this relation, a time-dependent analysis of the strain energy release rate, G, that accounts for the chemical degradation of the material is developed by considering of the balance of the energy

$$M_1 \delta\varphi_1 + M_2 \delta\varphi_2 = \frac{\partial U}{\partial a} \delta a + G h \delta a \tag{9}$$

where φ_1 and φ_2 are the rotation angles of the free ends of the two crack arms, δa is a small increase of the crack length, U is the strain energy in the beam. From (9), one obtains

$$G = \frac{M_1}{h} \frac{\partial \varphi_1}{\partial a} + \frac{M_2}{h} \frac{\partial \varphi_2}{\partial a} - \frac{1}{h} \frac{\partial U}{\partial a} \tag{10}$$

The strain energy is found as

$$U = \iiint_{(V_1)} u_{01} dV + \iiint_{(V_2)} u_{02} dV + \iiint_{(V_3)} u_{03} dV \tag{11}$$

where u_{01}, u_{02} and u_{03} are, respectively, the strain energy densities in the two crack arms and in the un-cracked beam portion, V_1, V_2 and V_2 are the volumes of the right-hand and left-hand crack arms and in the un-cracked beam portion, respectively. The density of the strain energy in the right-hand crack arm under chemical degradation is obtained as [14]

$$u_{01} = E\left(\sqrt{1+\varepsilon^2} - 1 \right) \tag{12}$$

It should be specified that (12) is time-dependent because the elasticity modulus changes continuously with time due to the chemical degradation according to (4), (5), (6) and (7). The Bernoulli's hypothesis for plane sections is applied for describing the distribution of the strains since beams of high length to thickness ratio is considered in the present paper. Thus, the distribution of strains in direction of the thickness of the right-hand crack arm is written as

$$\varepsilon = \kappa_1 (z_1 - z_{1n}) \tag{13}$$

where the curvature, κ_1, and the coordinate of the neutral axis, z_{1n}, are determined by using the following equations of equilibrium of elementary forces in the cross-section of the right-hand crack arm:

$$N_1 = \iint_{(A_1)} \sigma da, \quad M_1 = \iint_{(A_1)} \sigma z_1 da \tag{14}$$

where A_1 is the area of the cross-section, N_1 is the axial force, z_1 is the vertical centroidal axis of the cross-section. Apparently, $N_1 = 0$ (figure 1).

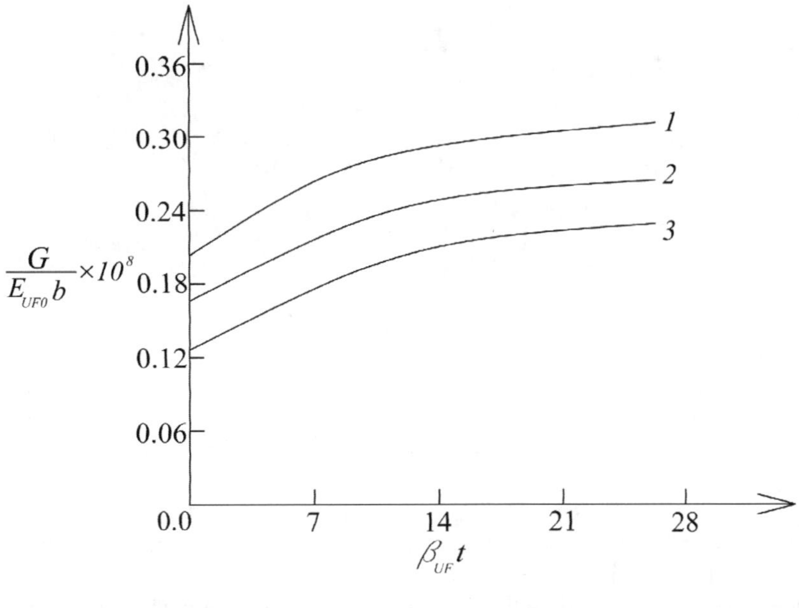

Figure 2. The non-dimensional strain energy release rate displayed as a function of non-dimensional time (curve 1 – at $a/l = 0.3$, curve 2 – at $a/l = 0.5$ and curve 3 – at $a/l = 0.7$).

After substituting of σ in (14), the two equations are solved with respect to κ_1 and z_{1n} at various values of time by using the MatLab computer program. The curvatures, the neutral axis coordinates and the strain energy densities in the left-hand crack arm and in the un-cracked beam portion are found in analogical way.

The angles of rotation which are involved in (10) are obtained by using the integrals of Maxwell-Mohr

$$\varphi_1 = \int_0^a k_1 dx_3 + \int_a^l k_3 dx_3 , \; \varphi_2 = \int_0^a k_2 dx_3 + \int_a^l k_3 dx_3 \tag{15}$$

where κ_2 and κ_3 are the curvatures of the left-hand crack arm and the un-cracked beam portion, respectively.

Equation (10) represents a solution of the strain energy release rate that depends on time since the elasticity modulus is function of time through (4), (5), (6) and (7) when the beam is under chemical degradation. Thus, (10) can be used to calculate the strain energy release rate at various values of time.

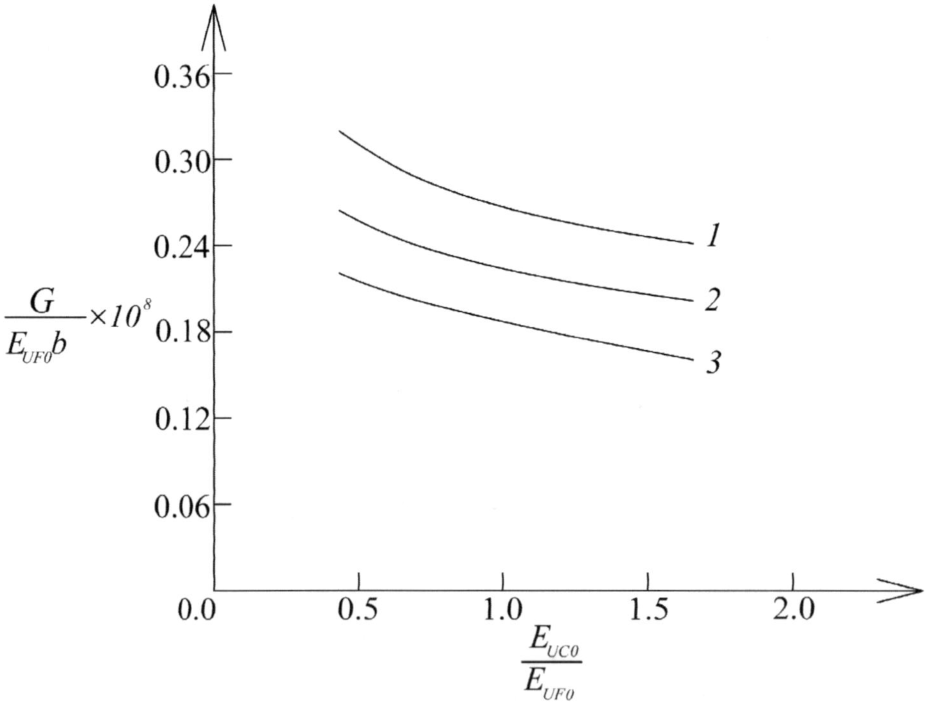

Figure 3. The non-dimensional strain energy release rate displayed as a function of E_{UC0} / E_{UF0} ratio (curve 1 – at $E_{LC0} / E_{LF0} = 0.5$, curve 2 – at $E_{LC0} / E_{LF0} = 1.0$ and curve 3 – at $E_{LC0} / E_{LF0} = 2.0$).

The *J*-integral approach with considering the chemical degradation of the material is applied to verify the strain energy release rate determined by using equation (10). In this relation, the *J*-integral is solved along the integration contour, Γ, shown by a dashed line in figure 1. It should be noted that the *J*-integral value is exact match of the strain energy release rate calculated by applying (10). This fact proves the correctness of the time-dependent analysis of the strain energy release rate that accounts for the chemical degradation of the material.

3. Parametric study

The aim of the parametric study is to evaluate the effects of the chemical degradation of the material on the longitudinal fracture behaviour of the continuously non-homogeneous beam shown in figure 1. For this purpose, calculations of the strain energy release rate are carried-out by applying the time-dependent solution (10). The strain energy release rate that takes into account the chemical degradation is presented in non-dimensional form by using the formula $G_N = G/(E_{UF0}b)$. The variation of the strain energy release rate with time induced by the chemical degradation is analysed. The influences of the location and length of the crack and the material non-homogeneity in the thickness and length directions on strain energy release rate under chemical degradation are analysed too. The following data are used: $b = 0.010$ m, $h = 0.012$ m, $l = 0.150$ m, $n = 0.7$, $m = 0.7$, $f = 0.7$, $M_1 = 6$ Nm and $M_2 = 4$ Nm.

The effect of the chemical degradation of the material on the longitudinal fracture in the beam is illustrated in figure 2 where the non-dimensional strain energy release rate is plotted against the non-dimensional time for three a/l ratios (the a/l ratio characterizes the relative length of the crack). It is evident from figure 2 that the chemical degradation leads to increase of the strain energy release rate with time. One can observe also in figure 2 that when a/l ratio increases, the strain energy release rate decreases.

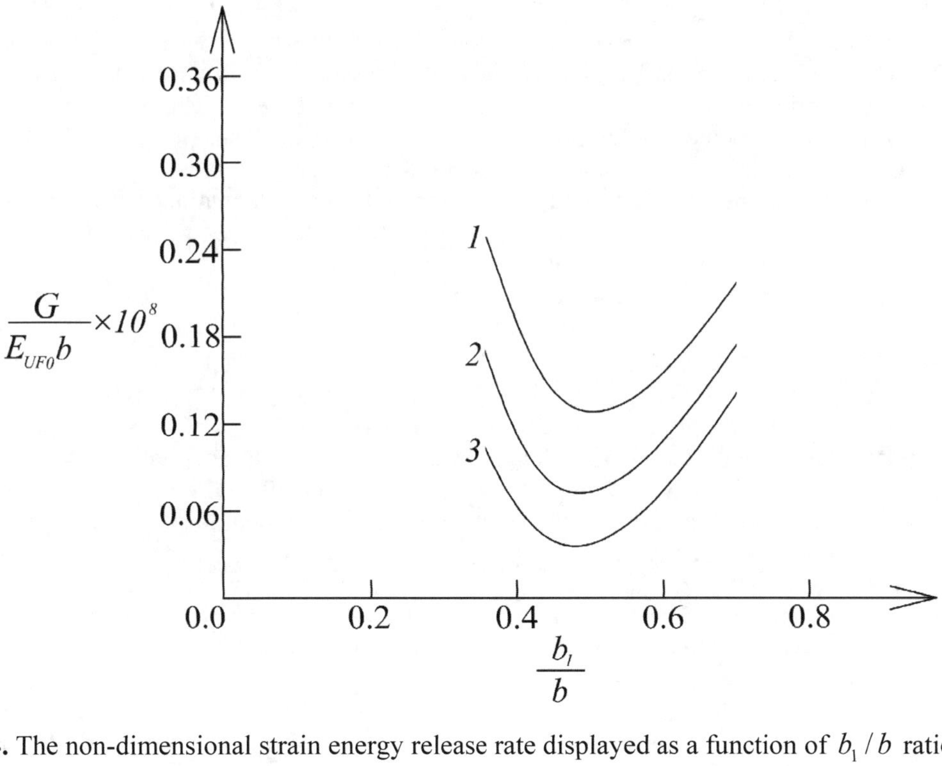

Figure 4. The non-dimensional strain energy release rate displayed as a function of b_1/b ratio (curve 1 – at $E_{LF0}/E_{UF0} = 0.5$, curve 2 – at $E_{LF0}/E_{UF0} = 1.0$ and curve 3 – at $E_{LF0}/E_{UF0} = 2.0$).

In order to evaluate how the material non-homogeneity along the length of the beam influences the longitudinal fracture behaviour under chemical degradation, the non-dimensional strain the energy release rate is displayed as a function of E_{UC0}/E_{UF0} ratio in figure 3 at three E_{LC0}/E_{LF0} ratios. The curves in figure 3 indicate that when E_{UC0}/E_{UF0} ratio increases, the strain energy release rate decreases. It can also be observed in figure 3 that the increase of E_{UC0} / E_{UF0} ratio leads to decrease of the strain energy release rate when the beam is subjected to chemical degradation.

The influences of the crack location in the width direction and the material non-homogeneity in the thickness directions on the longitudinal fracture under chemical degradation are also evaluated. In this relation, the non-dimensional strain energy release rate is plotted against b_1 / b ratio in figure 4 for three E_{LF0} / E_{UF0} ratios. The curves in figure 4 show that the strain energy release rate (when the beam is exposed to chemical degradation) is minimal at $b_1 / b = 0.44$. It can be seen in figure 4 that when E_{LF0} / E_{UF0} ratio increases, the strain energy release rate under chemical degradation decreases gradually.

4. Conclusions

The chemical degradation of the material and its effects on the strain energy release rate in a continuously non-homogeneous beam are analysed. The chemical degradation is described by decreasing the elasticity modulus with time. A time-dependent analysis of the strain energy release rate that accounts for the chemical degradation is developed. The *J*-integral approach with considering of the chemical degradation is used to verify the solution. The variation of the strain energy release rate with time due to the chemical degradation of the material is studied. The calculations performed show that the chemical degradation leads to increase of the strain energy release rate with time. The analysis reveals that when a / l, E_{UC0} / E_{UF0}, E_{LC0} / E_{LF0} and E_{LF0} / E_{UF0} ratios increase, the strain energy release rate in the beam under chemical degradation decreases. In general, the results obtained indicate that the strain energy release rate increases as a result of the chemical degradation. This finding is in a good agreement with prior studies [15]. The analysis presented in this paper is applicable mainly for functionally graded concrete beams. The degradation of concrete due to chemical causes such as sulphate reactions, alkali-reaction and carbonation affects the mechanical properties of concrete. Fracture behaviour of concrete is also deteriorated [16]. In this context, the paper helps for understanding of the influence of chemical degradation on longitudinal fracture behaviour of the functionally graded concrete beams.

5. References

[1] Naebe M and Shirvanimoghaddam K 2016 Functionally graded materials: A review of fabrication and properties *Applied materials today* **5** pp 223-245
[2] Udupa G, Shrikantha Rao S and Gangadharan K V 2014 Functionally Graded Composite Materials: An Overview *Procedia Materals Science* **5** pp 1291- 1299
[3] Kieback B, Neubrand A and Riedel H 2003 Processing techniques for functionally graded materials. *Materials Science and Engineering*: A **362** pp 81-106
[4] Müller E, Drašar Č, Schilz J and Kaysser W 2003 Functionally graded materials for sensor and energy applications *Materials Science and Engineering: A* **362** pp 17-39
[5] Pei Y T, De Hosson J Th M 2000 Functionally graded materials produced by laser cladding *Acta Materialia* **48** pp 2617-2624
[6] Robert C, Wetherhold Steven Seelman and Jianzhong Wang 1996 The use of functionally graded materials to eliminate or control thermal deformation *Composites Science and Technology* **56** pp 1099-1104
[7] Sola A, Bellucci D and Cannillo V 2016 Functionally graded materials for orthopedic applications – an update on design and manufacturing. *Biotechnology Advances* **34** pp 504-531
[8] Sobczak J and Drenchev L 2013 Metallic Functionally Graded Materials: A Specific Class of Advanced Composites *Journal of Materials Science & Technology* **4** pp 297-316
[9] Mortensen A and Suresh S 1995 Functionally graded metals and metal-ceramic composites: Part 1 Processing. *International Materials Reviews* **40** 239-265

[10] Groves J and Wadley H 1997 Fabrication of Functionally graded materials synthesis via low vacuum directed vapor deposition *Composites Part B: Engineering* **28** pp 57-69
[11] Chen Y and Lin X 2008 Elastic analysis for thick cylinders and spherical pressure vessels made of functionally graded materials. *Computational Materials Science* **44** pp 581-581
[12] Rizov V 2020 Analysis of Two Lengthwise Cracks in a Viscoelastic Inhomogeneous Beam Structure *Engineering Transactions* **68** pp 397-415
[13] Rizov V 2020 Viscoelastic inhomogeneous beams under step loading: a longitudinal fracture analysis. *IOP Conf. Ser.: Mater. Sci. Eng.***1031** 012008
[14] Rizov V 2020 Longitudinal fracture analysis of inhomogeneous beams with continuously changing radius of cross-section along the beam length *Strength, Fracture and Complexity: an International Journal* **13** pp 31-43
[15] Kishkilov M and Apostolov R 1994 *Introduction to theory of plasticity*, UACEG
[16] Wang Z, Zhu Z, Sun X, Wang X 2017 Deterioration of fracture toughness of concrete under acid rain environment Engineering Failure Analysis 77 pp 76-84.

IOP Conf. Series: Materials Science and Engineering 1141 (2021) 012017 doi:10.1088/1757-899X/1141/1/012017

Delamination analysis of multilayered viscoelastic beam exposed to chemical degradation

V Rizov[1]

[1]Department of Technical Mechanics, University of Architecture, Civil Engineering and Geodesy, 1046 - Sofia, Bulgaria

E-mail: v_rizov_fhe@uacg.bg

Abstract. Delamination of multilayered viscoelastic non-homogeneous beam structure subjected to chemical degradation is analysed in this paper. The creep behaviour is treated by using a linear viscoelastic model. The influence of chemical degradation is taken into account by decreasing the elasticity modulus and the coefficient of viscosity with time. The time-dependent delamination fracture under chemical degradation is studied by analysing the strain energy release rate. The strain energy is used to obtain the strain energy release rate that accounts for the creep and for the chemical degradation. The solution is verified by applying the compliance method with considering of the chemical degradation.

1. Introduction

Multilayered non-homogeneous structural materials are made of adhesively bonded layers of different non-homogeneous materials (for example, layers of fibre reinforced polymer composites). Very modern non-homogeneous materials are functionally graded materials [1, 2, 3, 4, 5]. These novel materials have a wide application in current engineering mainly due to their specific behaviour involving high strength-to-weight and stiffness-to-weight ratios [6, 7, 8, 9, 10, 11]. Therefore, multilayered non-homogeneous materials are used in advanced load-bearing structures where low weight is of special significance [12, 13, 14, 15, 16, 17].

It should be noted, however, that multilayered materials are employed with caution because of their poor delamination fracture behaviour [18, 19, 20, 21, 22, 23, 24, 25]. The delamination cracks drastically reduce the load-bearing capacity of the multilayered structure. Thus, the safety and reliability of multilayered non-homogeneous constructions depend largely on their delamination fracture behaviour.

The goal of the paper is to analyse the delamination fracture of a multilayered non-homogeneous viscoelastic beam configuration that is exposed to influence of aggressive chemical agent. It should be mentioned that such analysis is important since multilayered beam structures frequently are in contact with various chemical substances in their life-time. It should also be mentioned that previous publications do not consider the influence of chemical degradation on time-dependent delamination of non-homogeneous viscoelastic beam structural components [26, 27, 28]. Therefore, a time-dependent analytical expression of the strain energy release rate that accounts for the chemical degradation of the viscoelastic multilayered non-homogeneous material in a delaminated beam is presented in the paper.

2. Influence of chemical degradation on delamination of multilayered viscoelastic beam

The delamination crack in the multilayered non-homogeneous viscoelastic beam shown in figure 1 is under consideration. The crack has length, a. The lower and the upper crack arms have thickness, h_1 and h_2, respectively. The beam has width, b, and thickness, h. The beam is with length, l.

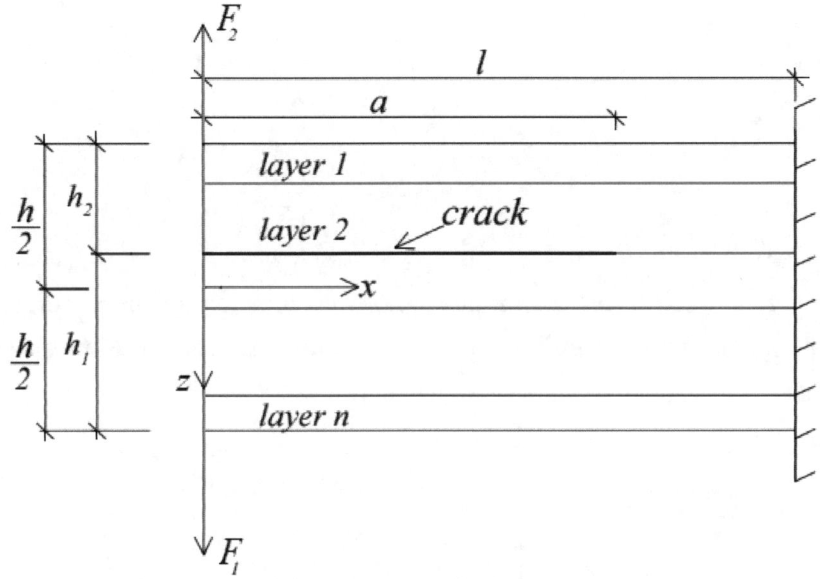

Figure 1. The multilayered non-homogeneous viscoelastic beam configuration.

The right-hand end of the beam is clamped. Two vertical forces, F_1 and F_2, are applied at the free end of the beam (figure 1). The creep behaviour of the i-th beam layer is treated by a linear viscoelastic model consisting of a spring in a series with a dashpot (damper) (figure 2). For this viscoelastic model, the stress-strain-time relationship under constant stress is written as [29]

$$\varepsilon = \frac{\sigma_i}{E_i} + \frac{\sigma_i t}{\eta_i} \tag{1}$$

where ε is strain, t is time, σ_i, E_i and η_i are the stress, the elasticity modulus and the coefficient of viscosity in the i-th layer, respectively. The layers are smoothly non-homogeneous along the beam width. Therefore, the elasticity modulus of the spring and the coefficient of viscosity of the dashpot in the i-th layer are distributed continuously along the width. The distributions are written as

$$E_i = \frac{E_{L_i}}{1 + p_{L_i} \dfrac{b + 2y}{b}} \tag{2a}$$

$$\eta_i = \frac{\eta_{L_i}}{1 + q_{L_i} \dfrac{b + 2y}{b}} \tag{2b}$$

for

$$-b/2 \leq y \leq 0$$

and

$$E_i = \frac{E_{L_i}}{1 + p_{L_i} \dfrac{b - 2y}{b}} \tag{3a}$$

$$\eta_i = \frac{\eta_{L_i}}{1 + q_{L_i} \dfrac{b - 2y}{b}} \tag{3b}$$

for

$$0 \le y \le b/2 .$$

In formulae (2) and (3), E_{L_i} and η_{L_i} are, respectively, the values of E_i and η_i at the lateral surfaces of the beam, p_{L_i} and q_{L_i} are material properties that control the non-homogeneity in the width direction, y is the horizontal central axis of the beam cross-section. According to formulae (2) and (3), E_i and η_i are distributed symmetrically with respect to $y = 0$.

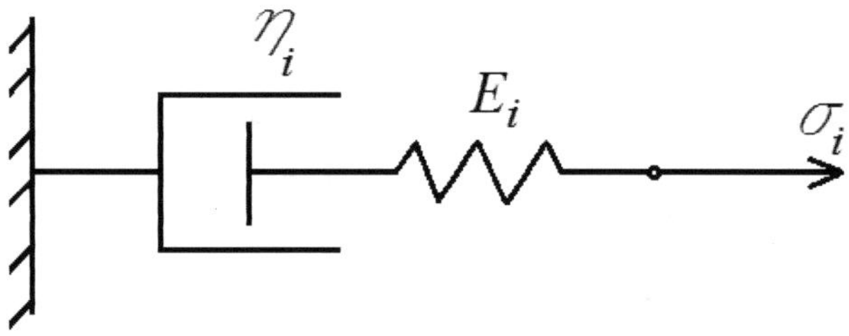

Figure 2. Viscoelastic model.

The multilayered non-homogeneous viscoelastic beam structure is exposed to the influence of aggressive chemical agent which leads to chemical degradation of the material in layers. Thus, E_{L_i} and η_{L_i} decrease with time on exponential laws [30]:

$$E_{L_i} = E_{L_i 0} e^{-\beta_{Ei} t - \psi_{Ei} \sqrt{t}} \tag{4a}$$

$$\eta_{L_i} = \eta_{L_i 0} e^{-\beta_{\eta i} t - \psi_{\eta i} \sqrt{t}} \tag{4b}$$

where $E_{L_i 0}$ and $\eta_{L_i 0}$ are, respectively, the values of E_{L_i} and η_{L_i} at the initiation of the chemical degradation, β_{Ei}, ψ_{Ei}, $\beta_{\eta i}$ and $\psi_{\eta i}$ are properties that describe the process of decrease of the elasticity modulus and the coefficient of viscosity in the layer with time.

The delamination fracture behaviour is analysed by investigating of the strain energy release rate with considering the chemical degradation of the material. The strain energy release rate, G, that takes into account the chemical degradation is obtained by using the formula

$$G = \frac{dU}{bda} \tag{5}$$

where U is the strain energy in the beam structure, da is an elementary increase of the length of the delamination. The strain energy is written as

$$U = U_1 + U_2 + U_3 \tag{6}$$

where U_1, U_2 and U_3 are the strain energies in the lower and upper arms of the delamination and in the un-cracked beam portion, $a \le x \le l$, where x is the longitudinal centroidal axis (figure1). The strain energy in the lower arm of the delamination is found as

$$U_1 = \sum_{i=1}^{i=n_1} \iiint_{(V_i)} u_{01i} dV \tag{7}$$

where u_{01i} is the density of the strain energy, V_i is the volume of the i-th layer of the lower arm of the delamination, n_1 is the number of layers in this delamination arm.

The density of the strain energy in the i-th layer is expressed as

$$u_{01i} = \frac{1}{2}\sigma_i \varepsilon \tag{8}$$

By substituting of (1) in (8), one obtains

$$u_{01i} = \frac{\sigma_i^2}{2E_i} + \frac{\sigma_i^2 t}{2\eta_i} \tag{9}$$

The distribution of strains along the thickness of the lower arm of the delamination is written as

$$\varepsilon = \kappa_1 (z_1 - z_{1n}) \tag{10}$$

where κ_1, z_{1n} and z_1 are the curvature, the neutral axis coordinate and the vertical centric axis of the cross-section, respectively. Formula (10) follows from the fact that validity of the Bernoulli's hypothesis for plane section is assumed since the beam under consideration has a high length to thickness ratio. The curvature and the neutral axis coordinate are determined by using the equations of equilibrium of the lower delamination arm.

The curvature, the neutral axis coordinate and the strain energy in the upper crack arm and in the un-cracked beam portion are found in analogical manner.

It should be noted that solution (5) is time-dependent since the strain energy densities in the layers are functions of the time.

The strain energy release rate that accounts for the chemical degradation of the viscoelastic multilayered beam is found also through the compliance method. According to this method, the strain energy release rate is written as

$$G = \frac{1}{2b}\left(F_1^2 \frac{dC_{F_1}}{da} + F_2^2 \frac{dC_{F_2}}{da} \right) \tag{11}$$

where the beam compliances, C_{F_1} and C_{F_2}, are expressed as

$$C_{F_1} = \frac{w_1}{F_1} \tag{12a}$$

$$C_{F_2} = \frac{w_2}{F_2} \tag{12b}$$

In formulae (12), w_1 and w_2 are the vertical displacements of the application points of the external forces, F_1 and F_2, respectively. These vertical displacements are found by applying the integrals of Maxwell-Mohr. The result is

$$w_1 = \int_0^a \kappa_1 x dx + \int_a^l k_3 x dx \tag{13a}$$

$$w_2 = \int_0^a \kappa_2 x dx + \int_a^l k_3 x dx \tag{13b}$$

where k_2 and κ_3 are the curvatures of the upper crack arm and the un-cracked beam portion, respectively.

The strain energy release rate obtained by (11) is in excellent agreement with that calculated by (5) which prove the correctness of analysis performed.

3. Numerical results

The time-dependent analysis described in section 2 of the paper is applied here in order to evaluate the influence of the chemical degradation of the multilayered viscoelastic material on the delamination fracture behaviour.

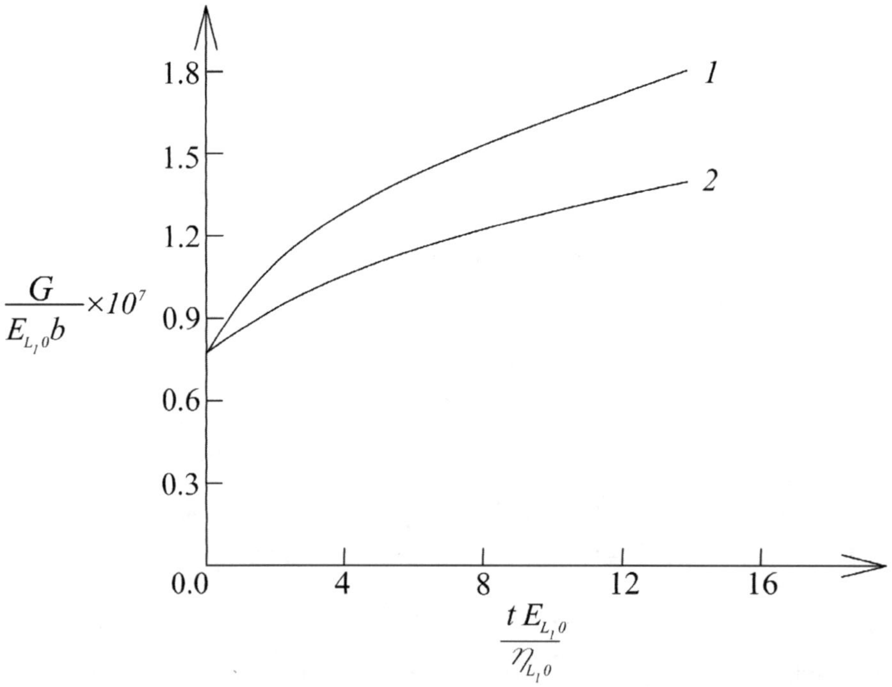

Figure 3. The non-dimensional strain energy release rate presented as a function of non-dimensional time (curve 1 – for multilayered inhomogeneous viscoelastic beam exposed to chemical degradation, curve 2 – for multilayered inhomogeneous viscoelastic beam that is not exposed to chemical degradation).

The strain energy release rate is expressed in non-dimensional form by using the formula

$G_N = G/(E_{L_1 0} b)$. Calculations of the strain energy release rate are carried-out for a three-layered cantilever beam exposed to chemical degradation. The thickness of each layer is h_d. The following data are used: $b = 0.020$ m, $h_d = 0.008$ m, $h = 0.024$ m, $l = 0.250$ m, $a = 0.100$ m, $F_1 = 2$ N and $F_2 = 3$ N.

The change of the strain energy release rate with time is analysed. For this purpose, calculations of the strain energy release rate are performed at various values of time. The results obtained are illustrated in figure 3 where the non-dimensional strain energy release rate is presented as a function of the non-dimensional time. The curves in figure 3 indicate that the strain energy release rate increases with time. This behaviour is due to the creep and to the chemical degradation of the multilayered viscoelastic material. The change of the strain energy release rate with time is analysed also assuming that chemical degradation does not occur. For this purpose, calculations are carried-out after substituting of $\beta_{Ei} = 0$, $\psi_{Ei} = 0$, $\beta_{\eta i} = 0$ and $\psi_{\eta i} = 0$ in (5). The results obtained are shown in figure 3. It can be observed in figure 3 that the chemical degradation leads to increase of the strain energy release rate. This finding indicates that chemical degradation has to be taken into account in fracture mechanics based structural design of multilayered non-homogeneous viscoelastic beams.

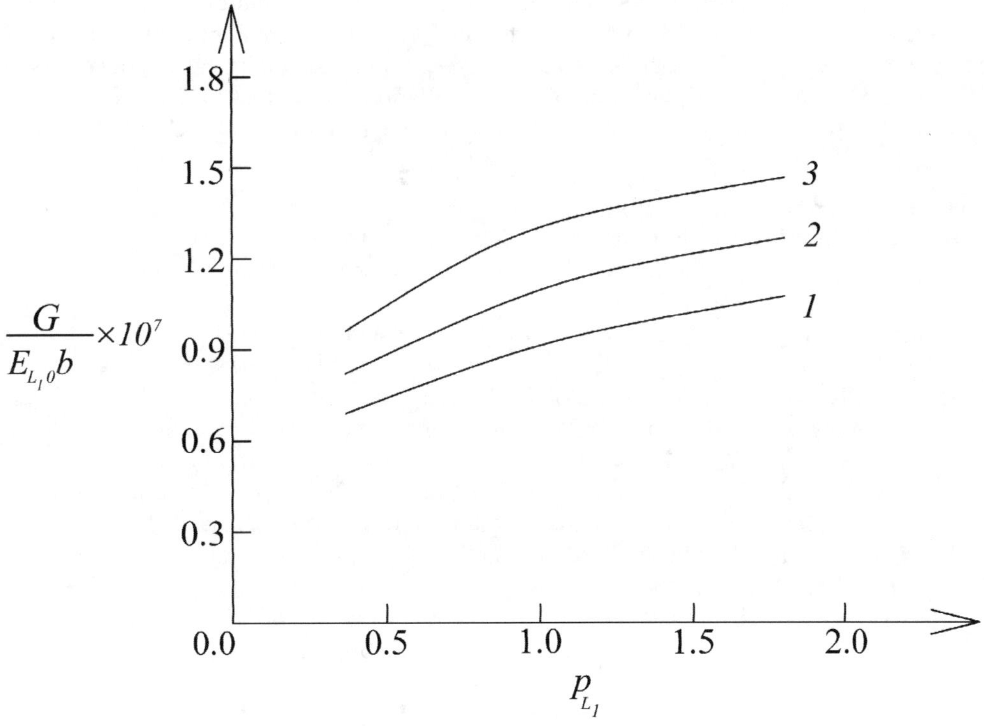

Figure 4. The non-dimensional strain energy release rate presented as a function of p_{L_1} (curve 1 – at $q_{L_1} = 0.5$, curve 2 – at $q_{L_1} = 1.0$ and curve 3 – at $q_{L_1} = 2.0$).

It is analysed also how the material non-homogeneity along the thickness of layers influences delamination of multilayered viscoelastic beam subjected to chemical degradation. For this purpose, calculations of the strain energy release rate are performed at various values of p_{L_1} and q_{L_1}. The influence of material non-homogeneity on the delamination is displayed in figure 4 where the non-dimensional strain energy release rate is presented as a function of p_{L_1} at three values of q_{L_1}. The curves in figure 4 show that the strain energy release rate increases with increasing of p_{L_1} when the beam is exposed to chemical degradation. The increase of material property, q_{L_1}, leads also to increase of the strain energy release rate under chemical degradation (figure 4).

4. Conclusions

A time-dependent delamination analysis of a multilayered non-homogeneous viscoelastic beam configuration subjected to chemical degradation is developed. The beam is made of adhesively bonded non-homogeneous viscoelastic layers of different thicknesses and material properties. The beam is exposed to influence of aggressive chemical agent. This leads to chemical degradation of the material in layers of the beam. As a result of the chemical degradation, the elasticity modulus and the coefficient of viscosity decrease with time. The time-dependent delamination fracture behaviour is studied by analysing the strain energy release rate with considering of the chemical degradation. For this purpose, the strain energy is used to derive the strain energy release rate. The strain energy is calculated by taking into account the chemical degradation. In this way, the solution obtained accounts for both the creep behaviour of the non-homogeneous structure and the chemical degradation of the material. The compliance method with considering of the chemical degradation of the multilayered non-homogeneous viscoelastic material is applied to verify the solution. The change of the strain energy release rate with time is analysed. The analysis reveals that the strain energy release rate increases with time (this behaviour is due to the creep and to the chemical degradation). Concerning the influence of material non-homogeneity, the calculations show that the strain energy release rate increases with increasing of p_{L_1} and q_{L_1} under chemical degradation.

References

[1] Lee W Y, Stinton D P, Berndt C C, Erdogan F, Yi-Der Lee and Mutasim Z 1996 Concept of Functionally Graded Materials for Advanced Thermal Barrier Coating Applications. *Journal of the American Ceramic Society* **79** pp 3003-3012

[2] Zhang Y, Ming-jie Sun and Zhang D 2012 Designing functionally graded materials with superior load-bearing properties. *Acta Biomaterialia* **8** pp 1101-1108

[3] Jedamzik R, Neubrand A and Rödel J 2000 Functionally graded materials by electrochemical processing and infiltration: application to tungsten/copper composites. *Journal of Materials Science* **35** pp 477–486

[4] El-Galy IM, Saleh BI and Ahmed MH 2019 Functionally graded materials classifications and development trends from industrial point of view. *SN Appl. Sci.* **1** pp 1378-1389

[5] Tejaswini N, Babu R and Ram S 2013 Functionally graded material: an overview. *Int J Adv Eng Sci Technol* **4** pp 183–188

[6] Nikbakht S, Kamarian S and Shakeri M 2019 A review on optimization of composite structures Part II: Functionally graded materials. *Composite Structures* **214** pp 83-102

[7] Dias CMR, Savastano Jr H, John VM 2010Exploring the potential of functionally graded materials concept for the development of fiber cement. *Construction and Building Materials* **24** pp 140-146

[8] Gandra J, Miranda R, Vilaça P, Velhinho A and Teixeira JP 2011 Functionally graded materials produced by friction stir processing.*Journal of Materials Processing Technology* **211** pp 1659-1668

[9] Toudehdehghan J, Lim W , Fool K E, Ma'arof M I N and Mathews J 2017 A brief review of functionally graded materials. *MATEC Web of Conferences* **131**, 03010

[10] Yoshinari Miyamoto 1996 The Applications of Functionally Graded Materials in Japan. *Materials Technology* **11** pp 1986-1998

[11] Wang J, Leon L and Shaw L L 2005 Fabrication of Functionally Graded Materials Via Inkjet

Color Printing. *Journal of the American Ceramic Society* **89** pp 3285-3289

[12] Hsueh C-H 2002 Modeling of elastic deformation of multilayers due to residual stresses and external bending. *J Appl Phys* **12** pp 9652-9689

[13] Drory M, Thouless M, Evans A 1988 On the decohesion of residually stressed thin films. *Acta Metall* **36** pp 2019-2028

[14] Townsend P, Barnet D, Brunner T 1987 Elastic relationships in layered composite media with approximation for the case of thin films on a thick substrate. *J Appl Phys* **62** pp 4438-4446

[15] Rizov V 2020 Analysis of Two Lengthwise Cracks in a Viscoelastic Inhomogeneous Beam Structure. *Engineering Transactions* **68** pp 397-415

[16] Rizov V 2020 Viscoelastic inhomogeneous beams under step loading: a longitudinal fracture analysis. *IOP Conf. Ser.: Mater. Sci. Eng.***1031** 012008

[17] Rizov V 2020 Longitudinal fracture analysis of inhomogeneous beams with continuously changing radius of cross-section along the beam length. *Strength, Fracture and Complexity: an International Journal* **13** pp 31-43.

[18] Zubchaninov VG 1990 *Theory of elasticity and plasticity*, Vishaya shkola

[19] Kishkilov M and Apostolov R 1994 *Introduction to theory of plasticity*, UACEG

IOP Conf. Series: Materials Science and Engineering 1141 (2021) 012018 doi:10.1088/1757-899X/1141/1/012018

Numerical modelling of point thermal bridge's impact on the thermal performance of facade systems

L Moga[1,*], I Moga[1]
[1] Technical University of Cluj-Napoca, 25 Bariţiu Street, *Romania*

*E-mail: ligia.moga@ccm.utcluj.ro

Abstract. Starting with 31 December 2020 all new buildings, as well as existing ones, must comply with the nearly Zero Energy Buildings levels, as defined by each member state. A very important design criteria is the one regarding an accurate assessment of both linear and point thermal bridges of the building envelope. Usually, point thermal bridges are either not considered in calculations or are assessed by using simplified methodology as described by the standard EN ISO 6946/2017. This is due to the modelling and calculation complexity required in assessing them. Therefore, in several situations the impact is underestimated, and the thermal performance results do not offer a clear image of the building's thermal performance. Thus, the paper focuses on the modelling and simulation of several constructive solutions for façade systems, aiming to identify a correlation between the geometrical and thermal characteristics of fixing components and the thermal performance of the facade. The studied case scenarios include two types of masonry walls, of 240 and 365 mm, that are thermally insulated in various thicknesses using an ETICS system. A sensitivity analysis is also performed to define a correlation between all the components that define the system and the point thermal bridge impact magnitude. The results of the study indicate an increased impact with a direct consequence on the energy performance of the building.

1. Introduction

The European Directive for the energy performance of buildings [1] gives the definition of a high performing energy efficient building, as being a nearly Zero Energy Building (nZEB). Two deadlines were set, one by 31 December 2018 and the second one for 31 December 2021, that mandates that all new buildings must reach nZEB levels in all European Union (EU) member states. Nevertheless, all member states had to define the nZEB levels than can be achieved at a national level, at optimum costs. A significant parameter in reaching nZEB levels is the reduction of the heat losses through thermal bridges. Depending on the constructive solution and the complexity of the architecture, these can count to about 30% or even more as mentioned by [2-3]. Several studies showed that a proper correction of the thermal bridges should result in a reduction of the energy consumptions for heating of at least 8.5%, that varying according to the climate type. [4]

In designing high performing energy efficient buildings, practitioners must carefully consider and asses the weak thermal areas in the structure of the building envelope. The weak thermal areas are known as thermal bridges and can be identified in the next cases, as given by the national thermo-technical design norm C107 [5]:

- when a material with a higher thermal conductivity λ [$W{\cdot}m^{-1}{\cdot}K^{-1}$], partially or fully penetrates a building envelope element (i.e. reinforced concrete column in a masonry wall);
- change in the thickness of the building envelope component.
- when differences exist between internal and external surfaces (i.e. intersection intermediate floor – exterior wall).

Therefore, thermal bridges can be point thermal bridges, linear thermal bridges, or spatial due to a spatial detail that defines such a weak thermal area. When calculating the thermal transmittance of building envelope components U [$W{\cdot}m^{-2}{\cdot}K^{-1}$], designers take into calculation only the value of the linear heat transfer coefficient denoted by ψ [$W{\cdot}m^{-1}{\cdot}K^{-1}$] (i.e. quantifies the magnitude of a linear thermal bridge), while the point heat transfer coefficient denoted by χ [W·K^{-1}] (i.e. quantifies the magnitude of a point thermal bridge), in many cases, is not considered in calculations.

Practitioners use atlases of thermal bridges [6,7,8] or the European standard [9] that gives 76 cases for thermal bridges details, as e.g. exterior corners, interior corners, intersection between slab on the ground and exterior walls, floors above unheated basements, attic roofs, roofs, intermediate floors – exterior walls intersection, window/door-wall intersection, balcony slabs-exterior walls, exterior wall with a column, and others. Designers can also address thermal bridges through modelling and simulation, but this requires knowledge which often lacks, and significant time for a proper simulation of the examined thermal bridge. Therefore, usage of thermal bridges atlases is the first choice in obtaining design values. The thermal bridges atlases offer design values for the ψ value, and not for the χ. On the same time, obtaining χ values requires advanced knowledge in modelling and simulation and a specific software for this aim. Due to that, the χ values are either not found in calculations or assessed using a simplified methodology defined in [10], although that for ventilated facades the point bridges can have a more significant impact on the thermal performance level compared the one assessed through a simplified approach. The scientific literature offers several studies in evaluating the negative impact of metal fasteners mainly for the case of ventilated facades or facades with integrated photovoltaics as mentioned by Theodosiou [11-12], curtain walls [13] or light steel frame structures [14].

Point thermal bridges are caused by fasteners (i.e. anchors, bolts, brackets and other similar components) that are needed for fixing the thermal insulation system or/and the ventilated façade system, or other types of systems to the support layer. The point thermal bridges can have a significant influence in establishing the real thermal performance of a building component, with a reach of 20% or higher, and implicitly with a same impact on the energy performance of a building [11,12,14]. Even in the case of envelope elements thermally insulated with aerogel blankets that are penetrated by metal anchors [15], the thermal resistance could decrease by 15% to 45% when using 3 to 6 anchors per m^2.

The aim of the paper consists in investigating the impact that the anchors, defined by their thermal properties and geometric dimensions, have on the thermal performance of the building envelope component. At the same time, the influence on the χ value due to the thermal insulation thickness and conductivity, number of anchors and type of support layer, are also assessed. Thermally insulated constructive details specific for External Thermal Insulation Composite Systems (ETICS) are modelled and simulated. Although that literature [12] mentioned that 3D thermal bridges (i.e. point thermal bridges) at ETICS can be considered as having a slightly less negative influence compared to other constructive solutions, the study will demonstrate that these should be considered in calculations.

2. Materials and methods

2.1. Parameters of ETICS and support layer

A typical ETICS system is defined by the following components: adhesive, thermal insulation material, anchors, base coat, reinforcement (i.e. usually glass fibre mesh) and a finishing layer: (i.e. finishing coat with a key coat and/or a decorative coat) and various accessories.[16] The anchors have the role of mechanically fixing the thermal insulation layer to the support layer (e.g. masonry, concrete). Anchors have a higher thermal conductivity λ [W·m^{-1}·K^{-1}] compared to the thermal insulating layer, thus conducting heat at a higher rate. Thus, the anchors become point thermal bridges in the structure of the building envelope due to their weak thermal performance.

The paper investigated the case for a support layer defined by a full brick masonry with λ =0.80 [W·m^{-1}·K^{-1}], in two situations, a 24 cm support layer, and a 36.5 cm layer. For the thermal insulating systems, the study gives the results for thicknesses starting from 50 mm up to 300 mm. Four situations for the number of used anchors, starting from 6 up to 12 anchors per m^2, were analysed. The length of the anchors, the radius of the anchors cap and the distance between the anchors was defined in accordance with ETICS standard [17]. The thermal conductivity for the thermal insulation was 0.040 [W·m^{-1}·K^{-1}] and 50 [W·m^{-1}·K^{-1}] for the steel anchors. The thermal insulation layer was chosen with an average value, without referring to a specific type.

Table 1. The investigated cases

Parameter	Range examined
Support layer (SL)	masonry
Thickness SL	240, 365 [mm]
Thermal conductivity SL	0.80 [$W \cdot m^{-1} \cdot K^{-1}$]
Anchor type (A)	Steel
Thermal conductivity A	50 [$W \cdot m^{-1} \cdot K^{-1}$]
Number of A	6, 8, 10, 12 /m^2
Thermal insulation (Tins) conductivity	0.04 [$W \cdot m^{-1} \cdot K^{-1}$]
Tins thickness (D_{ins})	50, 100, 150, 200, 250, 300 [mm]

Figure 1. Anchor placement - T pattern diagram

2.2. The modelling and simulation

The modelling and simulation of the point thermal bridges implies a 3D modelling approach for proper assessment of the heat losses, which is given by a higher complexity numerical approach compared to the case of linear thermal bridges. Therefore, the HISPAT 3D modelling and simulation software was used [18], which was developed by the authors of this paper.

The software knew three stages of national validation since its first development back in 1978:

- 1st stage: a validation of the program in corelation with the experimental data obtained at INCERC Iasi;
- 2nd stage: in the 90's together with IPCT, the program was used for the development of the thermal bridges annexes of C107/3 norm. The program was validated in accordance with the stipulations of Annex A, EN ISO 10211/1995;
- 3rd stage: when the Annex 1590, of the C107 Norm was developed, the program was again validated according to SR EN ISO 10211/2008.

- today the program is validated according to SR EN ISO 10211/2017, Annex C

The software is based on solving the spatial heat transfer differential equation in steady state thermal regime without internal heat sources, where T is the temperature and λ_x, λ_y and λ_z is the thermal conductivities of the material on x, y, z direction.

$$\frac{\partial}{\partial x}\cdot\left(\lambda_x\frac{\partial T}{\partial x}\right)+\frac{\partial}{\partial y}\cdot\left(\lambda_y\frac{\partial T}{\partial y}\right)+\frac{\partial}{\partial z}\cdot\left(\lambda_z\frac{\partial T}{\partial z}\right)=0 \tag{1}$$

The mesh is generated by the software following a grid dimension suitable for the studied case. The boundary conditions respect the European prescriptions [10]. The discretisation steps start from 1mm up to 25 mm, with an increased density in the anchor's mesh. The mesh is typical for metallic thermal bridges. The boundary conditions are R_{si}=0.13 $m^2 \cdot K \cdot W^{-1}$ and R_{se}=0,04 $m^2 \cdot K \cdot W^{-1}$, while the temperature difference has a unitary value according to Annex E of [10].

The distribution of the temperature field is calculated through an iterative process, that ends when the convergence criteria given by the difference between inner and outer heat flow is smaller than 0.001 W and heat flow between two consecutive iterations is smaller than 0.000001W.

To determine the influence of the anchors on the wall's thermal transmittance, the case scenarios were considered with and without the anchors. First, the 3D modelling and simulation was done with the 3D HISPAT software for the first case scenario, and a 1D calculation was done for the latter case. The value of the difference between the two heat flows indicates the amount of heat flow through the point thermal bridge. The formula for calculating χ was used according to [10]:

$$\chi = L_{3D} - U \cdot A \quad \left[\frac{W}{K}\right] \tag{2}$$

where L_{3D} [$W \cdot K^{-1}$] is the thermal coupling coefficient calculated by a 3D approach, the U [$W \cdot m^{-2} \cdot K^{-1}$] is the thermal transmittance and A [m^2] is the area of the examined building element.

The changes in the examined detail for each case scenario is followed by a new redefined mesh, in order to ensure the accuracy in calculation. As it can be observed in figure 2 at the anchors mesh, when the geometry of the element is more complex, the density of the mesh will increase.

3. Results and discussions

The research was made for different types of support layers, starting from full brick masonry, reinforced concrete, aerated cellular concrete. However, the paper will present the results for the masonry case. Results are summarised for the case scenarios mentioned in table 1, the unidirectional thermal resistance R and the adjusted thermal resistance R' [$m^2 \cdot K \cdot W^{-1}$] are pointed out as main indicators. R was calculated in accordance with the layers that defined the analysed constructive details. The temperature field distribution is presented in figure 3, for one of the studied cases.

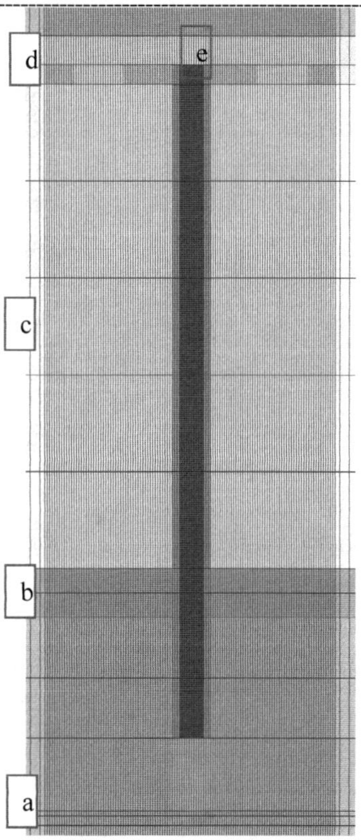

Figure 2. Discretization network for the anchor (a) masonry, (b) adhesive, (c) thermal insulation, (d) finishing, (e) anchor

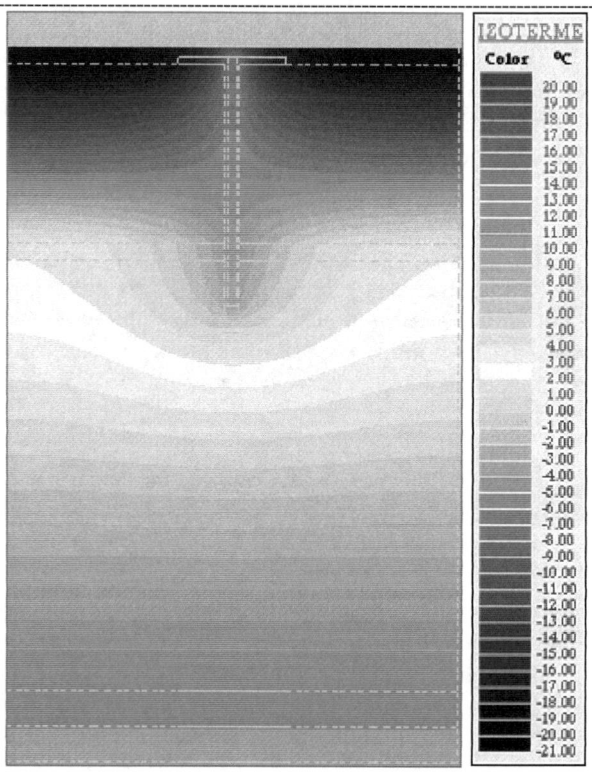

Figure 3. Isotherms of the studied detail – D_{ins}=100 mm, SL=240 mm

As it can be observed, the steel anchor is causing a disturbance in the temperature curves distribution and implicitly a weaker thermal point, which leads to a decreased thermal performance.

3.1. Impact of the number of anchors

The results are given starting from the case with 50 mm to 300 mm of thermal insulation for both case scenario, i.e. full brick masonry 240 mm – case scenario 1 (CS1) and full brick masonry 365 mm – case scenario 2 (CS2). The simulations were made with steps of 1 cm for thermal insulation thickness (i.e. 26 variants), but for a clean presentation the scale only indicates the six thicknesses values, from 50 to 300 mm.

As it can be observed in figure 4 and figure 5, χ has a greater value, and implicitly impact, in the case of a thinner masonry compared to a thicker one. For CS1 the values range from the lowest value of 0,012 [$W{\cdot}K^{-1}$] for the case of 6 anchors/m^2 and 300 mm Tins to an almost 0.043 [$W{\cdot}K^{-1}$] for the case of 12 anchors/m^2 and 70 mm Tins.

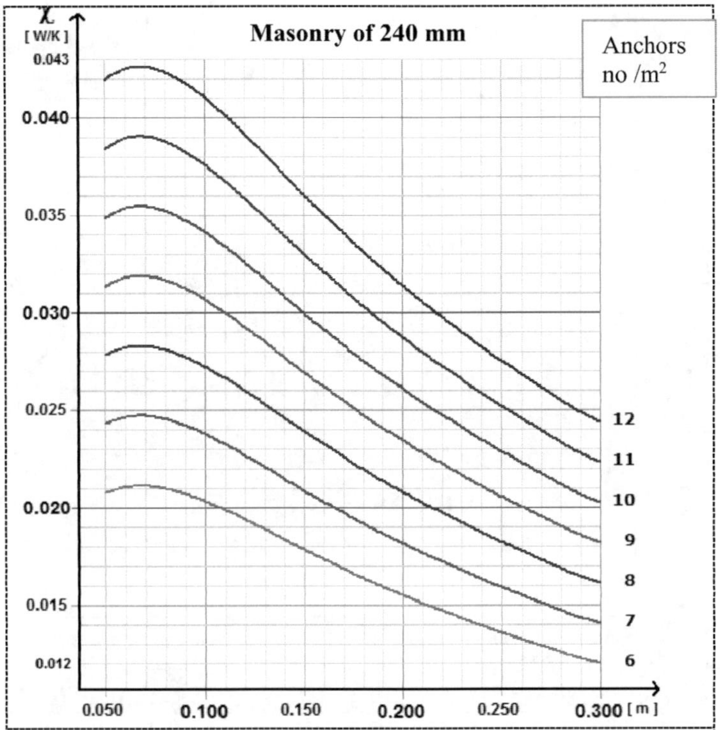

Figure 4. The χ value – full brick masonry 240 mm

For CS2 the values range from the lowest value of 0,010 [$W\cdot K^{-1}$] for the case of 6 anchors/m^2 and 300 mm Tins to an almost 0.036 [$W\cdot K^{-1}$] for the case of 12 anchors/m^2 and 70 mm Tins. At the same time, one can see that the step for the variation of χ increases with the increase of the anchors number but decreases with the increase of the support layer thickness.

3.2. The impact of the anchor's length

For CS1, the growth for the R value starts by a 41% and ends with a 78% compared with the reference case, the one for 50 mm. For CS2, the growth is smaller, starting from a 39% to a 76%. Furthermore, both R and R' are smaller compared to the reference value for exterior walls R'_{min}=1.80 [$m^2\cdot K\cdot W^{-1}$] [19] in the case of C1. For CS2, all R and R' values are greater than the R'_{min} reference value.

In the light of the NZEB criteria, the proposed R' values will be for residential buildings R'_{min}=4.00 [$m^2\cdot K\cdot W^{-1}$] and for non-residential buildings R'_{min}=3.00 [$m^2\cdot K\cdot W^{-1}$] [20]. Thus, for CS1 the values will not be met for residential buildings for cases 50 mm to 150 mm, respectively for non-residential buildings for cases 50 to 100 mm Tins. For CS2 the values will not be met for residential buildings for cases 50 mm to 100 mm Tins, respectively for the 150 mm Tins, 10 to 12 anchors/m^2 cases. In case of non-residential buildings, the values will not be met for 50- and 100-mm Tins, especially 8 to 12 anchors/m^2 case scenarios.

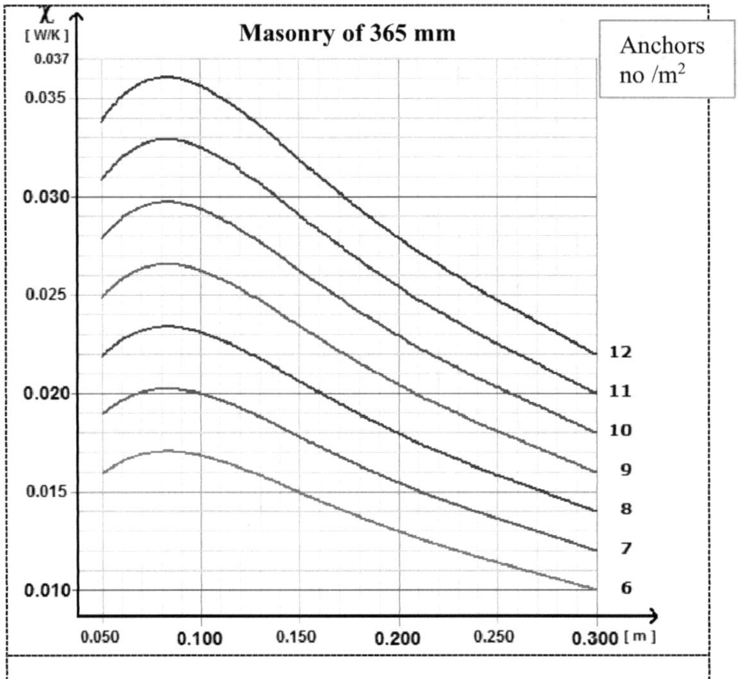

Figure 5. The χ value – full brick masonry 365 mm

Table 2. The adjusted thermal resistance - no of anchors - masonry 240 mm

D_{ins} [mm]	R $\left[\frac{m^2 \cdot K}{W}\right]$	R' $[m^2 \cdot K/W]$ Number of anchors 6	8	10	12
50	1.776	1.711	1.691	1.671	1.652
100	3.026	2.847	2.792	2.740	2.689
150	4.276	3.967	3.873	3.784	3.699
200	5.526	5.080	4.947	4.820	4.700
250	6.776	6.190	6.017	5.853	5.698
300	8.026	7.300	7.087	6.886	6.695

Regarding the reduction of R' value with respect to the number of anchors and the R value, the reduction percentage increases with the increase of the thermal insulation layer, starting form 4% (i.e. for 6 anchors/m^2, 50 mm Tins) to 17% (i.e. for 12 anchors/m^2, 300 mm Tins) for CS1 and 3% (i.e. for 6 anchors/m^2, 50 mm Tins) to 16% (i.e. for 12 anchors/m^2, 300 mm Tins) for CS2. The increased percentage value could be also associated to a longer dimension of the steel anchors.

Table 3. The adjusted thermal resistance - no of anchors - masonry 365 mm

D_{ins}	R	R' $[m^2 \cdot K/W]$			
[mm]	$\left[\frac{m^2 \cdot K}{W}\right]$	Number of anchors			
		6	8	10	12
50	1.931	1.864	1.847	1.827	1.807
100	3.182	3.003	2.947	2.893	2.842
150	4.432	4.122	4.029	3.939	3.853
200	5.682	5.238	5.105	4.978	4.858
250	6.932	6.345	6.170	6.005	5.849
300	8.182	7.452	7.237	7.033	6.841

The obtained results are consistent with results from reference papers [11,12]. As it can be observed in [12], the effect of the thermal insulation on the χ value is given with a similar allure, with a clear spike in between 50 – 100 mm of thermal insulation. Sadauskiene in [21] obtained continuous reductions of χ values for increased thickness of the SL, similar to the findings presented in figures 4 and 5.

3.3. The impact of the anchor on the thermal performance and energy consumption

In order to assess the impact that the anchors have on the overall thermal performance of a building, the case of a residential building is considered. The building is in Cluj-Napoca, location defined by the IIIrd climatic zone and an exterior temperature θe=-18°C. The height regime is basement partially heated+ ground floor + 1st floor. The heated useful area is 166 m^2 and the area of the building envelope is 435 m^2. In table 4 are presented the R' values for a part of the building envelope components.

Figure 6. The investigated building

Table 4. R' for the building envelope elements (extract) [$m^2 \cdot K \cdot W^{-1}$]

Exterior wall	5.29
Exterior glazing	1.10
Attic roof	5.48
Floor in contact with the ground	6.37

The assessed thermal bridges were just the linear ones, without addressing the point thermal bridges. The final energy consumption for heating resulted in a value $q_{heating}$=67.38 [$kWh/m^2 \cdot yr$], which means an A energy class for heating, based on the national legislation which requires a value ≤70 [$kWh/m^2 \cdot yr$]. By including the effect of the point thermal bridges the $q_{heating}$ will rise up to a value of 74.49 [$kWh/m^2 \cdot yr$] which will lead to a B energy class for heating. Therefore, the impact of the anchors should be investigated to not underestimate the actual heat flows through the building's façade.

4. Conclusions

The nZEB requirements come with a lot of pressure on both energy auditors for buildings and designers, mainly the one specialised in the thermal design of the building envelope. Nevertheless, reaching the required levels in design and operation, will be translated in a proper assessment of all type of thermal bridges.

In most cases, point thermal bridges are not considered in calculations due to their complexity and lack of knowledge regarding their simulation. Also, a general perception is that for single-skin façade, their input is not significant, although that the present papers demonstrates that not always this applies. The constructive details for nZEB must have increased thicknesses of the insulation layers which will translate in an increased impact of the anchor's presence. The results obtained for the 240 mm and 365 mm full brick masonry walls, thermally insulated in various thicknesses, demonstrate the need of considering the impact of χ values. The magnitude of the point thermal bridges increases with an increased anchors length needed in case of high thermal insulations thicknesses, resulting in a R' reduced of up to 17% compared to R value.

For more complex details (i.e. ventilated façade), the heat flows associated with point thermal bridges can represent more than 25% from the total heat flows through the building envelope, as mentioned by literature. The results of the study can be used as input in design work, for similar cases with the ones investigated here.

5. References

[1] Directive 31/UE 2010 *Directive on the energy performance of buildings* European Parliament, Council of EU

[2] Erhorn H, Erhorn-Kluttig H, Citterio M et al 2010 *An effective handling of thermal bridges in the EPBD context* Asiepi Project Fraunhofer Institute for Building Physics (Germany) p 30

[3] Ascione F, Bianco N, de Rossi F, Tutni G and Vanoli GP 2012 Different methods for the modelling og thermal bridges into energy simulation programs: Comparisons of accuracy for flat heterogeneous roofs in Italian climates *Appl.Energy* **97** 405-18

[4] Evola G, Margani G and Marletta L 2011 Energy and cost evaluation of thermal bridge correction in Mediterranean climate *Energy and Buildings* **43** 2385–93

[5] C107 2005 *Norm for thermotechnical calculation of construction elements of buildings* **3** (Romania)

[6] Moga L and Moga I 2013 *Punţi termice specifice clădirilor cu pereţi structurali din zidărie* ed U.T.Press (Cluj-Napoca) p 134

[7] Moga L and Moga I 2017 *Punţi termice specifice planşeelor terasă, de pod, deasupra subsolului şi*

plăcilor pe sol la clădiri cu pereţi din zidărie ed U.T.Press (Cluj-Napoca) p 162
[8] Order 1590 2012 *Catalogue of specific thermal bridges for buildings* Annex (Romania)
[9] ISO 14683 2007 *Thermal bridges in building construction — Linear thermal transmittance — Simplified methods and default values* The International Organization for Standardization
[10] ISO 10211 2017 *Thermal bridges in building construction — Heat flows and surface temperatures — Detailed calculations* The International Organization for Standardization
[11] Stonkuviene A, Bliudzius R et al 2021 The impact of connector's thermal and geometrical characteristics on the energy performance of facade systems *J. of Build. Eng.* **35**
[12] Theodosiou T, Tsikaloudaki K et al 2019 Thermal bridging problems on advanced cladding systems and smart building facades *J. of Cleaner Prod.* **214** p 62-9
[13] Song JH, Park SH et al 2015 Influence of Thermal Bridges on the Insulation Performance of Curtain Wall Panel Systems *J. of Asian. Arch. and Build. Eng.* **14** (3) p 741–48 doi:10.3130/jaabe.14.741
[14] Gomes AP, de Souza et al 2013 Impact of thermal bridging on the performance of buildings using Light Steel Framing in Brazil *App. Therm. Eng.* **52** (1), p 84–9. doi:10.1016/j.applthermaleng.2012.11.015
[15] Berardi U and Akos L 2019 Thermal bridges of metal fasteners for aerogel-enhanced blankets *Energy Build* **185** p 307–15
[16] External Thermal Insulations Composite Systems *About ETICS* accessed at https://www.ea-etics.eu/etics/about-etics/
[17] *European Guideline for the Application of ETICS* ed European Association for External Thermal Insulations Composite Systems
[18] HISPAT software 2021 *Three-Dimensional Heat-Transfer for Buildings Modelling and simulation software for calculating point thermal transmittances values* Building Physics Collective (Technical University of Cluj-Napoca)
[19] Order 2641 2017 *Order regarding the modification and completion of the technical regulation "Methodology for calculating the energy performance of buildings", approved by the Order of the Minister of Transport, Construction and Tourism no. 157/2007* (Romania)
[20] Mc001 2021 *Methodology for calculating the energy performance of buildings- updated version* Preprint (Romania)
[21] Šadauskienė J, Ramanauskas J et al 2015 A Simplified Methodology for Evaluating the Impact of Point Thermal Bridges on the High-Energy Performance of a Passive House *Sustainability* **7** (12) p 16687-702

IOP Conf. Series: Materials Science and Engineering 1141 (2021) 012019 doi:10.1088/1757-899X/1141/1/012019

Computation and experimental considerations on dynamic testing of cement mortar

S M Alexa-Stratulat, D Covatariu, I O Toma, G Covatariu
Faculty of Civil Engineering and Building Services, "Gheorghe Asachi" Technical University of Iasi, nr. 1 Blvd. Prof. Dimitrie Mangeron, Iaşi 700050 Iasi, Romania

Email: mihai.s.stratulat@gmail.com

Abstract. Non-destructive tests to assess the dynamic characteristics of the materials are important for both in-situ and laboratory situations. Elastic modulus, rigidity modulus and Poisson ratio can be determined through dynamic testing, by employing resonant frequency measurements. The ASTM C-215 standard contains the necessary information for the experimental set-up and subsequent computation using the longitudinal, transversal and torsional frequencies, for both prismatic and cylindrical specimens. Nevertheless, some experimental particularities are not explained and their influence on the final results might prove important. In this paper, the experimental and computational insights on performing the dynamic Young modulus test for cement mortar prismatic samples, with emphasis on the importance of interpretation of the results are presented. The type of specimen support and the actual data manipulation, when extracting results, are taken into consideration. The results may prove important for the correct employment of dynamic testing.

1. Introduction

Elastic modulus of building materials is a key characteristic, with the structural design process heavily relying on its value. For concrete, it is usually determined by a static method, which implies repeated measurements of the strain when increasing the compressive force from zero to maximum 40% of the compressive strength [1][2]. This could also be done for mortar but it is much more challenging due to its much lower compressive strength [3].

Non-destructive, dynamic, methods have been used for both concrete and mortars [3]–[7]. Vibrational tests can be employed for determining the modulus of elasticity, modulus of rigidity and the damping properties by measuring the resonance frequency of the specimen and the propagation of a signal through it. These tests can be done by:

- forced oscillations, where a signal of constant amplitude and increasing frequency is induced in the specimen by means of a shaker; the natural frequency to be used in computing Young's modulus will be noted when the response of the sensor attached to the specimen is maximum.
- impact resonance method, where an impact on the specimen produces a signal, and the free, decaying, vibration is used to identify the resonance (FFT analysis).

The values obtained via this route for Young's modulus are significantly higher than those obtained through static tests but empirical correction equations have been determined in order to correlate the results from both dynamic and static approaches [8].

The purpose of the present research is to show the particularities of employing the impact resonance dynamic test approach when used on mortar prismatic samples and how to correctly asses the elastic modulus from the obtained data during the experiment. The impact method presents some advantages over the forced oscillations method. Firstly, it does not need a special device to create the oscillations (usually a shaker powered by a power amplifier and a frequency generator), which can be quite expensive and can induce errors due to the large frequency range needed in the experiments on concrete and mortar specimens (0-2kHz). Secondly, the measured decaying signal can be used to assess the damping properties of the material.

2. Experiment

2.1. Materials

Mortar samples have been prepared with a classical mix to accommodate a M10 strength class. The mortar was cast in metal moulds, following EN 1015-11 guidelines, obtaining prismatic specimens with 40 × 40 × 160 mm dimensions (width × height × length). The specimens have been further put in water and cured for 28 days.

2.2. Dynamic characterisation

The impact resonance dynamic test, as described in the ASTM C215-14 standard [9] was used. This type of test can be easily performed both in the laboratory and in-situ. Nevertheless, the data interpretation can be challenging, especially when not using a commercially developed apparatus for this specific purpose.

The focus was on the longitudinal geometry. The specimen was supported in the middle. However, ASTM C215-14 code does not give any details related to the width of the support. For this reason, two support widths were used to see the influence of this parameter: t_{narrow}=2cm and t_{wide}=5cm. Also, an innovative approach to reduce unwanted influences from outside sources was tested. The specimen was placed in a vertical position on a material with damping properties. The set-up layouts are presented in figure 1.

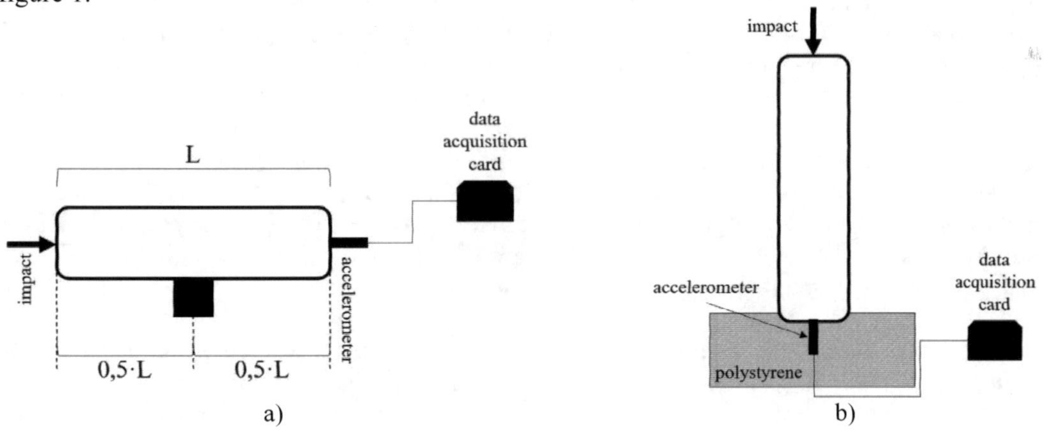

Figure 1. Experimental setup for longitudinal resonance test: a) classic, b) vertical.

Following this experiment, Young's modulus can be computed using equation (1). The result is expressed in [Pa].

$$E_d = D \cdot m \cdot f_n^2 \tag{1}$$

where:

- $D = 4\frac{L}{b \cdot t}$, in [m^{-1}], is a constant related to the length (L) of the specimen and the cross-sectional dimensions (b, t); this equation is applied for a prismatic specimen only.
- m, in [kg], is the mass of the specimen.
- f_n, in [Hz], is the natural longitudinal frequency that is measured during the experiment.

3. Results

The geometric dimensions and the mass of the prismatic sample have been measured (table 1), in order to perform the calculations of the elastic modulus as shown in equation (1).

Table 1. Specimen size and mass

Property	value
mass [kg]	m=0.533
length [m]	L=0.160
cross-section dimensions [m]	b=t=0.04

The longitudinal impact experimental setup is described to use a support at the middle of the specimen, as shown in figure 1. Nevertheless, the characteristics of this support are not given (width, material etc.). The position of the support is chosen to coincide with the node of the longitudinal vibration, in order to allow the signal to pass freely through the specimen. Due to practical considerations, it is unlikely to effectively use a thin enough support, as it would be desired. Also, the position of the node at the geometric middle of the specimen depends on two things that are supposed to be true but might not be: i) the impact is done exactly in the middle of the cross-section; ii) the specimen is a perfect rectangular prism.

During the first experiment, the specimen on a narrow wood support (t=2cm) was tested. This width represents around 12% of the length of the prism. It was the smallest width that did not present significant challenges in practice. Wood was chosen as the support material (and it was used for the next experiments also) since it has vibrational properties quite different from those of the cement mortar.

In figure 2a), an example of the resulting impact for the first experiment is presented. There is a vibration that modulates the traveling signal through the specimen for more than half the time. This can be explained by the relatively low weight of the mortar prism when compared to: i) the accelerometer and its connector (which were around 80g); ii) the impactor (close to 100g). The wave could be attributed to the mechanical movement (displacement) induced in the specimen during the impact and to the resonance of the impactor. Towards the end of the signal, the natural vibration can be observed unaltered, thus the resonant frequency can be inferred. Nevertheless, the damping properties of the material are obstructed by the damping of the modulating wave when the full signal would be considered and the unaltered part is too small to be usable for this purpose.

Subsequently, to reduce any possible hinge effect, a wider support was used. The width of 5cm accounts for more than 30% of the specimen length. The result in figure 2b) shows a big improvement, but the modulating signal is still very important in terms of damping characteristics.

Following this improvement, the specimen was fully rested on a wooden support. The result presented in figure 2c) shows a considerable degradation with the initial modulation existing throughout most of the signal. Moreover, the support starts to act as a secondary source of vibration, as it can be observed after the first complete decay of the impact signal.

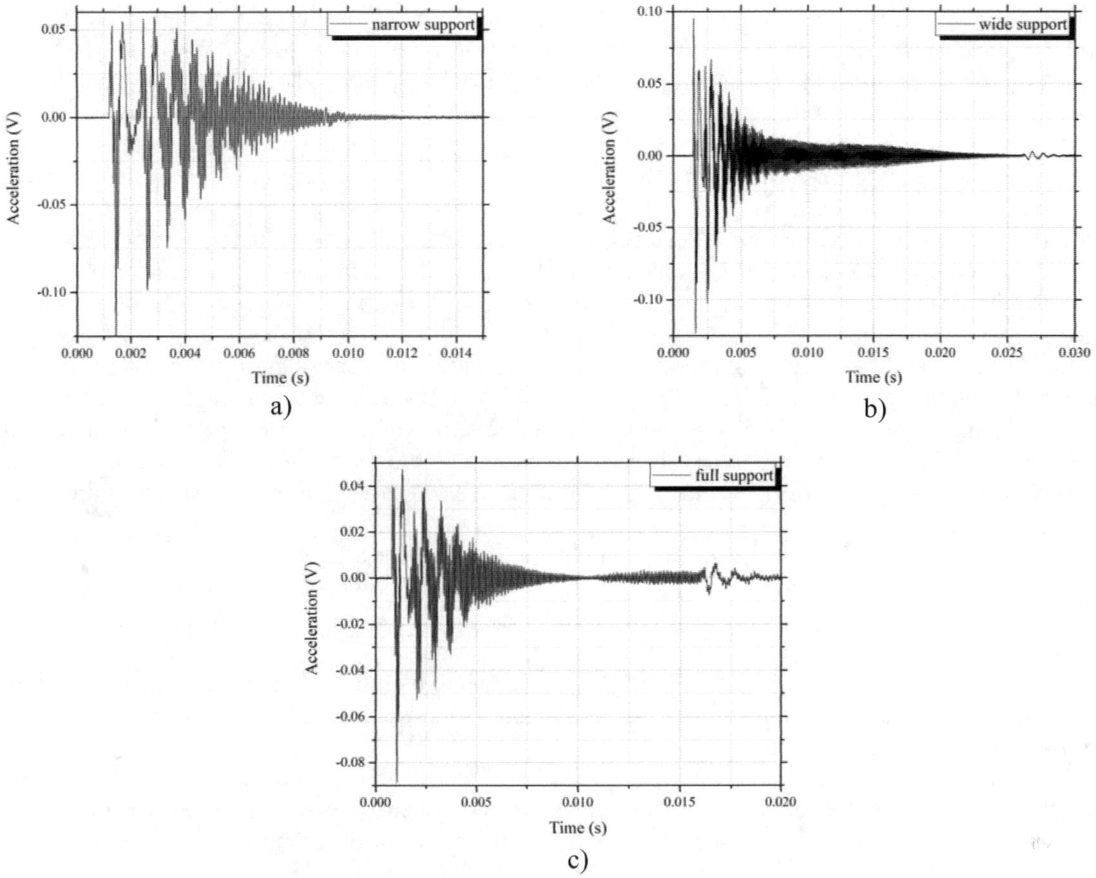

Figure 2. Impact signals through the specimens: a) narrow support; b) wide support; c) full support.

The experimental program pursued to improve these tests by vertically positioning the specimen, thus completely removing the influence of the support. The result is superior to all previous tests, with the damping of the impact signal **unmodified** by the modulating wave, as shown in figure 3.

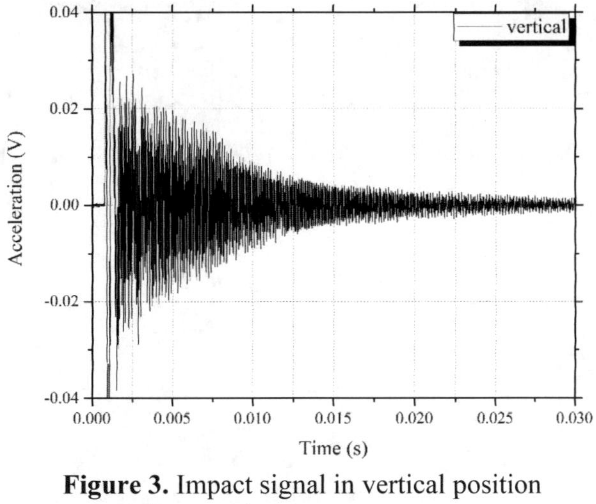

Figure 3. Impact signal in vertical position

The signal in all situations was analysed by fast Fourier transform (FFT). In every case, both the whole signal and the final, unmodulated, part were considered. It is important to compare the two types

IOP Conf. Series: Materials Science and Engineering 1141 (2021) 012019 doi:10.1088/1757-899X/1141/1/012019

of assessments, as it may be a source of error to examine too much or too little of the signal. The FFT was used to quantify both the position and the width of the resonance peaks. While the position is important for computing the dynamic modulus of elasticity, the width is related to the damping properties of the material, with a wider peak indicating more damping.

The analyses of the experiments based on supporting the specimen are presented in figure 4.

The FFT of the experiment involving the narrow support shows that the full signal contains a lot of frequency components (figure 4a). Two big peaks can be observed, at around 2 kHz and 12 kHz, respectively, and many smaller ones. Performing the analysis on the unmodulated part of the signal, at the end, two components are clearly visible (figure 4b). The 12 kHz frequency remains the higher peak in the plot, indicating it as the longitudinal resonant frequency (as expected for cement mortar specimens this size and composition). The second highest peak that can be observed is situated at around 5 kHz and can be related to the transversal mode of vibration (additional experiments – not shown here – have been performed to check for this fact). The presence of the transversal component can be related to the error in hitting the specimen. The selected signal can be seen to retain a small part of the modulating signal at around 2 kHz. An important observation related to the width of the longitudinal resonance peak, which differs greatly in figure 4a) and figure 4b), making the damping assessment unreliable in this case.

The wide support setup, which results in a cleaner signal over a longer period of time, shows the resonance to be positioned close to the same value as before. As expected, the modulating signal is less well established in the FFT plot, even when considering the full signal. The profiles in figure 4c) and 4d) are both slim, pointing towards a viable damping assessment.

Resting the specimen fully on the support does dampen the modulating signal, but the error is higher when compared to the wide support. Also, the peaks have somewhat different profiles when comparing the full signal and a selected one.

The narrow support and the full support presented the same problems when comparing peak profiles. This can be assigned to the fact that the selected signals in both cases were a small portion of the whole measurement. Restricting the analysed part was due to the big error induced by the modulating signal.

The vertical placement has the most consistent results in terms of FFT analyses (figure 5). Almost all of the measurement contains the natural vibration within the specimen. The transversal frequency was present, but not dominant. It can be removed with more careful positioning of the specimen but it was not detrimental to the fundamental longitudinal value.

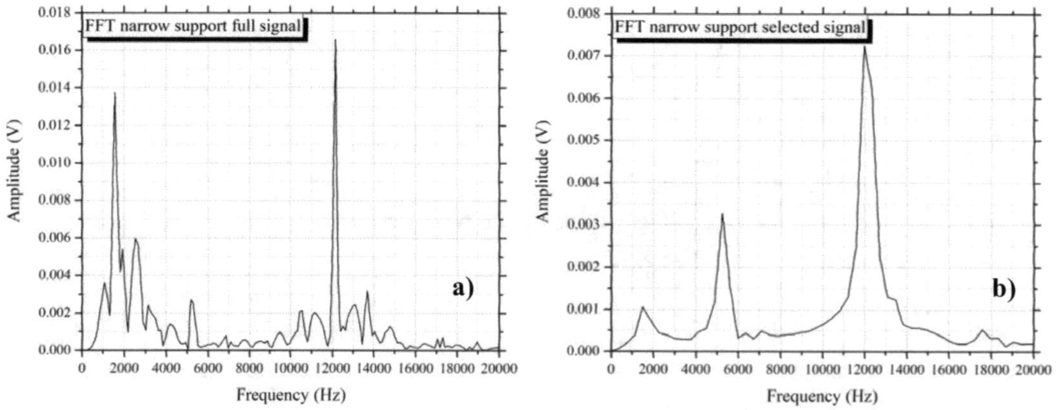

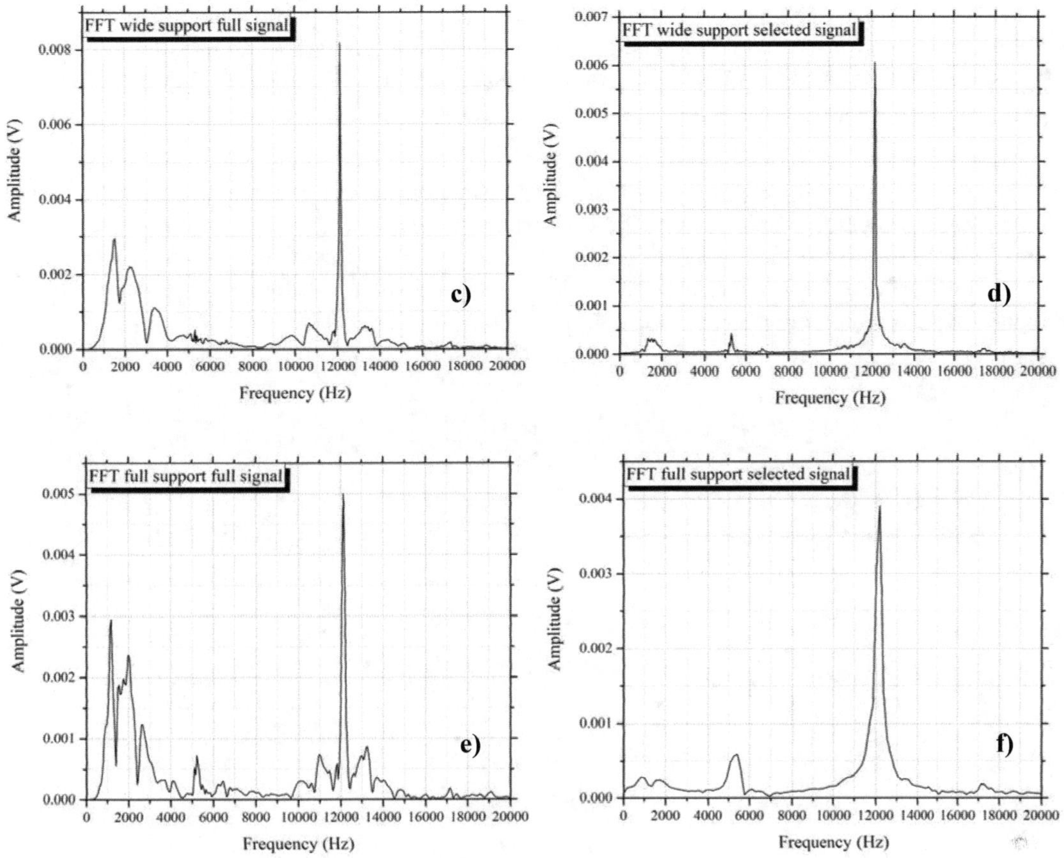

Figure 4. FFT results of experiments employing a wooden support: a) narrow support (full signal); b) narrow support (selected signal); c) wide support (full signal); d) wide support (selected signal); e) full support (full signal); f) full support (selected signal).

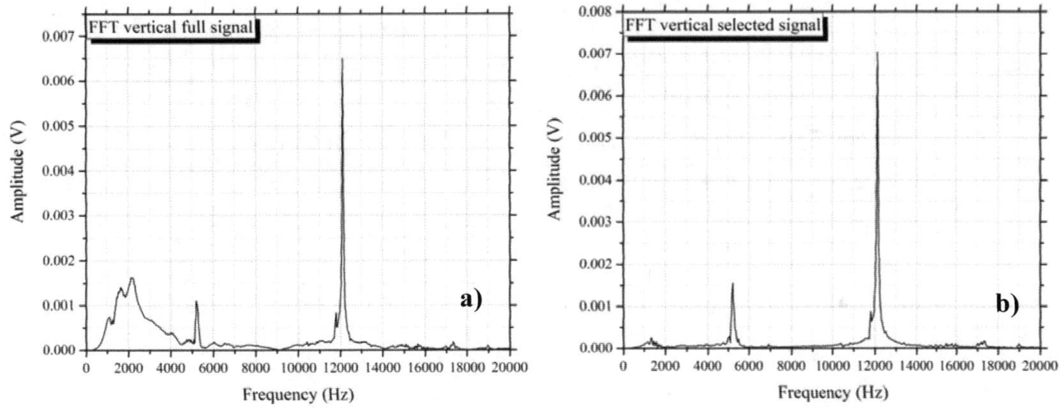

Figure 5. FFT analyses for the vertical position experiment: a) full signal; b) selected signal.

All FFT plots have been used to obtain the fundamental longitudinal frequency by analysing the peaks. A peak fitting operation has been performed, in order to reduce the errors given by the measuring devices. The fit has been done using a Voigt profile in order to take into account both machine and process peak widenings and displacements.

An example is given in figure 6 and the results for peak positions are given in table 2. Moreover, the full width at half maxima (FWHM) is obtained during this fitting, and it can be used to approximate the damping coefficient (noted *dc*) of the material applying equation (2).

$$dc = \frac{FWHM}{2 \cdot f} \tag{2}$$

The results are consistent in terms of elastic modulus. Most experiments provide a value within the 31.6 – 31.7 [GPa] interval. Notably, the wide support and vertical experiments are the most coherent, as expected. The damping coefficient varies significantly in the experiments employing a narrow support and a full one. The vertical measurement shows the most promising result, since the analysis on the full signal and the selected one gives almost the same value.

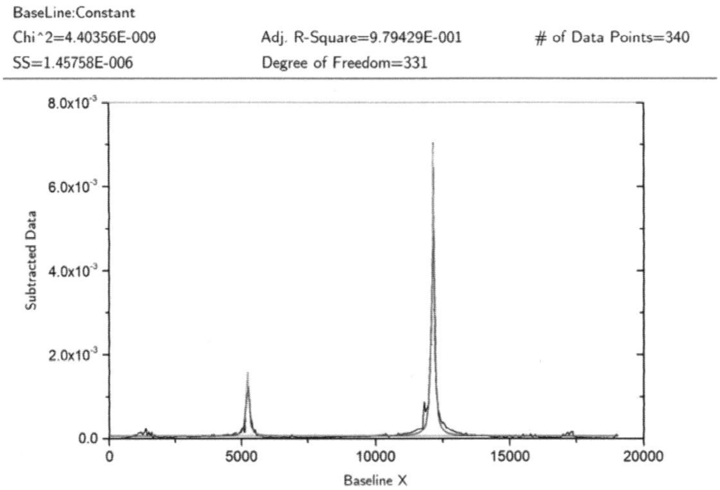

Fitting Results

Peak Index	Peak Type	Area Intg	FWHM	Max Height	Center Grvty	Area IntgP
1	PsdVoigt1	0.28895	129.09156	0.00152	5224.99545	20.87871
2	PsdVoigt1	1.09501	84.41538	0.00693	12123.54589	79.12129

Figure 6. Example of peak fitting result

Table 2. Longitudinal resonance test results

Support	Signal	f [Hz]	*FWHM [Hz]*	E [GPa]	*dc [%]*
narrow	full	12114	165	31.63	0.68
narrow	selected	12119	595	31.65	2.45
wide	full	12128	72	31.7	0.29
wide	selected	12128	92	31.7	0.37
full	full	12124	185	31.68	0.76
full	selected	12138	306	31.75	1.26
vertical	full	12117	81	31.6	0.33
vertical	selected	12123	84	31.6	0.34

4. Conclusions

An experimental program was conducted in order to assess the established procedure for the impact test method in the longitudinal configuration. Different supporting options are evaluated and an alternative to the prescribed arrangement is proposed. The validity of the proposed lay-out is demonstrated through data analysis of the impact signal. Testing the specimen in the vertical position gives the best results since no source of error and interference are present. For all other tests, the signal has been shown to have various components and proper investigation requires choosing the relevant segment of the recorded signal.

The proposed method presents advantages on both "fronts": experiment and analysis.

First of all, the experiment is easier to perform (there is no need to choose a support – material, width, a.s.o.).

Analysis-wise, since most of the signal is "clean" there is no need for in-depth examination of the measurements. This makes it promising for future industrial applications, providing a facile procedure.

Due to the reliable nature of the experiment, it can also be viable in the investigation of the damping properties of materials.

References

[1] Zhou Y, Gao J, Sun Z, and Qu W 2015 A fundamental study on compressive strength, static and dynamic elastic moduli of young concrete, *Constr. Build. Mater.*, vol **98**, pp 137–145

[2] Lee B J, Kee S H, Oh T and Kim Y Y 2017 Evaluating the Dynamic Elastic Modulus of Concrete Using Shear-Wave Velocity Measurements," *Adv. Mater. Sci. Eng.*, vol 2017

[3] Marques A I et al 2020 Modulus of elasticity of mortars: Static and dynamic analyses, *Constr. Build. Mater.*, vol **232**, p 117216

[4] Breysse D et al 2017 Non-destructive assessment of in situ concrete strength: comparison of approaches through an international benchmark, *Mater. Struct. Constr.*, vol **50**, no 2

[5] Tomsett H N 1981 Non-destructive testing of in situ concrete structures, *NDT Int.*, vol **14**, no 6, pp 315–320

[6] Hassan A M T and Jones S W 2012 Non-destructive testing of ultra-high-performance fibre reinforced concrete (UHPFRC): A feasibility study for using ultrasonic and resonant frequency testing techniques, *Constr. Build. Mater.*, vol **35**, pp 361–367

[7] Jebli M, Jamin F, Malachanne E, Garcia-Diaz E and El Youssoufi M S 2018 Experimental characterization of mechanical properties of the cement-aggregate interface in concrete, *Constr. Build. Mater.*, vol **161**, pp 16–25

[8] Popovics J S, Zemajtis J,and Shkolnik I 2008 A study of static and dynamic modulus of elasticity of concrete, *ACI-CRC Final Report, Civ. Environ. Eng. Univ. Illinois, Urbana,* no. October, p 16

[9] Test C C, Drilled T, Ag C, and Statements B 2014 *Standard Test Method for Fundamental Transverse, Longitudinal, and Torsional Resonant Frequencies of Concrete Specimens - ASTM C215 - 14* pp 1–7

Asphalt mixtures for bituminous coatings with BF slag filler

D R Nanea[1)]
[1)] "Gheorghe Asachi" Technical University of Iasi, Faculty of Civil Engineering and Building Services, "Dimitrie Mangeron" Blvd., no 43, 700050, Iasi, Romania

E-mail: ramona.nanea@gmail.com

Abstract. Asphalt mixtures, as a construction material intended to produce bituminous coatings, consists of a homogenous mixture made from natural aggregates and bitumen. Replacing one of the natural rocks, the filler with artificial filler, a metallurgical industry waste material, will determine a saving of material and financial resources. This can be used as an actual method of environmental protection by conserving the relief, protecting localities from noise produced in stone quarries and reusing a waste material. The processing of granular blast furnace slag from the Liberty Galati plant by grinding it exceptionally fine, into filler and their subsequent use in the preparation of asphalt mixtures, leads to obtaining favourable results on two of the most used currently in the country, AC16 surf50/70 and AC 22.4 bin50/70.

These values will provide an image on the behaviour resistance and their deformability under traffic. A high-water absorption can indicate a porous mixture and favours the penetration of water into the layer, water that can change its volume during freezing-thawing and can lead to cracks in the bitumen film, stripping, dislocation of pieces in the layer. A high creep can indicate a mixture which will deform under traffic, will form slopes, or will favour the phenomenon of exudation.

High stability may indicate a rigid mixture and under traffic does not have an elastic conduct, which can lead to cracks in the road surface.

Framing of the results and the appreciation of this conduct of asphalt mixture made with bf slag filler compared to a standard, leads us to the finding that the mixture is accurate and will withstand environmental factors.

1.Introduction

Asphalt mixtures, as a construction material, are obtained by mixing mineral aggregates with a bituminous binder, based on predetermined dosages, to obtain their best properties and good operating behaviour. Considering that the properties of the materials of which the asphalt mixtures are composed are different, depending on the physical-mechanical characteristics of the rocks, the size and the granularity, the properties of the bituminous binder, then the asphalt mixture formed by these components may have different properties. influenced by the properties of the component materials.
Regardless of the type of asphalt mixture to be prepared, the component materials must be:

- The natural aggregates should have the size and shape of the maximum layer size to be made of the respective asphalt mixture.
- The natural aggregates should have the shape as close as possible to the cubic one, so as not to crumble under the action of traffic.
- Its aggregates have a good mechanical strength, to give strength to the layer.
- The binder can coat all the granules, to unite them in a compact mass, in which the granules adhere to each other.
- Its filler does not react with the binder, to increase its viscosity, to give the mixture a higher resistance and a small deformation.

All these properties are directly influenced both by the characteristics of the component materials from the asphalt mixture, but also by the percentages with which each material participates in the mixture. This is an important aspect of a recipe study, to design an asphalt mixture, on this criterion depending mostly on the success of the recipe.

Given that the goal of an asphalt mixture is to be used in a communication path, as a road layer, the best behaviour will be observed both by laboratory study, in the design and testing stage. material, but also by compacting information in situ.

These are the starting premises based on which this study is made.

2.Design

To design the asphalt mixture recipe, respectively to establish the best proportions of the component materials to obtain its best characteristics, all its components were studied, both mineral and artificial aggregates and binder according to the provisions of norms, [7], [8], [9].

2.1. Mineral aggregates

Different properties of aggregates influence their performance in the composition, [14].

Regarding the maximum size, this feature is defined for the destination of the layer: a mixture will be made with the aggregates and they should be evenly distributed throughout the mass, compact, with a remaining void volume, without segregated granules or agglomerations of particles end, [15].

To prepare an asphalt mixture we used mineral aggregates with continuous granularity, which will form asphalt mixtures on the principle of concrete, respectively small granules to fill the gaps between large granules, so that the volume of gaps is minimal. This involved the use of mineral aggregates with a wide granularity, distributed on each sieve that falls into the field d min – d max, and which includes a whole suite of fractions but also rich in exceptionally fine fractions, with a size below 0,063 mm. This distribution of the granules, characterized by the granulometric curve of the mineral aggregate, will also influence the granularity of the total mineral mixture in the composition of the asphalt mixture.

In the present study, mineral aggregates from the Suseni quarry - Harghita county and natural sand from the Tupilați gravel pit - Neamt county will be used. The aggregates consist in: Grade 0/4 mm crushed stone – Tupilati gravel pit, grade 0/4, 4/8, 8/16 and 16/22,4 mm crushed stone -Suseni quarry.

2.2. Bitumen

To obtain an asphalt concrete, this being a mixture of agglomerated mineral materials with a bituminous binder, plain petroleum bitumen for roads, non-paraffin, class 50/70 was used.

2.3. Filler

Added to the asphalt mixture, the filler will influence its properties, [10], [11], such as:

- stability and creep of asphalt mixtures varies significantly with filler dosage.
- improves binder viscosity, this being given primarily by its granularity.
- reduces the volume of gaps in the asphalt mixture.
- the filler ratio influences significantly the physical-mechanical properties of the mixture, as follows:
    - if the filler-binder ratio is increased by keeping the percentage of binder constant, then the obtained mixture will be more rigid, will have a high stability and low creep, and during the execution the mixture will require mixing and a more energetic compaction. The increased rigidity is given by the increase of the softening point of the fillerised bitumen as well as by the reduction of the volume of apparent gaps. Excess filler generally leads to a hard-to-work mixture, possibly even with uncoated granules, which also favours water penetration.
    - if the filler-binder ratio decreases while keeping the percentage of binder constant, then the mixture will be more elastic, will have a low stability and a high creep. The mixture will deform under the action of a low force, and especially at high temperature. Also, a low filler content favours the leakage of the binder from the surface of the granules due to its low viscosity.

For the present study we used blast furnace slag filler from the Liberty Galati plant and limestone filler.

3.The proportion of mineral aggregates in asphalt mixtures.

A high-performance asphalt mixture was made with a mineral skeleton consisted of sorts of chipping (Suseni quarry) and bf slag filler from Liberty Galati plant, the types as followings:

- AC 16 surf50/70 and AC 22.4 bin50/70, of asphalt concrete type with chipping - hereinafter referred to as standard asphalt mixtures for surface and binder layers of road systems.
- AC 16 surf50/70 and AC 22.4 bin50/70, asphalt concrete type with chipping - hereinafter referred to as asphalt mixtures with slag filler for surface and binder layers of road systems.

The granulometric curves obtained for the four types of asphalt mixtures are shown in figures 1, 2, 3, 4.

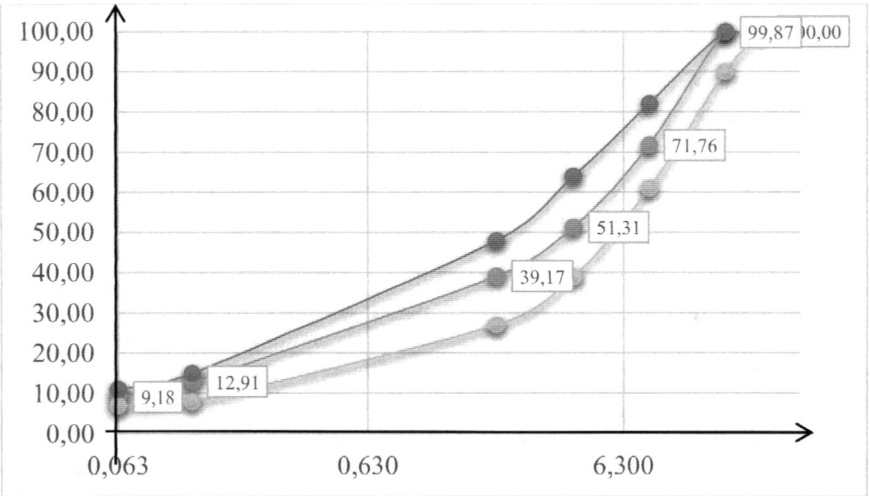

Figure 1. Granulometric curve of AC16 surf50/70 standard

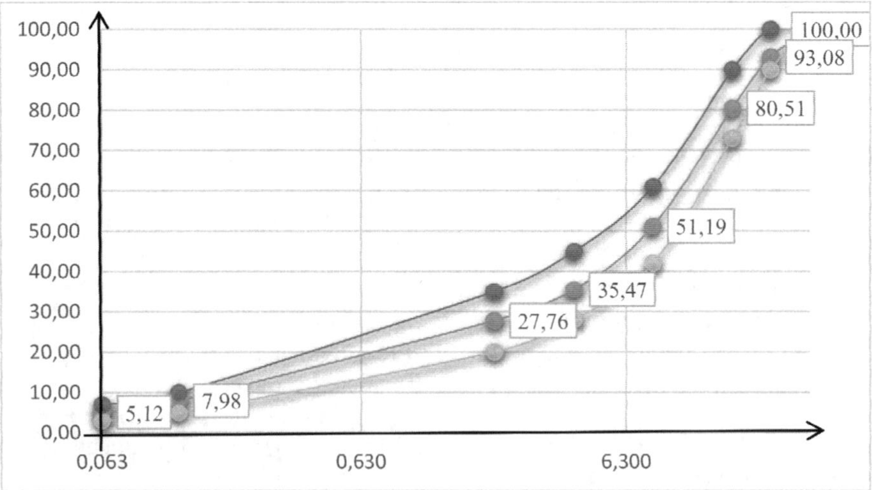

Figure 2. Granulometric curve of AC22,4 bin50/70 standard

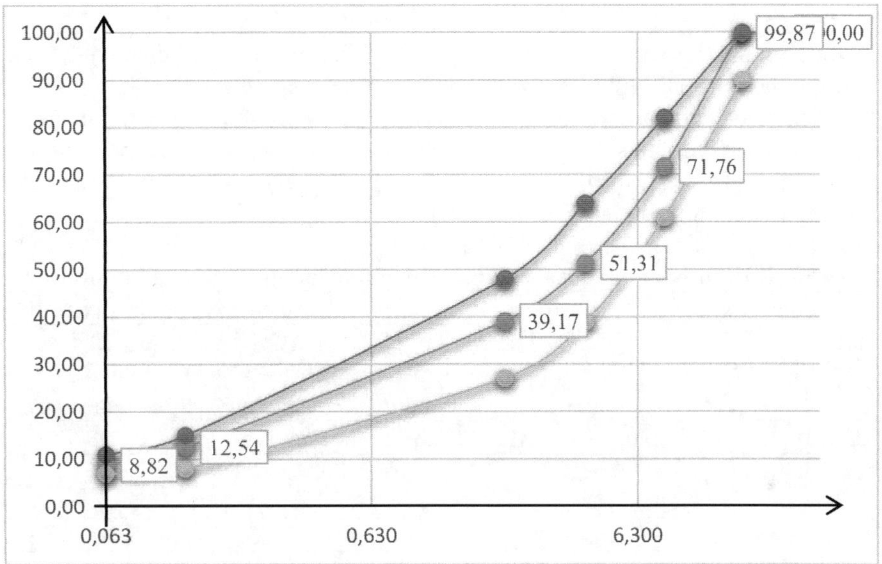

Figure 3. Granulometric curve of AC16 surf50/70 with bf slag filler

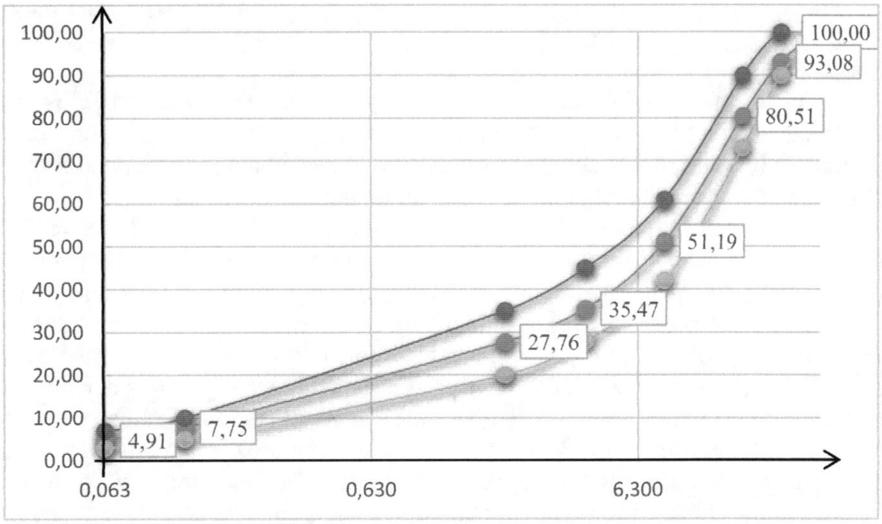

Figure 4. Granulometric curve of AC22,4 bin50/70 with bf slag filler

The asphalt mixtures prepared in the laboratory consists in the following batches:

- AC16 surf50/70 standard for surface layers with different bitumen dosage: 5.7%, 5.9% and 6.1%.
- AC22.4 bin50/70 standard for binder layers, with different bitumen dosage: 4.3%, 4.5% and 4.7%.
- AC16 surf50/70 with slag filler for surface layers, with different bitumen dosage: 5.7%, 5.9% and 6.1%.
- AC22.4 bin50/70 with slag filler for binder layers, with different bitumen dosage: 4.3%, 4.5% and 4.7%.

4.Tests

Tests were performed and the values obtained of the physical-mechanical characteristics on Marshall cylinders of the standard asphalt mixtures compared to those prepared with slag will be compared and interpreted. On each batch and each series of cylinders in the batch, the following determinations of the physical-mechanical characteristics were performed, in accordance with the provisions of the actual norms, to highlight the performances of the asphalt mixtures, [1]. Characteristic variations of the asphalt mixture and the regulated limits are shown in table 1 below.

Table 1. Results of physical-mechanical characteristics on asphalt mixture

Physical-mechanical characteristics	Requested values[1), [9]	Results on asphalt mixture			
		AC 16 surf50/70 standard	AC 16 surf50/70 BF slag	AC 22,4 bin50/70 standard	AC 22,4 bin50/70 BF slag
Apparent density, kg/mc	**Not imposed**	2,376	2,375	2,368	2,324
Stability at 60 degree C, KN	**6,5...13**	10,30	9,70	8,30	9,30
Marshall flow, mm	**1,5...4,0**	3,70	3,70	3,30	3,20
S/I report (rigidity) min, KN/mm	**1,6**	2,80	2,70	2,50	2,90
Water absorption, % vol	**1,5...5,0**	1,80	1,70	2,90	4,20

([1) Values regulated by the norm, [9])

5. Conclusions

The results interpretation was performed based on the provisions of the norm, [9], by reporting to its limits and by comparing the two types of asphalt mixtures and issuing opinions and interpretations. This interpretation represents a first stage in the evaluation of the behaviour of asphalt mixtures in traffic, compared to the current requirements on the Romanian market.

Based on the performed measurements, a first difference was observed between the apparent densities obtained on the asphalt mixture with artificial filler (bf slag filler) compared to those with natural filler, this being due to the difference in density between artificial rock and natural rock. The apparent density of filler from natural rock is higher than the filler obtained from high furnace slag filler.

There was a relatively small decrease in the stability of mixtures with artificial filler compared to those with natural filler, this being due to the fineness of the filler but also to its density.

The asphalt mixtures were deformed relatively in the same way, the flow values falling mainly in the same interval, of 3 ... 4%. However, given that the flow of asphalt mixtures prepared with artificial filler is located at the minimum end of this range, respectively close to 3% compared to those with natural filler, which have a value close to the maximum end of the range, and by reporting creep stability, it can be concluded that the stiffness is approximately constant regardless of the type of mixture used. The S / I ratio in all cases is in the range of 2 ... 3%.

A significant difference is also observed in the water absorption process, this may be due to the granularity and porosity of the filler.

Regarding with the compliance with provisions of current standards, we conclude:

- the values obtained comply with the provisions of current technical regulations according to the information related in table 1, which leads us to assess their quality, up to this point of testing, as compliant.

Regarding with comparative analysis of the two types of asphalt mixture, we conclude:

- The density of the mixtures increases as the binder dosage increases.
- Water absorption decreases as the binder dosage increases.
- Stability increases as the binder dosage increases.
- The creep increases as the binder dosage increases.

Analysing in parallel the standard asphalt mixtures and those prepared with slag filler, the same tendency of increasing / decreasing the characteristic values can be observed, along with the variation of the binder dosage which will lead us to the conclusion that bf slag mixture will have a behaviour in traffic similar with a standard one, made with natural filler.

Acknowledgments
Thanks to professor coordinator, Prof. PhD. Gugiuman Gheorghe for support in making this study possible.

References

[1] Gaman D-V, Bulau C E 2015 *Supervisor for laboratories construction tests*
[2] Lucaci G, Costescu I, Belc F 2020 *Road constructions,* Technical publishing house, Bucuresti
[3] USIRFF 2005 *Les enrobes bitumineux* – Tome 1+2, Doutes de France, RGRA, Paris
[4] Andrei R, Stoica Ct 2014 *Roads in the conception of the current generation*, Ed Soc Acad "Matei Teiu-Botez" Iasi, ISBN 978-606-582-055-5
[5] Cristea N, Cioroi L M 2007 *Metallurgical slag recycling potential*
[6] Faculty of Metallurgy and Material Science of Galati 2007 *Advanced Technologies and Materials International Conference*
[7] EN 13043:2002, *Aggregates for bituminous mixtures and surfaces treatments for roads, airfields, and other trafficked areas.*
[8] EN 13108:2016, *Asphalt mixtures. Material specifications. Part 1: Asphalt concrete.*
[9] Indicative AND 605-2016, *Technical conditions for design, preparation, and commissioning of asphalt mixtures.*
[10] Diab A, Enieb M 2018 Investigating influence of mineral filler at asphalt mixture and mastic scales, *International Journal of Pavement Research and Technology,* vol **11**, iss 3, pp 213-224
[11] Fallah Tafti M, Khabiri M M , Khani Sanij H 2016 Experimental investigation of the effect of using different aggregate types on WMA mixtures, *International Journal of Pavement Research and Technology,* vol **9**, Iss 5 pp 376-386
[12] Wang G C 2016 *The utilisation of slag in Civil Infrastructure Construction,* ISBN 978-0-08-100994-9, pp 115-129, 131-153 201-238
[13] Netinger Grubesa I, Barisic I, Fucic A, Bansode S S 2016 *Characteristics and Uses of Steel Slag in Building Construction*, Woodhead Publishing, Pages 67-82
[14] Zumrawi M M E, Khalill F O A 2017 Experimental study of Steel Slag used as Aggregate in Asphalt Mixture, *American Journal of Construction and Building Materials*, vol **1**, iss 1, pp 12-18
[15] Swathi M, Andiyappan T, Guduru G, Reddy M A, Kuna K K 2021, Design of asphalt mixes with steel slag aggergates using the Bailey method of graduation selection, *Construction and Building Materials*, vol **279** iss 0

Database assessment of the reinforced concrete water tanks built fund in Iasi county

D Covatariu[1], A Filip[2], G Covatariu[1], I O Toma[1]
[1] Faculty of Civil Engineering and Building Services, “Gheorghe Asachi” Technical University of Iasi, no. 1, Blvd. Prof. Dimitrie Mangeron, 700050, Iasi, Romania
[2] S.C. “Apavital” S.A. Iasi, no. 6, Str. Mihai Costăchescu, 700495, Iasi, Romania

Email: daniel.covatariu@academic.tuiasi.ro

Abstract. Reinforced concrete water tanks are special structures and important from a strategically point of view. In particular, they are degraded both at the exterior (aggressive environmental agents) and interior (chlorinated water). A case study based on a group of water tanks (over-ground, semi-buried, underground) which have been investigated / analysed in the period 2016-2020 was conducted based on the degradations / non-conformities identified (on site visits, in monitoring / technical expertise reports, etc.). In this regard, for each constructive type (also considering the exploitation period and storage capacity), certain types of degradations were identified. Taking into account the state of degradation of each objective studied / analysed, a scoring system was assigned. The aim of this article is to conceive an algorithm (following the analysis of the database) regarding the level of degradations for each constructive system and to propose the suitable rehabilitation procedures corresponding to the investigated water tanks.

1. Introduction

In Romania, especially in the area of Moldova, the constructive composition of reinforced concrete tanks must lead to the construction of structures that can take over the demands during operation, so as to ensure and comply with the fundamental requirements of strength, stability and durability.
In the case of drinking water tanks, water-tightness and water quality retention are essential.

Water quality regulations and standards have required operators/distributors of drinking water in major cities/metropolitan areas to monitor and ensure a programme that must include the control of the efficiency of the treatment process, in particular the disinfection and quality of drinking water produced, distributed and used.

In Iasi County the water treatment process is carried out by using a complex set of procedures and installations. Chlorination is the most widely used method of water treatment (for technical and economic reasons). Treatment of water by chlorination has positive effects on the potability character by improving the qualitative parameters, respectively, can have negative effects.

Inside, the attack of chlorine ions (due to the chlorination process) has a negative impact on reinforced concrete water tanks (by decreasing the service life/ durability of the construction) often leading to the degradation of the structure by the expulsion of the coating with concrete and corrosion of the reinforcement (Figure 1).

On the outside, water tanks are mainly degraded due to the action of aggressive environmental agents (temperature, humidity, etc.) frequently manifested by the removal of plasters and/or finishing layers (on different areas) (Figure 1).

Figure 1. Exterior wall degradation: a) interior, b) exterior

Regardless of the constructive type (over-ground, semi-buried and underground), storage capacity and operating time, the structural elements of reinforced concrete water tanks (perimetral wall, horizontal plates) are "attacked", both inside and outside (Figure 2.a, 2.b).

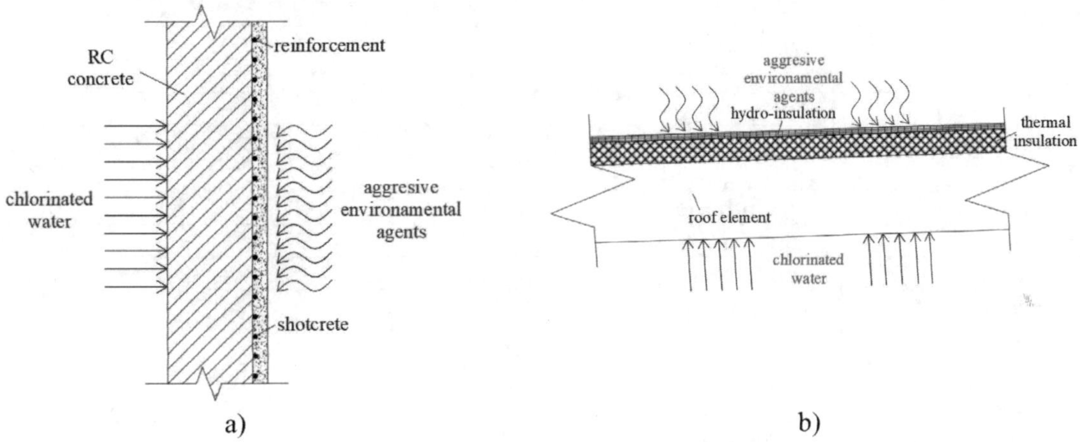

Figure 2. Water tank structure:
a) vertical wall, b) covering horizontal plate

In this respect, a thorough inventory of the types of degradation (depending on the constructive type) may highlight the need (or not) for the performance of structural and/or non-structural rehabilitation works to increase operational safety and the prescribed duration of service.

2. Identification of types of degradation

The identification of deteriorated water tanks was carried out as a result of site investigations carried out between 2016 and 2021, in this respect a database (photographic characteristics and relevance) consisting of 29 objectives investigated (Table 1) has been created.

Table 1. The list of water tanks investigated

Water tank's location	Constructive type	Volume capacity [m3]	Building year	Rehabilitation year
1. Sărărie	underground	100	2006	2017
2. Aeroport	semi-buried	200	1981	-
3. Boureni	over-ground	200	2002	2017
4. Satu Nou	semi-buried	200	2005	2015
5. Strunga	over-ground	500	1982	2016
6. Miroslava	over-ground	5.000	1981	2020
7. Păcurari	underground	5.000	1974	-
8. Târgu Frumos	over-ground	5.000	1985	-
9. Aurora	over-ground	10.000	1974	2017
10. Păcurari	over-ground	10.000	1980	2018
11. Valea Lupului	semi-buried	5.000	1974	2015
12. Deleni	underground	100	1991	-
13. Tomeşti	semi-buried	300	1995	2014
14. Bucium	underground	500	1995	-
15. Hlincea	over-ground	150	2006	2015
16. CUG	over-ground	1.000	1995	-
17. Aroneanu	semi-buried	200	2006	2016
18. Agronomie	underground	2.000	1966	-
19. Leţcani	semi-buried	500	1995	2018
20. Galata	semi-buried	200	1981	2012
21. Vişan	underground	500	1981	2019
22. Andrieşeni	underground	150	2014	-
23. Gorban	underground	500	1984	-
24. Roşu	semi-buried	1000	1984	2017
25. Răducăneni	over-ground	100	1982	2016
26. Munteni	over-ground	200	2013	2017
27. Belceşti	over-ground	2500	1982	2013
28. Tansa	semi-buried	300	2004	-
29. Uricani	underground	500	2013	-
30. Podu Iloaiei	semi-buried	500	1981	-

Statistically, from the constructive type point of view, the built fund is equilibrated:

- over-ground – 11 (37%);
- semi-buried – 10 (33%);
- underground – 9 (30%).

Taking into account the constructive type of water tanks, the existing degradations/non-conformities at the operational objectives (of different types and degrees of intensity) were quantified following the investigations carried out (Table 2).

Table 2. Types of degradation / non-conformities of the investigated tanks

Constructive type	Structural Element	Degradation / Non-conformities
Over-ground water tank	Covering Plate	- cracks/detachments/dislocations of the concrete coating; - corroded reinforcements (significant surfaces); - reduction of the cross-section of the armature (accelerated corrosion due to the action of chlorine ions); - deformations of the element's soffit (in some cases); - moisture / dampness / efflorescence (insufficient ventilation); - detachment/ out-of-plane of waterproofing, metal sheet (at the attic's level); - local water accumulations (due to execution defects - slightly incorrect slopes); - indoor water infiltration (caused by microcracks of waterproofing); - coats of protection detached / exfoliated.
	Vertical Wall	- 0.3-0.5 mm wide cracks / segregations / detachments of the shotcrete's coating; - unprotected reinforcement bars (local); - reinforcements section's significant reduction (corrosion); - settlements (the base of the elements); - expulsion of the shotcrete coating on extended surfaces (strongly corroded reinforcements); - removal of plasters / finish layers on small / extended areas; - high humidity (on the interior surface); - out-of-plan/detaching of the thermo-system; - mould / efflorescence / dampness on extensive surfaces.
	Support plate	- cavitational erosion / abrasion (reduced areas); - cracks in the finish layers (hydro-insulation); - very high humidity values / general water accumulations; - dampness stains / mould (extensive areas)
	Foundation	- visible settlements of the perimetral pavement; - cracks / areas with crushed concrete (local surfaces); - leaky joint (between the vertical wall and the sidewalk); - infiltrations (caused by meteoric waters); - concrete pavement detachments (local areas).
Semi-buried water tank	Covering Plate	- cracks/detachments/dislocations of the concrete coating; - corroded reinforcements (significant surfaces); - reduction of the cross-section of the armature (accelerated corrosion due to the action of chlorine ions); - moisture / dampness / efflorescence (insufficient ventilation); - detachment/ out-of-plane of waterproofing, metal sheet (at the attic's level); - local water accumulations (due to execution defects - slightly incorrect slopes); - indoor water infiltration (caused by microcracks of waterproofing); - coats of protection detached / exfoliated.
	Vertical Wall	- cracks / segregations / detachments of the shotcrete's coating; - unprotected reinforcement bars (local); - reinforcements section's significant reduction (corrosion); - expulsion of the shotcrete coating on extended / reduced surfaces; - high humidity (on the interior/exterior surface); - out-of-plan / detaching of the thermo-system; - mould / efflorescence / dampness on extensive surfaces.

	Support plate	- cavitational erosion / abrasion (reduced areas); - cracks in the finish layers (hydro-insulation); - very high humidity values / general water accumulations; - dampness stains / mould (extensive areas).
Underground water tank	Covering Plate	- dislocations of the concrete coating (local); - corroded reinforcements (significant surfaces); - reduction of the cross-section of the armature (accelerated corrosion due to the action of chlorine ions); - moisture / dampness / efflorescence (insufficient ventilation); - local water accumulations (due to execution defects - slightly incorrect slopes); - indoor water infiltration (caused by microcracks of waterproofing); - dislocations of the plaster / protection's layers of the waterproofing significantly degraded (due to the action of water resulting from precipitation, aggressive environmental agents, fillers settlements, etc.);
	Vertical Wall	- microcracks / segregations / local detachments (in some cases) of the concrete coating; - partial reduction of the reinforcements section (corrosion); - tamping of the earth filling; - removal of plasters/ finishing layers on reduced areas (inside the structure) - high humidity (on the inner surface); - significant degradations (precipitation, frost-thaw cycles, etc.) of the waterproofing system;
	Support plate	- cavitational erosion / abrasion (reduced areas); - cracks in the finish layers (hydro-insulation); - very high humidity values / general water accumulations; - dampness stains / mould (extensive areas).

Most of the water tanks are over-ground, Iasi County being a seismically active zone (ag = 0.25g), due to the value of the peak acceleration in conjunction with the constructive type, they may be susceptible to seismic action.

The objectives analysed present relatively high periods of exploitation (50 years), some bearing the effects of numerous earthquakes, which requires the obligation and/ or the need to establish a rigorous programme of special monitoring of in-time behaviour (for all types of construction, but especially in the case of over-ground water tanks).

Most of the objectives analysed are at an early stage of degradation (requires hydro- and/or thermal rehabilitation), showing superficial non-conformities that did not affect the structure's strength.

At the same time, during visual inspections, several tanks have been identified which require intervention/structural rehabilitation measures, due to the lack of maintenance/maintenance work in conjunction with the action of treated water (indoor) and aggressive environmental agents (outdoor).

In this respect, the study of factors influencing the durability of a particular structure may be of major importance in subsequent decision-making of intervention on the structure.

The Ishikawa Diagrams (cause-effect diagrams) figure 3 was used as a tool for identification of the important causes that lead to the most important problems that have been observed.

Even with simple graphics, this kind of tool can help to define procedures in monitoring of the life-cycle of water tanks and brings several benefits to the organization and to the team using it.

Some of the benefits can be:

- identification and ranking of possible causes
- a better perception of the problems,
- a better understanding of the causes of the problem and their relationships,
- to highlight other causal relationships (potential risks or causes of the desired effects),
- visualization that makes the future analyses easier, a.s.o.

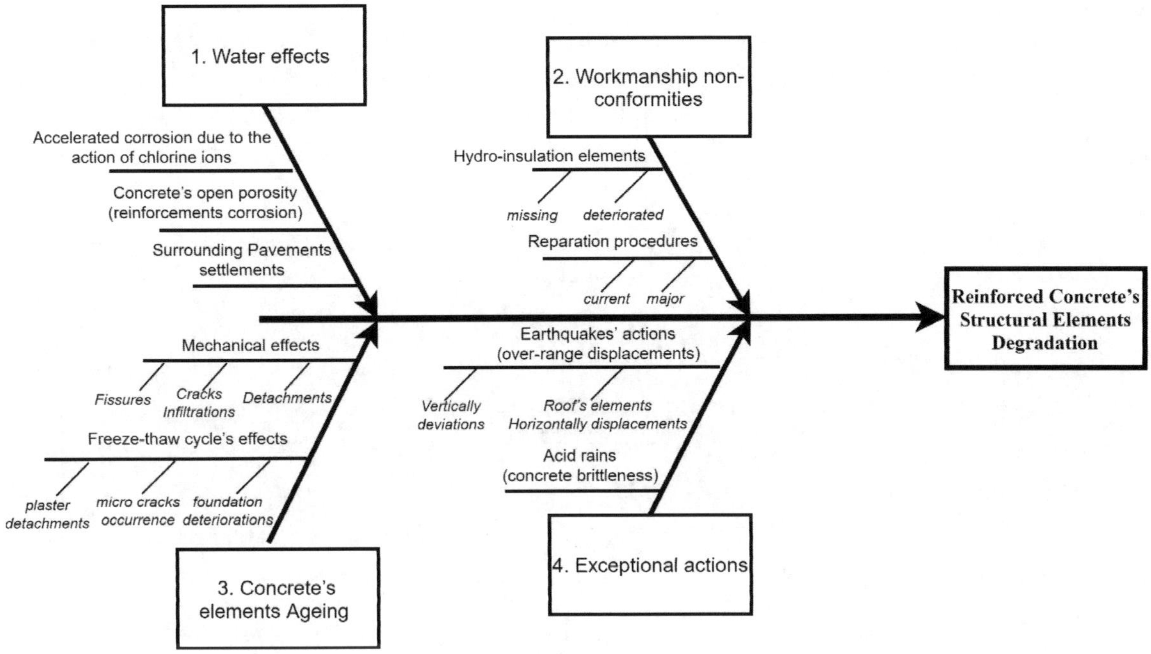

Figure 3. Ishikawa diagram for finding the causes of the problem

3. Evaluation of the water tank database – Iași County

The rehabilitation of special water storage buildings is a complex process. Technical intervention procedures must be designed in such a way that the structural system will be not affected while the level of safety in service will be increased.

From the point of view of the degradations/non-compliances found and the specific actions on these types of special structures, the following types of rehabilitation works have been identified:

- Structural rehabilitation;
- Water-proof rehabilitation;
- Operational rehabilitation;
- Thermal rehabilitation.

In this respect, the classification of the technical condition of the construction and the intervention measures for each situation have been carried out for the development of non-structural/structural rehabilitation procedures (Table 3).

Table 3. Classification of tanks investigated according to the state of degradation found

Degradation level	Technical state	Intervention procedures
1	Very Good	Proper use (preventive maintenance)
2	Good	Performing current repair work (normal maintenance)
3	Acceptable	Non-structural rehabilitation work
4	Week	Need for consolidation work (structural rehabilitation)
5	Very Week	Structural instability of the tank (replacement)

Taking into account the technical condition of the analysed database (according to Table 2) and knowing that more than 400 drinking water tanks (in various stages of degradation) are located in the area of Moldova County, it was highlighted the need to develop an algorithm to streamline the rehabilitation processes of reinforced concrete water tanks (Figure 4).

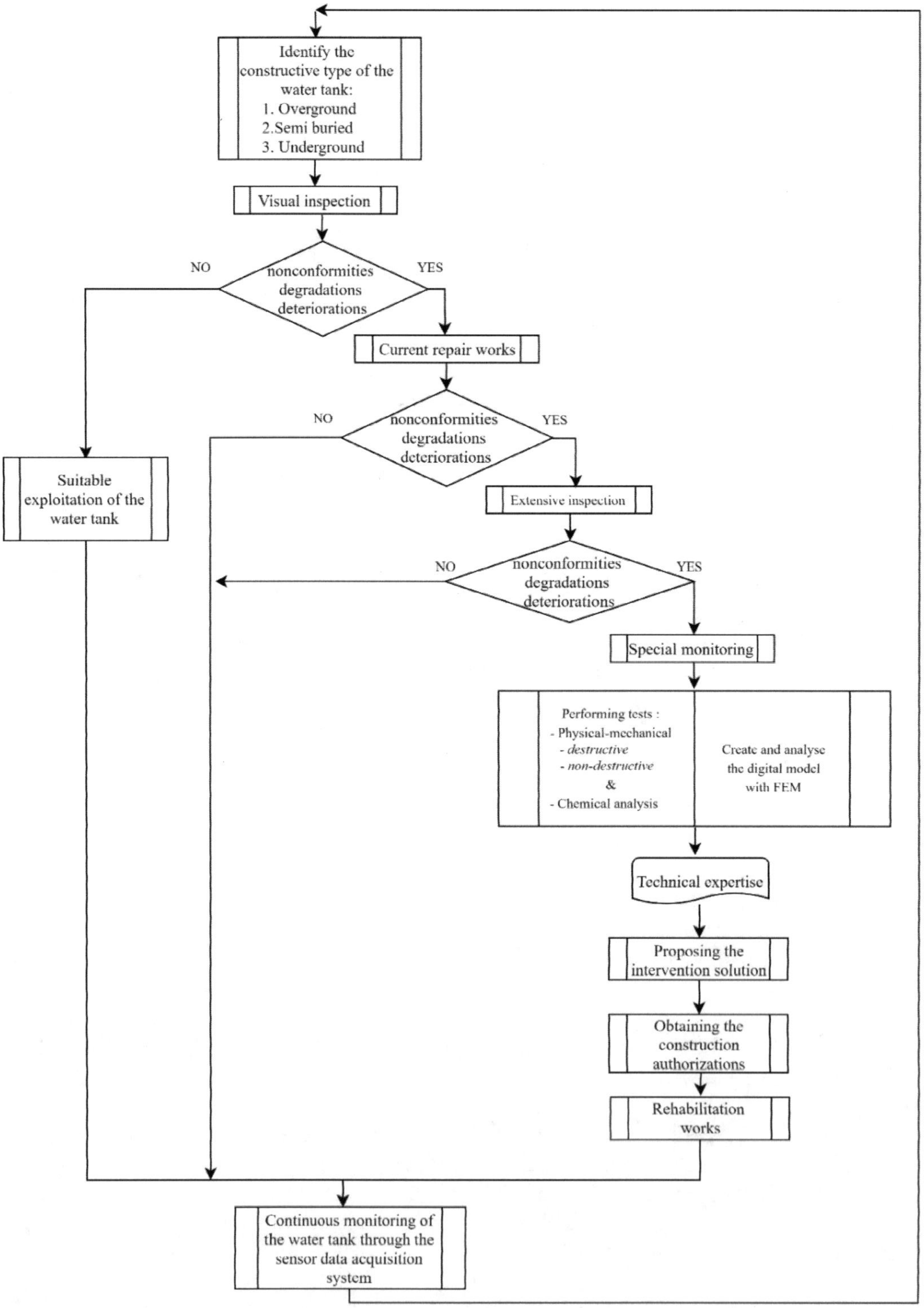

Figure 4. Efficiency Algorithm for the rehabilitation processes of reinforced concrete water tanks

Following the investigation of the tanks (according to the algorithm Figure 4), the detected degradations (Table 2) and the classification of degradation levels (Table 3) an inventory of the tanks was carried out on the principle of priority of rehabilitation/consolidation intervention (Table 4).

Table 4 – Degradation level classification (rehabilitation priority)

	Constructive type	Building / rehabilitation year	Degradation level
12. Deleni	underground	1991 / -	5
5. Strunga	over-ground	1982 / 2016	4
6. Miroslava	over-ground	1981 / 2020	4
21. Vişan	underground	1981 / 2019	4
29. Uricani	underground	2013 / -	4
2. Aeroport	semi-buried	1981 / -	3
14. Bucium	underground	1995 / -	3
16. CUG	over-ground	1995 / -	3
19. Leţcani	semi-buried	1995 / 2018	3
20. Galata	semi-buried	1981 / 2012	3
22. Andrieşeni	underground	2014 / -	3
23. Gorban	underground	1984 / -	3
27. Belceşti	over-ground	1982 / 2013	3
30. Podu Iloaiei	semi-buried	1981 / -	3
4. Satu Nou	semi-buried	2005 / 2015	2
9. Aurora	over-ground	1974 / 2017	2
10. Păcurari	over-ground	1980 / 2018	2
13. Tomeşti	semi-buried	1995 / 2014	2
15. Hlincea	over-ground	2006 / 2015	2
18. Agronomie	underground	1966 / -	2
24. Roşu	semi-buried	1984 / 2017	2
25. Răducăneni	over-ground	1982 / 2016	2
26. Munteni	over-ground	2013 / 2017	2
28. Tansa	semi-buried	2004 / -	2
1. Sărărie	underground	2006 / 2017	1
3. Boureni	over-ground	2002 / 2017	1
7. Păcurari	underground	1974 / -	1
8. Tărgu Frumos	over-ground	1985 / -	1
11. Valea Lupului	semi-buried	1974 / 2015	1
17. Aroneanu	semi-buried	2006 / 2016	1

4. Final remarks

The Ishikawa Diagram brings a lot of benefits due to creating procedures for identifying (very fast) the causes of failures/flosses and improving the monitoring and maintenance of the facilities. Also, if there are complementary used others methodologies (such as brainstorm meetings, which potentiate meeting output and help the team to present their ideas and points of view) the benefits are fulfilled.

The database created to monitor the operation's behaviour of reinforced concrete drinkable water-tanks structures important information on their technical condition. The information contained and updated can form the basis of decisions regarding the imminence of repair/rehabilitation, the justification of these decisions can bring great benefits to the organization.

Also, an algorithm (following the analysis of the database) regarding the level of degradations for each constructive system of the built fund of water tanks is compulsory, in order to identify the suitable rehabilitation procedures and to maintain the fundamental requirements of strength, stability and durability.

The creation, updating and operation of such a database (with regional and/or national inputs) thus becomes mandatory in order to optimize the management of the drinking water supply network in Romania.

5. References

[1] GT 052/2002 – *Guide to investigate the technical condition of operating tanks, decanters and meta-tanks*;
[2] Law 458/2002 on the quality of drinking water.
[3] PMI 2019 *The Standard for Risk Management in Portfolios, Programs and Projects*
[4] Covatariu D, Filip A, Covatariu G 2019 The Impact of Thermal Rehabilitation of the Reinforced Concrete Water Tanks Romanian Built Fund over its Lifecycle, *International Scientific Conference CIBv2019*, 1 –2 November, Braşov, România, *IOP Conf. Ser.: Mater. Sci. Eng.* **789** 012014
[5] Covatariu G, Covatariu D, Filip A 2019 Planning Risk Management in case of a Drinkable Water Tank Facility Operation, International Scientific Conference CIBv2019, 1–2 November, Braşov, România, *IOP Conf. Ser.: Mater. Sci. Eng.* **789** 012015;
[6] Covatariu D, Filip A 2018 A comprehensive approaching of the water tank durability's affecting factors in different stages, *Buletinul Institutului Politehnic din Iaşi, secţiunea Construcţii – Arhitectură*, **64** (**68**) (4): 142 – 156

IOP Conf. Series: Materials Science and Engineering 1141 (2021) 012022 doi:10.1088/1757-899X/1141/1/012022

Indirect determination of the measurement accuracy of the FMK-004 track geometry measuring car used on Hungarian rail network

R Nagy[1], F Horvát[1]
[1] Department of Transport Infrastructure and Water Resources Engineering, Széchenyi István University, Győr 9026 Hungary

E-mail: nagy.richard@sze.hu

Abstract. During the examination of the Hungarian national network for the period from the 2nd half of 1999 to the 2nd half of 2013, had to answer to the reason why the geometrical errors are able to decrease, despite the fact that no regulation intervention was done on the line section during the study period. This anomaly leads us to the procedure in which it became important to determine the accuracy of the FMK-004 measuring car (FMK in English: railway superstructure measuring car). With the VBA code written for the research task, the data set was optimized, then deterioration functions were fitted to the data, and where outliers were detected, the extent of the outlier was analysed on a statistical basis. This determines the measurement accuracies for each of the three metrics separately. Furthermore, it can be clearly demonstrated by the examination of the national network that in 90% of cases there is a series of geometrics that have a value that decreases compared to the previous half-year, while other geometrics value still is increasing.

1. Introduction

Our age is characterized by major changes that affect not only our everyday lives, our tools but also our machine analysis and data processing. In scientific life, a significant part of the results of publications is produced by some method assisted by artificial intelligence and then analysed [1]. An examination was needed whether higher-level mathematical models can be used more effectively in the case of models describing railway track degradation than the various simpler prediction procedures used so far [2], [3] a procedure is required for its design that predicts with good certainty the change of the state of the track geometry under a given traffic load [4]. This is necessary because intervention timing based on correct and reliable data would significantly improve cost-effectiveness at the national level. Several Hungarian researchers [5] and international [6] publications have long drawn attention to the fact that the reduction of the cost of construction works, be it new construction or maintenance, can be achieved with thorough preparation and proper organization. But in order for this prediction to be described with sufficient certainty, reliable data is needed, data whose accuracy and magnitude of error are known. This short article summarizes the procedure used to determine the measurement accuracy of a measuring train.

2. Theoretical background

The deterioration process of railway track geometry is influenced by several factors, which mainly consist of the traffic passing through it and the non-negligible environmental impacts. Deterioration of the non-intervention track can be properly described by the following function:

$$C = C_0 * e^{\alpha * m * v^2} \tag{1}$$

where:

C - general parameter characterizing the geometric condition of the railway track,
C_0 - general parameter characterizing the initial geometric condition of the railway track,
α - "track sizing" factor depending on the track structure,
m - the rolling traffic load [br. tons],
v - the equivalent speed [m * s-1].

In the deterioration process shown in Figure 1, the values of C on the αmv^2-axis in the range 0 to 1 are considered. According to the Hungarian regulations, $\alpha mv^2=0$ is the construction state, and the geometry cannot reach the state 1 value because it must be preceded by an intervention.

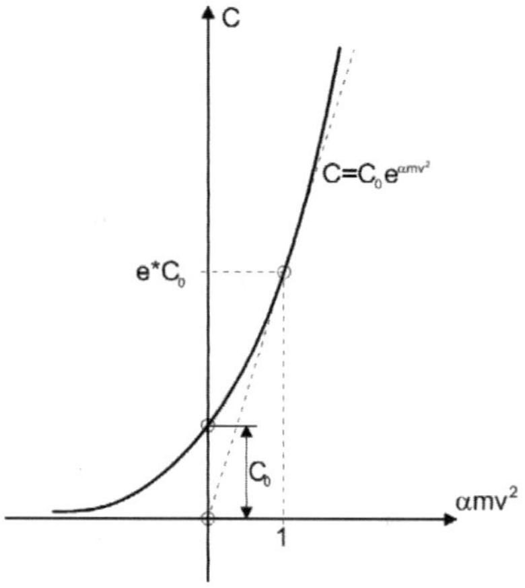

Figure 1: The exponential deterioration of the rail geometry according to Dr. Pál Vaszary[7]

From a technical point of view, the deterioration of the general geometric quality can be attributed primarily to the effect of railway trains on the track, and secondly to the effects of weather and on-site circumstances. The condition of the railway substructure and superstructure is continuously deteriorating geometrically and structurally from the date of construction as a result of rolling traffic loading. Rails are worn by wheels; fastenings are loosened due to lowering and ascending movements in the load cycles. The sleepers are increasingly pressed into the ballast; the crushed stone bedding grains are pressed into the additional layer or into the earthwork; and while the subsoil does not reach final consolidation, the earthwork also sinks. In the case of superimposition of all these effects, different dimensional deviations and then position errors develop in the track. The relative geometry of the railway track is characterized, by three quantities:

- alignment; (Y),
- longitudinal level; (Z)
- track twist (X).

The units of measurement of these numbers are the distortion-free measurement results of the calculation in the original wavelength range, which is done according to MSZ EN 13848-2.

The general track geometry classification is based on qualification lengths, the station spacing, line and line network are classified on the same basis. There are several methods using measured data, which are collected in each 0,25 m: area-based evaluation, error-maximum (peak-to-peak) evaluation, and projection-based evaluation. During the general qualification, the measuring car forms measuring numbers and qualification numbers for the qualification lengths, in the case of my work 500 m. In general, the sign of longitudinal level is Z (average of the values measured on the left and right rails), the alignment characteristic is Y (the average of the values measured on the left and right rails), and that of the track twist is X (see fig. 2). The measuring numbers are:

•Longitudinal level $$Z = \frac{Z_{left}+Z_{right}}{2} \tag{2}$$

•Alignment $$Y = \frac{Y_{left}+Y_{right}}{2} \tag{3}$$

•Track twist $$X = X_{(for\ base\ of\ 2,5m)} + X_{(for\ base\ of\ 6m)} \tag{4}$$

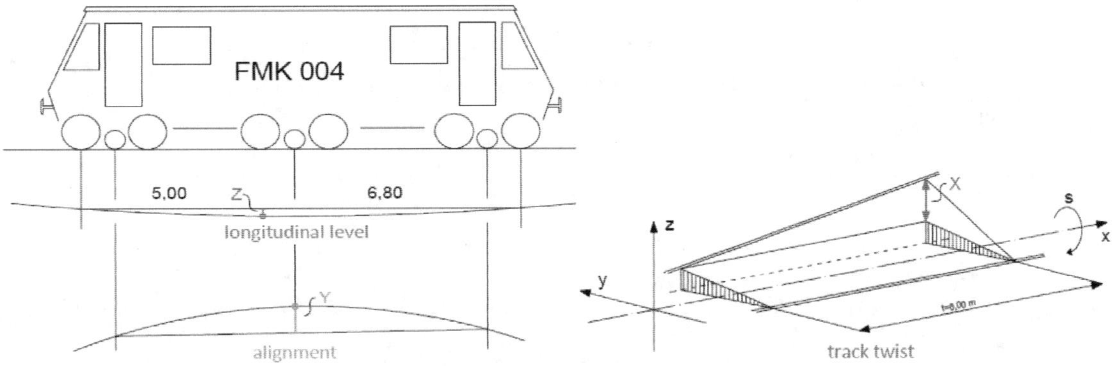

Figure 2: FMK - 004 measuring car

Measuring numbers are used to create SAD qualifying number (Summary qualifying number weighted with ADded track twist), which describes the general geometric condition of the track. It is calculated to a given qualifying length (earlier 500 m, currently 200 m).

$$SAD = (Z + Y + X) * \frac{1}{3} \quad (5)$$

3. Analytical program

Besides the evaluation graphs and tables, the railway engineers can use MS Excel as well, to analyse all the geometrical evaluation data of the railway track. It was clear that if a new program is needed it must be created in Visual Basic for Applications. Other programs had been used like MATLAB, which is a special program system developed to perform numerical calculations, as well as a programming language. The software system developed by The MathWorks is capable of performing matrix calculations, representing functions and data, implementing algorithms, and creating user interfaces. Although the software is purely numeric, it can also display mathematical expressions graphically by adding the MuPAD package. Furthermore, in the artificial neural network procedure were assisted. MATLAB is able to work with and use data from a well-prepared Excel workbook. A procedure (see fig 3) had been developed that can be easily analysed and modelled from the existing data table but can also be clearly interpreted for others.

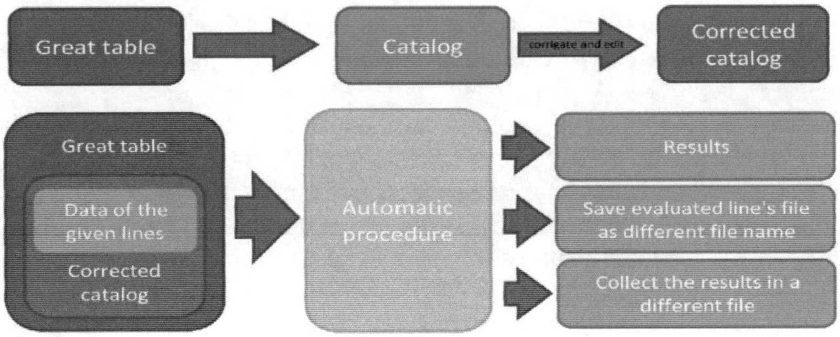

Figure 3: The automatic procedure of the analytical program (own work)

The basic idea was to find characteristic sections suitable for the purpose of the study and then to analyse the geometric deterioration by comparing different measuring and qualification numbers. As the number of subtasks to be solved increased, a whole set of procedures was finally born as a result of our work. A computer procedure had been created that automatically processes, analyses and uses the measurement and qualification numbers provided by the FMK-004 measuring car:

- collects the specified number of semi-annual data sets which have not been repaired,
- fits linear, exponential, logarithmic and power-based regression descriptive functions to the collected data sets,

- collects, organizes the function parameter and deterministic coefficient of each regression equation and writes the equation specific to the given line from all four types, which is plotted on a graph,
- the parameters of the equations are plotted separately on a histogram, and new ranges can be selected according to the distribution, for which it recalculates the parameters of the given function type,
- calculates the correlation between the selected measuring and rating numbers, i.e. the magnitude of the linear relationship between two values,
- calculate the deviation of the missing or outliers from the regression-matched equation of the data set,
- calculate the time difference function and spare time function of the selected lines,
- summarizes the predictions made by the seven models [8] and analyses their accuracy based on the mean squared root deviation.

This program can provide a real processing system for programming in the industry for knowledgeable engineers. Because it is open-source, there is scope for development and customization.

At the following links (QR-codes, see fig. 4), the running of the program can be observed at a slow speed, for demonstration purposes:

Figure 4: The presentation of certain parts of the program

4. Measurement accuracy of FMK-004 measuring car

With the help of the program which performs the evaluation of millions of data with various settings, research on the entire national network of Hungary has been carried out. The examined period lasted from the 2nd half of 1999 to the 2nd half of 2013. In the course of the performed examinations, there are 14434 sections of 500 m where there is a time interval of 6 half-year when the SAD qualification number is constantly increasing (there was no regulatory intervention) taking into account the pre-set error limit of 5%. The length of the deterioration period was adjusted to six half-years based on preliminary studies.[2], [4]

The extent to which the alignment was needed, level and twist in these cases determine the magnitude of the SAD rating. With the help of another program written for this purpose, analysation has performed on this issue on the national network. After the successful examination 1402 of the 14434 sections with 500 m length remained, where both the measuring and the qualification number increased for 6 consecutive half-years, i.e. on these sections the track deterioration is monotonous in the given period.

This article attempts to answer the question of why the value of the three measuring numbers decreases in the additional 13023 pieces of 500 m sections.

Programming was again used to help us conduct the study. The first really interesting question was how much of the weight of the different measuring numbers in the rating? Then, analyses had to be made around the anomaly of time series showing geometrical improvement without regulation intervention, since the geometric condition of the track could not improve "by itself".

The calculations of the dominance studies did not provide an answer as to the reason why the measuring numbers are able to decrease, despite the fact that no work was done on the line section during the study period.

It is important to note that the measuring car is calibrated as required in all cases. Furthermore, the measurement inaccuracies published in this article determine the measurement accuracy based on data from the entire national network. It does not take into account the differences in the condition of the railway sections. On the other hand, it can be clearly demonstrated at the national network level that in 90% of cases, for some reason, there is a series of measuring numbers that have a value that decreases compared to the previous half-year, in the case of no intervention.

It is important to emphasize that the data that were examined belonged to the whole national network, and the 13032 data series is a series of six half-yearly measurement and qualification numbers for a 500 m section, where the SAD number always increases while one of the measuring numbers decreases compared to the previous half.

Furthermore, the determination of how much the improving value deviated from the value calculated by linear regression had much importance. This is the reason why another program was written that calculates the difference from the pre-and post-average of the decreasing metric number between a metric and a linear regression value. In the case where the decreasing measuring number is at the beginning or end of the timeline, enter the expected rate of deterioration backwards from the next two values or forward from the two values preceding it. This is how the difference could had been calculated. It should be noted that the outliers were not considered when constructing the descriptive equation by the regression procedure.

The analyses had to be continued with newer studies that clearly showed at the national network level that in 90% of cases there is, for some reason, a series of measuring numbers that have a declining value compared to the previous half-year without intervention. For these examined data sets, see Figure 5. and Table 1.

Table 1. Formatting sections, subsections and subsubsections

	Sum	SAD	SAD + Z	SAD + Y	SAD + X	Z	Z + Y	Z + X	Y	Y + X	X	SAD + Z+ Y	SAD + Z + X	SAD + Y + X	Z + Y + X	SAD + X + Y + Z
Piece	13023	0	14	7	14	2613	130	442	5233	538	3866	20	104	29	13	0
Percentage	100,0%	0,0%	0,1%	0,1%	0,1%	20,1%	1,0%	3,4%	40,2%	4,1%	29,7%	0,2%	0,8%	0,2%	0,1%	0,0%

where:

Sum - 13032 data series are the series of six half-yearly measurement and qualification numbers for a 500 m section, where the SAD number always increases, while one of the measuring numbers decreases compared to the previous half-year

SAD - Shows case numbers where SAD increases but SAD decreases. By definition, there is no such case, it is necessary to check.,

Z - 2613 means that of the 13023 sections there are 2613 pieces of 500 m sections where the SAD numerical value increases, of the three measuring numbers, the Y and X numbers increase, but the Z value decreases.

Y+X - 538 means that out of the 13023 sections, there are 538 are pieces of 500 m sections, where the numbers of the SAD qualification number and the Z measuring number increase compared to the previous half-year, while the Y and X numbers decrease.

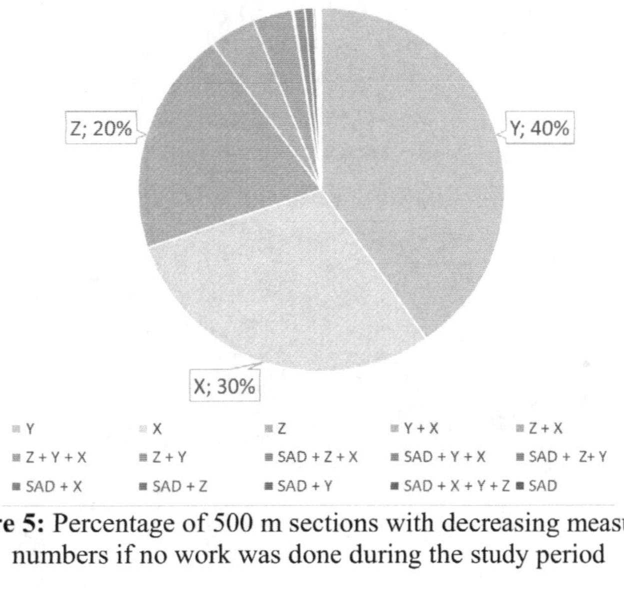

Figure 5: Percentage of 500 m sections with decreasing measuring numbers if no work was done during the study period

It is worth to dwell a little on the result that one of the measuring numbers (Y, X, and Z) decreases while there was no intervention, and the number of these types of 500m sections are altogether 90%. Thus, the probability of a combined decrease in measuring numbers from non-working data sets is less than 10%.

The multiplicity and the ratio have already been shown, now let us look at the magnitude of the difference. The difference of the improving value from the value calculated by linear regression had to be determined. In contrast to the theoretical exponential model, linear regression was used because it is for short periods (max. 6 semesters) and the linear approximation could be quite accurate here. That was the reason why the program that calculates the difference from the average before and after the decreasing measuring number was written. Give in advance the degree of expected deterioration, against which the difference can be calculated. The outlier was not considered when creating the descriptive equation generated by the regression procedure.

In figure 6-8 the difference can be seen between the given measure and the linear regression calculation. The diagrams show the values of the measuring number on the horizontal axis and the determined deviation from the regression approximation on the vertical axis. The number of error sizes on the histograms were plotted. This is how the results for all three measuring numbers were reported.

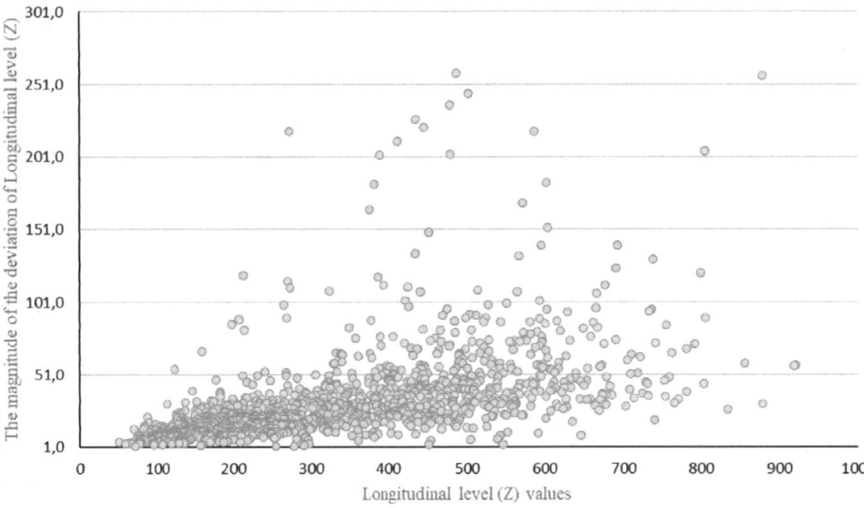

Figure 6. Longitudinal level (Z) deviation from regression fit diagram

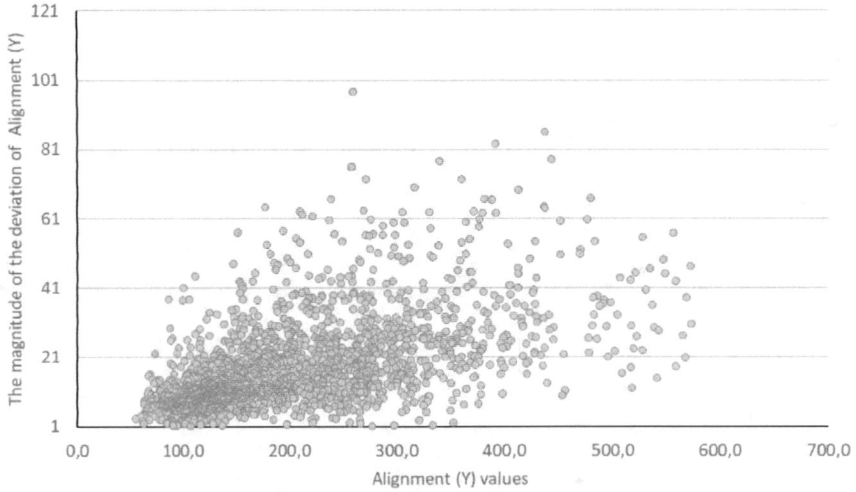

Figure 7. Alignment (Y) deviation from regression fit diagram

IOP Conf. Series: Materials Science and Engineering 1141 (2021) 012022 doi:10.1088/1757-899X/1141/1/012022

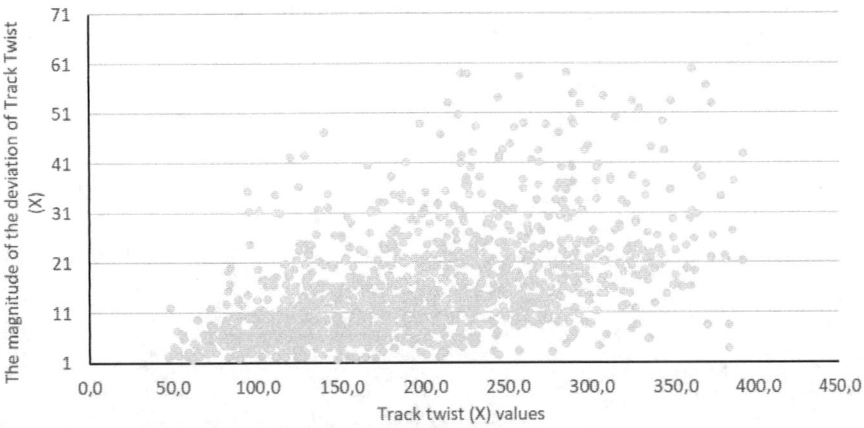

Figure 8. Track twist (X) deviation from regression fit diagram

In the following, figures 9-11 the results according to a different idea have been plotted. The percentage of the deviation is assigned to the given measuring number, because this makes the data easier to analyse. Furthermore, the percentage of error rates can thus be compared to the condition of a given 500 m section of a given railway line.

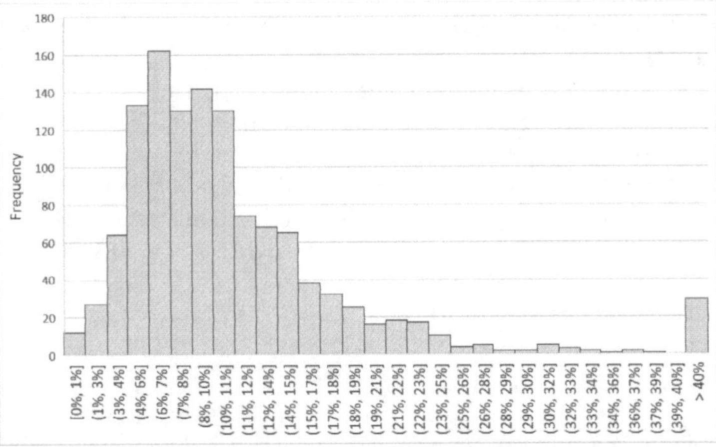

Longitudinal level (Z) deviation values from the regression fit, descripting statistic	
Mean	1,12
Standard Error	0,00
Median	1,09
Mode	1,08
Standard Deviation	0,12
Sample Variance	0,01
Kurtosis	34,44
Skewness	5,15
Range	1,27
Minimum	1,00
Maximum	2,27
Sum	1363,61
Count	1221,00

Figure 9. Longitudinal level (Z) deviation from the regression fit diagram and this description statistic

The deviation of the Longitudinal level (Z) and the regression fit shows a lognormal distribution, with an expected value of 12%.

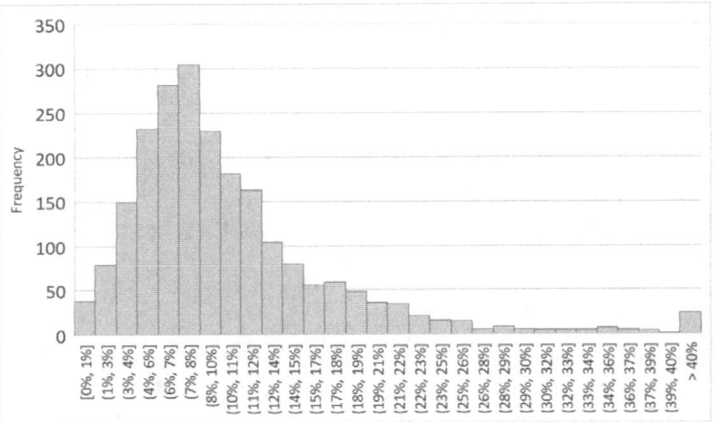

Alignment (Y) deviation values from the regression fit, descripting statistic	
Mean	1,10
Standard Error	0,00
Median	1,08
Mode	1,12
Standard Deviation	0,08
Sample Variance	0,01
Kurtosis	12,61
Skewness	2,75
Range	0,76
Minimum	1,00
Maximum	1,76
Sum	2428,70
Count	2201,00

Figure 10. Alignment (Y) deviation from the regression fit diagram and this description statistic

The deviation of the Alignment (Y) and the regression fit shows a lognormal distribution, with an expected value of 10%.

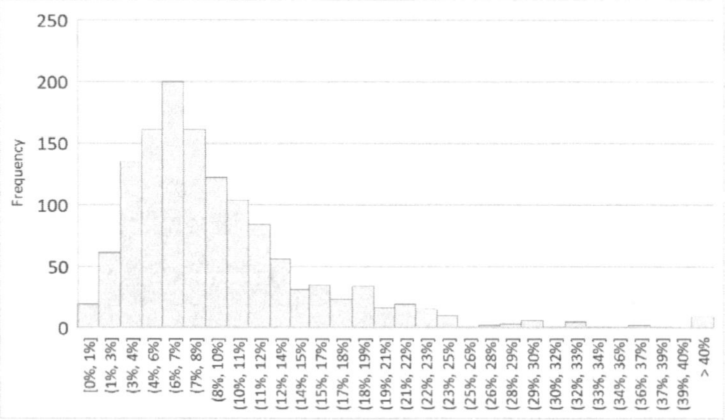

Track twist (X) deviation values from the regression fit, descripting statistic	
Mean	1,09
Standard Error	0,00
Median	1,08
Mode	1,02
Standard Deviation	0,07
Sample Variance	0,00
Kurtosis	11,16
Skewness	2,58
Range	0,59
Minimum	1,00
Maximum	1,59
Sum	1440,35
Count	1318,00

Figure 11. Track twist (X) deviation from the regression fit diagram and this description statistic

The deviation of the Track twist (X) and the regression fit shows a lognormal distribution, with an expected value of 9%.

5. Conclusion and comparison of results with the literature

In summary, the main indicators for the statistical evaluation of the results are:
Longitudinal level (Z): Expected value: 1.117, median: 1.092, standard deviation: 0.119,
Alignment (Y): Expected value: 1.103, median: 1.085, standard deviation: 0.075,
Track twist (X): Expected value: 1.093, median: 1.075, standard deviation: 0.067

In this part of the research, examining the entire nationwide network in the 27 half-years it turns out that in almost 90% of the cases without intervention assumed on the basis of a monotonous increase in the number of qualifications, at least one of the indicators shows improvement. By processing metric increments of this improving nature, are:

- Longitudinal level (Z) accuracy 11.7%
- Alignment (Y) accuracy: 10.3%
- Track twist (X) accuracy: 9.3%

These numbers characterize the geometrical measurement accuracy of the FMK-004 measuring car.

In the light of the results, more accurate and reliable data provision will be available for the FMK-004 measuring car, which may also improve the accuracy of track deterioration forecasts. There is no research in the literature that tries to answer the question of the severity of the rating numbers in the SAD rating by processing such a large amount of data. There are articles in the international literature, including [9], [10], [11], which deal with similar topics and calculations, but because each country evaluates and analyses its own railway network differently, although with similar evaluation data (alignment, longitudinal level, twist) it was not possible to find any research comparable to our results.

6. References

[1] Rajan K and Saffiotti A 2017 Towards a science of integrated AI and Robotics, *Artif. Intell.*, vol **247** pp 1–9, doi: 10.1016/j.artint.2017.03.003

[2] Nagy R 2017 Description of rail track geometry deterioration process in hungarian rail lines no. 1 and no. 140, *Pollack Period.*, vol **12**, no 3, doi: 10.1556/606.2017.12.3.13

[3] Nagy R 2021 A vágánygeometria romlási modelljeinek összehasonlító elemzése,"*Sinek világa*, vol 2, pp 25–30

[4] Nagy R,2016 A vasúti pályageometria romlási folyamatának leírása, *Sinek világa*, vol **6**, pp 12–18

[5] Gáspár L, Horvát F, and Lublóy L 2011 *Közlekedési létesítmények élettartama*

[6] Atkinson R 1999 *Project management: cost, time and quality, two best guesses and a*

phenomenon, its time to accept other success criteria
[7] Vaszary P 1992 *A vasúti pálya geometriai állapotértékelésének elméleti alapjai* - Kandidátusi értekezés
[8] Nagy R 2017 Analytical differences between seven prediction models and the description of the rail track deterioration process through these methods, *Intersections/Intersecţii*, ISSN 1582-3024, vol. **14**, no. 1, pp. 14–32, 2017, [Online]. Available: http://www.intersections.tuiasi.ro
[9] Luber B, Haigermoser A, and Grabner G 2010 Track geometry evaluation method based on vehicle response prediction, *Veh. Syst. Dyn.*, vol. **48**, no. SUPPL. 1, pp. 157–173 doi: 10.1080/00423111003692914.
[10] Raymond L 2011 *Track geometry recording and usage: notes for a lecture to Network Rail*
[11] Li H and Xu Y 2010 A method to correct the mileage error in railway track geometry data and its usage, in *Proceedings of the Conference on Traffic and Transportation Studies, ICTTS*, vol. **383**, pp. 1130–1135, doi: 10.1061/41123(383)106

The concept of an automatic system for preliminary entry of hazardous facility into emergency mode with an approaching earthquake

V Zhukovsky [1], I Rusu [2], M Zavoloka [3], I Grynyova [4]

[1] Department of Physics; South Ukrainian National Pedagogical University named after K. D. Ushynsky; Staroportofrankovskaya str. 26, 65029, Odesa, *Ukraine*

[2] Technical Faculty of Architecture and Urban Planning, University of Moldova, Faculty of Architecture and Urban Planning, Ştefan cel Mare şi Sfînt Blv.168, Chişinău 2004, *Technical University of Moldova*

[3] Department of production of building products and constructions; Odesa State Academy of Civil Engineering and Architecture, Didrihson, 4, 65029, Odesa, *Ukraine*

[4] Department of Architectural structures; Odesa State Academy of Civil Engineering and Architecture; Didrihson, 4, 65029, Odesa, *Ukraine*

E-mail: grynyova@ogasa.org.ua

Abstract. The article discusses the concept of introducing an automatic system for preliminary input of hazardous facilities into emergency mode in the event of an approaching earthquake, developed at the South Ukrainian National Pedagogical University named after K. D. Ushinsky. Theoretical prerequisites for the study of heterogeneities, using a passive method based on the location of secondary waves reflected from the inhomogeneities under the basements and foundations of buildings of already operated building structures, were taken as the basis of the proposed system. Due to the finite speed of propagation of seismic waves and having a sensor in the focus of an earthquake, it is possible for distant objects to learn about an impending disaster in a few tens of seconds. The effective use of the time window from the focus of the earthquake to hazardous facilities requires the development of: measuring equipment, data transmission systems, automatic event recognition, elements for automating the process of putting hazardous objects into emergency mode - all this is the goal and objectives of this article.

1. Introduction

Among natural disasters, earthquakes are the most destructive. Unlike hurricanes and floods, which can be predicted by meteorological services, earthquakes occur suddenly for no apparent reason, leaving no opportunity to prepare for a disaster. The influence force of nature finds trains in motion, people in elevators, cranes with lifted loads, hazardous production in the midst of a shift, etc. It is obvious that this force of nature can provoke a number of technogenic accidents, which lead to an increase in human casualties and economic losses. Until now, all attempts at early prediction of earthquakes have been unsuccessful [1-3]. However, due to the finite speed of propagation of seismic waves, it is still possible for distant objects to learn about an impending disaster in a few tens of seconds [4]. Effective use of this time window, which requires the development of: measuring equipment, data transmission systems, automatic event recognition, elements of automation of the process of putting hazardous facilities into emergency mode - all this is the goal and objectives of this project.

2.Theoretical information and main idea concept

The velocities of seismic waves in the earth's crust are known - they are about 6 km / s for longitudinal waves and 3.5 km / s for shear waves, which have the main destructive effect. Consider an element of the Google map of the Vrancea area (Figure 1), which is known for the periodic epicentre of earthquakes and adjacent earthquake-prone regions (investigation in Greece region [5]). This area is located 110 km from Bucharest, 250 km from Chisinau and 350 km from Odessa. With the help of a special manual for the speed of transverse waves, it is not difficult to calculate that they will reach Bucharest in ≈30 s, Chisinau in ≈70 s, and Odessa in ≈100 s.

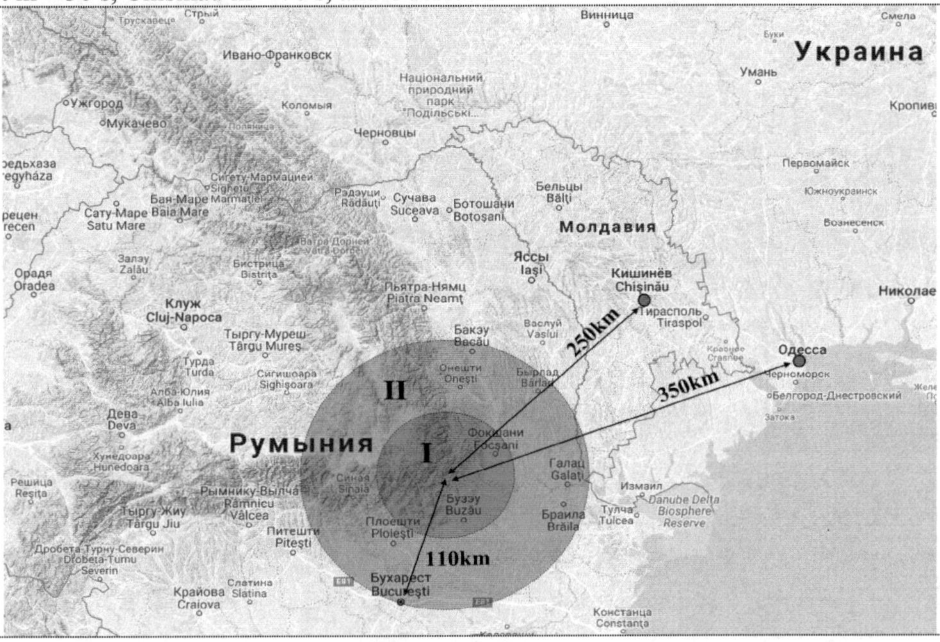

Figure 1. Google map of Vrancea area and adjacent earthquake-prone regions

Obviously, having a sensor in the center of the earthquake, which is connected with the indicated cities by a radio channel, it's possible to know about the approaching element in advance for those very few tens of seconds. Naturally, notifying the population at such time intervals is out of the question. Mass evacuations from apartment buildings, schools and public institutions will only exacerbate the situation. However, this time may be sufficient to lower elevators to the first floor, stop the movement of goods in ports and construction sites, start emergency braking of trains, stop conveyors and shut off pipelines of hazardous chemical industries, cut off gas supply, cut off ammonia pipelines at refrigeration plants, transfer hospitals for autonomous power supply, etc.

For this task, special seismic receivers were developed, which are distinguished by high performance characteristics and are easily integrated into modern telemetry systems.

When designing low-frequency seismic receivers, an optical method was chosen for converting mechanical vibrations into electrical ones, which does not depend on the frequency of ground vibrations. The signal-to-noise ratio of optical converters is much higher than that of capacitive and inductive ones. In works [6, 7], successful results are given on the use of interferometric laser seismic receivers. Such seismic receivers are distinguished by high sensitivity and a relatively simple converter circuit, but their weak point is the laser, which has a relatively short service life (about 10,000 hours) and requires a rather complex temperature stabilization circuit. The high-power consumption of such seismic receivers causes problems with autonomous power supply. Converters using an optocoupler (LED / photodiode) are less sensitive to pendulum oscillations, but their service life is 10 times longer, they are more economical and do not require additional temperature and current stabilization circuits.

In works [8, 9], a method is given for calculating an integrator and a differentiator for adjusting the amplitude-frequency response of seismic receivers assembled on analogue microcircuits. However, the modern architecture of microcontrollers allows to completely abandon analogue control circuits, solving the problem of adjusting the amplitude-frequency response by changing the coefficients directly in the control equation, which in discrete representation has the form:

$$U(t) = Px(t) + I\sum_i \left[\frac{x(t_i) + x(t_{i-1})}{2}\right]\Delta t + D\left[x(t+\Delta t) - x(t-\Delta t)\right]/2\Delta t \tag{1}$$

where P, I, D – weight coefficients, x(t) – deviation from the equilibrium position on the seismogram, Δt – time between measurements.

The principle of operation of a digital seismic receiver (Figure 2) is that the signal of the converter is digitized by means of an analogue-to-digital converter (ADC) and the control signal is calculated using the formula (1). The control signal from the pulse width modulation (PWM) generator is applied directly to the electromagnetic feedback coil.

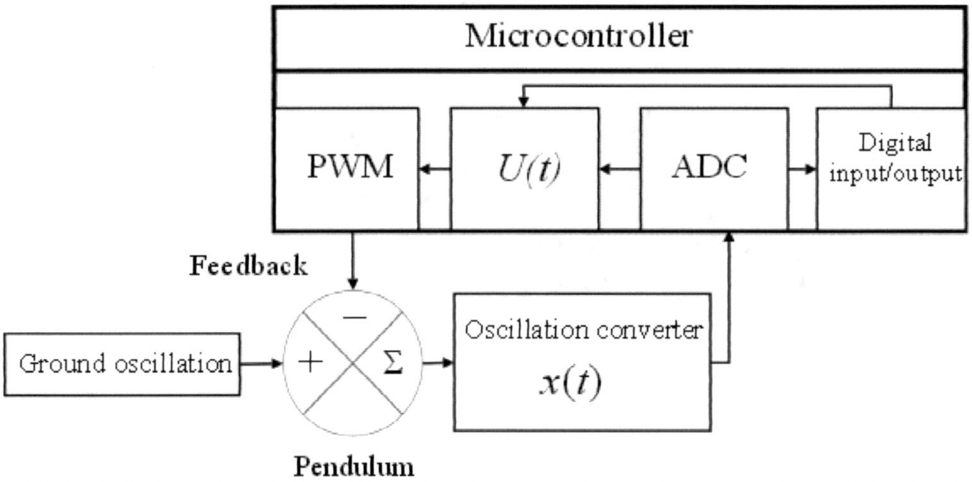

Figure 2. Block diagram of a digital seismic receiver

Through the digital input / output interface, the seismic receiver is tuned by entering the weighting coefficient and a digital signal is output for recording and visualizing the data.

Obviously, this approach greatly simplifies calibration and allows the seismic receiver to be tuned remotely via digital communication lines and even via the Internet. Figure 3 shows a photo of the system, which includes a digital optical seismic receiver, an optical microbarograph and a minicomputer for registration, automatic data processing and external communication via the Internet or other digital lines.

Figure 3. Digital seismic receiver and microbarograph connected to the Internet by mean of a microcomputer

Provided that an optical conversion method is used, where a powerful magnet and an inductor are not required, such a device is no larger than a matchbox with batteries and wireless digital communications.
This miniaturization of devices allows you to solve a number of problems, for example:

- reduction in the cost of measuring equipment due to lower consumption of materials;
- reducing the cost of installing seismic receivers due to the smaller diameter of filling wells;
- delivery of equipment to hard-to-reach places by unmanned aerial vehicles;
- creation of reconnaissance and signal systems with the recognition of seismic images by the spectrum of seismic vibrations, etc.

3. Experimental technique.
Seismic receiver will be located around Vrancea in two zones (Fig. 4):
I Focsani, Buzau, Ploiesti;
II Bacau, Bârlad, Galati, Bucharest.

As an example, Figure 7 shows a diagram of the installation of digital seismic receivers developed at Ushinsky University with data transmission over the Internet.

By Internet channel, with a frequency of 1 s, the seismic receivers will transmit to the server the maximum amplitude values for the last 30 seconds. The time of 30 seconds is the maximum value for which a seismic wave propagates from zone I to zone II.

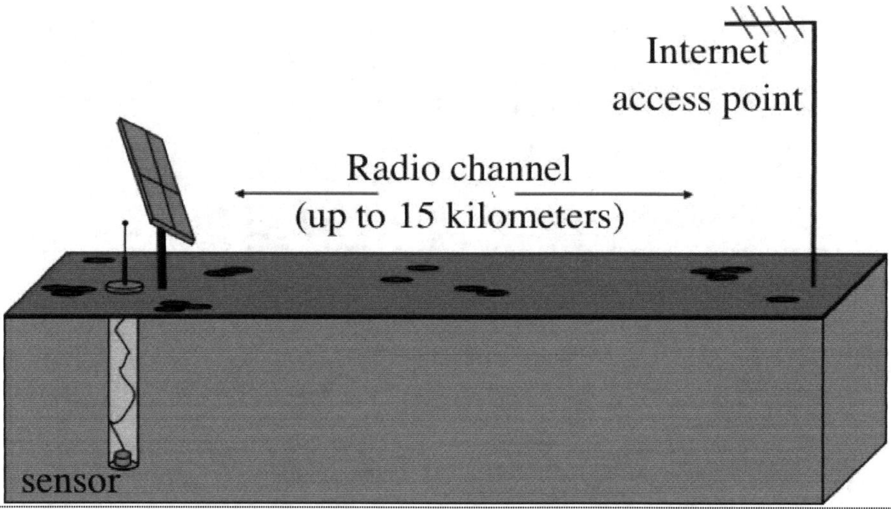

Figure 4. Installation diagram of digital seismic receivers with mixed data transmission to the server

On the server, automatically in on-line mode, according to the amplitudes of seismic waves in zones I and II, the amplitudes of seismic waves will be extrapolated using an exponential approximating function, and a hazard level code will be generated, which will be received by the actuators via the radio channel. In turn, each actuator can only be activated at a certain level of danger. At the same time, in Chisinau, it will take about 40 seconds to take security measures, and about 70 seconds for Odessa.

Figure 5 shows the general structure of an automatic system for preliminary input of hazardous facilities into emergency mode in the event of an approaching earthquake.

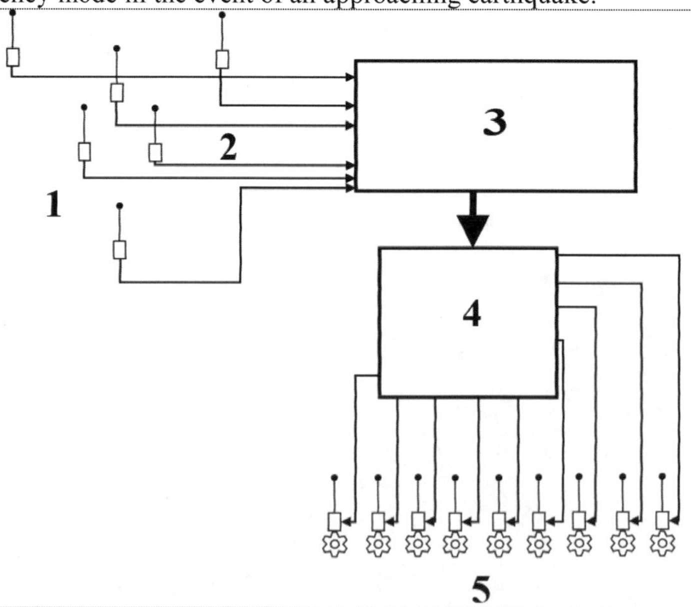

Figure 5. The structure of an automatic system for preliminary input of hazardous facilities into emergency mode in the event of an approaching earthquake, where 1 - seismic receivers, 2 - Internet lines, 3 - server and decision-making center, 4 - dedicated radio channel, 5 - actuators.

4. The discussion of the results.

During 2020 - 2021, a prototype of the proposed system was installed at the Ukrainian Antarctic Station "Akademik Vernadsky", which successfully operates to this day in harsh polar conditions. The system includes 4 spaced seismic receivers and 4 microbarographs combined into a seismic and acoustic antenna (Figure 6), which, against the background of strong natural noise, record seismic and atmospheric phenomena with high reliability.

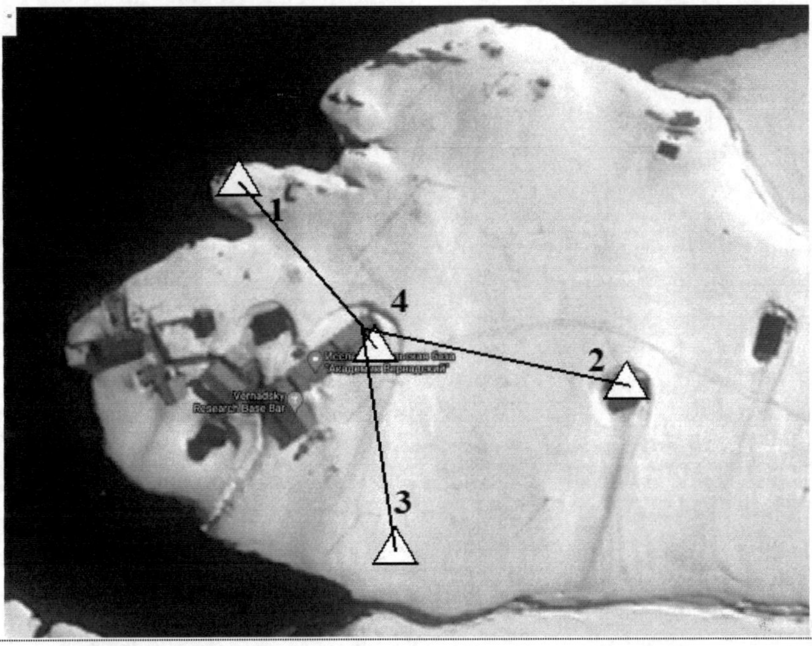

Figure 6. Scheme of installation of optical seismic receivers and acoustic sensors on Galindez Island in the vicinity of the Antarctic station "Akademik Vernadsky"

Specially developed on-line software detects, locates and recognizes signals, as well as classifies them using a neural network and generates a report, which is sent via the Internet to the Antarctic Center in Kiev.

5. Conclusions

Today, there are all the prerequisites for the creation of an automatic system for preliminary input of hazardous facilities into emergency mode in the event of an approaching earthquake. For this, the main blocks of the system have already been created:

- sensors;
- means of communication;
- software;
- executive devices of the "smart home" type;

which need to be combined into a single system.

6. References

[1] Savin M G and Rokityansky I I 2011 A method for monitoring earthquake foci *Geofiz. zhurn.* **33 (3)** pp 129-143

[2] Haishan Z. and Leiv-J. 2009 An Open-Source Passive Seismic Data Processing and Visualization Package: Passeis *CSPG CSEG CWLS Convention* pp 408–411

[3] Sheng Yi, Kong Qi and Beroza Gr C 2021Network analysis of earthquake ground motion spatial correlation: a case study with the San Jacinto seismic nodal array *Geophysical Journal International* **225 (3)** pp 1704–1713 https://doi.org/10.1093/gji/ggab058

[4] Song Ch, Alkhalifah T and Waheed U B 2021 Solving the frequency-domain acoustic VTI wave equation using physics-informed neural networks *Geophysical Journal International* **225 (2)** pp 846–859 https://doi.org/10.1093/gji/ggab010

[5] Kourouklas C, Console R, Papadimitriou E, Murru M and Karakostas V 2021 Modeling the large earthquakes recurrence times along the North Aegean Trough Fault Zone (Greece) with a physics-based simulator *Geophysical Journal International* ggab085, https://doi.org/10.1093/gji/ggab085

[6] Zhukovskiy V K, Karyagin E V and Popov A Yu 2010 Application of an electrooptical converter of mechanical vibrations with a holographic optical element in seismometry *Technical Physics* **55 (4)** pp 532-536

[7] Berger, Davis, Widmer-Schnidrig and Zumberge 2014 *Performance of an Optical Seismometer from 1 μHz to 10 Hz* BSSA (Preprint: *https://eost.unistra.fr/websites/eost/eost/Observatoires/Gravimetrie/Colloque_G2/G2_Widmer.pdf*)

[8] Havskov J and Alguacil G 2002 Instrumentation in Earthquake Seismology (Modern Approaches in Geophysics) p 313

[9] Senigov P N 2000 Automatic control theory: Lecture notes (Chelyabinsk: SUSU) p 93

IOP Conf. Series: Materials Science and Engineering 1141 (2021) 012024 doi:10.1088/1757-899X/1141/1/012024

Methodological approaches to creating the electronic databases of building operation safety

S Kolesnichenko[1], I Seliutin[1] and Y Grytsuk[1, 2]
[1]Donbas National Academy of Civil Engineering and Architecture, Kramatorsk, Donetska oblast, Heroyv Nebesnoy Sotni str.,14, 84333, Ukraine

E-mail: [2] ccit@donnaba.edu.ua

Abstract. One of the tasks of ensuring the safety of construction sites is to maintain or increase the design reliability of buildings and structures during their renovation and reconstruction. The reliability of the object is calculated at all stages of its life cycle and for kinds of work corresponding to these stages. An effective method of determining quality indicators is the registration method, which is based on observations of the actual condition of the building structure. The registration of the condition of the elements and the construction is carried out through complex characteristics allowing to estimate the rate of imperfections (defects and damages - DD) progression throughout time, and to define their criticality level. For building structures (facilities), the registered DD allows performing the analysis of types, consequences, and the criticality of possible failures (FMECA classification). The proposed strategy includes the process approach to resource management through developing a system of accounting and functional controlling, analysis, and regulation of safety. One of the alternate solutions of the problem may be the establishment of information systems based on registration methods, namely electronic databases (DB) of defects and damages of structures. Information databases in the conditions of an actual enterprise may become the first but still very important step on the way to generalizing and controlling the data about the actual state of building structures and facilities for further decision-making concerning their operation and maintenance. The database is to be implemented in the network form, with a clear hierarchical access framework and editing capabilities for the officials, the ability to generate reports and extracts. The obtained results in the framework of the reliability-assurance program should provide the ability to calculate the probability of failure-free operation, the residual resource with assigned level of risk.

1. Relevance of the issue

The condition for the safe operation and modernization of buildings and structures is ensuring systemic control of possible changes in the technical condition parameters. Decision-making on restoration, calculation of degradation models or assessment of operating risks in the presence of non-critical imperfections (defects and damage) is possible only if there is a history of object operation. The information contained in the project, reports and other documents requires system analysis, which can be performed using electronic digital resources - databases - DB.

Under technological safety we mean a partial component of the general safety of the industrial enterprise (industrial safety), concerning maintenance of absence of unacceptable risk of structural failures and collapses that characterizes the system of measures for working capacity maintenance, prevention of decrease in operational properties of building structures, constructions and engineering networks which have completely or largely exhausted their regulatory resources, and are the source of potential danger in relation to technological functions during operation, including within the assigned extended residual life of the facility. Ensuring technological safety is the simultaneous combination and calculation of indicators of reliability of building structures and indicators of organizational and technical safety - OTS, aimed at safe maintenance of industrial buildings and structures.

Technological safety is an integral part of the overall safety of the enterprise along with economic, financial, environmental etc., and characterizes the system of measures to maintain the acceptable level of its operation in terms of depletion by building structures and engineering networks of their regulatory (design) resource. Thus, the main task of technological safety is to increase or maintain the operational properties of structures and networks in the design state during their modernization, technical re-equipment, and reconstruction.

Technological security requires a process approach, which supposes, inter alia, the creation of organizational and technical means, based primarily on the company's management wanting to systematically implement the principles of quality management and reliability. The reliability of the object must be calculated at all stages of its life cycle and at the types of work corresponding to these stages. The need for reliability calculations, goals and objectives, the range of calculation indicators, timing and order of results are set in the reliability program - RP [1, 2].

2. Research methods

Organizational and technical measures to maintain and restore the performance indicators of industrial facilities are regulated by the requirements of ISO 9001, ISO 2394, ISO 3100. To ensure normal operation, measures are determined depending on the actual (current) parameters of the technical condition of structural elements of buildings and structures. The preventive maintenance management system complies with the sectoral structure of maintenance and repair management. The control system is built without the use of modern requirements for reliability and safety based on ISO 9001 standards. The lack of procedures for accumulating and analysing data on the technical condition of structures causes difficulties for timely and professional decision-making to prevent emergency situations.

One of the methods of determining quality indicators is the registration method, which is carried out based on observations of the actual condition of the building structure. Registration of the condition of the structure (element) is carried out through complex characteristics allowing to estimate the rate of the damage progression in time, to determine the degree of its criticality. For structures, buildings (structures) in general, registered DD allow performing the analysis of types, consequences, and criticality of possible failures - FMECA [3]. FMECA is performed according to the plan, which is directly included in the RP or in the form of a separate document attached to the RP.

Also, the condition of the structures of the building (construction) is determined based on regular inspections performed by maintenance departments – typically these are the heads of shop units and specialists of (departments) of technical supervision service - TSS. The results of such inspections are formalized by acts and instructions. The above-mentioned documents are required for compiling statements on the scope of repair work to eliminate the identified DD. The problem is that the information obtained during investigations and inspections often cannot be analysed and summarized comprehensively due to its large volume and is usually provided for performing urgent one-time works on accident-hazardous facilities. In addition, the current state documents on the activities of TSS contain general instructions on decision-making procedures, they are not connected to the organizational and managerial structure of the enterprise, which has been developing for years.

Accident-free operation of building structures (BS) is possible only based on decision-making with the constant accumulation, systematization and processing of information about their condition, analysis of the progression of DD over time. Given the need to store and process a very large amount of data, one way to solve the problem may be to create information systems based on registration methods - databases (DB) of defects and structural damage.

The need for the creation of the information systems has been repeatedly pointed out in the studies by the specialists that are engaged in the investigation issues and TSS specialists who make decisions based on the obtained data.

3. Databases of operational safety of buildings

The implementation of RP for BS at the enterprise can be carried out according to the scheme "information collection" - "development of information databases" - "creation of expert systems" - "calculation of reliability indicators". Information databases in the conditions of an actual enterprise can be the first, but very important step in the generalization and control of information about the actual state of the BS for further decision-making procedures on their operation and maintenance. The methodology for creating a database is quite fully developed in the works [4, 5, 6]. The database is the basis for further development of expert systems - systems involved in the general monitoring of BS structures. However, it should be noted that expert systems are quite complex information systems [7, 8], the implementation of which requires a sufficiently high level of technological culture of the enterprise.

Given all the above-mentioned, for the conditions of enterprises with a sufficient number of BS, which are in conditions of long-term operation, we can recommend, as a priority measure, the implementation of electronic databases of the actual state of the BS.

The main sections of the database are:

- passport data on the building (structure). The information is entered based on data of the architectural and technical passport, technical state passport, the passport of the Technical Inventory Bureau, data of technical reports and the available design documentations. The categories of buildings and individual structures for fire hazard;
- spatial planning and design solutions. Data on existing structures, plans, facades, sections, units are entered. Data on the existing technology, equipment, operating conditions of the building as a whole and its individual structures are also provided;
- data on engineering networks (conventional and technological). Their location schemes, state of networks, impact of the networks on the condition of the structure. Probability of destruction of the structure during an emergency with the network;
- data on the organizations that performed the inspections, data on the technical supervision service, on the persons who performed the inspections. The date of each event, the technical condition of the object, which was determined after the inspection and possible features of the structures are recorded. The terms for the elimination of the DD are provided, the residual resource and the term of the next investigation / inspection are determined;
- information on existing defects and damages - DD. To describe them, the data obtained during the inspections and surveys of the DD are entered. The nomenclature of DD is adopted in accordance with the existing norms [9, 10, 11]. Each DD should be described and clearly attached to the structure in plan and height. The detection time for each DD is recorded. It is also necessary to provide for the possibility to assess the progression of each DD over time. Based on the data by categories of DD, we can prepare reports (extracts) for each category. The report is formed based on the request of the official concerned both for an individual building and for all (groups) of buildings. To ensure the possibility of making preliminary decisions on repairs, a base of options for reinforcement or repair is formed for each individual DD. There are several basic options for solutions, from which the project designer can choose a solution and implement it for each situation;
- reference information. This includes reference data on organizations - suppliers of construction products (metal, paints, welding materials, etc.), project designers, manufacturers, repairing companies. It is also possible to provide information on the categories of hazards, fire hazard, hazard of technological processes, basic regulatory documents or extracts from them.

The principles of database creation should be such that the data on the identified DD, even after their elimination (repair), are permanently stored with the indication of the person who made the decision to repair, the date of repair and information about the post-repair condition. The proposed strategy includes a process approach to the management of operational safety by setting up a system of accounting and functional controlling, analysis, and regulations, which accounts for the actual technical condition of construction objects and can be presented in the form of a diagram (Fig. 1).

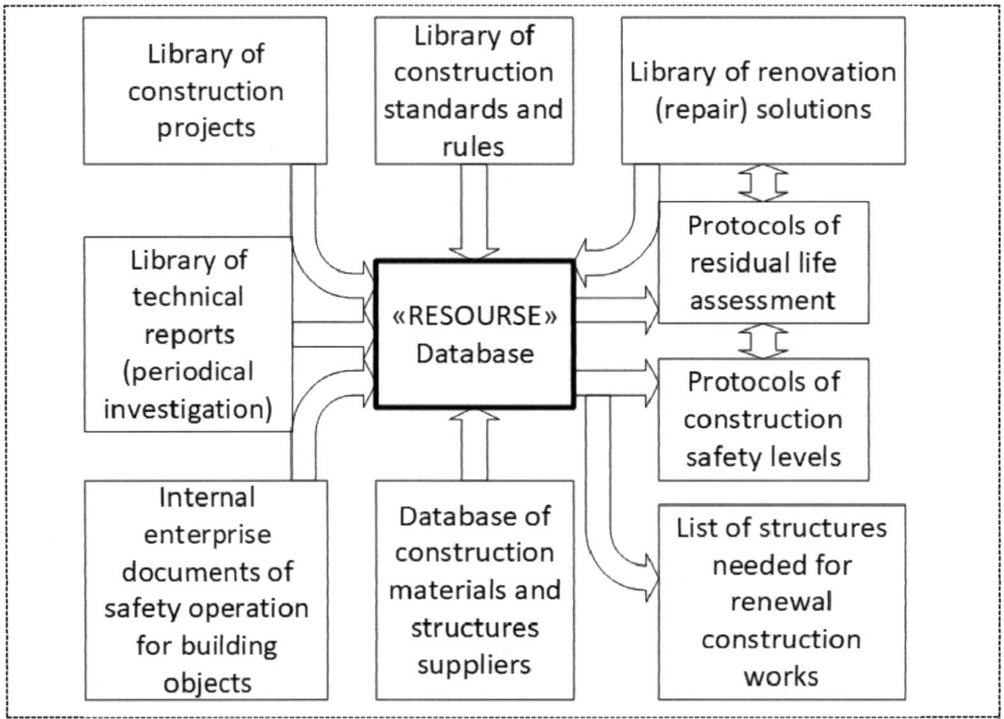

Figure 1. Scheme of organizational, methodological, and regulatory support of the "Resource" database.

The database should be implemented in a network form, with a clear hierarchical structure of access and editing capabilities for officials, the ability to generate reports and extracts (the authors used a free platform for database management systems FirebirdSQL). The list of persons is usually formed from TSS specialists who have the right to enter data based on received technical reports based on the results of investigations and regular inspections.

Due to the availability of tracking the progression of the DD parameters over time, it is possible to forecast the transition of DD from one category to another. Registration of DD allows to perform FMECA with the development of appropriate design solutions for their elimination. The FMECA plan should ensure the interconnection and coordination of the facility's RP with the terms and scope of restoration works, ensuring further post-repair accident-free work.

The developed method of analysis of safety indicators is the basis for the formation of the structure of the description of tasks by blocks and modules of the "Resource" information-analytical electronic database (Fig. 2). The implementation of the process approach to safety at the facility level is aimed at ensuring the risk assessment when extending the service life, considering the levels of vulnerability and threat, maintainability when servicing facilities according to their actual state.

Access and editing of the database should be performed by qualified and trained personnel. The scheme proposed in Fig. 2 for editing the database needs to be developed for each company separately and set in company orders and references to the enterprise standard.

Constant control over the possible change in the parameters of the detected DD allows to concentrate the attention of the shop staff on the most dangerous places during periodic scheduled inspections.

The results obtained in the framework of RP should provide the ability to calculate the failure-free operation probability, the residual life with ensured risk level.

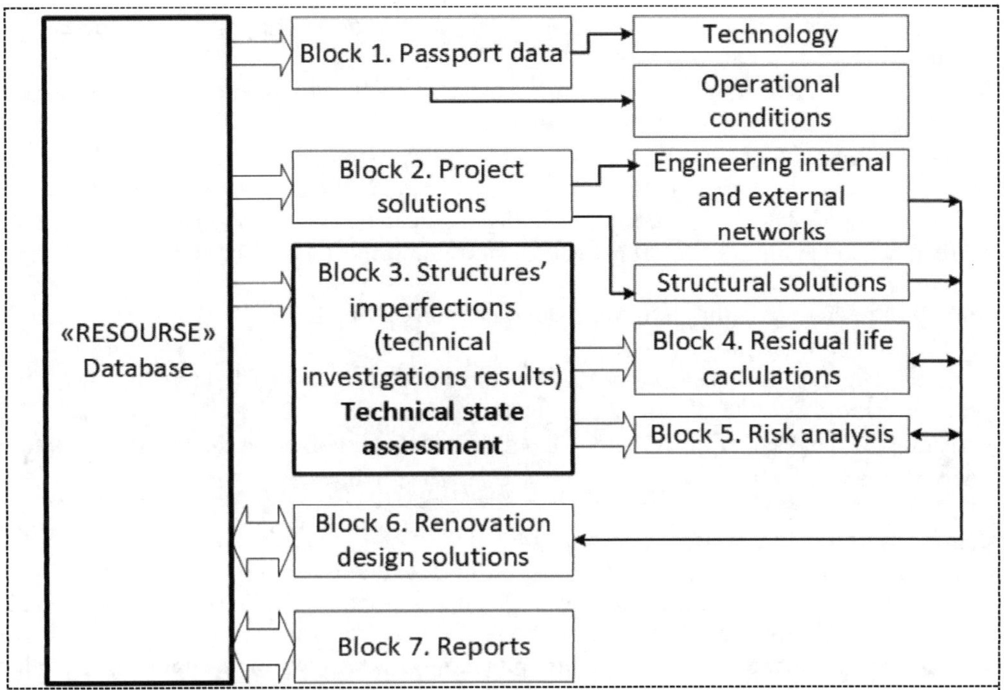

Figure 2. Block diagram of the "Resource" database

The database should be developed and implemented at the enterprise by a research institution. During the installation, there should be constant contact between the developers and those responsible for further database work and provisioning. The first objects are entered directly with the representatives of the institution. Later, it is necessary to organize the training of production staff and to establish the scientific support for the DB use.

Given the fact that the database is the property of the enterprise, further calculations of the residual life, forecast models of operation, the risk of operation can be performed only by employees of the supporting research institution jointly with the company's specialists.

4. Database elements - Blocks 1 and 2

These blocks of the "Resource" database contain information for:

- maintaining technical passports of buildings and structures, storing project documents and data of regulatory and technical requirements for quality, structural reliability assurance and technological safety of structures;
- providing information support of materials and structures quality indicators basing on the results of control in the process of their production, installation and operation;
- accumulating data on the maintenance and operating conditions of structures of industrial facilities;
- developing schedules for current, periodic inspections, technical supervision and engineering inspections of buildings and structures by specialized organizations;
- registering acts, protocols of registration of the nature, size and degree of progression of defects and damages;
- streamlining and improving the quantitative accounting of operational indicators of the technical condition of construction projects based on the comparison of the identified defects and damages with the established tolerance limits;
- ensuring the requirements of technological safety during the operation of the key production assets according to the actual state;

- managing basic data on the parameters of the operating conditions of structures, resources, maintenance, and repair processes;
- collecting and analysing data on actual costs for their subsequent use at the stages of risk reduction and investment decision making;
- controlling over the fulfilment by the enterprise structural divisions of the requirements of regulatory and sectoral regulatory and guidance documents, decrees, decisions, orders and instructions of state and trade union bodies, orders of state supervision bodies, orders and directions concerning the technical operation of buildings and structures.

5. Structural imperfections and assigning technical state - Block 3

The block is intended for:

- creating the DD database, both for the current one and for all previous inspections and investigations;
- creating the database of the repaired DD of different categories for the object's structures;
- determining the technical condition of the object's structures.

6. Categories of defects and damages

When assessing the bearing properties of structures, defects and damages are categorized as critical – Ad, significant Bd and insignificant Cd. According to the estimation, the DD depending on the cause of their occurrence, can be divided into two groups:

AD – imperfections that characterize quality indicators that have normalized numerical values;

BD – imperfections associated with irregularities in the work practices or damages that do not have standardized numerical values.

For defects and damages of the AD group, their hazards are determined by the amount of excess or understatement of the actual value of the controlled parameter Xi in comparison with its limit value according to the formula:

$$\Delta = \frac{X_i - X_{\min(\max)}}{X_{\min(\max)}}, \tag{1}$$

where $X_{min/(max)}$ - limit values determined in accordance with the standards.

In this case, the DD is critical at $\Delta > 40\%$, significant at $\Delta \leq 40\%$ and insignificant at $\Delta \leq 10\%$.

For defects and damages of the BD group, the assignment of one or another imperfection to a certain class is based on the analysis of its consequences, the degree of impact on the main indicators of the functional performance of the building element in question.

By the number (zone of progression) of imperfections in the element or in the area under consideration, they are distinguished as:

a) single imperfections, occupying up to 10% of the area, linear size or quantity;
b) numerous – from 10% to 40%;
c) mass – over 40%.

Such categories for DD were accepted based on the fact that the total value of the reliability indexes gives a margin of about 40% when determining the bearing capacity of the element. Also, it is seen as convenient to assess the hazard of DD, when a periodic site survey is performed by operating organization representatives who are not experts in the field of construction. However, when determining the category of DD hazard for putting it into the DB, it must be assigned by an expert with the account for the actual hazard of the identified DD.

7. Technical state determination

In accordance with the requirements of regulatory documents, the main task of the structural survey is to assign the technical condition of the object. During monitoring, the assessment of the technical condition of building structures or engineering systems, as well as the structural performance of the building, is carried out according to separate groups of performance indicators. According to their technical condition, structures may belong to the following categories:

I – normal condition – minor defects are eliminated during maintenance;

II – satisfactory condition – defects are eliminated in the process of maintenance and repair;

III – limited operational condition – there is no danger of collapse. All operating requirements are to be met. Limitations to individual operating parameters are possible. Repair required;

IV – emergency condition – withdrawal of people from the danger zone, urgent unloading of structures and (or) the arrangement of temporary fasteners with subsequent disassembly and replacement of structures is required.

For steel structures, the technical condition is also determined depending on the reliability index β [12, 13] and the CC consequence class (Table 1). Here, the technical condition is determined according to three categories. Categories II and III correspond to category 2.

Table 1. Assigning a technical condition of structures depending on β and CC.

Consequence class	β Design	β 1 - state satisfactory	β 2 - state, in need of repair	β 3 – state emergency
CC1	>3.3	β ∈ [3.3…2.8[	β ∈ [2.8…1.8[	β ≤ 1.8
CC2	>3.8	β ∈ [3.8…3.3[	β ∈ [3.3…2.3[	β ≤ 2.3
CC3	>4.3	β ∈ [4.3…3.8[	β ∈ [3.8…2.8[	β ≤ 2.8

8. Residual Life Calculation - Block 4

The block is designed to solve the calculation of the residual life - RL, depending on the number of identified DD in each of the categories [14]. The number of repaired DD is also considered according to previously performed inspections. Residual life is defined as the time from the transition from state I, II and III to state IV for structures of all types of materials and for steel structures from states 1 and 2 to state 3.

All structures are considered partially recoverable systems. Depending on the amount of information available on the life cycle of each structure (data from inspections and surveys), the calculation is carried out either by deterministic or semi-probabilistic models.

When calculating on deterministic models, a simple dependence of determining RL by the term of decreasing certain technical condition parameters - TCP. All necessary TCP must be available after the inspection, calculation, and assessment of technical condition. The general formula for determining the RL is as follows:

$$T_i = f\left(R_{i,t} / R_{i,0}\right) \text{ or } T_i = f\left(R_{i,t+1} / R_{i,t}\right), \tag{2}$$

where R is the strength characteristic or the TCP characteristics;

i = 1,2,3 - the number of times of the survey; Ri, 0 - design strength.

When calculating using semi-probabilistic models, statistical characteristics are calculated from the data on DD, loads and characteristics of materials determined from the results of previous technical investigations and inspections.

For steel structures, the calculation of RL is carried out according to the reliability index β. The calculation principle is shown in Fig. 3.

9. Risk analysis – Block 5

After calculating the RL, another task appears – to analyse the possible risk of operating structures within the RL. This DB does not consider the economic consequences of a probable accident. Only the probability of an adverse event occurring is analysed [15, 16].

Analysing the hazards, it is possible to identify the following risks of operation of building structures from the standpoint of reliability and organizational and technical safety.

IOP Conf. Series: Materials Science and Engineering 1141 (2021) 012024 doi:10.1088/1757-899X/1141/1/012024

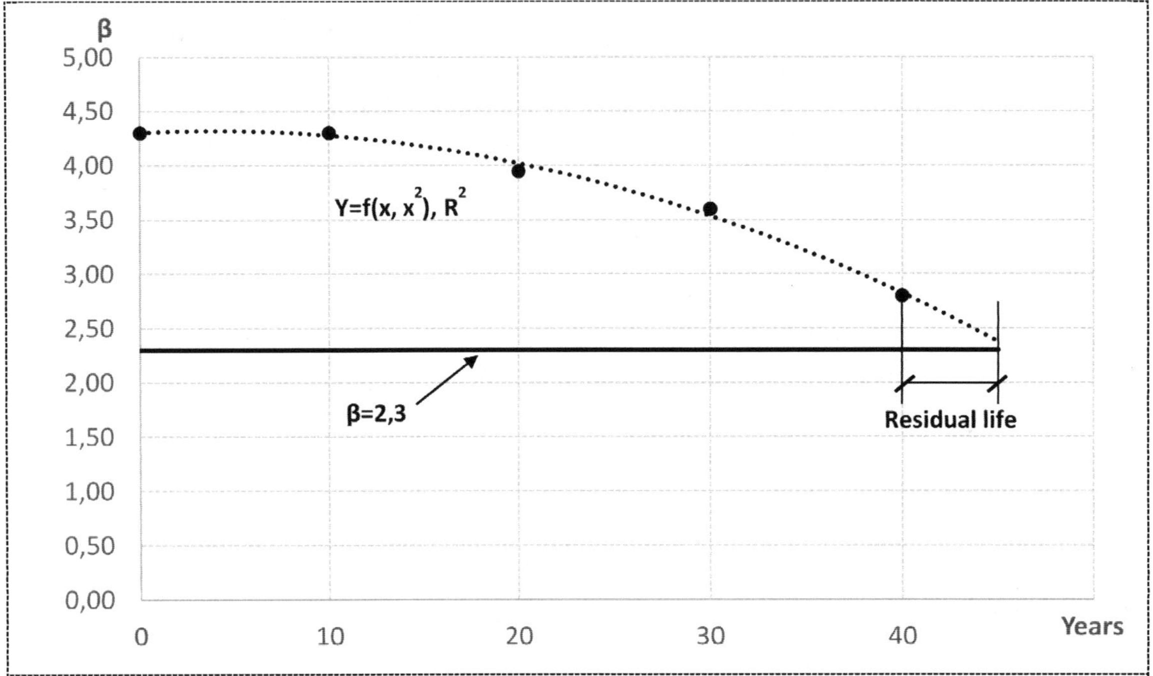

Figure 3. Residual life assessment with reliability index β. Y, x – the parameters of approximative function; R^2 - significance level.

For reliability, the following risks can be identified:

Group K1 - the risk of non-identifying during the survey of the imperfections that have an increased risk and may lead to DD for the transition to category Ad:

Group K2 - the risk of situations that were not considered during the design and did not arise during the operation. The most common risk is increased workload.

For organizational and technical hazards:

OTS Group - the risk of non-compliance or termination of organizational and technical measures that make up the OTS indicators. Possible consequences and potential creation of emergency situations of OTS risks can be uncontrolled change of design diagrams, occurrence of technological accidents.

To forecast the existence of unidentified imperfections of the K1 group, it is necessary to use the information on the DD determined during the survey. There are two possible options:

Option A: Damages are found, and they are identified in the relevant zones of the structure.

Option B: No damage detected. But, based on theoretical calculations (for example, the number of load cycles), their existence is possible.

The following are taken as design parameters:

- the ratio of the survey area to the entire structure area;
- the number of DD of each type of each category;
- the number of DD of each type of each category, which were found during the previous surveys;
- the number of DD of each type of each category, which were repaired according to the results of previously performed surveys.

The probability of emergency occurrence is calculated according to the corresponding dependencies. The basis for assessing the risk of R, we take the functional F, which connects the probability P of an adverse event and the expectation value of damage U from this adverse event:

$$R = F_R\{U,P\} = \sum_i \left[F_{R_i}(U_i,P_i) \right] = \int C(U)P(U)dU = \int C(P)U(P)dP, \quad (3)$$

where: i is type of adverse event; C is a weight function that considers the overall impact of risks. U values are not calculated.

To calculate the OTS risk, a list of $N_{i,OTS}$ (Block 1 and Block 2) indicators of safety assurance by technical operation services is formed. Their corresponding ranking and weighting rates are entered. Existing indicators from the entire list of the required ones are considered. The probability of implementation of OTS measures for a certain residual life can be defined as the sum of the percentage of all remaining indicators, divided by the maximum possible at 100%:

$$P_{OTS} = \left(\Sigma N_{i,OTS}\right)/100 . \quad (4)$$

Also, based on the obtained RL and the operating risk, the recommended timing of the next inspections for different types of structures is determined.

All the results obtained are formed into a block of design solutions for the restoration of structures (Block 6). The database contains a library of standard solutions for reinforcement and repair, which are proposed depending on the type, material, and rate of damage to the structures.

For developing plans for repair and restoration works, the forms of corresponding reports have been created (Block 7).

10. Conclusions

The creation of the 'Resource' database provides a vertically integrated safety management system for the structures of buildings and constructions of industrial facilities based on the principles of ISO 9001 management. The main tasks of information-analytical construction of blocks and modules contribute to:

- reduction by 10-15% of costs associated with the operation of buildings' structures and constructions during their technical maintenance according to their actual state. This reduction value is determined based on the economic pre-analysis (decreasing expenses for season inspection works, planned reparation and technical investigation works) for enterprises where this DB has been installed.
- maintenance of industrial facilities based on the requirements of reliability assurance programs and working documentations based on the calculated residual life during operation with an acceptable risk;
- further training of specialists of enterprise services and personnel, based on the training instructions of the database;
- creation of the internal monitoring system for the quality of works of technical maintenance services of buildings and constructions;
- introduction of materials and technologies to ensure long-term and trouble-free operation of steel building structures.

11. References

[1] IEC 61511. Edition 1.0 2003-03. International standard. Functional safety – Safety instrumented systems for the process industry sector – Part 3. Guidance for determination of the required safety integrity levels. International electrotechnical commission. IEC. p. 19

[2] ISO 2394:2015. International standard. General principles on reliability for structures. ISO 2015. p. 112

[3] Rausand M and Høyland A 2003 System Reliability *Theory: Models, Statistical Methods, and Applications*. Second edition. (New York: Wiley–Interscience) p. 672

[4] Garcia-Molina H, Ullman J and Widom J 2014 *Database Systems the Complete Book* (New Jersey: Pearson Education). 2nd edition, p.1139

[5] Coronel C and Morris S 2018 *Database Systems: Design, Implementation, & Management*, (Cengage Learning, 13th edition), p. 816

[6] Hernandez M 2014 D*atabase Design for Mere Mortals: A Hands-On Guide to Relational*

Database Design, (Addison-Wesley Professional; 3rd edition), p. 672
[7] Mitchell T 1997 *Machine learning* (McGraw-Hill Education; 1st edition), p. 432
[8] Expert Systems in Engineering Applications 1993 Editor: Spyros Tzafestas (Springer Berlin Heidelberg) p. 383
[9] DSTU B V.2.6–210:2016. Assessment of technical conditions for steel structures under operation. Ukrainian State standards (Kyiv: 2016). 80 p. (in Ukrainian)
[10] DSTU B V.1.2-18:2016. Recommendations for buildings and structures technical investigation for technical state assessment (Kyiv: 2017). 44 p. (in Ukrainian)
[11] NPAOP 45.2-1.01-98. The rules for technical investigations, technical state assessment and industrial buildings and structures passporting (Kyiv: 1999) 35 p. (in Ukrainian)
[12] Eurocode 0: ENV 1990:2002+A1. Basic of structural design. CEN, Brussels. 2002
[13] Eurocode 1: EN 1991-1-1. Actions on structures - Part 1-1: General actions - Densities, self-weight, imposed loads for buildings. CEN, Brussels, 2002
[14] Kühn B, Lukić M, Nussbaumer A, Günther H-P, Helmerich R, Herion S, Kolstein M, Walbridge S, Androic B, Dijkstra O and Bucak Ö 2007 Assessment of Existing Steel Structures: Recommendations for Estimation of Remaining Fatigue Life, JRC Scientific and Technical reports. Joint Report Prepared under the JRC – ECCS co-operation agreement for the evolution of Eurocode 3 (program of CEN/TC 250), Editors: G. Sedlacek, F. Bijlaard, M. Géradin, A. Pinto and S. Dimova, (Luxembourg: Office for Official Publications of the European Communities) p. 89
[15] ISO 31000:2009. International standard. Risk management. Principles and guidance. ISO 2009. p. 32
[16] Kumamoto H and Henley E 1996 Probabilistic Risk Assessment for Engineers and Scientists (New York: IEEE Press). p. 597

Seismic hazard mapping using R-CRISIS

E Cutia[1], V Ţurcan[1],
[1] Technical University of Moldova, Faculty of Construction, Geodesy and Cadaster, *Republic of Moldova.*

E-mail: evgheni.cutia@cms.utm.md

Abstract. This article presents methods and software used to develop a seismic hazard map. As an example, the territory of the Republic of Moldova is being studied, which is affected by subcrustal earthquakes of medium depth in the Vrancea seismic area. The elaboration of the map is based on the theoretical model of probabilistic seismic hazard assessment established and described in detail in the specialized literature. In order to facilitate and minimize errors, specialized software was used, one of this software is R-CRISIS, a software created and presented for premier in 1986 by Professor Mario Ordaz at the Institute of Engineering UNAM, Mexico and which is continuously being developed and improved. R-CRISIS allows the users to perform a big range of deterministic and probabilistic analyses through a user-friendly interface.

1. Introduction
Natural hazards represent a real challenge not only for engineers, but for entire society as well. In order to manage and have a good emergency management policy in case of natural catastrophe, every country relies on different kinds of documents, statistics and maps. In this paper is presented how a seismic hazard map could be developed using specialized software. A hazard map represents graphical delimitation of affected or vulnerable areas to a particular phenomenon. As for seismic hazard map, the definition remains unchanged, only the phenomena are known – earthquake; so, one can state that a seismic hazard map is a graphical representation of a particular territory that is divided in small areas where the ground shaking, caused by earthquake, have the same behaviour and characteristics. Ground shaking is defined in terms of some parameter, usually peak ground acceleration (PGA), spectral acceleration or macroseismic intensity (e.g., EMS-98) [1].

1.1. Uses of seismic hazard maps
Generally speaking, hazard maps could be divided into two groups: maps for regulatory purposes and maps for information purposes.

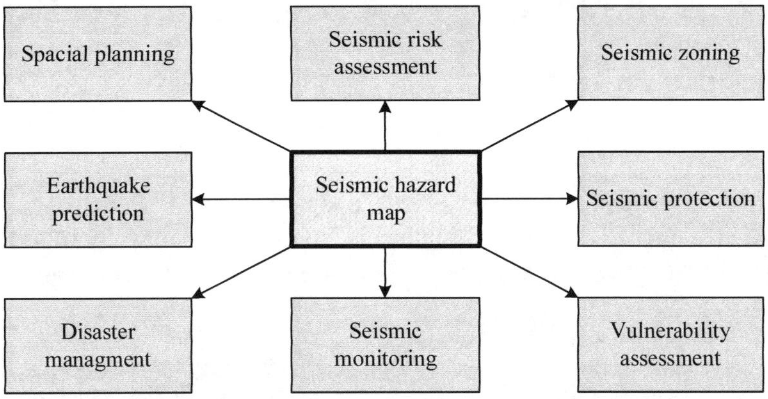

Figure 1. Applicability of seismic hazard map [2]

1.2. Seismic hazard evaluation

In present there are two methods of seismic hazard assessment: probabilistic seismic hazard assessment (PSHA) and deterministic seismic hazard assessment (DSHA). DSHA is a method for estimating location-specific seismic hazard that is affected by the maximum earthquake from the controlling sources that can influence the specific location. Probabilistic seismic hazard is detailed described in [3,4,5,6,7]. PSHA for a site is developed considering all ground motions that can occur in the specified seismic zones of any possible magnitude and/or any possible distance between source and site. The main advantage of PSHA is that it combines all of accidental uncertainties that comes from the analysis of seismic zone parameters from a future earthquake at particular site [6]. PSHA provides a probabilistic approach of a characteristic unit such as ground acceleration.

One should understand that the mapping process is the last step in hazard assessment. In order to get to this final level, we need to perform a seismic hazard assessment.

2. Software R-CRISIS

R-CRISIS is a software that works on Windows Operating System and has capability to carry out probabilistic seismic hazard analysis using a fully probabilistic approach [8]. The first version of software was issued in 1986 by Professor Mario Ordaz - the lead author and was software with a command line interface developed in FORTRAN [8]. In 2007 the program was upgraded and now provides a user-friendly environment to perform seismic hazard assessment. R-CRISIS software is logically organized in six main tabs: File, Input, GMPE Analyzer, Run, Hazard and Tools.

The work in this software can be described in few steps as follows:

- **Adding Map Data** – at this step user should have a reference map in *.sph shapefile format of region that is analysed. It should be pointed out that R-CRISIS uses the World Geodetic System 84 (WGS-84) projection system and does not make any changes of coordinate projection. In addition to the map, the program offers user to add a text file with cities that could be subjected to hazard assessment (see Figure 2).
- **Computation Site Data** – is used to define a site where the PSHA is computed. Computation can be performed for a grid with a given step or for a specific site and output is required in terms of hazard curves (see Figure 3).
- **Source Geometry Data** – intend of this panel is to provide a fast and easy way of inserting the geometry of seismic sources. Various geometric models are realized in R-CRISIS software that characterize seismic sources. In this article will be used area-plane sources that are inserted as a set of vertexes which create a plan. Line, point and grid sources are available also (see Figure 3).
- **Spectral Ordinates Data** – in this window user can define number of spectral ordinates and their fundamental periods (see Table 1)
- **Seismicity Data** – for each seismic source defined in *Source Geometry Data*, user should specify seismicity parameters. R-CRISIS, by default, assigns modified G-R seismicity model to each source (see Table 2).
- **Attenuation Data** – in R-CRISIS, ground motion prediction model could be added from a pre-defined list of existing attenuation laws or could be used a custom *.atn file, created by user.

2.1. Adding Map Data

For map data, authors had prepared 2 types of file to add in software:

- A shapefile (*.shp) of Republic of Moldova in WGS-84 system of coordinates;
- A text file (*.txt) with 5 biggest cities as follows (Name of state, name of city, longitude, latitude):
    - Chisinau District, Chisinau City, lat $N\ 47°0'59''$ and long $E\ 28°52'0''$.
    - Balti District, Balti City, lat $N\ 47°45'41''$ and long $E\ 27°55'44''$.
    - Cahul District, Cahul City, lat $N\ 45°54'18''$ and long $E\ 28°11'47''$.
    - Stefan-Voda District, Stefan-Voda City, lat $N\ 46°30'55''$ and long $E\ 29°39'47''$.
    - Edinet District, Edinet City, lat $N\ 87°10'14''$ and long $E\ 27°18'15''$.

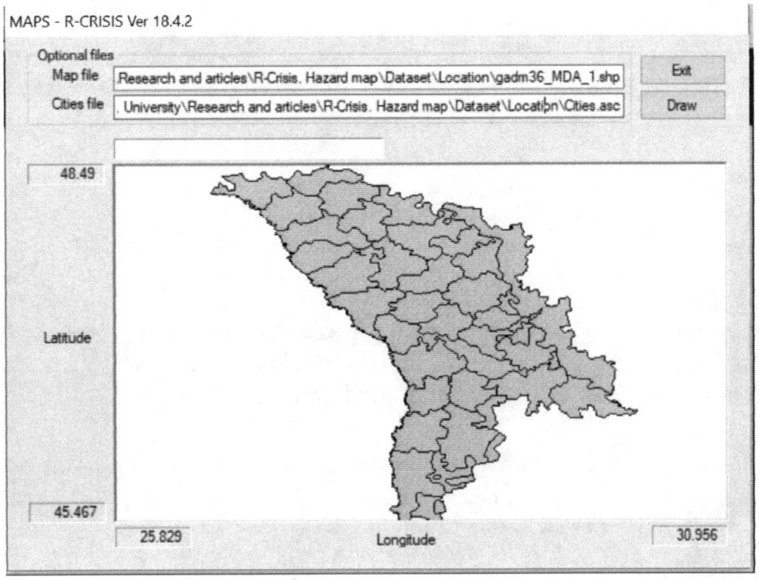

Figure 2. Inserting map of Republic of Moldova.

2.2. Computation Site Data

In site data computation dialog window was chosen the Grind of sites option for performing PSHA. Was defined origin of the grid and incremented with the number of needed lines to cover all territory in two orthogonal directions. As the base point was chosen coordinates at latitude $N\ 45°20'24''$ and the longitude $E\ 26°29'23''$. The grind size was chosen to be equal of 40 vertical and 36 horizontal lines with the increment of $0.1°$ or $11.1\ (km)$. The resultant map with grid point is presented in Figure 2.

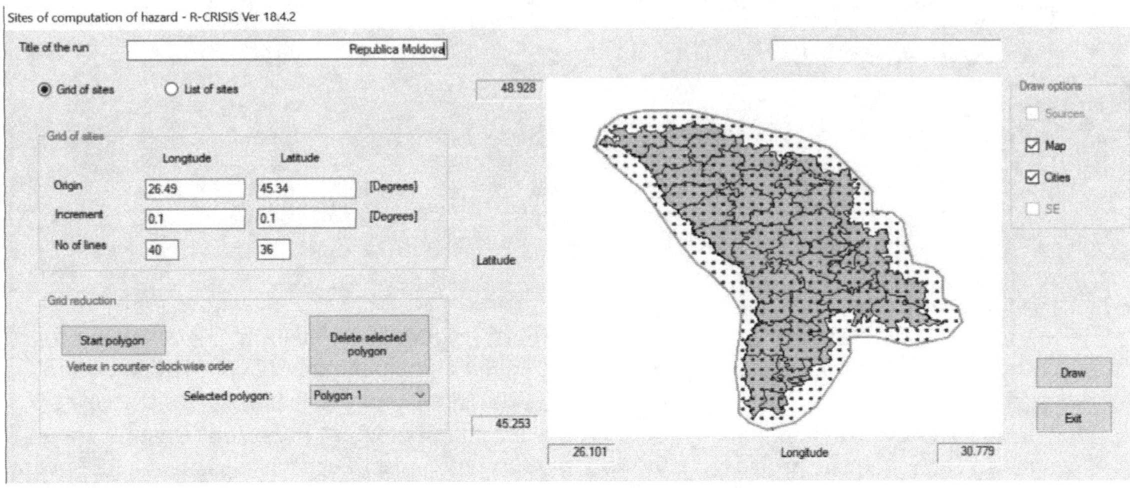

Figure 3. Division in grids of size $0.1°\ by\ 0.1°$.

2.3. Spectral Ordinates Data

This screen allows entering the parameters for each spectral ordinate (or, in general, intensity measure) for which seismic hazard will be computed. The values of spectral ordinates for defining fundamental periods for each spectral ordinate are entered. The measure unit is gal or cm/s^2. It this case study spectral ordinates are assigned for time periods of $0.01\ sec$ (i.e., PGA), $0.2\ sec$ and $0.4\ sec$.

Table 1. Spectral ordinates.

Period index	T (s)	Lower limit	Upper limit	Units
1	0.01	1	1000	gal
9	0.2	1	3000	gal
10	1	1	3000	gal

2.4. Source Geometry Data

In the geometry data window, users can define the geometry of seismic sources. There are different geometrical models implemented in R-CRISIS, in this article only the area source model is used. They are defined as polygons by coordinates for each of their vertexes.

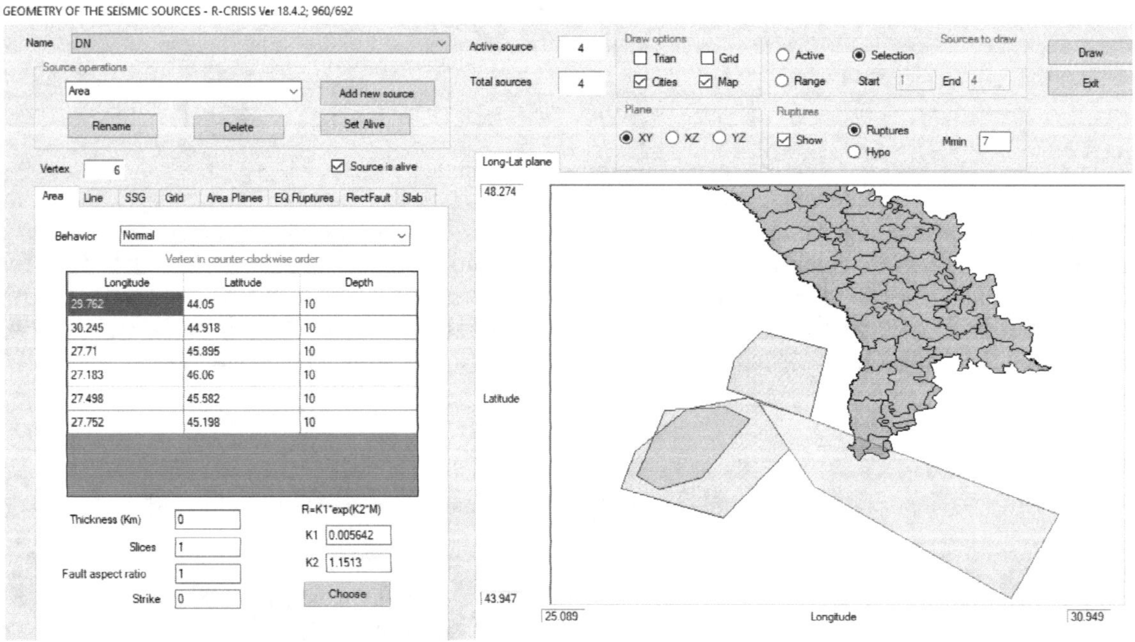

Figure 4. The area source geometry.

The analyzed sources are:

- Vrancea crustal seismic source – located in South-Eastern Carpathians arc, spread over a stripe area delimited to the north by Pecenega-Camena fault and to the south by the Intramoesian fault [9].
- Vrancea intermediate seismic source – is the Romania's highest hazardous seismic zone with the epicentral area confined to about **40 x 60 km** with most of earthquakes that occur at depths between **60 km** and **180 km** that are felt over the large areas [10].
- Bârlad depression – located in the North-East of the Vrancea region in the Scythian platform, represents the extension to the North-West of North Dobrogea source zone. The observed earthquakes are of moderate magnitude and do not exceed $\mathbf{M_w = 5.6}$ [11].
- Predobrogian Depresion – is part of the southern boundary of the Predobrogian Depression, following alignment of the Sfintul Gheorge fault. From seismotectonic perspective, it is a part of Scythian platform with the maximum observed magnitude of $\mathbf{M_w = 5.3}$.

2.5. Seismicity Data

Seismic parameters for all defined seismic sources should be assigned in "Seismicity Data" button. The modified Gutenberg-Richter seismicity model is used to describe every seismic source. This model is associated to Poisson occurrences, so the probability of exceeding intensity level a in the next T_f years, given that an earthquake with magnitude M took place at a distance R from the site, is given by:

$$P_e(a,T|M,R) = 1 - \exp[-\Delta\lambda(M)Tp_1(a|M,R)] \quad (1)$$

where $\boldsymbol{p_1(a|M,R)}$ is the exceedance probability of intensity level $\boldsymbol{a}$, given that a magnitude $\boldsymbol{M}$ event occurred at a distance $\boldsymbol{R}$ from the site, and $\boldsymbol{\Delta\lambda(M)}$ is the Poissonian magnitude exceedance rate associated to the magnitude range characterized by magnitude $\boldsymbol{M}$.

Table 2. Seismic sources seismicity parameters.

Seismic source	Vertex number	Coordinates	Average depth	M_{min}	M_{max}	b	$\lambda_{M_{min}}$	*Std of b*
	1	26.823 / 44.870						
	2	27.498 / 45.582						
	3	27.183 / 46.057						
VRC	4	27.084 / 46.116	20	4.5	7.0	1.1	0.16	±0.20
	5	26.463 / 45.998						
	6	25.936 / 45.561						
	7	25.793 / 45.174						
	1	26.180 / 45.166						
	2	26.611 / 45.284						
	3	27.100 / 45.893						
VRI	4	26.482 / 46.012	160	4.9	8.1	0.87	1.20	±0.08
	5	26.238 / 45.983						
	6	25.955 / 45.307						
	1	27.710 / 45.895						
	2	27.880 / 46.609	10					
DB	3	27.218 / 46.790		4.5	5.8	1.10	0.07	±0.42
	4	26.938 / 46.496						
	5	26.867 / 46.197						
	1	29.762 / 44.050						
	2	30.245 / 44.918						
	3	27.710 / 45.895						
DN	4	27.183 / 46.060	10	4.5	5.8	1.84	0.06	±0.31
	5	27.498 / 45.582						
	6	27.752 / 45.198						

2.6. Attenuation Data

Selection of attenuation model is the key step in every seismic hazard assessment. The ground motion prediction models (GMPM) are developed after a lot of research. Generally speaking, the GMPM establishes probabilistic relations between earthquake characteristics, intensities and distances at the computation site.

The attenuation relationship developed by Youngs [12] for subduction zone earthquake is used in case study described in this article. Youngs provides attenuation relationships for rock and soft soils and the relationships are used for estimating both, peak ground acceleration (PGA) and spectral acceleration (SA) for 5% damping value. The attenuation relation for rock site is following:

$$ln(y) = 0.2418 + 1.414M + C_1 + C_2(10 - M)^3 + C_3\, ln\left(r_{rup} + 1.7818 \cdot exp(0.554M)\right) + 0.00648H + 0.3643Z_T \quad (2)$$

standard deviation is:

$$\sigma = C_4 + C_5 M \quad (3)$$

where y – is PHA or SA in g, M – moment magnitude, r_{rup} – closest distance to rupture in km, H – is focal depth in km, Z_T – source type (0 for interface and 1 for interslab) and σ – standard deviation.

Based on the research [13] it is decided to use hybrid attenuation models. A hybrid GMPM is the result of the weighted combination of two or more distributions that can have different means values and standard deviations [14]. In general form, the conditional probability of exceeding an intensity measure A is calculated using following relationship [8]:

$$P(A > a) = \sum_{i=1}^{N} w_i \left\{1 - \Phi\left[\frac{a-\mu_i}{\sigma_i}\right]\right\} \quad (4)$$

where w_i is the weight assigned to the ith base GMPM, Φ – is normal distribution and μ_i, σ_i – are the mean and standard deviation values of the ith base GMPM. For crustal seismic sources are used Cauzzi and Faccioli – CF08 [15] and Akkar and Bommer – AB10 [16] attenuation relationships. The final weighing scheme is given in Table 3.

Table 3. Weighing scheme for PSHA of the Republic of Moldova.

GMPM	Weighting factors
Crustal	
CF08	0.6
AB10	0.4

3. Results

Having all input information described above in section XX, for all 4 seismic sources that affect the most territory of the Republic of Moldova was computed seismic hazard for the return period of 475 years, which corresponds to the probability of exceedance of 10% in 50 years. The seismic hazard map of the Republic of Moldova in terms of Peak Ground Acceleration and Spectral Acceleration for periods of 0.2 sec and 1 sec as well as spectral acceleration graphs for analysed cities are shown in the figures below. The mean hazard curves for the five analysed cities in the Republic of Moldova is shown in Figure 6.

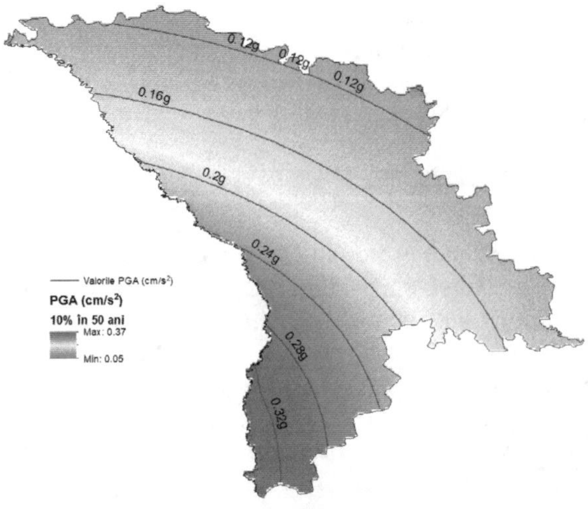

Figure 5. Seismic hazard map of the Republic of Moldova (Return Period 475 years).

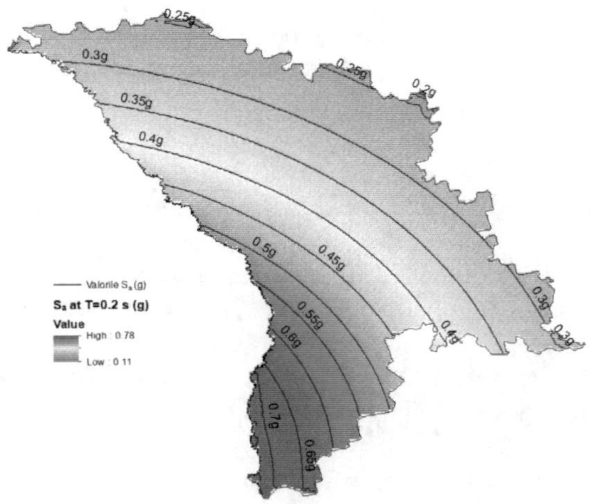

Figure 6. Seismic hazard map of the Republic of Moldova for Spectral Acceleration of 0.2 sec.

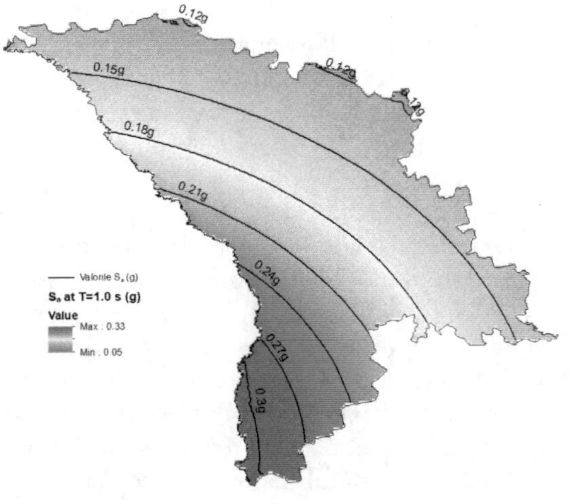

Figure 7. Seismic hazard map of the Republic of Moldova for Spectral Acceleration of 1.0 sec.

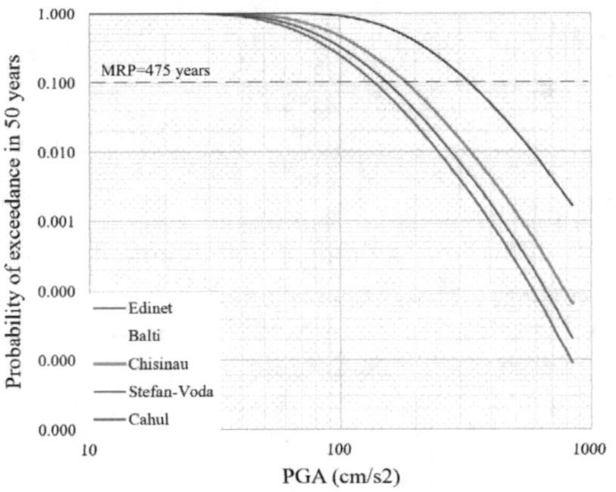

Figure 8. Mean hazard curves for PGA for the five analyzed cities.

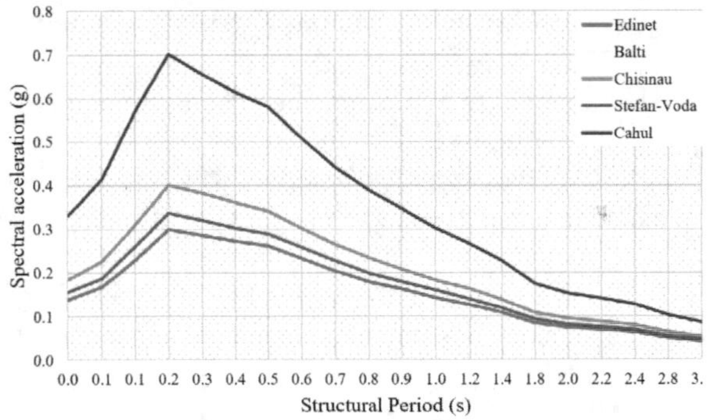

Figure 9. Spectral acceleration for five analyzed cities.

4. Conclusion

In this article is presented a probabilistic seismic hazard assessment of the Republic of Moldova territory. Using R-CRISIS software, firstly developed by professor Mario Ordaz, it was possible to evaluate probabilistically seismic hazard and obtain mean seismic hazard curves as well as spectral acceleration graphs. Wide possibilities of use of R-CRISIS software was shown along with large predefined tools for hazard evaluation. Definitely, the software is useless if key data entered are wrong or missing, therefore to obtain a veridic output one should comprehend the importance of work that lays before software, i.e. identification of seismic sources, determination of seismic parameters for each source, choosing the adequate attenuation law, etc.

5. References

[1] Musson R and Sargeant 2007 *Eurocode 8 seismic hazard zoning maps for the UK*, British Geological Survey, Keyworth, Nottingham

[2] GSHAP 2000 Global Seismic Hazard Assessment Program [Interactiv]. Available: http://static.seismo.ethz.ch/GSHAP/index.html. [Accesat 20 February 2021].

[3] Cornell A C 1968 Engineering seismic risk analysis, *Bulletin of the Seismological Society of America,* vol 58, pp 1583 - 1606
[4] Kramer S 1996 *Geotechnical Earthquake Engineering*, New Jersey: Upper Saddle River
[5] McGuire R 2004 *Seismic hazard and risk analysis*, Earthquake Engineering Research Institute
[6] Văcăreanu R, Pavel F, Aldea A, Arion C and Neagu C 2015 *Elemente de analiză a hazardului seismic*, București: Conspress București, pp 79-80
[7] Reiter L 1990 Earthquake hazard analysis, *Issues and Insight*
[8] Ordaz M and Salgado-Galvez M A 2019 *R-CRISIS Validation and Verification Document, Mexico City*: ERN Technical Report
[9] Moldovan I A, Diaconescu M, Popescu E C and Radulian M 2016 Input parameters for the probabilistic seismic hazard assessment in the Eastern part of Romania and Black Sea Area, *Romanian Journal of Physics*
[10] Ardeleanu L, Leydecker G, Bonjer K P, Busche H, Kaiser D, and Schmit T 2005 Probabilistic seismic hazard map of Romania as a basis for a new building code, *Natural Hazards of Earth System Science,* vol **5**, no 5, pp 679-684
[11] INFP, Institutul național de cercetare - dezvoltare pentru fizica pământului, 2020. [Interactiv]. Available: http://www.infp.ro/index.php?i=hzd. [Accesat 5 Mai 2020].
[12] Youngs R R, Chiou S J, Silva W J and Humphrey R J 1997 Strong ground motion attenuation relationships for subduction zone earthquakes, *Seismological research letters,* vol **68**, no 1, pp 58-73
[13] Florin P, Vacareanu R, Douglas J, Radulian M, Cioflan C and Barbat A 2016 An updated probabilistic seismic hazard assessment for Romania and comparison with the approach and outcomes of the SHARE project, *Pure and Applied Geophysics,* vol **173**
[14] Scherbaum F, Bommer J J, Bungum H, Cotton F and Abrahamson N A 2005 Composite ground-motion models and logic trees: methodology, sensitivities and uncertainties, *Bulletin of the Seismological Society of America,* vol **5**, no 95, pp 1575-1593
[15] Cauzzi C and Faccioli E 2008 Broadband (0.05 to 20 s) prediction of displacement response spectra based on worldwide digital records, *J Seismol,* vol **12**, no 4, pp 453-475
[16] Akkar S and Bommer J J 2010 Empirical equations for the prediction of PGA, PGV and spectral accelerations in Europe, the Mediterranean Region and the Middle East, *Seismological Research Letters,* vol **81**, no 2, pp 195-206

IOP Conf. Series: Materials Science and Engineering 1141 (2021) 012026 doi:10.1088/1757-899X/1141/1/012026

Bio product used in the self-sealing process of microcracks in hydrotechnical concretes

C Sbirlea[1], D N Isopescu[1], D Ungureanu[1], S G Maxineasa[1]
[1] "Gheorghe Asachi" Technical University of Iasi, Faculty of Civil Engineering and Building Services, Department of Civil and Industrial Engineering, 1, Prof. Dr. docent Dimitrie Mangeron Blvd., Iasi, 700050, Romania

E-mail: catalin_sbirlea@yahoo.com

Abstract. This paper is focused on quantifying the influence of the Bio materials that are used in the self-healing process of concrete. The Bio material that was studied has a high content of calcium carbonate, which, after precipitation, may lead to the sealing of the microcracked surface of concrete. This nucleation process is possible only in humid environments; thus, water has a paramount role in ensuring the mobility of the molecules. The difference between a complete and a partial seal is given by the amount of the Bio product that is available in the cracked area, the size of the calcium carbonate particles and the amount of water. For this study, the SEM Quanta Dual Beam device was used, through which scans were performed by electron microscopy, in order to observe the precipitation of the Bio product on the cracked surface. The chemical process throughout the sealing takes place has been presented through various illustrations and reports. By analysing previous experimental studies, it might be stated that the use of calcium carbonate in various forms is a sustainable and an environmentally friendly repair solution that, also, may improve the durability of cement-based materials. The use of ecological products in new cement-based materials is one of the main topics of worldwide interest for various research teams.

1. Introduction

Due to the extreme phenomena that occurred more and more often after the global warming, the need of population to adapt to severe environmental conditions is accentuated. In order to consolidate the existing constructions techniques and to develop new constructions technologies, various products are created, with the main role of increasing the mechanical and the physical resistance of the construction materials. Nowadays, an important emphasis has been placed on public opinion, namely that these materials must use less raw resources, and thus reducing the emission of harmful emissions. In recent years, researchers have presented in various publications, studies focused on the use of organic raw materials that can either slow down the phenomenon of degradation, or completely remediate the degraded construction elements. This process is referred to as the autogenic self-healing. Among the newly developed and even patented organic materials brought to the public opinion, the most advantageous ones are bacterial and fungal spores that can survive in extreme environments, inside the concrete construction elements. Bacteria survive in the PH 13 environment and, they can stay in latency for long periods of time (over 200 years). Removal from hibernation is possible only at the moment when the degradation phenomenon appears on the construction element and the organic matter comes into contact with oxygen and water. In this frame, the hydrotechnical constructions, due to their constructive nature of being in contact with water (permanent or infrequent), represent one of the main applications of the autogenous self-healing process. However, researchers encountered several issues in choosing the protective layer of organic matter for hydrotechnical concrete elements, due to the fact that the sealing materials activates as soon as they come in contact with water. The incipient phase through which a concrete construction element begins to lose its initial strength is represented by the appearance of microcracks. At the moment of micro-cracking, water and various impurities starts to infiltrate. The longer is postponed the intervention time, the more expensive are the works and the materials used for rehabilitation. For these reasons, the premature sealing of microcracks with Bio materials is considered to be a viable alternative to the traditional rehabilitation methods. The two autogenous self-healing processes that use bacteria and fungi as organic raw materials have a common denominator, namely the fact that both precipitate calcium carbonate ($CaCO_3$) in different amounts, on the cracked surface. The higher the amount of organic material is present in the microcracked area, the more complete is the sealing.

2. Stable connections in micro-cracked concrete

2.1. The chemical compounds of the Bio concrete matrix

The processes underlying autogenic self-healing are of a physical, chemical and mechanical nature, as illustrated in figure 1.

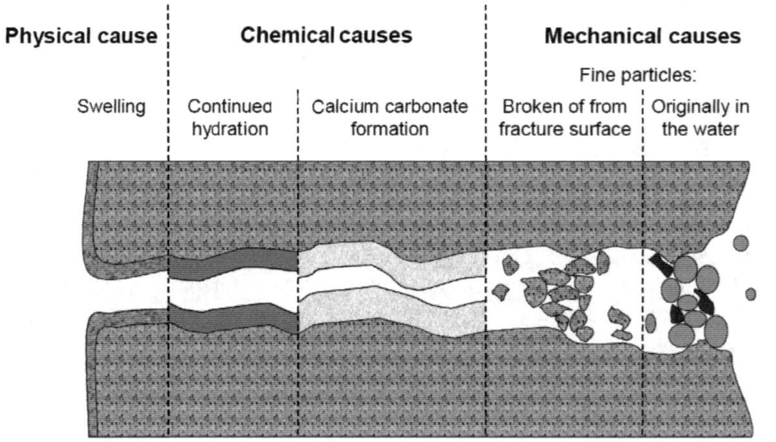

Figure 1. The processes of autogenic self-healing [1].

The physical processes consist in hydrating the cement stone by the absorbed water. The latter induce a swelling inside the crack and, further on, the expansion of constituents cause the closure of about 10% of the cracks (according to statistics) [1].

By precipitating the calcium carbonate ($CaCO_3$), the calcium ions in the cement matrix react with the carbon dioxide or the carbon ions in water and, they form crystals. [2]. The chemical reactions highlighted by the equations (1, 2, 3), [2], represent the most important mechanism in autogenic healing. Edvardsen [3] stated that the process of crystallization of calcium carbonate involves two distinct phases. In the initial phase, the kinematics of crystallization is generated by the reactions that took place on the surfaces of the cracked concrete element. In the second phase, there are involved several reactions with diffusion ions, through the layer already created in the first stage by the precipitation products. Yang [4] evaluated the efficiency of the above-mentioned chemical mechanism, and concluded that there are two important aspects which influence the outcome. Firstly, the presence of calcium ions is mandatory, in a relatively high concentration, and secondly, the humidity is essential and it can be ensured by alternating the water – air amounts or by repeated immersions. These aspects are highlighted in figure 2.

$$H_2O + CO_2 \leftrightarrow H_2CO_3 \leftrightarrow H^+ + HCO_3^- \leftrightarrow 2H^+ + CO_3^{2-} \quad (1)$$

$$Ca^{2+} + CO_3^{2-} \leftrightarrow CaCO_3 \ (pH\ water > 8) \quad (2)$$

$$Ca^{2+} + \mathrm{HCO}_3^- \leftrightarrow \mathrm{CaCO}_3 + H^+ \ (7.5 < \mathrm{pH\ water} < 8) \quad (3)$$

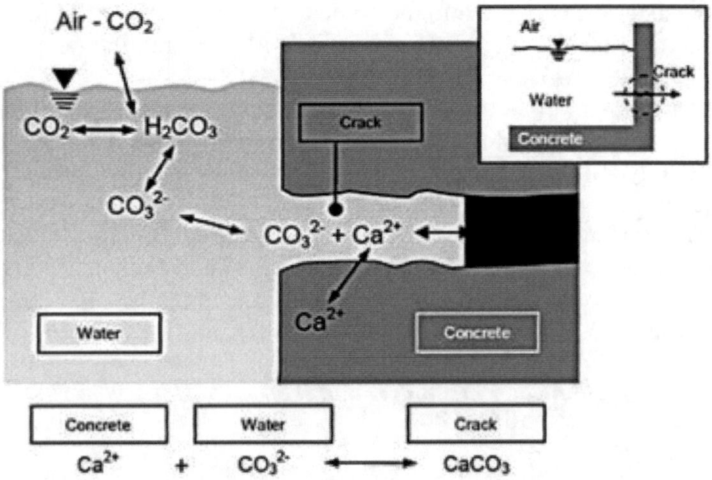

Figure 2. The activation of the calcium carbonate precipitation due to the interaction of the cement matrix with water and air [2].

The mechanical processes occur by blocking the crack with fine particles (impurities), resulted from the cracking process and residues that exist in the water and infiltrate the crack. The mechanical process can produce effects only in combination with the chemical mechanisms described above [2].

Starting from the two autogenous self-healing processes based on biological compounds, namely the use of bacterial and fungal spores, it can be identified the repetitive use of calcium carbonate ($CaCO_3$) precipitation. Through this specific process, the biological compounds feed on nutrients and they transform these substances into calcium carbonate ($CaCO_3$). A complete sealing of a microcrack is possible if there is a sufficient concentration of curing agent in the respective area.

The experimental outcomes reported so far demonstrate that the autogenous healing of microcracks by precipitation of the calcium carbonate ($CaCO_3$) can be very effective due to an adapted bacterial suspension. Indeed, due to the calcite layer formed on the crack wall, the limiting component is not the carbonate ions, that can be generated by bacterial activity or supplied by air, but the available calcium ions [5].

Whether we are talking about bacteria that activates when they come in contact with oxygen and water, or we refer to the fungal spores which germinate, both processes are based on the same kinetics, namely the precipitation of the calcium carbonate ($CaCO_3$) on the cracked concrete surface.

Kim Van Tittelboom [6] mentions that the calcium carbonate ($CaCO_3$) is highly compatible with concrete and, thus, it represents a sustainable and an environmentally friendly repair solution. In 1938, Bessey et al. [7] identified the formation of calcium-carbo-aluminate in the hydration process of cement, by the time calcium carbonate was incorporated. This phenomenon was named "*the chemical effect of calcium carbonate*".

Mingli Cao [8] stated that when calcium carbonate is introduced into the matrix its density will increase due to its filling effect and, the hydration process can be accelerated due to its nucleation effect. Thus, the mechanical properties, workability and hydration process of cementitious materials are directly influenced by the size of calcium carbonate particles. The following sizes have been identified by Mingli Cao:

- Macro-Calcium carbonate (> 1 mm) such as coarse limestone sand or limestone aggregates. At this scale the chemical and the nucleation effects are not significant and the effect on the hydration process is negligible; the substance behave like an inert filler in the cementitious compositions [8].
- Micro-Calcium carbonate (1 μm - 1 mm), such as limestone powder and limestone dust, are often used in the manufacture of cement, and due to the fact that they have smaller dimensions than the cement grains they participate in the hydration process and can affect the kinetic factors. The

incorporation of calcium carbonate into the cemented composites may lead to an increase both in the matrix density and their resistance to acid attacks. In general, the nucleation effect is directly proportional to the lime powder content [8].

- Nano-calcium carbonate (<1 μm) in cemented composites can significantly improve their mechanical properties and durability [10]. Nano-calcium carbonate (50 - 120 nm) is very effective in accelerating the hydration of cement, especially for the induction period of tricalcium silicate (C_3S), due to its nucleation effect [8].

For this work, a Bio product with a concentration of over 90% of micro-calcium carbonate ($CaCO_3$) was used, without adding any other biological compounds. The bonding reactions between the Bio product and the components of the concrete matrix are triggered by the chemical reactions specific to the autogenous self-healing.

2.2. The materials used in the Bio concrete matrix and their chemical compositions

2.2.1. Cement

Cement is a building material provided in the form of a fine powder that is made by grinding the clinker and which, in contact with water, sets and hardens. Cement is the most widely used material in the field of civil, industrial, hydrotechnical, road and bridge engineering. Addition agents are used, primarily, to lower the cost of production and secondly to improve some properties [11]. The chemical composition of normal Portland cement is shown in table 1 [11].

Table 1. The chemical composition of normal Portland cement [11].

Chemical composition of Portland cement	Amounts (%)
Calcium oxide (CaO)	60-67 %
Silicon dioxide (SiO_2)	19-24 %
Aluminum oxide (Al_2O_3)	4-7 %
Iron oxide (Fe_2O_3)	2-6 %
Magnesium oxide (MgO)	4-5 %
Sulfur trioxide (SO_3)	3 %
Alkaline compounds (K_2O, Na_2O)	< 1%

2.2.2. Sand

Sand is an unconsolidated sedimentary rock, derived from the crushing of minerals, rocks or organisms, which is found in the form of an accumulation of fine grains (0,063 – 2 mm) [12]. The sands used in the preparation of mortars and concretes can be of two types, depending on their origin, namely:

- natural (quarry, river, sea, dunes);
- artificial (obtained by breaking, crushing);

The chemical composition of river sand is presented in table 2 [12].

Table 2. The chemical composition of river sand.

The chemical composition of river sand.	Amounts (%)
Silicon dioxide (SiO_2)	51.00 %
Aluminium oxide (Al_2O_3)	6.83 %
Iron oxide (Fe_2O_3)	0.32 %
Calcium oxide (CaO)	0.48 %
Potassium oxide (K_2O)	0.40 %
Titanium dioxide (TiO_2)	0.58 %

2.2.3. Bio product

The Bio product is made of micronized powder which is subsequently granulated in dimensions between 2 - 6 mm; this material is strongly reactive. On the outside it is protected with a layer of paraffin (1/2 mass of the Bio product). The chemical composition of the Bio product is presented in table 3 [13].

Table 3. Chemical composition of the Bio product.

Chemical composition of granular Bio product	Amounts (%)
Calcium carbonate ($CaCO_3$)	*91%*
Magnesium carbonate ($MgCO_3$)	*2 %*
Paraffin ½ mass from Bio product	*~ 47%*

Paraffin is a white solid, used as a raw material in various industries, with a melting point between 52°C - 54°C. Calcium carbonate is the salt of calcium with carbonic acid. The former is a solid substance, of white colour, with the chemical formula ($CaCO_3$), and the ratio of Ca^{2+} and CO_3^{2-} ions being 1:1 (4, 5).

Calcium carbonate is widely spread in nature, in minerals (calcite, aragonite, vaterite), in bones, teeth, shells, corals and shellfish crust. In rocks it is found in the form of limestone (where it is almost pure), an example being the dolomites, which are a mixture of calcium and magnesium [14].

$$CaO + H_2O \leftrightarrow Ca\,(OH)_2 \tag{4}$$

$$Ca(OH)_2 + CO_2 \leftrightarrow CaCO_3 + H_2O \tag{5}$$

This product has been approved to be used in agriculture, zootechnics, the paint industry, the glass industry, the plastics industry, the mortar and sandstone industry, the steel industry and the rubber industry. [13]. Given the fact that this product is also found in the pharmaceutical industry, it might be stated that it can be used in different dosages to treat people who lack calcium [15].

2.3 Autogenous self-healing of microcracks under ideal conditions

In order to establish the intermolecular reactions at the matrix level of Bio concrete, a very important element is mandatory, namely water (H_2O). The smaller is the environment in which the microcracking occurs, the shorter is the reaction time of the Bio product. This is due to the fact that there is a smaller area, strongly saturated with ($CaCO_3$), and the nucleation between components is almost instantaneous. Figure 3 illustrates a scenario through which, under the effect of some stresses, a micro-crack of size 0.4 μm appears in a concrete element. The two granules of Bio product cracks, and the water penetrates inside, ensuring the mobility of the molecular species.

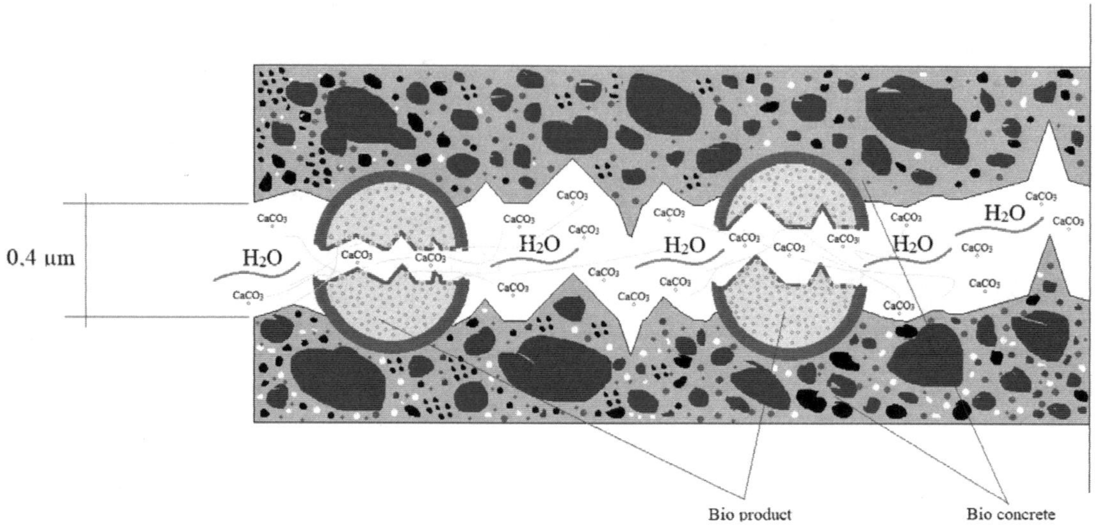

Figure 3. Micro-cracking and water penetration into the Bio concrete element.

The more material of Bio product is found in the microcrack, the more complete is the sealing. Figure 4 shows the moment of decomposition of the two granules of the Bio product and the migration of molecules in a humid environment. The molecules detached from the Bio product seek to create stable relationships mainly with the other molecules in micronized form (SiO_2, Al_2O_3, Fe_2O_3). The chemical process will continue until the molecules in the Bio product are consumed.

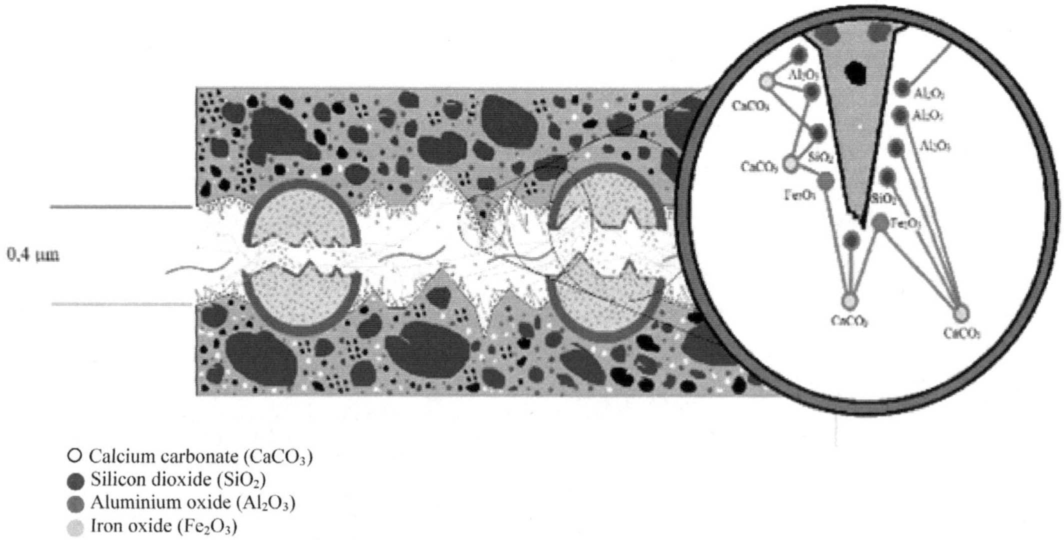

Figure 4. Stable bonds in the Bio concrete matrix.

An important parameter in this chemical process is represented by the amount of paraffin that is used to create the protective layer (that conserve the Bio product during concrete mixing). The more the amount of paraffin is absorbed by the Bio product, the less micronized compound ($CaCO_3$) is obtained; thus, the temperature of the paraffin melting process is important. From what has been observed so far, the

paraffin layer being in a relatively small amount, on long-term contacts with water (total immersion) it remains in the form of a spherical shell and the Bio product diffuses into the water. It can also be removed by increasing the temperature of the matrix, phenomena that can be easily reached in common operating environment.
All this kinetics underlies the autogenic self-healing or rather the self-sealing of the concrete micro-cracks up to 0.4 μm. This whole process can be observed with special high precision equipment (SEM). The chemical reactions that occur after nucleation are presented in the relationships below (6, 7, 8)

$$CaCO_3 + Al_2O_3 = Ca_3Al_2O_6 \quad (6)$$

$$CaCO_3 + SiO_2 = Ca_2SiO_5 \quad (7)$$

$$CaCO_3 + Fe_2O_3 = Ca_4Al_2Fe_2O_{10} \quad (8)$$

Following the chemical nucleation process, stable bonds were obtained (highlighted in figure 5) and sealing took place by deposits on the cracking areas.

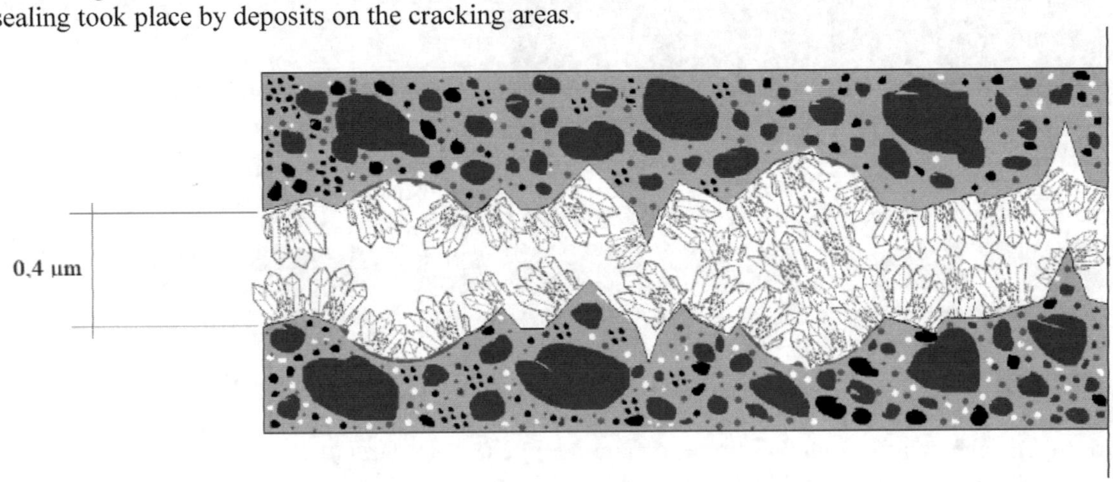

Figure 5. Sealing of 0.4 μm microcracks.

The total or partial sealing of the 0.4 μm microcracks presented in the above scenario is influenced by the amount of the Bio product, the amount of existing water and the amount of the protective layer.

3. Scanning electron microscopy (SEM). Case studies

Microbial calcium carbonate has been widely investigated in the self-healing process of concrete cracks. In this chapter are reviewed 2 studies related to the phenomenon of precipitation of the calcium carbonate using bacteria. The outcomes of these studies are in a general good agreement with the ones obtained throughout the investigations presented below, in the 3.2 section.

3.1. Study 1

Hassan A. [16] presents in a study, a comprehensive investigation carried out on bacterial growth and factors influencing the evolution of urea hydrolysis, aiming to accurately describe the precipitation of the calcium carbonate inside concrete. The results obtained by Hassan A. [16] indicate that the bacteria managed to survive in a latent state without any reproduction at a pH level of 12-13. A decrease of almost 75% of the hydrolysis efficiency inside the concrete pores was observed, mainly due to the high pH value (pH = 13) [16]. Moreover, the complete curing of the 0.4 μm microcrack was achieved after 70 days, while the cure ratio was 15% lower inside the concrete. This fact was highlighted by the SEM analysis (figure 6). Hassan A., [16] concluded that these bacteria can self-heal cracks, and can lead to a significant increase in the life cycle of concrete elements.

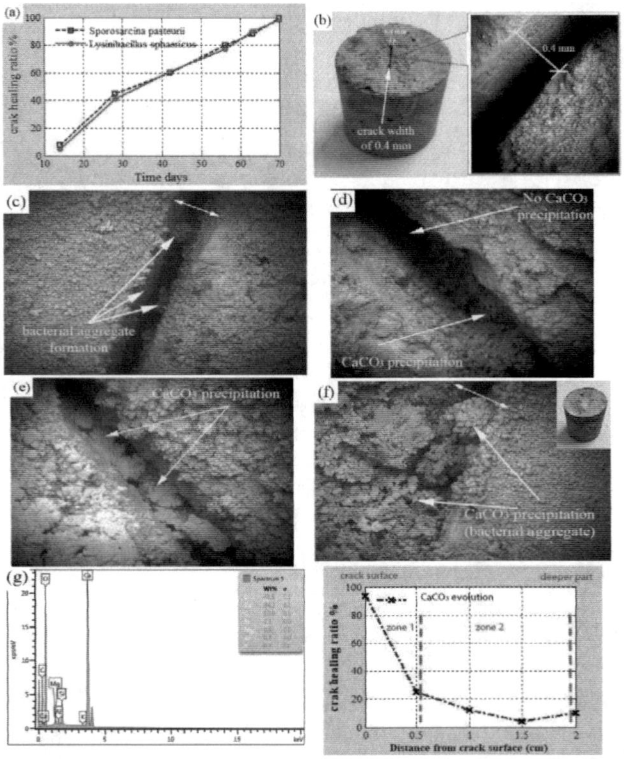

Figure 6. Evolution of crack healing with the help of bacteria (a), healing ratio at a different time (b), 0 days (c), 14 days (d), 28 days (e), 56 days (f), 70 days (g), SEM-EDS (h), distribution of $CaCO_3$ along the crack area [16].

3.2. Study 2

Shekhar [17] stated that the presence of pores and cracks in the structure allows the penetration of harmful chemicals, gases and moisture, endangering the life of the structure. The basic remedy for these issues is to fill these gaps and cracks with synthetic polymers. The quoted research focuses on the development of a cement bio-mortar using bacteria, with the role of sealing cracks through the process of inducing calcium carbonate precipitation. The results obtained by Shekhar [17] indicated a 28.3% increase in mortar strength and a 12% reduction in water absorption. This biotechnological approach, if adopted, could provide sustainable concrete protection by filling cracks and pores. SEM and XRD analyses were also performed, in order to have a better perspective on the overall process and on the precipitation of the calcium carbonate from the pores, in particular (figure 7).

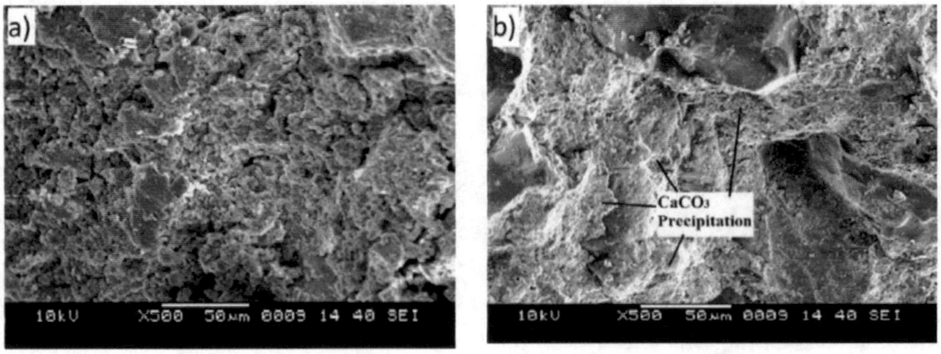

Figure 7. $CaCO_3$ precipitation inside the mortar [17].

The results obtained by Shekhar [17] were as follows:

- The biotechnological approach using the bacterial solution helps to improve the properties of the cement mortar;
- An improvement in the compressive strength of mortar by 28.3% is achieved when isolated urea is produced;
- The use of the bacterial solution in the preparation of the mortar reduces the water absorption of the mortar by 12%, at 28 days compared to the control mortar;
- SEM images of mortar prepared with bacterial solution show the precipitation of ($CaCO_3$) which fills the gaps and makes it dense and compact, which ultimately increases durability;

3.3 Study 3

In order to highlight the existing nucleation process inside a microcrack between the Bio product and the rest of the materials existing in the mortar matrix, the SEM Quanta 200 3D Dual Beam device was used, through which scans were performed by electron microscopy. For this case study, a sample consisting in a slice of mortar, cut from a prism measuring 40x40x160mm, which has a content of about 10% Bio product, was investigated. Figures 8 - 9 shows the fixing of the sample in the clamping system of the device and the adjustments that were performed according to the sample dimensions.

Figure 8. Fixing the sample in the clamping system of the device.

Figure 9. Adjusting the device according to the sample size.

In figures 10 - 13 the precipitation ($CaCO_3$) at different distances from the observation point is highlighted. A sealing of the microcrack can be observed in these images, due to the fact that the Bio product is decomposed, and the molecules have entered into nucleation with the others, existing on the surface of the crack.

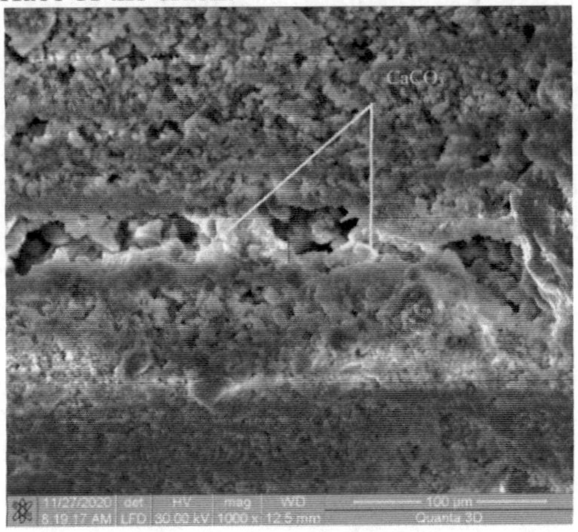

Figure 10. Visualization of $CaCO_3$ precipitation in the microcrack at a distance of 100 µm.

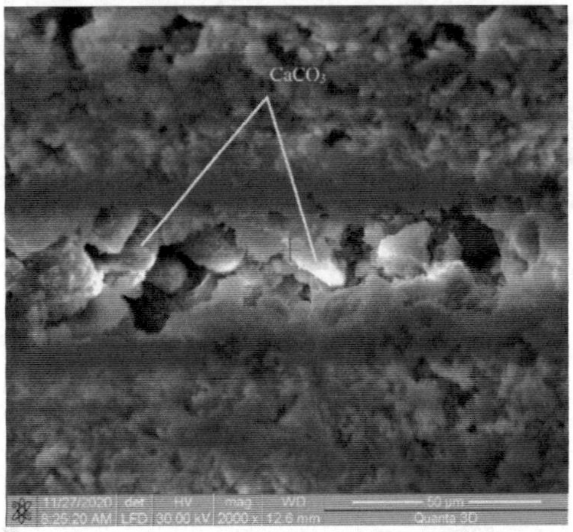

Figure 11. Visualization of $CaCO_3$ precipitation in the microcrack at a distance of 50 µm.

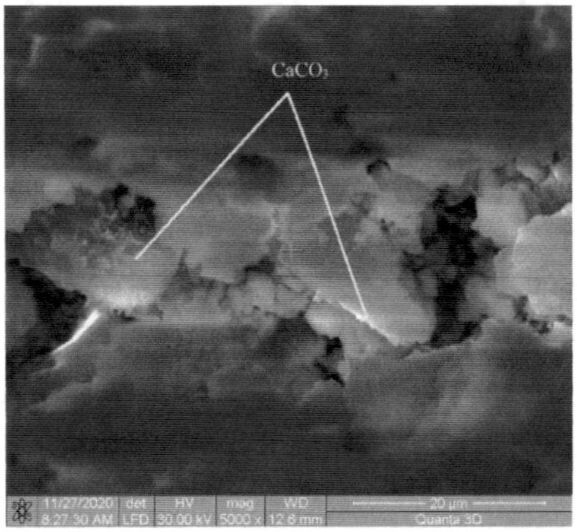

Figure 12. Visualization of $CaCO_3$ precipitation in the microcrack at a distance of 20 μm.

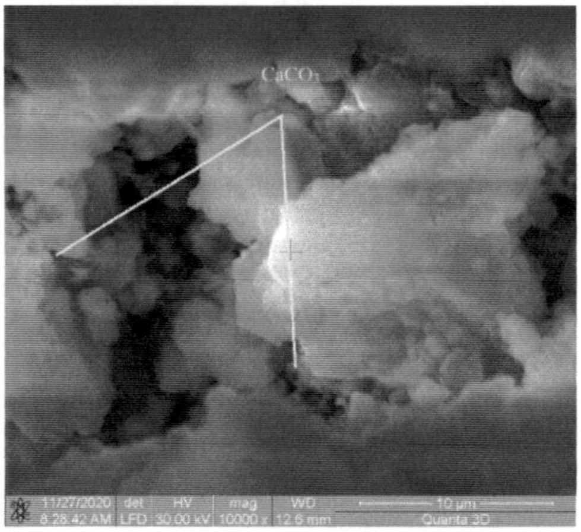

Figure 13. Visualization of $CaCO_3$ precipitation in the microcrack at a distance of 10 μm.

4. Conclusions

This work presents the chemical reaction process of Bio product and cement, in a scenario in which the concrete/mortar microcracks are limited to 0.4 μm. The optimal environmental conditions for the Bio product to work at the molecular level with the other molecules present in the concrete matrix were identified. Based on the presented laboratory studies, the following conclusions can be formulated:

- The chemical reactions presented in these studies can be identified only by scanning microscopy;
- Water has an important role in achieving stable bonds, due to the fact that it ensures the mobility of components from the BIO product to the cracked area;
- The reaction time is determined by the dimensions of calcium carbonate (macro, micro and nano);
- The higher the amount of calcium carbonate and the moisture in the cracked area, the more complete sealing can be achieved;
- The working temperature is a key parameter in the activation of the Bio product (the material is protected with a paraffin layer);
- The amount of the protective layer, namely paraffin, must be as small as possible, thus avoiding a partial closure without making stable connections.

The main conclusion regarding the chemistry of the self-healing process of microcracks is that both in the case of using bacteria and the Bio product, the raw material reacts through durable and stable bonds only in the presence of water. As described by other researchers, the presence of calcium carbonate on the cracked surface has a significant role in the self-sealing process. Thus, it can be stated that the use of products based on biological raw materials and Bio product in hydrotechnical construction elements can lead to immediate results in self-healing of microcracks. This fact, also, implies an increase in the lifespan of the concrete element and a reduction in its maintenance costs.

References

[1] De Rooij M R and E Schlangen 2011 *Self-healing phenomena in cement-based materials* Draft of State-of-the-Art report of RILEM Technical Committee

[2] Baera C 2016 *The self-curing capacity of engineering advanced cement composites*, PhD thesis Technical University Cluj Napoca, Romania

[3] Edvardsen C 1999, Water pemeability and autogenous healing of cracks in concrete, *ACI Materials journal*- American Concrete Institute, Vol. **96**, No.4, 448-454

[4] Yang EH 2008 *Designing added functions in Engineered cementitious composites*, PhD thesis,

University of Michigan, Department of Civil Engineering, Michigan, USA
[5] Lors C et all 2017 Microbiologically induced calcium carbonate precipitation to repair microcracks remaining after autogenous healing of mortars, *Construction and Building Materials* **141**, 461- 469
[6] Van Tittelboom K et all 2010 Use of bacteria to repair cracks in concrete, *Cement and Concrete Research*, Vol **40**, Issue 1,157-166
[7] Wang D et all 2018 A review on use of limestone powder in cement-based materials: Mechanism, hydration and microstructures, *Constr. Build. Mater.* **181**, 659–672
[8] Cao M et all 2019 *Effect of Macro-, Micro- and Nano-Calcium Carbonate on Properties of Cementitious Composites-A Review, Materials*,12(5), 781
[9] Thongsanitgarn P et all 2014, Heat of hydration of Portland high-calcium fly ash cement incorporating limestone powder: Effect of limestone particle size, *Constr. Build. Mater.*, 66, 410–417
[10] Silvestre J et all 2016, Review on concrete nanotechnology, *Eur. J. Environ. Civ. Eng.* **20**, 455–485
[11] Harja M, *Materiale oxidice convenționale și avansate. Tehnologia Materialelor de Constructii* (in Romanian) “Gheorghe Asachi” Tehnical University of Iasi, Romania – course notes
[12] Nurul I R R et all 2016 *Physical and Chemical Properties of Coal Bottom Ash (CBA) from Tanjung Bin Power Plant,* IOP Conference Series Materials Science and Engineering **160** 012056
[13] Published on the website, https://www.omya.com (accessed 04.03.2021)
[14] Published on the website, https://ro.wikipedia.org/wiki/Carbonat_de_calciu (accessed 04.03.2021)
[15] Published on the website, https://www.carbonatdecalciu.ro/utilizari/ (accessed 04.03.2021)
[16] Hassan A et all 2020 Insight into the role of microbial calcium carbonate and the factors involved in self-healing concrete, *Construction and Building Materials*, **254**
[17] Shekhar S, Ajay R 2020 Microbiological Induced Calcium Carbonate Process to Enhance the Properties of Cement Mortar, *Materials Today: Proceedings* **21** 1350–1354

IOP Conf. Series: Materials Science and Engineering 1141 (2021) 012027 doi:10.1088/1757-899X/1141/1/012027

Ecological and economic impact of an innovative bioproduct used to consolidate the concrete elements of hydrotechnical constructions

C Sbirlea[1], D N Isopescu[1], D Ungureanu[1], S G Maxineasa[1]
[1] "Gheorghe Asachi" Technical University of Iasi, Faculty of Civil Engineering and Building Services, Department of Civil and Industrial Engineering, 1, Prof. Dr. docent Dimitrie Mangeron Blvd., Iasi, 700050, Romania

E-mail: catalin_sbirlea@yahoo.com

Abstract. The present paper takes into observation the evaluation of the environmental and economic impact of a innovative bioproduct that can be used in order to reduce the volume of rehabilitation works needed during the life cycle of different structural reinforced concrete elements utilized for hydrotechnical constructions. In order to determine the performances of the analysed material, seven scenarios that imply the use of different rehabilitation solutions have been considered. In order to achieve the goal of the study, the Cradle-to-Gate Life Cycle Assessment type of study was used. The amount of materials, the distances from the extraction point to the site of the materials and the implementation technology are detailed in all proposed scenarios. In recent years, researchers carried out various studies on the use of raw materials in different compositions, with different quantities, in order to improve the sustainability and mechanical characteristics of constructions, as well as to increase the users' level of satisfaction. The research and development of new types of materials that can be used in the construction sector with a low ecological influence will have a tremendous positive impact over the present state of the natural environment. The authors conclude that by considering the resulted values, the analysed solution based on the use of calcium carbonate can be considered a suitable solution with respect to the environmental and economic dimensions of sustainability.

1. Introduction

In the last decades, the needs and demands of the ever-growing global population have significantly increased the need of developing new solutions in order to reduce the global environmental impact. Taking into account the fact that the construction sector is considered one of the most pollutant economic activities, it is required that civil engineering specialists develop new materials and structural solutions in order to significantly reduce the negative influence of the built environment over the natural one. Therefore, the development of innovative materials that increase the physical and mechanical performances of the constructive elements of hydrotechnical works should be analysed from an ecological point of view [1-4]. It is important to note that it is increasingly required for a new material to have both good results in laboratory tests, from a physical and mechanical point of view, as well as a low environmental impact.

Taking into consideration the importance of hydrotechnical constructions, the development and use of new materials which have self-healing properties and which can be used for increasing the lifespan of these structures should be considered as extremely important in the construction sector. Currently, specialized personnel closely check hydrotechnical constructions with a high degree of importance through various methods, with the goal of preventing but also providing prompt intervention in case it is necessary. Hydrotechnical works are a crucial part of our development, seeing as they directly influence water quality and protection against flood phenomena. At the same time, besides the fact that these structures consume tremendous amounts of traditional building materials that have an important environmental impact, they are also responsible for negatively affecting the natural aquatic ecosystems. Thus, it is paramount to take under observation various solutions that can help reduce these negative impacts.

In the last years, the number of tools that can be considered in order to correctly asses and interpret the environmental impact of a product or of a service has significantly increased. For example, there are different software applications like GaBi Software (PE International, Germany), openLCA (GreenDelta GmbH, Germany), Sima Pro LCA (PreConsultants, The Netherlands), that can enable a correct environmental impact analysis for different stages of the life cycle of a product. In order to use the tools, the specialist must take into consideration the Life cycle assessment (LCA) methodology. In order to determine the impact on the environment, LCA studies help identify the flow of materials as well as the energy and waste of a product over its life cycle [5]. Depending on the stages from the life cycle of a product used in an analysis, there are different type of LCA, such as Cradle-to-Gate; Cradle-to-cradle; Gate-to-Gate [6, 7]. Taking into account the above, the goal of the present study is to analyse a solution, with respect to the environmental and economic dimensions of sustainability, based on the use of calcium carbonate in order to achieve self-healing properties for reinforced concrete (RC). The environmental and economic performances of the considered bio base product are compared with those of different other rehabilitation solutions. Both the assessment of the impact over the natural environment and the economic influence are determined and compared from a Cradle-to-Gate point of view. The ecological performances are determined by using the GaBi software.

2. Definition of the bioproduct

The bioproduct, shown in Figure 1, is made of micronized calcium carbonate ($CaCO_3$) powder which is subsequently granulated in small sizes (2-6 mm), being strongly reactive with water. On the outside, it is protected by a layer of paraffin (1/2 of the bioproduct) that prevents $CaCO_3$ from reacting with water in the mixing process. Paraffin is a solid, white substance used as a raw material in various industries, with a melting point between 52°C and 54°C.

The chemical composition of the bioproduct [8] expressed in percentages is presented in Table 1. The bioproduct is added to the concrete composition in various quantities compared to the amount of cement and has the role of providing stable bonds at the molecular level, thus sealing the micro-cracks that may occur during operation. Calcium carbonate is widely spread in nature in the form of minerals (calcite, aragonite, vaterite), in bones, teeth, shells, corals and shellfish crust [9]. In rocks, it comes in the form of limestone, where it is almost pure, an example being the dolomites, which are a mixture of calcium and magnesium [9].

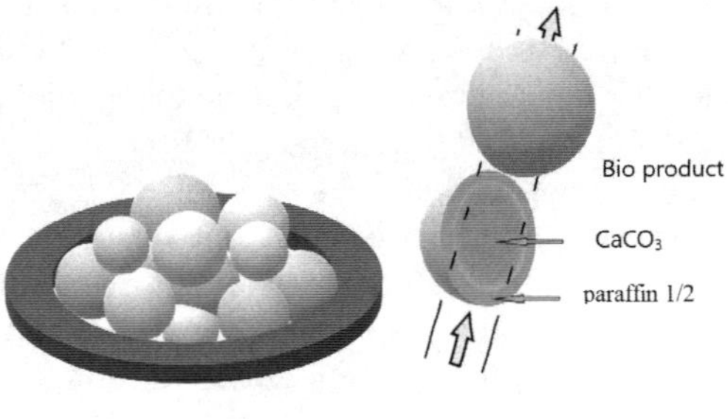

Figure 1. Bioproduct.

Table 1. Chemical composition of the bioproduct. [8]

Chemical composition of granular bioproduct	Amounts (%)
Calcium carbonate ($CaCO_3$)	91%
Magnesium carbonate ($MgCO_3$)	2 %
Paraffin ½ from bioproduct	~ 47%

IOP Conf. Series: Materials Science and Engineering 1141 (2021) 012027 doi:10.1088/1757-899X/1141/1/012027

3. Cradle-to-Gate LCA analysis

As stated before, the Cradle-to-Gate LCA type of study has been used in order to determine the environmental impact of the considered construction solutions. This type of study takes under observation the life cycle of a product from the extraction and processing of raw materials ("cradle") to the factory gate ("gate"), before the assessed product is utilised, as shown in Figure 2. In each stage of its life cycle, a product is responsible for a different impact over the natural environment. These ecological influences can be translated into a high level of negative emissions into the atmosphere, soil and water, or in a certain amount of consumed natural resources [10]. Thus, even if the LCA analysis cannot clearly show that the use of a product or service is safe, it provides a number of indicators, depending on the type of the environmental impact categories considered, in terms of impact assessment scores [5].

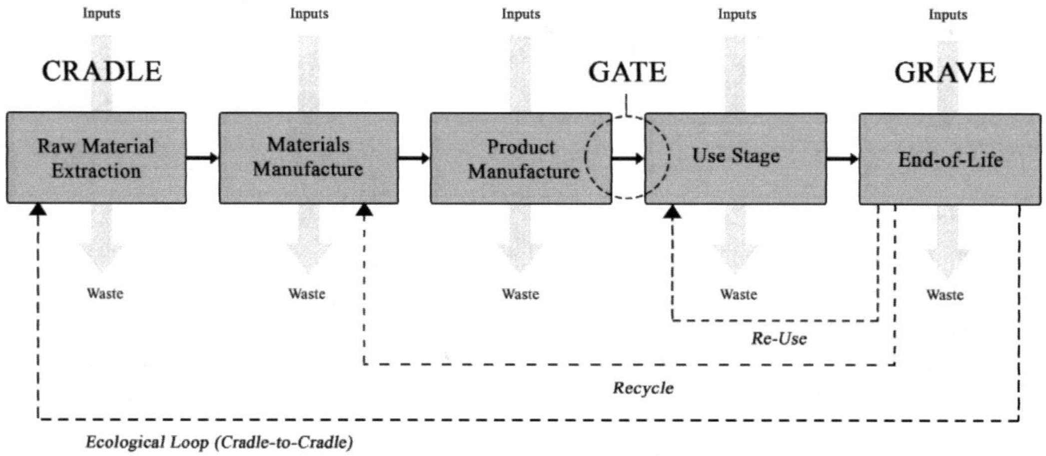

Figure 2. LCA (Life Cycle Assessment) [11].

Concrete is one of the most widely used materials in the world, being employed in the construction of civil and industrial buildings, hydrotechnical constructions, roads and other types of infrastructure. On average, about 1 tonne of concrete is produced each year for every human being in the world [12]. In recent years, concrete has been subjected to detailed analysis given that this material has a significant impact on the environment. Out of the composition of the aggregates used in its manufacture, cement is responsible for the most important part of the greenhouse gas emissions that are related to the use of concrete. It is considered that at the global scale, the cement industry is responsible for approximately 8% of total CO_2 generated [12]. The scope of the Bio concrete concept is to replace a volume of cement with other materials or to make changes to the composition parameters in its design to obtain a concrete that is sufficiently durable, resistant, and sustainable. It is necessary to make an environmental impact comparison between bio and traditional concrete to clearly identify the ecological efficiency of replacing traditional raw materials that have a high carbon footprint, a toxic impact on humans and an important negative influence over the stratospheric ozone layers, with new materials. Research and development of new raw materials has an important role in our society, seein as there are many analyses worldwide that show us that nature goes through a continuous process of change, one that is directly influenced by human activities. All these negative influences on nature and climate change are caused by the alarming growth of the global population, which in turn caused the high demand and need for industrialization. Raw material consumption is continuously increasing, endangering the existing natural stocks. Therefore, by adding that the construction sector is one of the most important consumers of natural resources, it can be stated that it is necessary to develop new solutions that will significantly help tackle this environmental problem. In addition to the consumption of large amounts of natural resources and the emission of polluting gases into the atmosphere, this sector is responsible for 25% to 30% of the total waste produced in Europe [13].

The Cradle-to-Gate LCA type of study was used in the present paper in order to assess the environmental impact of traditional as well innovative materials used in construction by considering seven scenarios. Table 2 shows the impact categories used to quantify the effects on the environment [14].

Table 2. Impact categories used to quantify the effects on the environment [14].

Global Warming	Global Warming Potential – GWP	kg. CO2 eq.
Human Toxicity	Human toxicity potential, cancer effects – HTPc	CTUh
Ozone Depletion	Ozone Depletion Potential – ODP	kg. CFC 11 eq.

Within the seven scenarios, a Diesel truck with Euro 5 pollution norm and a transport capacity of 3.3 t was considered for the transport of materials. The distances adopted in the LCA studies for the transport stages of the life cycle of the analyzed products are listed in Table 3.

In order to take into account the mixing of concrete stage, an electricity consumption of 39.672 MJ was considered [14]. In the proposed scenarios, for the LCA analysis, only one concrete recipe C16/20 was used; the quantities of raw materials are presented in tables 4. The quantities of materials from the seven proposed scenarios correspond to a concrete beam measuring 120 mm x 90 mm x 1500 mm, presented in Table 5 as primary materials. The quantities of materials needed to repair cracks and strengthen the concrete element are defined as secondary materials.

In the scenarios considered in the analysis, the same type of reinforcement was used, 2 steel reinforcing bars with a diameter of 10 mm (0.616 kg/ml), which were cut to the size of 1490 mm arranged at 30 mm from the sides of the formwork at the time of casting. For each element, the compositions of secondary materials used to repair cracks and/or strengthen the degraded element and the technology used are mentioned. The support concrete beams used in the analysis are numbered from Gr.0 to Gr. 6, with the same technical and compositional characteristics.

Table 3. Distances used in the LCA analysis.

Material	Distance (km)	Starting point - Arrival point
Cement Steel	165 25	- manufacturing unit - place of installation
Sand Gravel	30 30	- ballast - concrete station
Concrete	15	- concrete station
Bioproduct	400	- manufacturing unit - place of installation
Construction lime	400	- manufacturing unit - place of installation
Paraffin	400	- manufacturing unit - place of installation
Fiber fabric of glass /carbon Epoxy resin	1850	- manufacturing unit - place of installation

IOP Conf. Series: Materials Science and Engineering 1141 (2021) 012027 doi:10.1088/1757-899X/1141/1/012027

Table 4. Recipe for 1 m^3 C16/20 concrete used in the LCA analysis.

Raw material used	Amounts (kg/m^3)
cement I 42.5 R	270
sand 0-4 mm	716
fine gravel 4-8 mm	345
coarse gravel 8-16 mm	850
water	170

Table 5. Primary materials related to the size of the concrete beam 120 mm x 90 mm x 1500 mm.

Raw material used	Amounts (kg/m^3)
cement I 42.5 R	4.4
sand 0-4 mm	11.6
fine gravel 4-8 mm	5.6
coarse gravel 8-16 mm	13.8
water	2.8
steel reinforcement	1.8

4. Evaluation of the environmental impact

4.1. Scenario 1

The present scenario entails the determination of the ecological impact resulted from the replacement of an existing reinforced concrete beam which no longer meets structural requirements with a new linear concrete element Gr.0 (Figure 3). Table 5 presents the type and amount of materials considered for the evaluation. By analyzing Figures 4 and 5, it can be observed that the mixing stage from the life cycle of the assessed product is responsible for the highest impact over the carbon footprint and human toxicity impact category. At the same time, the negative influence over the stratospheric ozone layer is mostly influenced by the steel rebars (Figure 6).

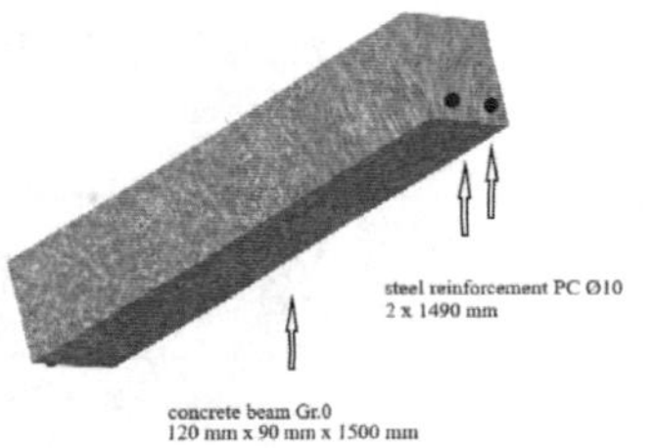

Figure 3. RC beam details Gr.0 in scenario 1.

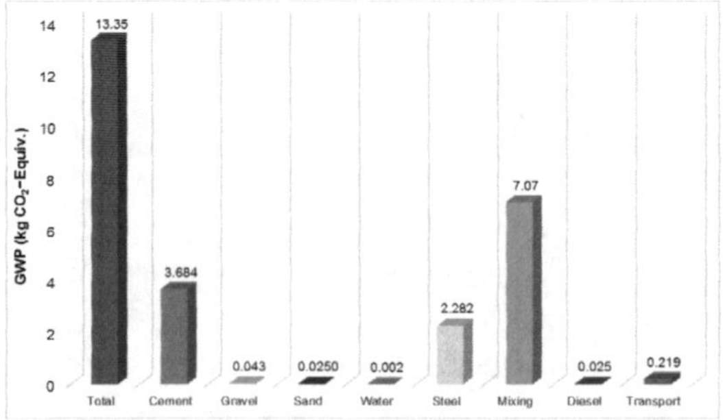

Figure 4. Global warming potential in scenario 1.

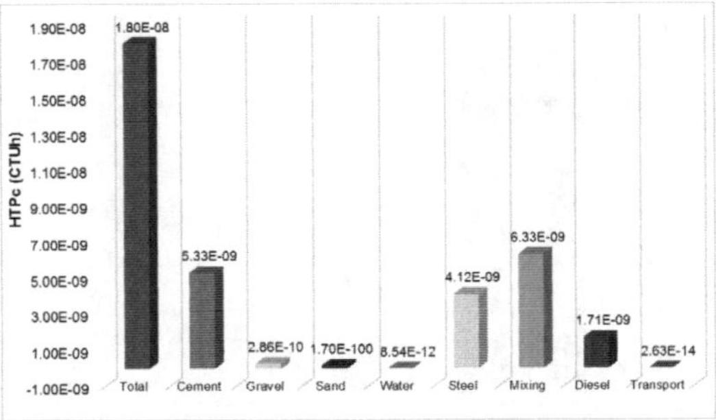

Figure 5. Human toxicity potential, cancer effects in scenario 1.

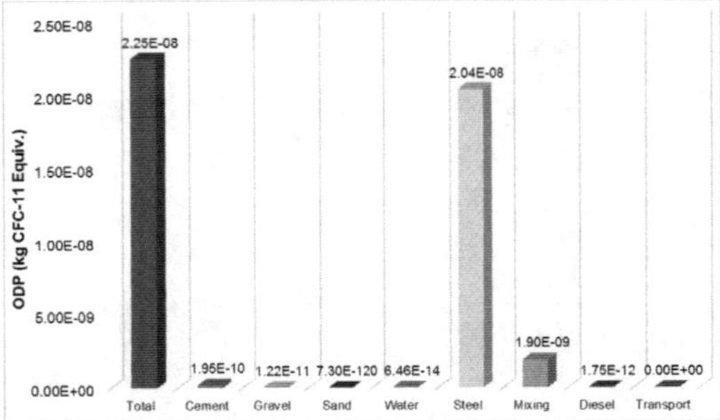

Figure 6. Ozone Depletion Potential in scenario 1.

4.2. Scenario 2

In this scenario, in order to fix the cracks on the surface of the Gr.1 concrete beam, an M100T mortar was used; the quantities of materials are mentioned in Table 6.

Table 6. Quantities of materials used in scenario 2.

Secondary materials	Amounts (kg/m^3)
cement II AM 32.5 R	2.19
sand 0-3 mm	0.052
hydrated lime	0.599
water	0.502

The first step was to clean the crack with a spatula and wire brush. After removing the residue from the damaged surface and vacuuming the surface, the area was moistened to attach the filler. After the preparation of the mortar, with the help of the spatula, it is applied on the entire surface of the damaged element, as well as finishing layer (Figure 7). By analyzing Figures 8, 9 and 10, it can be observed that the amount of cement consumed is significantly influencing the environmental performances in all three considered impact categories.

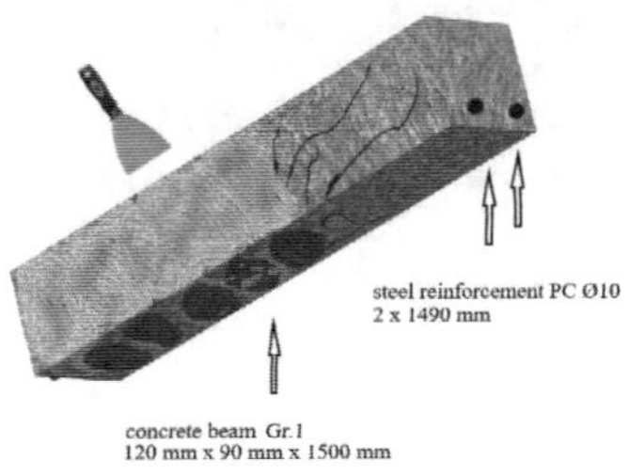

Figure 7. RC beam details Gr.1 in scenario 2.

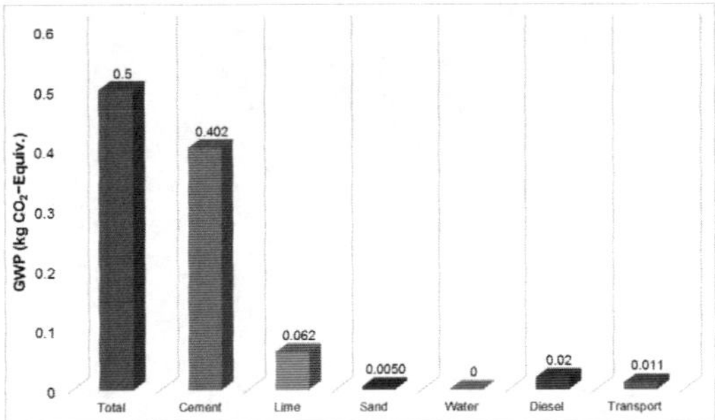

Figure 8. Global warming potential in scenario 2.

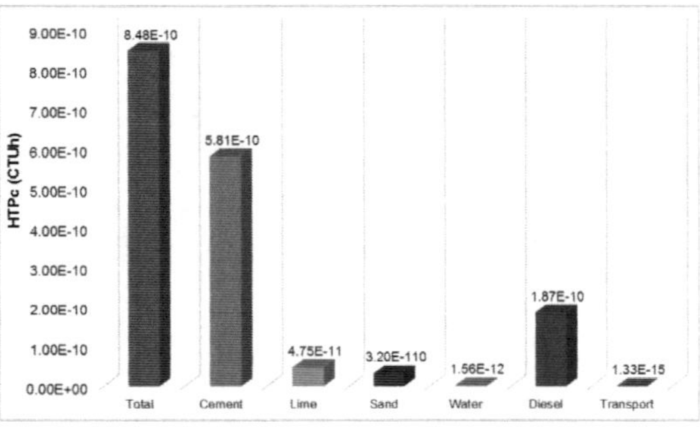

Figure 9. Human toxicity potential in scenario 2.

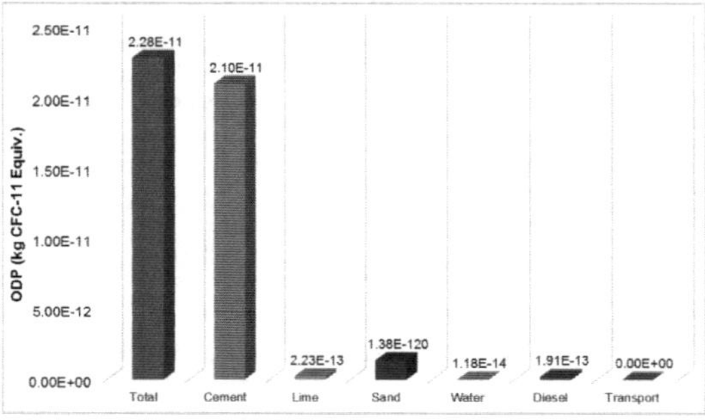

Figure 10. Ozone Depletion Potential in scenario 2.

4.3. Scenario 3

In this scenario, the percentage of bioproduct is 10% of the mass of cement (Figure 11). The materials were dosed and then mixed until the desired consistency was obtained. Before casting, the moulds were prepared by cleaning and lubricating with stripping oil. The casting was performed in several layers, after the first layer the reinforcement was introduced, and the continuation of the casting was performed in small quantities by vibrating the mould. The quantities of materials used in this scenario are presented in Table 7. By analysing the resulted values, it can be observed that in the case of the GWP and HTPc indicators, the amount of paraffin has the highest environmental impact, while in the case of ODP impact category, calcium carbonate is responsible for more than 99% of the total negative effect.

Table 7. Quantities of materials used in scenario 3.

Secondary materials	Amounts (kg/m^3)
Bioproduct	0.437
Paraffin	0.219

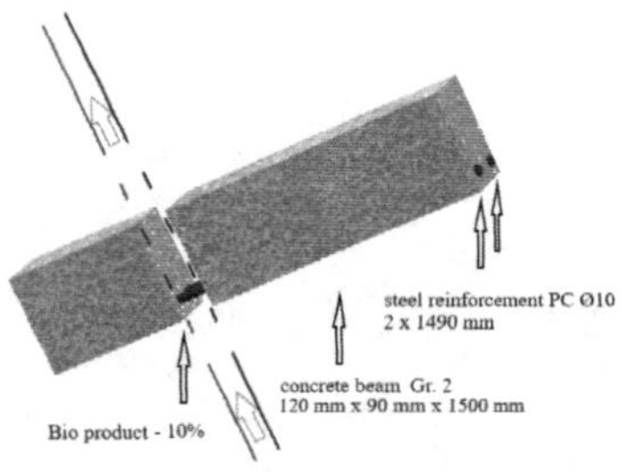

Figure 11. RC beam Gr.2 with 10% bioproduct in scenario 3.

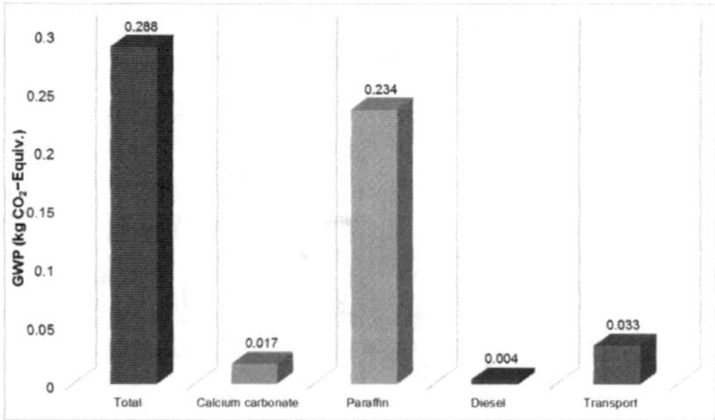

Figure 12. Global warming potential in scenario 3.

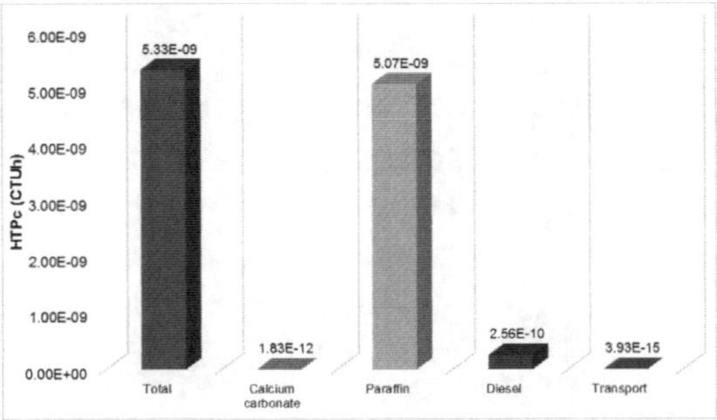

Figure 13. Human toxicity potential in scenario 3.

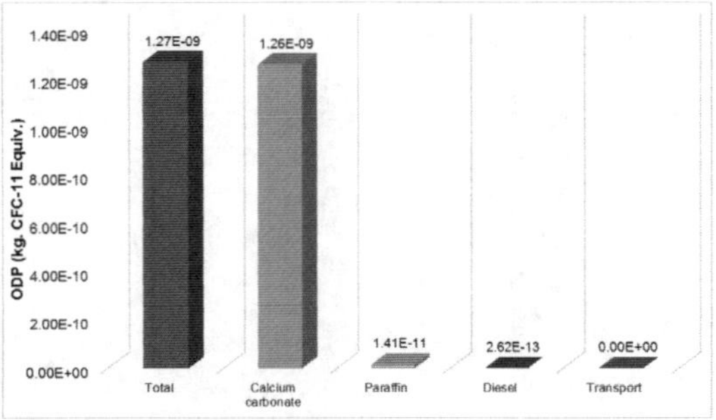

Figure 14. Ozone Depletion Potential in scenario 3.

4.4. Scenario 4

The LCA analysis was performed by taking into account the use of 20% bioproduct compared to the cement mass. The installation technology is the same as explained in scenario 3, the difference being the percentage of used bioproduct (Figure 15). The quantities of secondary materials used in this scenario are mentioned in Table 8. As in the previous scenario, in the case of the GWP and HTPc impact categories, the amount of paraffin used has the highest negative influence, while the $CaCO_3$ has the most important impact in the case of the last considered environmental indicator (Figures 16, 17, 18).

Table 8. Quantities of materials used in scenario 4.

Secondary materials	Amounts (kg/m^3)
Bioproduct	0.875
Paraffin	0.437

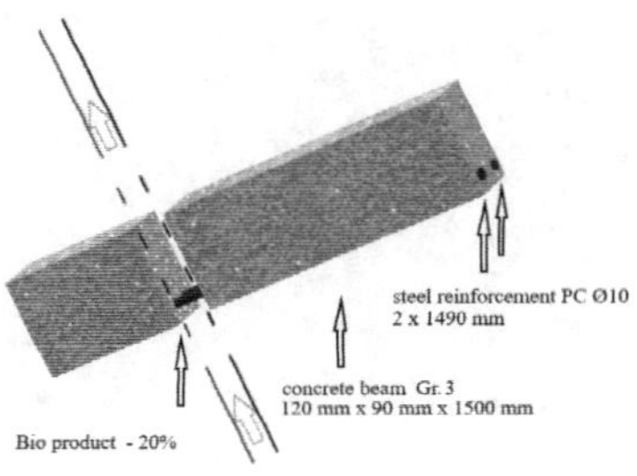

Figure 15. RC beam Gr.3 with 20% bioproduct in scenario 4.

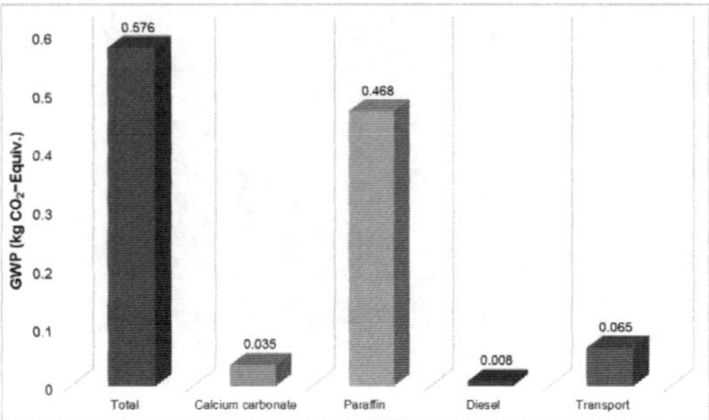

Figure 16. Global warming potential in scenario 4.

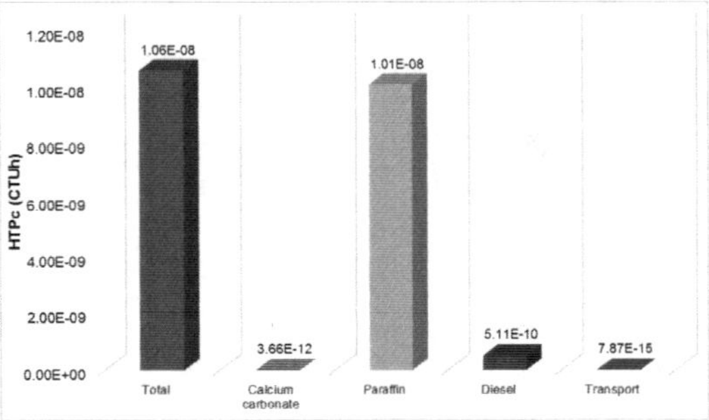

Figure 17. Human toxicity potential in scenario 4.

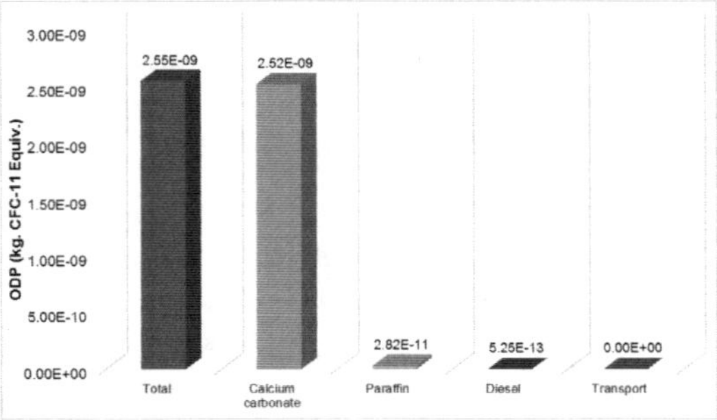

Figure 18. Ozone Depletion Potential in scenario 4.

4.5. Scenario 5

In this scenario, EPOMAX-L20 [15] was used, a two-component, injectable epoxy resin, used to repair cracks of 0.1-1.0 mm, with a consumption of about 1.1 kg/litre. The quantities of secondary materials used to repair cracks in the Gr.4 concrete beam are presented in Table 9.

Table 9. Quantities of materials used in scenario 5

Secondary materials	Amounts (kg/m^3)
epoxy resin	0.111

After removing the degraded material with a spatula and a wire brush, the cracked surface was vacuumed. The epoxy resin was injected into the cracks using a special device. After hardening, this surface was cleaned of excess material as shown in Figure 19. In Figures 20 and 22, it can be noticed that epoxy resin has the most important carbon footprint and the highest negative effect over the stratospheric ozone layer. At the same time, the amount of diesel consumed accounts for almost 89% of the total negative impact over human health (Figure 21).

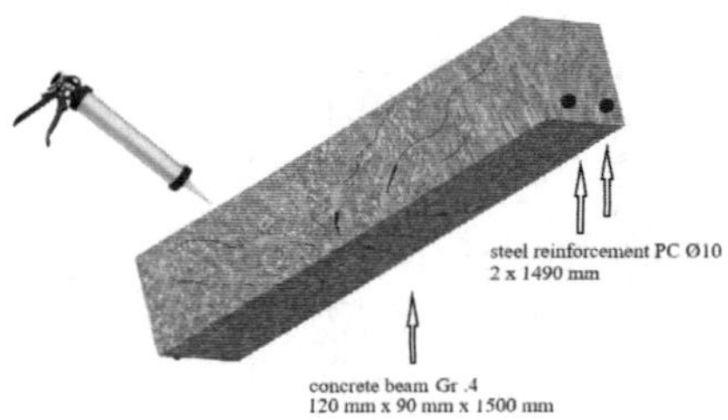

Figure 19. Injection of epoxy resin into cracks.

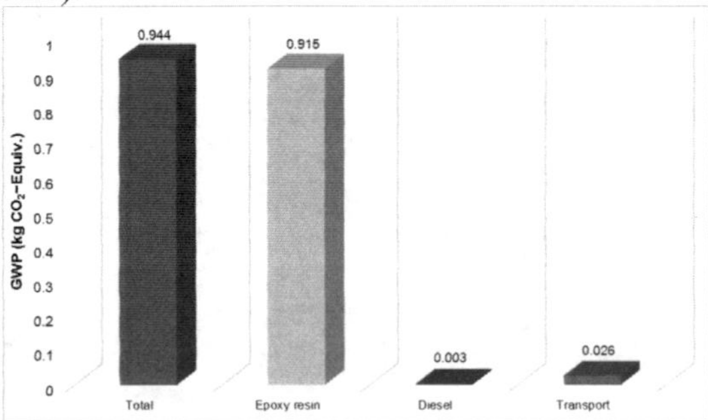

Figure 20. Global warming potential in scenario 5.

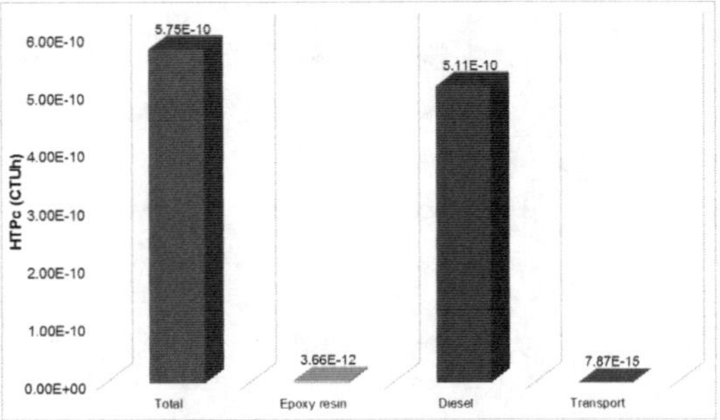

Figure 21. Human toxicity potential in scenario 5.

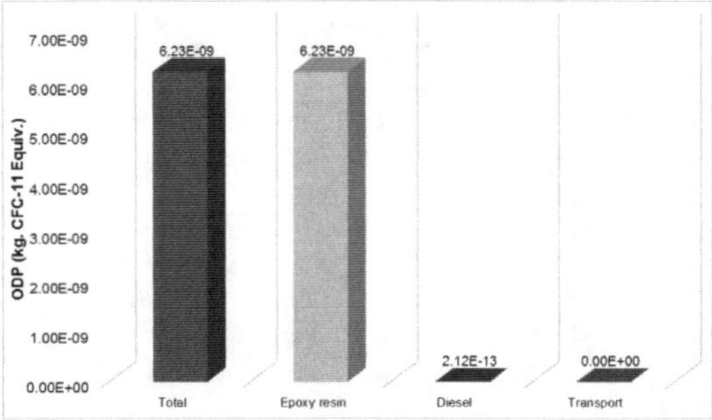

Figure 22. Ozone Depletion Potential in scenario 5.

4.6. Scenario 6

This scenario analyses the impact on the environment resulted from using fiberglass fabric to strengthen the concrete beam Gr.5 (Figure 23). Glass fabric type E-Glass [16] with a width of 100 cm was applied on the entire beam by using EPOMAX-LD [17], an epoxy resin that has a consumption of 0.7-1.2 kg/m² as specified in the data sheet. The quantities of secondary materials used in this scenario are mentioned in Table 10. Analysing the resulted values leads to the conclusion that epoxy resin has the most important negative effect over the natural environment in all considered environmental impact categories.

Table 10. Quantities of materials used in scenario 6

Secondary materials	Amounts (kg/m^3)
fiberglass fabric	0.047
epoxy resin	0.288

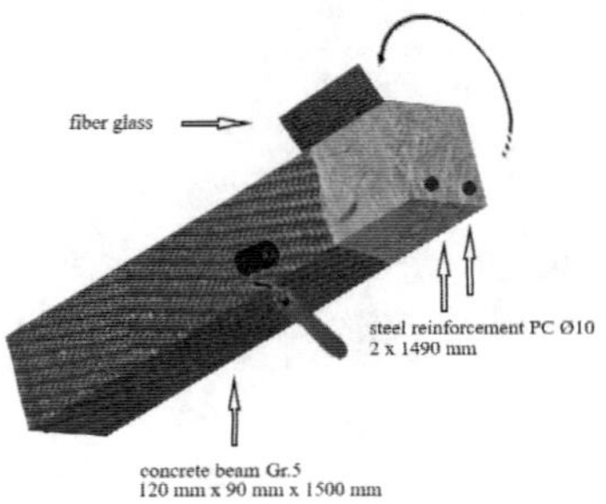

Figure 23. RC beam with fiberglass fabric in scenario 6.

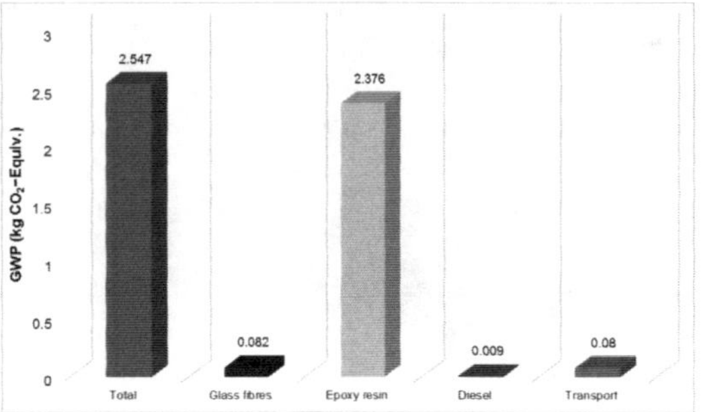

Figure 24. Global warming potential in scenario 6.

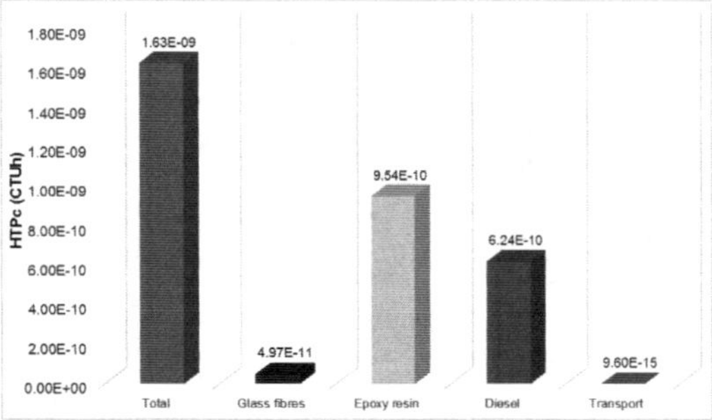

Figure 25. Human toxicity potential in scenario 6.

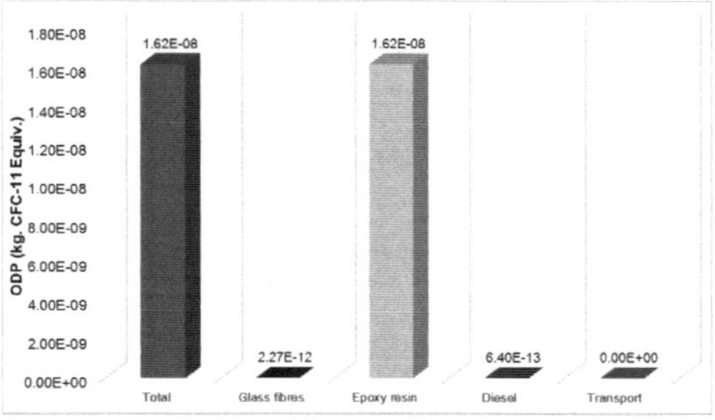

Figure 26. Ozone Depletion Potential in scenario 6.

4.7. Scenario 7

The last considered scenario assesses the impact resulted from using carbon fiber fabric for strengthening concrete beam Gr.6 (Figure 27). The MEGAWRAP-200 carbon fiber [18] with a thickness of 0,11 mm and a width of 60 cm has been applied on the entire beam by using the EPOMAX-LD [17] epoxy resin. Table 11 presents the amounts of materials considered. By analysing the resulted values, it can be observed that in the case of the HTPc environmental impact category, carbon fibers have the most significant influence (Figure 29). At the same time, the overall impact in the case of the other two environmental indicators is mainly produced by epoxy resin.

Table 11 Quantities of materials used in scenario 7.

Secondary materials	Amounts (kg/m^3)
carbon fiber fabric	0.054
epoxy resin	0.288

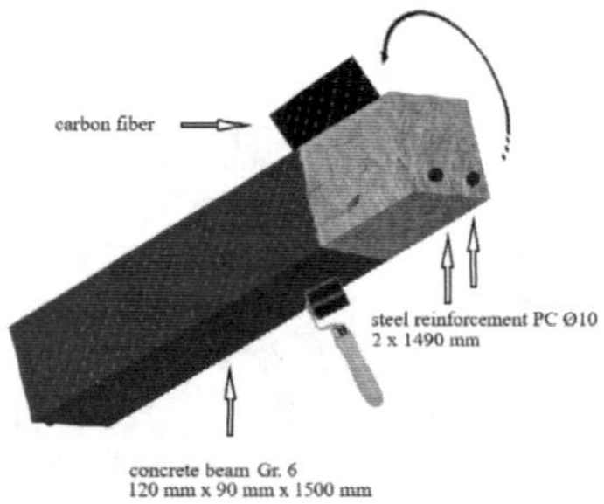

Figure 27. RC beam reinforcement with carbon fiber fabric.

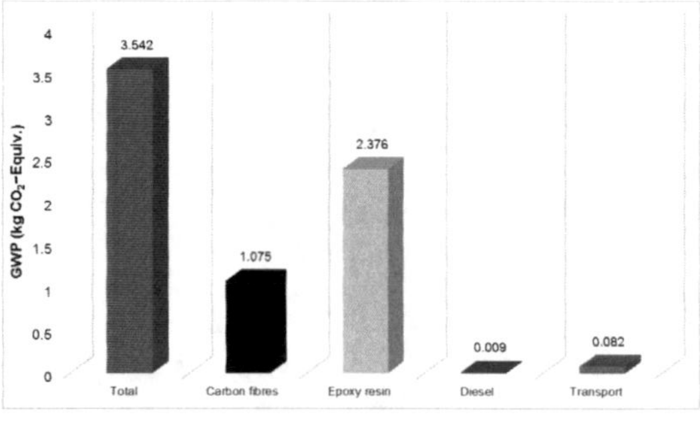

Figure 28. Global warming potential in scenario 7.

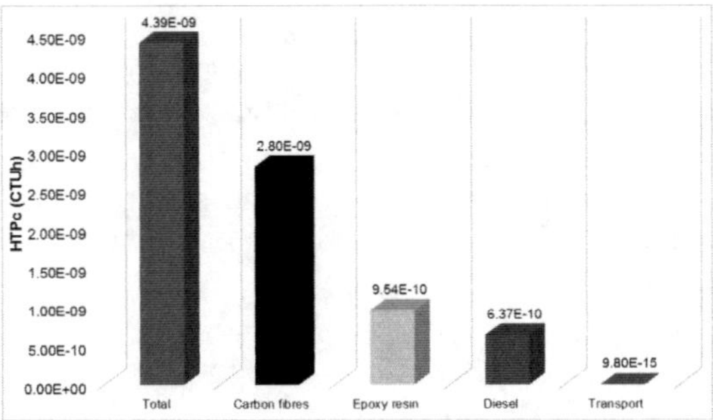

Figure 29. Human toxicity potential in scenario 7.

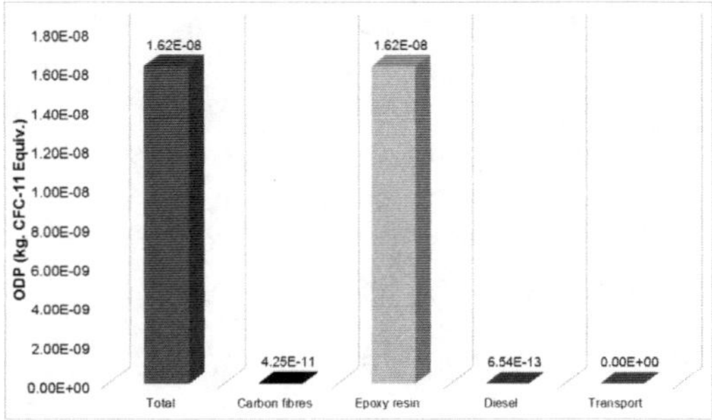

Figure 30. Ozone Depletion Potential in scenario 7.

4.8. Comparing the environmental performances

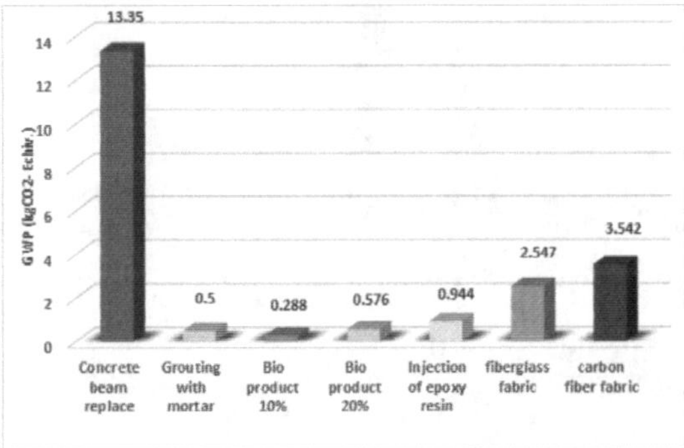

Figure 31. Comparing the results for the Global Warming Potential environmental indicator.

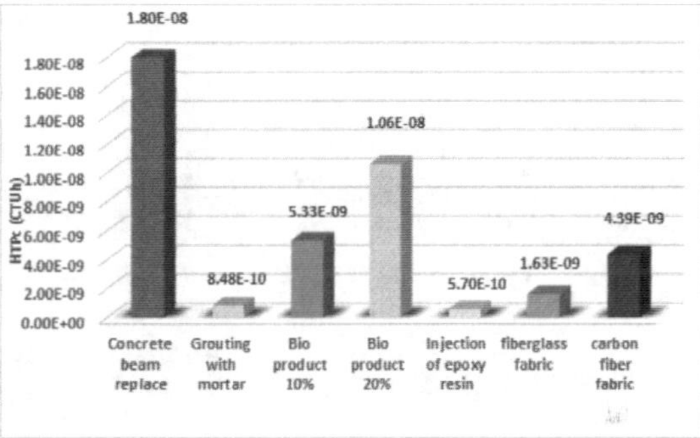

Figure 32. Comparing the results for the Human toxicity potential environmental indicator.

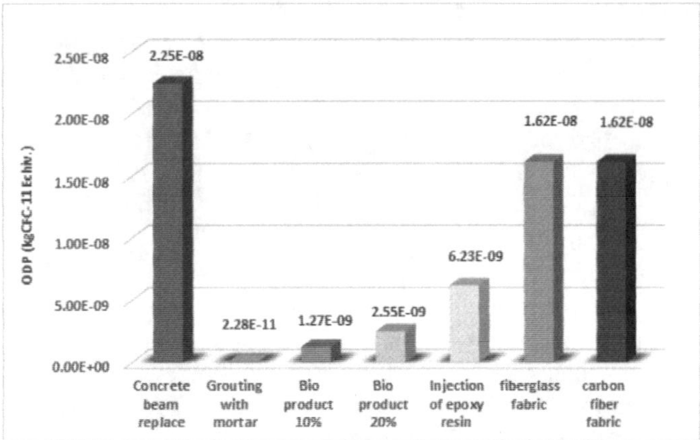

Figure 33. Comparing the results for the Ozone Depletion Potential environmental indicator.

By analysing the resulted values, presented in Figures 31, 32, and 33, it can be observed that:

- in the case of the Global Warming environmental impact category, the results reflected by the use of 10% calcium carbonate represent the most suitable application, followed by the

solutions studied in scenario 2 and 4;
- for the Human toxicity potential, cancer effects environmental indicator, the new reinforced concrete element and the use of the analysed bioproduct have the highest negative impact over human health, followed by the solution considered in the last scenario;
- the values resulted in the last environmental impact category show that the solution which entails repairing works from the first scenario has the lowest impact over the stratospheric ozone layer, closely followed by the use of 10% and 20% of bioproduct.

5. Evaluation of the economic impact

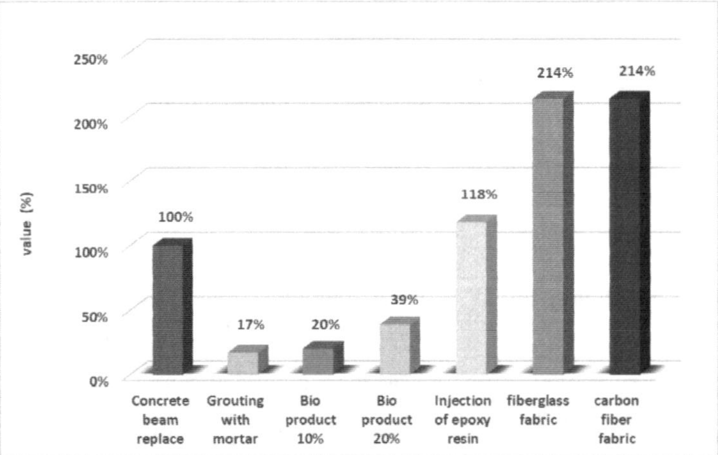

Figure 34. Comparison of the economic impact of the analysed materials.

In this chapter, the authors have analysed, from a financial point of view, the sustainability of the materials used in the considered scenarios. The economic analysis was performed only in terms of material costs, the cost of labour not being included. Both the dimensions of the concrete beams of class C16/20 (120 mm x 90 mm x 1500 mm) and the quantities used are the same as in the case of LCA studies. By analysing Figure 34, it can be stated that:
- in scenario 1, the total replacement of the degraded beam was established. In order to compare the costs resulted in the other six scenarios, the authors decided the value achieved in this case to be considered as reference;
- in scenario 2, the costs of the repairing works of the considered cracks of the linear RC element only represent 17% of the value resulted in the first scenario;
- in scenarios 3 and 4, the bioproduct was used in a proportion of 10%, respectively 20%, and the resulted costs account for only 20%, and 39% respectively, of the cost resulted in the case of building a new RC element;
- the injection with two-component epoxy resin that was analysed in the fifth scenario implies costs that are higher by about 18% compared with the ones from the first scenario;
- the use of fiberglass and carbon fabrics on the entire surface of the concrete beam were considered in scenarios 6 and 7. By analysing Figure 31, it can be observed that in both considered scenarios, the strengthening schemes are more expensive then the solution from scenario 1 by approximately 114%.

6. Conclusions

The goal of the present study is to assess if the use of calcium carbonate can be considered a sustainable solution with respect to the environmental and economic dimensions of the sustainable development concept. This bioproduct is considered to achieve self-repairing capacity for the reinforced concrete products, mostly for those elements that are used in hydrotechnical constructions. The assessment of the environmental performances of the considered construction products has been completed by using the Life Cycle Assessment methodology from a Cradle-to-Gate perspective, the studies having been performed by using the GaBi software. As shown in the present study, the use of bioproduct represent a solution that has a low impact over the natural environment.

The economic assessment has revealed that the solutions studied in scenarios 2, 3, and 4 are the only ones which imply costs lower that the ones resulted in the case of constructing a new RC element (scenario 1). By taking into account all the above, it is justified to conclude that the use of calcium carbonate in different reinforced concrete elements with the goal of achieving a self-repairing capacity represents a highly suitable solution that can be successfully used in order to accomplish the main objectives of sustainable development in the construction sector.

7. References

[1] Bribian I Z, Capilla A V and Uson A A 2011 Life cycle assessment of building materials: comparative analysis of energy and environmental impacts and evaluation of the eco-efficiency improvement potential, *Build. Environ.* **46** 1133-1140

[2] Maxineasa S G and Taranu N 2018 Life cycle analysis of strengthening concrete beams with FRP, in: Pacheco-Torgal F, Melchers R E, Shi X, De Belie N, Van Tittelboom K and Saez A (Eds.) *Eco-efficient Repair and Rehabilitation of Concrete Infrastructures*. Woodhead Publishing, Duxford, 673-721

[3] Romano E, Cascini L, D'Aniello M, Portioli F and Landolfo R 2020 A simplified multi-performance approach to life-cycle assessment of steel structures, *Structures* **27** 371-382

[4] Kamali M, Hewage K and Sadiq R 2019 Conventional versus modular construction methods: A comparative cradle-to-gate LCA for residential buildings, *Energ. Build.* **204** 109479

[5] Toniuc L and Musuroaea V 2014, *Analiza ciclului de viață în domeniul gestionării deșeurilor din construcții și demolări*

[6] Tundrea H, Maxineasa S G, Simion I M, Taranu N, Budescu M and Gavrilescu M 2014 Environmental impact assessment and thermal performances of modern earth sheltered houses, *Environ Eng Manag J.* **13(10)** 2363-2369

[7] Luca A 2018, *Cercetări privind evaluarea ciclului de viață al produselor de încălțăminte din perspectiva asigurării durabilității*, teza doctorat, Iasi, Romania

[8] Brochure Product Calciprill, published on the website https://www.omya.com/calciprill

[9] https://ro.wikipedia.org/wiki/Carbonat_de_calciu

[10] Draft thematic document 2010, *LIFE ENVIRONMENTAL INFORMATION on daily household products*, published on the website, https://ec.europa.eu/environment/industry

[11] https://en.wikipedia.org/wiki/Life-cycle_assessment#/media/File:Example_Life_Cycle_Assessment_Stages_diagram.png

[12] Bujor O 2018 *The massive source of pollution that very few are aware of*, published on the website https://www.descopera.ro/natura/17790065-sursa-masiva-de-poluare-pe-care-foarte-putini-o-constientizeaza (accessed 04.03.2021)

[13] JRC-IES - Joint Research Centre, 2011, *Institute for Environment and Sustainability Institute for Environment and Sustainability Supporting Environmentally Sound Decisions for Construction and Demolition (C&D) Waste Management European Commission Joint Research Centre, Luxembourg*

[14] Maxineasa S G 2015 *Soluții compozite și hibride pentru dezvoltare durabilă în construcții*, (in Romanian) PhD thesis, «Gheorghe Asachi» Technical University of Iasi, Romania

[15] Data sheet, epoxy resin - EPOMAX-L20, published on the website, https://www.isomat.ro/product/epomax-l20-2-ro/

[16] Data sheet, fiber glass, published on the website https://compositesplaza.com/ro/product/plain-woven-glass-fabric-195-g_m2-100cm/

[17] Data sheet, epoxy resin - EPOMAX-LD, published on the website

https://www.isomat.ro/product/epomax-ld-2-ro/
[18] Data sheet, fiber carbon- MEGAWRAP-200, published on the website https://www.isomat.ro/product/megawrap-200-2-ro/

Remote optical measurements of cracks in concrete structures using computer image processing

Y Vashpanov[1,2], G Heo[1], J-Y Son [1] and T Podousova[3]
[1] Department of Civil Engineering, Public Safety Research Center, Konyang University, Nonsan, Chungnam, 32992, Republic of Korea
[2] Department of Physics, [3] Department of Applied Mathematics, Odessa State Academy of Civil Engineering and Architecture, Odessa, 65029, Ukraine

E-mail: yuriy.vashpanov@gmail.com

Abstract. This report develops non-contact optical methods for measuring concrete cracks for remote diagnostics of hard-to-reach building surfaces using drones. The drone's high-resolution digital camera captures remote close-up photography of the concrete surface under high light intensity levels. The digital image of the crack converted in a data matrix, each element of which contains information about the level of reflected light from the concrete surface and the crack. Using the formulas obtained by us earlier, with the help of computer processing of the image, the morphology of the crack in real geometrical dimensions can be calculated. Using special tools computer programs, the crack's geometrical parameters can be measured maximal accurately as possible. The proposed remote optical method for non-contact detection of cracks in concrete when photographing with a drone from a close distance allows achieving an accuracy of measuring the length and width of a crack of 5%, and its depth of 10%.

1. Introduction

Cracks in heavily loaded concrete structures are dangerous because their enlargement can lead to destruction and therefore require constant monitoring to ensure public safety. Recently, the number of publications on optical methods for detecting cracks in concrete structures has significantly increased [1-14]. The advantages of optical methods for registering problem areas are a non-contact, non-destructive method of detecting cracks and its low cost and ease of implementation, based on the latest achievements in image processing and their acquisition by mega pixels' cameras. Previously, we proposed a method for determining the depth of a crack by processing images obtained at close range and at high light intensity, by mathematical processing of the intensity of image pixels in the OriginLab program [15]. This method allows, in the presence of a high light intensity and a close distance to the object of study, to obtain sufficiently accurate results of measuring the geometric dimensions (length, width and depth) of cracks in concrete. However, the application of this method at a considerable distance from the camera to the object of study is limited.

Huge reinforced concrete structures (hydraulic structures, multi-kilometer bridges, etc.) are of great interest for operational monitoring of the state of concrete structures, since their destruction leads to significant material losses and the detection of dangerous areas is not always possible or expensive using conventional diagnostic methods. For example, the condition of the dam near the city of Nonsan in South Korea (Figure 1) requires such constant monitoring, since the dam is under significant stress and its destruction could lead to flooding of a large area. Constant contact with water, its freezing in winter has a destructive effect on the concrete surface, causing the appearance of cracks, the size of which can increase. The use of standard contact methods for diagnosing concrete structures requires the installation of equipment on their surface, which is time-consuming and expensive. There are a significant number of the similar heavy structures that require constant monitoring of its behaviour [16].

Machine vision technologies by National Instrument have a specific tools and hardware for computer image processing [17]. Since mathematical image processing requires a powerful computer with LabView program installed, the drone can store high-resolution images in its memory and then, after the drone returns, computer will be processing image by the method [15]. A big data can be transferred to a control computer using a connection under the newest cellular communications. In this case the image can be processed in real time.

The concrete surface reflects light diffusely [18] and has a reflection coefficient (albedo) of the order of 36-69% [19]. A feature of the photographic image of the concrete surface is that the reflected light rays from the surface elements are collected by the lens into images. When using digital cameras, a matrix of photodetectors that form image pixels. The light intensity registered by each pixel of the image corresponds to the intensity of the reflected radiation of certain areas of the real surface. In the presence of a crack in concrete, a sharp contrast of reflected light appears. Therefore, it is possible to describe the real surface of the material from their images.

Figure 1. Photo of the dam near the city of Nonsan (South Korea).

Recently, the direction of surface monitoring with the help of drones (small aircraft, radio-controlled from the surface of the Earth) has been intensively developing. In particular, National Instrument (USA) [20] has created software and hardware for controlling the flight of a drone and receiving video data on a computer and capturing high-resolution images in the electronic memory of the drone. In this case, it is possible to examine the surfaces of hard-to-reach places of the monitoring object from a close distance and obtain images of the maximum surface quality of the investigated damage to the concrete structure. Then it seems possible to use the image processing technology to measure the geometric dimensions of cracks, proposed in [15].

2. Imaging processing for crack's width, length and depth using computer tools

The digital images of the cracks were converted into a dataset, each element of which contains information about the level of reflected light from the concrete surface and cracks. In Ref. [15] is obtained formulas for width, length and depth of crack $x = x_i \frac{L-f}{f}$; $y = y_i \frac{L-f}{f}$; $d \cong L(\frac{1}{\sqrt[4]{\xi}} - 1)$,

where $\zeta = L/f$, L the distance between the objective of camera and the concrete surface, f is camera focal length. These formulas have to convert to discrete form. The pixel data of the digital image was converted into real geometric parameters (width and length) by the formulas:

$$x_i = n_i\delta(\zeta - 1), \quad y_j = n_j\delta(\zeta - 1), \tag{1}$$

where δ is camera pixel size, i and j are the number of pixels in the width and length of the image captured by the camera (Figure 2).

To obtain crack depth data, the data matrix was recalculated according to the formula:

$$d_{i,j} \cong L(1/\sqrt[4]{\xi_{i,j}} - 1), \tag{2}$$

where $\xi_{i,j} = I_d^{i,j} / I_s$, I_s and $I_d^{i,j}$ are experimental values of luminous intensity from undamaged concrete surface and from different depths of the crack.

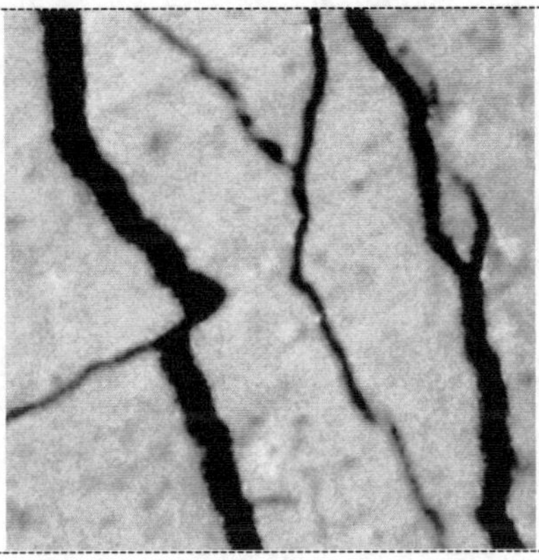

Figure 2. A photo of a crack on a concrete surface with a known properties (etalon) that was captured by the drone photo equipment

Figure 2 shows a test photo of a crack on a concrete surface that was captured by the drone's optical camera. A concrete sample with a crack was selected specifically for this experiment in order to compare the known parameters of the crack with the calculated data obtained using the proposed optical method.

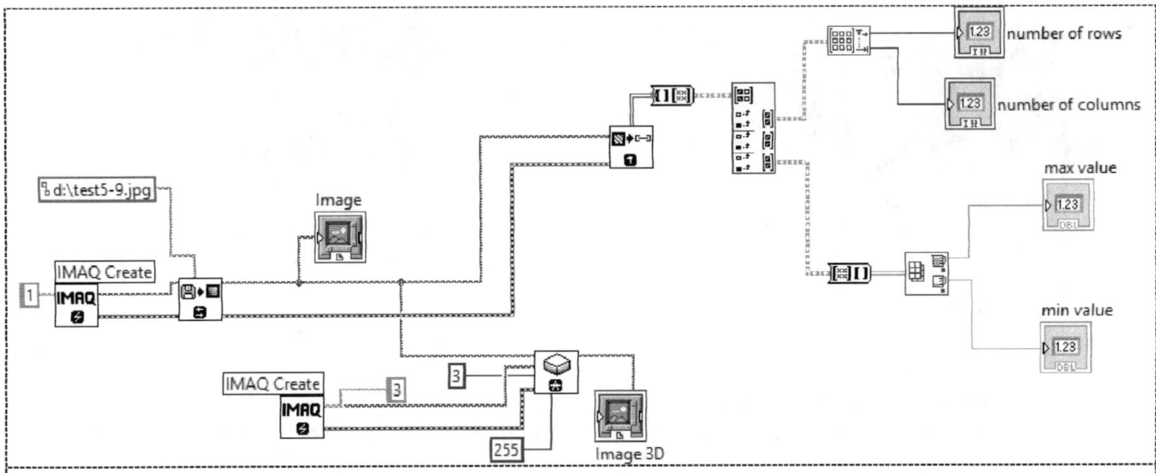

Figure 3. LabVIEW program for reading an image and convert it to matrix and 3D graph.

Figure 3 shows a LabVIEW program for reading an image and converting it to a matrix and 3D graph. The IMAQ Create module reads data written to the computer disk. The image data is displayed on the front panel of LabView as a regular photograph using the Image module. The image data is then processed by a data matrix transform unit that displays the intensity data of each pixel in the image (Figure 4). The image data is displayed on the front panel of LabView using the special software module as a 3D graph (Figure 5). Converting the image to 3D format is useful for direct visualization of the object of study with cracks (Figure 5). It is also important to know the maximum and minimum values of the image matrix, which are determined by the tools min value and max value.

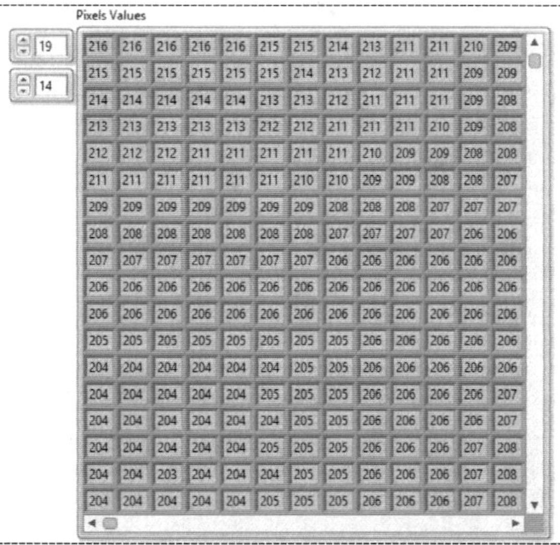

Figure 4. The image data as a matrix of photo 3 in front panel of LabVIEW program.

Figure 5. 3D graph of the crack image in front panel of LabVIEW program.

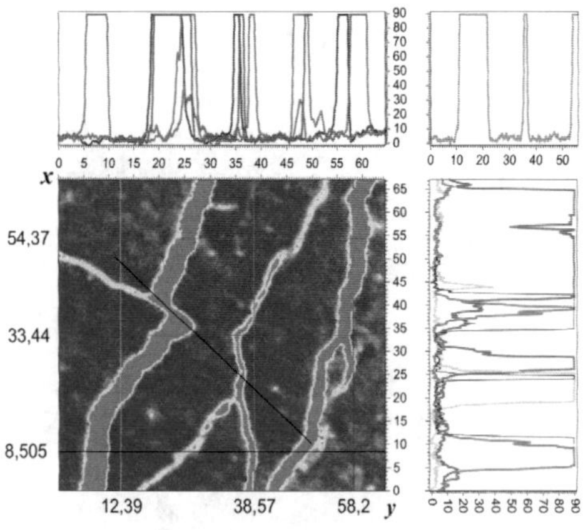

Figure 6. Profile of the image with recalculated data for real geometric parameter of crack in concrete.

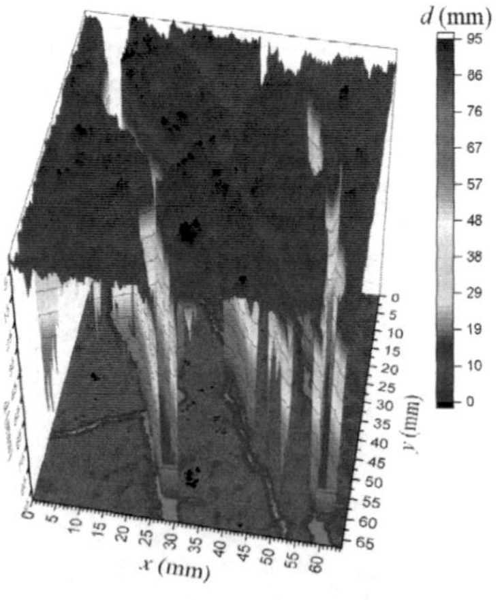

Figure 7. 3D graph with recalculated data for real geometric parameter of crack in concrete.

The real *x*, *y* coordinates and the crack's depth value *d* from the image data are calculated according to formulas (1) and (2). Results of this calculation are shown in Figures 6 and 7. Profiles of images data (Figure 6) allow to find the crack's geometrical size in *xy* plane, that is, the real visible dimensions of cracks on the surface of a concrete structure. In this case, it is possible to obtain profiles of both vertical, horizontal axes, and oblique (top right distribution of Figure 6). For 36-megapixel camera with pixel size δ= 4.88 μm when photographing from a close distance, the resolution in measuring the width and length of the crack reaches 0.03 mm, and the measurement error is ± 0.1 mm. More accurate results can be obtained by decreasing pixel sizes and increasing their number. The accuracy of measuring the length and width of a crack equals 5%. The data in Figure 6 also contains the data for the crack depth along the concrete surface. The accuracy of crack's width and length measurements have to depend on the resolution of the camera (pixel size δ).

Calculations for the depth of cracks along the surface of the sample are shown in Figure 7. Based on these data, the maximum depth and distribution of the crack depth along the cracked concrete surface can be found. The error in measuring the crack depth depends on the accuracy of measuring the distance between the camera and the concrete surface. In our case, the error is 10%.

3. Conclusion

Remoted optic measurements using drones with an installed high-resolution camera can be effectively used for monitoring the condition of concrete structures in hard-to-reach places and allows you to get high-quality images of the damaged concrete surface at close range. In this case, the use of computer image processing using the method proposed in [15] makes it possible to calculate the geometric dimensions of cracks in concrete to assess the critical state of the structure.

Using the latest, higher resolution megapixel cameras with advanced optics can improve measurement accuracy. However, increasing the size of the image leads to problems with computer processing of big data. Further development of this optical method will be associated with the development of new software and hardware.

4. References

[1] Dong C-Z and Catbas F N 2020 A review of computer vision-based structural health monitoring at local and global levels *Structural Health Monitoring* **1-52**

[2] Mei Q and Gül M 2020 A cost effective solution for pavement crack inspection using cameras and deep neural networks *Construct. Build. Mater.* **256** 119397

[3] Gehri N, Mata-Falcón J; Kaufmann W 2020 Automated crack detection and measurement based on digital image correlation, *Construct. Build. Mater.* **256** 119383

[4] Ji X, Miao Z and Kromanis R 2020 Vision-based measurements of deformations and cracks for RC structure tests, *Eng. Struct.* **212** 110508

[5] Khuc T, Anh Nguyen T, Dao H 2020 Swaying displacement measurement for structural monitoring using computer vision and an unmanned aerial vehicle, *Measurement* **159** 107769

[6] Al-Mistarehi Bara W, Obaidat T M 2020 Detection of Crack on Concrete Surface Using Semi-Automated Digital Image Processing Technique *International Journal of Advanced Science and Technology*, **29**, 2465-2480.

[7] Rui L, Zappa E, Collina A 2020 Vision-based measurement of crack generation and evolution during static testing of concrete sleepers. *Eng. Fract. Mech.* **224** 106715.

[8] Słonski M and Tekieli M 2020 2D Digital Image Correlation and Region-Based Convolutional Neural Network in Monitoring and Evaluation of Surface Cracks in Concrete Structural Elements *Materials* 13 3527.

[9] Song W, Jia G, Zhu H, Jia D and Gao L 2020 Automated Pavement Crack Damage Detection Using Deep Multiscale Convolutional Features *Journal of Advanced Transportation*, Article ID 6412562, 11 pages.

[10] Kim B and Cho S 2020 Automated Multiple Concrete Damage Detection Using Instance Segmentation Deep Learning Model *Applied Sciences*, **10** 8008.

[11] Feng C, Zhang H, Wang H, Shuang W and Yonglong L 2020 Automatic Pixel-Level Crack

Detection on Dam Surface Using Deep Convolutional Network *Sensors* **20** 2069.

[12] Ahmed H, La Hung M and Gucunski N 2020 Review of Non-Destructive Civil Infrastructure Evaluation for Bridges: State-of-the-Art Robotic Platforms, Sensors and Algorithms *Sensors* **20** 3954.

[13] Won K and Sim C 2020 Automated Transverse Crack Mapping System with Optical Sensors and Big Data Analytics *Sensors* **20** 1838.

[14] Spencer Jr B F, Hoskere V, Narazaki Y 2019 Advances in Computer Vision-Based Civil Infrastructure Inspection and Monitoring *Engineering* **5** 199–222.

[15] Vashpanov Y, Son J-Y, Heo G, Podousova T and Kim Y S 2019 Determination of Geometric Parameters of Cracks in Concrete by Image Processing *Advances in Civil Engineering*, Article ID 2398124, 14 pages.

[16] Stochino F, Bedon C, Sagaseta J and Honfi D 2019 Robustness and Resilience of Structures under Extreme Loads. Review. *Advances in Civil Engineering*, Article ID 4291703, 14 pages.

[17] National Instrument Corporation (USA). Available from: http://www.ni.com/vision/.

[18] Ryer Alex Light Measurement Handbook 1998 *International Light Inc.* http://www/intl-light.com

[19] Levinson R and Akbari H 2001 Effects of Composition and Exposure on the Solar Reflectance of Portland Cement Concrete, *Lawrence Berkeley National Laboratory Report LBNL-48334, Berkeley*, CA. http://www-library.lbl.gov/docs/LBNL/483/34/PDF/ LBNL-48334.pdf; American Concrete Pavement Association www.pavement.com.

[20] AR Drone Toolkit for LabVIEW–LVH http://sine.ni.com/nips/cds/view/p/lang/ru/nid/211837#resources.

IOP Conf. Series: Materials Science and Engineering 1141 (2021) 012029 doi:10.1088/1757-899X/1141/1/012029

Analytical methods for selection of demolition technology

O Popov, A Chystiakov, A Petrovsky
Odessa State Academy of Civil Engineering and Architecture, Odessa, Didrihsona str., 4, *Ukraine*,

E-mail: artemchis@gmail.com , oleg.a.popov@gmail.com

Abstract. Demolition of buildings and structures, or their structural elements, is primarily aimed in designing stage – for the reconstruction or renovation of a building or complex of buildings, updating the appearance, design or philosophical meaning of an object. Therefore, demolition must be approached no less responsibly than the construction itself. The article deals with data on the economic feasibility of a correct approach to demolition, which allows the maximum extraction of raw materials from construction waste for further construction. The article provides a brief overview of the technologies for demolition structures to select the optimal and most effective. For the best use of the existing knowledge on this subject, and as a result – obtaining the greatest economic benefit, it is proposed to classify demolition technologies, their description and recommendations for use. For this, it was proposed to create a certain database that can, in an interactive form, help to choose the optimal technology for demolition a building or a separate structure.

1. Relevance

Nowadays, the relevance of topics related to the recycling and reuse of building structures and materials is becoming more valuable. This is mainly due to the end of the service life of residential buildings that were massively built in the 60s of the last century. These buildings are currently undergoing for reconstruction and modernization. In prestigious areas they are completely demolished. As a result of dismantling huge volumes of construction and demolition waste remain, which is expensive to dispose of and, moreover, is unfavourable for the environment. Considering the fact that the dismantling of a building is an expensive process, it becomes obvious the need for competent demolition so that the resulting raw materials can be sold or used and automatically solve the problem of waste disposal. There are many technologies for demolition of buildings and structures, which are more or less appropriate to apply. A competent approach to the choice of demolition technology will allow to reduce the cost of the process as much as possible and, as a result, to obtain the highest quality building materials from waste.

2. The main idea

The main potential reserves for saving material and energy resources in the construction industry is the use of waste concrete scrap, today in Ukraine the use of this resource is small. Analysis of indicators of waste disposal and their use as secondary raw materials indicates the economic attractiveness of this area. So, according to statistical reporting data for 53 types of resource-valuable construction waste, the volume of their use can be 58.5% of the total volume of this waste. Construction waste contains a large amount of metal and concrete scrap. In some cases, replacing natural resources by construction waste can save up to 50% of material [1].

Waste recycling, including construction and demolition waste, is a promising business, since the cost of recycled rubble is only 50 Euro per ton [2]. Secondary crushed stone from concrete constructions of demolished buildings is much cheaper than natural, since energy consumption for its production is 8 times less, and the cost of concrete with it is reduced by 25% [3]. Moreover, the absence of certain standards may allow manufacturers of secondary non-metallic building materials to sell almost all components. But, for example, the ingress of gypsum contained in finishing materials and partitions can lead to significant corrosion of new concrete, up to its destruction [3]. To obtain a high-quality product, it is necessary to strictly observe the technology of demolition buildings, sorting materials and preparing waste for recycling. Therefore, for the widespread introduction of complexes for the destruction of substandard concrete goods by a mechanical method and the production of crushed stone from crushed concrete, a number of issues should be resolved on its rational use in construction as a filler in the production of concrete work. It is especially important to choose the correct technology for demolition structures [4].

The problem of construction and demolition waste is of particular importance due to the fact that the service life of buildings massively erected in the 60s of the twentieth century is being exhausted. So, when a five-storey residential building is demolished, about 3 thousand cubic meters of construction and demolition waste are formed that needs to be processed, not stored. In the conditions of urban development, next problems arise: how to demolish a building, destroy a worn-out structure of a reinforced concrete or stone bridge or an industrial facility that has become unnecessary and where to subsequently use the resulting secondary resource. In general, the technological process of utilization of construction and demolition waste can be represented in two stages:

- preliminary disassembly or destruction of the most valuable and complex concrete and reinforced concrete structures.
- recycling waste into secondary crushed stone; removal and disposal of reinforced concrete and construction and demolition waste.

The advantages of processing reinforced concrete waste into secondary crushed stone include the following: at each construction site where structures are dismantled, crushed stone will soon be needed during the construction process (the presence of secondary crushed stone eliminates the need to purchase crushed stone, and pay the costs of its delivery, removal and storage).

The main components of reinforced concrete are rubble, sand, cement and metal reinforcement. These materials are relatively stable in the environment: their total volume and weight change little even after prolonged use. After crushing and separation of the metal, the obtained secondary aggregates compared to natural ones have lower density, bulk density and percentage of volume in a solid body, higher water absorption, weight loss in the test for resistance to weathering and losses due to abrasion. All these changes are associated with the presence of a mortar component in the concrete scrap. A great importance for predicting the behaviour of the aggregate in mixtures and in hardened concrete is the quantitative content of components in various fractions of concrete scrap. Knowledge of the patterns of changes in its component composition allows it to optimize the use of various fractions.

Today there are many methods of destruction of building structures – static (splitting, crushing, cutting and expansion) and dynamic (shock, vibration, explosive). Currently, the greatest results have been achieved in improving the technology of destruction of building structures by impact methods, splitting, cutting, crushing and expansion. However, at the same time, each of the methods has restrictions on the use, or vice versa, recommendations for use for the destruction of buildings and structures made of a certain material. In this regard, the issue of their classification remains relevant.

2.1. Impact methods

The most widespread are hydraulic and pneumatic hammers on self-propelled units, which are characterized by high productivity, mobility and the ability to accurately apply a blow. Compared to pneumatic hammers, hydraulic hammers have less noise, vibration and dust generation. The best proven hydraulic hammers with a single impact energy of 9000 J and hydropneumatic installations with a load of up to 3000 J. Pneumatic breakers with an impact energy of 80-90 J are used for dismantling concrete and reinforced concrete structures and rubble, rubble concrete and brick structures and semi-rocky rocks. For disassembly and destruction of various types of structures, they are supplied with replaceable working bodies (lance and shovel). Pneumatic jackhammers with an impact energy of 30-45 J are used for demolition concrete and asphalt concrete pavements, brick walls. During operation, pneumatic manual machines are turned on by pressing the handle with a certain effort and turned off when the effort is removed from the handle. To provide energy to manual pneumatic machines, stationary compressor units and air distribution networks of the reconstructed enterprise are used. In the absence of such a possibility, portable compressors are used. Electric manual impact machines have a lower energy of a single blow compared to pneumatic ones, however, during their operation, the noise level is significantly lower, which leads to a decrease in fatigue of workers.

It is advisable to use electric hand hammers and concrete breakers for element-wise disassembly of a structure of medium and low strength, as well as when working at height, where, in cases of using pneumatic manual machines, workers need to make additional efforts to lift and hold the air hose, which leads to rapid fatigue and, accordingly decreased productivity. Pneumatic manual impact machines are effectively used to dismantle more durable concrete, reinforced concrete and brick structures.

2.2. Cracking
When concrete and reinforced concrete structures are destroyed by splitting, hydraulic wedges are used, which allow working without harmful effects of vibration, noise and dust formation. The hydraulic wedge consists of a hydraulic cylinder and a propping device inserted into the drilled hole and creating a force of up to 130 tons, as well as a pumping station that creates pressure in the hydraulic cylinder. The average performance of hydraulic wedges is approximately 510 times higher than manual breakers.

For splitting concrete foundations, installations are used, consisting of an oil pump station and several (up to 5) wedge devices. To separate parts of concrete, boreholes are drilled in it with a pitch that depends on the strength of the concrete and is 400-800 mm. The diameter of the holes is 3-5 mm larger than the diameter of the working body. The working body is introduced into the hole, then pressurized oil into the hydraulic cylinder. Chipping off pieces of concrete occurs without the scattering of fragments, accompanied by a weak crack. Installation performance 0,25-0,5 m3/h.

2.3. Cutting
When structures are destroyed, cutting methods are used, which make it possible to dismember a structure into separate elements (blocks) suitable for reuse. In this case, diamond cutting wheels and thermal cutting using oxygen blast, plasma or electric arc are used. Modern machines with diamond wheels allow cutting reinforced concrete to a depth of 400 mm and a mechanical feed rate of up to 2 m/min.

2.4. Splitting up
Crushing is carried out using teeth that are installed on the concrete breaker or separately mounted on the excavator. Replaceable working equipment allows crushing reinforced concrete structures up to 700 mm thick and foundations up to 1200 mm thick. One of the main advantages of a crushing plant is the ability to use it directly at the site of construction and demolition waste generation. In this case, the mobile crushing and screening complex is delivered to the construction site, where it immediately starts to work.

2.5. Destruction by expanding compositions
For the destruction of building structures with the help of expansion, cartridges of liquid carbon dioxide (cardox) are most often used, the action of which is based on an increase in volume as a result of the transition of carbon dioxide from a liquid to a gaseous state, while the developed pressure varies from 125 to 275 MPa. Recently, other expanding compounds have appeared, the action of which is based on various chemical processes that take from several hours to 30 minutes. The destruction of structures occurs as a result of the expansion of the mixture of powder and water poured into the drilled holes, but the resulting pressure is much lower than when using a frame (within 30-40 MPa). Therefore, with the help of this, as a rule, light reinforced concrete structures are destroyed.

When all production processes are carried out near the demolished building, a mobile or self-propelled recycling equipment is used, placed on a mobile construction and demolition waste recycling site. The set of equipment includes: a tower crane (when building is demolished), which forms stacks of building elements with different characteristics; excavator with removable working equipment (bucket, hydraulic hammer and hydraulic shears); a loader for removing destroyed elements of buildings prepared for primary crushing from a stack, moving these elements to the primary crushing unit and loading the unit's primary device (a bulldozer can be used in these processes); units for primary and secondary crushing; screen for separating crushing products by size; conveyors for placing products of several fractions, processing waste and reinforcement, feeding into stacks. Loading of products and waste is carried out by loaders, and valves by excavators, less often by loaders.

2.6. Electro-hydraulic method of destruction of structures

It is carried out without the formation of a blast wave and scattering of fragments, which is a fundamental factor when performing work in places with the release of dust or the likely appearance of gas. This method is completely safe for people working near people and installed equipment, so it can be successfully applied not only on open construction sites, but also inside industrial premises. The use of the installation of the electrohydraulic effect for the destruction of stone and concrete massifs, rubble concrete and brickwork make it possible to increase labour productivity tenfold and even completely exclude the use of physical labour in these works.

2.7. Drilling and blasting method of destruction

The method uses the energy of an explosion generated when an initial impulse from a spark or impact is applied to an explosive. This method has long been used in construction; therefore, it is considered one of the first. In justified cases of destruction of structures by this method, blasthole charges and camouflage explosions are used. To reduce the scattering of pieces, explosion localizers of various designs are used. Their advantages over other means are the absence of fragments and noise, a large number of simultaneously filled boreholes, which in a day cause cracking of massifs unlimited in volume. Destruction of concrete blocks of grade 300 and more, as well as densely reinforced blocks is carried out with preliminary drilling of vertical or inclined boreholes. The disadvantages of explosion generators are the large spread of fragments, significant noise (up to 108 dB in a radius of 50 m) and the release of toxic gases.

An analytical comparison experiment was carried out of all methods of demolition structures to establish a correspondence between building structures and technology in order to obtain the most efficient extraction of raw materials. It was found that it is impossible to strictly envisage the regulations for the selection of demolition technology. This is due to the fact that even the slightest changes in the design feature or its environment can change the efficiency of using the demolition technology, compared to other similar structures. However, for all buildings and structures, for the most efficient extraction of high-quality raw materials from construction and demolition waste, it is important to perform a phased demolition of structures and coatings made of different materials. It is especially important to separate materials that could subsequently cause corrosion of concrete made from the raw materials obtained.

For the most efficient and rational use of existing knowledge about demolition of buildings, structures and their individual structures and, as a result, to obtain the greatest economic benefit, we propose a classification of demolition technologies, their description and recommendations for use in tabular form (Table 1). This will make it possible to help in a convenient form to choose the optimal technology for demolition of a building or a separate structure.

Table 1. Recommendations for the selection of demolition technology.

Name of technology	Used equipment	Recommendations for use
Impact methods	Self-propelled hydraulic and pneumatic hammers, hand hammers	For demolition of concrete and reinforced concrete structures, rubble concrete and brick structures, semi-rock and asphalt pavements
Cracking	Hydraulic wedges	Used in situations where the harmful effects of vibration, noise and dust formation are unacceptable
Cutting	Saws with diamond cutting discs, thermal cutting devices using oxygen blast, plasma or electric arc 23.9	If necessary, to dismember a structure into separate elements suitable for reuse
Splitting up	Teeth that are mounted on concrete breakers or separately mounted on an excavator	Used when it is necessary to crush and process concrete scrap directly at the place of formation
Destruction by expanding compositions	Cartridges of liquid carbon dioxide (cardox)	For the destruction of light reinforced concrete structures
Electro-hydraulic method of destruction of structures	Installation of electrohydraulic effect	Can be used inside industrial premises, where the use of a blast wave and scattering of fragments, dust and gas excretion is unacceptable. The method is safe for people working near and installed equipment.
Drilling and blasting method of destruction	Initial impulse explosive (blasthole charges and camouflage explosion)	The method is effective for quickly demolition a large volume of structures.

3. Conclusions and results

Effective demolition of buildings and structures allows performing a number of tasks, namely: elimination of landfills and burials of construction and demolition waste; creation of resource-saving technologies for the processing of construction and demolition waste, allowing to save building materials; reducing the cost of building structures, saving natural resources. For the most efficient extraction of useful resources from construction and demolition waste, it is important to choose the right technology or a set of technologies for demolition structures. To do this, it is necessary to analyse the above technologies for dismantling structures and select the optimal one based on the recommendations. When choosing methods for disassembling and destroying structures, one of the main indicators is the labour intensity and timing of work, however, the effectiveness of using one or another method also significantly depends on the yield of materials ready for reuse. The use of a competent approach to the choice of demolition technology will allow a responsible and most rational approach to the task of demolition of old, morally and physically obsolete buildings. The proposed

IOP Conf. Series: Materials Science and Engineering 1141 (2021) 012029 doi:10.1088/1757-899X/1141/1/012029

classification of technologies for demolition buildings and structures and their individual parts and recommendations for use will make it possible to competently and most efficiently selection of optimal technology. This will maximize the benefit of the generated waste and to save money.

The classification of technologies for demolition buildings and structures and their individual parts in a tabular form should be regarded as the basis for the further creation of a database in a more interactive form.

4. References

[1] Erukhimov A L 1999 Building materials. Prospects for the use of construction waste. *Publ. group Press*, no **12** p 23 - 28

[2] Gusev B V, Zagursky V A 1988 *Recycling of concrete*. Stroyizdat, 95 p.

[3] Pentegov I V 2001 Secondary resources. *Congress on waste management* Moscow, Publ. group Press, no **1** p 25 - 30

[4] Arsentiev V A, Marmandyan V Z, Dobromyslov D D 2006 Construction materials. *Modern technological lines for construction recycling* no **8** p 64–66

IOP Conf. Series: Materials Science and Engineering 1141 (2021) 012030 doi:10.1088/1757-899X/1141/1/012030

Investigation of parameters affecting the soil amplification by means of 1-D analysis

C C Gürtürk[2], M E Selçuk[1]
[1] Yıldız Technical Univ., Dept. of Civil Eng., Geotech. Div., 34220 İstanbul, *Turkey*
[2] Yıldız Technical Univ., Dept. of Civil Eng., Geotech. Div., 34220 İstanbul, *Turkey*

E-mail: f0919017@std.yildiz.edu.tr[1], meselcuk@yildiz.edu.tr[2],

Abstract The increase in the amplitude of earthquake waves as they pass through layers of soft soil close to the surface can have significant effects on ground amplification and structural damage. The aim of this study is; ground response analysis was performed using 1D (One Dimensional) Equivalent Linear Analysis method with DeepSoil software for specific soil profiles. The analyses were carried out with 3 different strong ground motions selected in accordance with the response spectrum. Strong ground motion records selected according to the regulation were scaled in accordance with the spectrum in the frequency domain. Analysis results were compared with the Turkish Earthquake Response Regulations 2018 (TEBR 2018). It is explained how the soil transfer functions, namely how the amplification functions vary depending on the parameters of the soil layer thickness and shear wave velocity. Depending on the characteristics of the soil profile, it has been observed that the incoming earthquake waves only amplify those in a certain frequency range. By performing ground response analysis, it should be known which frequency content of the ground amplification and therefore which types of structures will affect and the design should be made by considering these criteria.

1. Introduction

It is important to determine the characteristics of the layers that make up the foundation soil in the ground studies carried out before the project design and construction, but it is also necessary to determine how the soil layers will behave in a possible earthquake. Many different methods have been used in studies to determine the effects of local soil conditions on strong ground motion, the factors affecting this and the values of these effects. These methods are in the form of one, two- and three-dimensional ground response analysis, microtremor measurements, classical spectral ratio and horizontal to vertical spectral ratio. Among these, one-dimensional site response analysis provides the advantage of easily creating a mathematical model of the soil to be analysed, with the assumption that each layer of the soil environment can be defined precisely with shear modulus, critical damping ratio, density and thickness values.

The significant changes in the degree of damage in settlements based on different soil conditions in the same region prompted researchers to investigate the effect of local soil conditions on earthquake waves. As a result of the observation and analysis of the shear wave velocity during the ground motion, they stated that the average value of the velocity has a significant effect on the amplification of the soil layers that waves travelling from the bedrock through to the soil layers up to the ground surface [1]. In studies on the subject, the amplification factors depending on the period for the analysis made in certain regions and soils have been revealed [2]. Erdal Şafak worked on seismic field amplification types and generation mechanisms and the models and methods used to characterize them [3]. Yalçınkaya investigated the effect of local ground conditions on earthquake waves by using 5 different earthquake records at 6 stations established within the scope of the BYTNET project. Classical spectral ratio and horizontal vertical spectral ratio methods are used in the calculation of local ground effects. It is observed that stations located on soft soil layers show significant amplification in periods close to the dominant vibration periods that engineering structures can have [4].

2. The Effect of Local Soil Conditions on Soil Amplification

Soil amplification effect is the change in the amplitude, frequency content and duration of earthquake motions depending on the physical and dynamic properties of the soil layers through which the waves pass. The most important factors that cause soil amplification are the density and stiffness of the soil layers through which the earthquake waves pass. As an earthquake wave travelling from the bedrock to a less rigid soil layer, the earthquake wave amplitude increases, changing the frequency content and duration. Another factor due to the conservation of energy principle is the impedance ratio. The ratio of the product of the shear wave velocity and density of the bedrock to the product of the shear wave velocity and density of the soil layer on the surface is called the impedance ratio. The fact that the impedance ratio is greater than one indicates that the bedrock is passed to softer ground and the wave amplitudes will increase. The main reason for the increase in the amplitude of earthquake motion in soft soils is the seismic impedance difference between the ground and the bedrock beneath it. Seismic impedance can be considered as a measure of the resistance of the environment to particle motion [5]. For a vertical S wave propagating in a soil layer, the seismic impedance (z) is the product of the density (ρ) and the S wave velocity (V_S),

$$Z = \rho V_s \tag{1}$$

Generally, seismic wave velocities and intensities are smaller in regions close to the surface of the earth. If the losses due to scattering and damping are neglected, the elastic wave energy must remain constant until the earth surface, according to the energy conservation principle. Seismic energy, E (t),

$$E(t) = \frac{1}{2}(\rho V_s) v_{(t)}^2 \tag{2}$$

the particle velocity ($v_{(t)}$) must increase in order to conserve energy because the density (ρ) and S-wave velocity (VS) decrease as the waves approach the earth surface, that is, the seismic impedance decreases [6].

For a one-dimensional single layer model, the fundamental resonance frequency is given by the following relation, respectively [7].

$$f_0 = \frac{V_s}{4H} \tag{3}$$

In equation (3) V_s is the S-wave velocity of the layer and H is the thickness of the layer on the bedrock. Soil pre-dominant period T_0 is defined as $\frac{1}{f_0}$. The earthquake characteristics that will affect the responses in earthquakes are significantly affected by the ground conditions in the regions where the responses are located. For this reason, determining the effect of soil layers is an important step in the studies conducted to predict the damages that may occur in the responses. The response of layers differs depending on their thickness, soil type and properties. Likewise, soil layers of different thicknesses and properties may be different from one point to another. For this reason, it should be determined how the dynamic behaviour characteristics such as soil fundamental periods and soil amplification change from one point to another in earthquake resistant response design.

The structure of the soil layers has a great influence on the earthquake behaviour of structures. During an earthquake, seismic waves are transmitted from the bedrock to the foundation by the amplifying effect of the ground through the soils between them. This may cause an amplification or deamplification effect. This shows the soil amplification that can be effective with the presence of the structure due to the soil-structure interaction, which is the result of the difference between the free field motions that will occur at the point where there is no construction and the motion that will occur on the basis of the response [8]. The changes that seismic waves undergo in the soil layers are called "local ground effect".

Since this change is generally observed as increasing amplitudes, the term local ground effect; It is also called soil amplification, transfer function or ground response.

3. 1-D Equivalent Linear Analysis

The aim of this study is to obtain the parameters of the motion on the surface by performing ground response analysis under the effect of 3 different real earthquake records will be analysed using one-dimensional equivalent linear analysis. 1-D analysis can be done with different software that include different material behaviour models. Deepsoil program was used in this study. Parametric values have been chosen for the equivalent linear analysis selected soil profiles.

3.1. Obtaining Data to be Used in Ground Response Analysis

This study is based on Turkish Earthquake Response Regulations 2018 (TEBR 2018) [9]. Response spectrum will be construction by taking site specific spectrum coefficient according the seismic hazard studies from Turkey Earthquake Hazard Maps Interactive Web Applications design by AFAD (Disaster Emergency Management Presidency) [10]. The earthquake records to be used in the analysis were obtained from the PEER (Pacific Earthquake Engineering Research) [11]. The simple scaling of the strong ground motions was done in accordance with the TEBR 2018, and 1-D Equivalent Linear analyses were performed by Deepsoil Program.

3.1.1. Selection and Scaling of Earthquake Records

Web-based Pacific Earthquake Engineering Research (PEER) strong ground motion database; Provides tools to search, select, and download ground motion data. Different spectrum models are available in PEER (ASCE Code, PEER NGA-West2, User Defined). Filtering can be made by entering the desired earthquake parameters. In this study, filtering has been done by selecting the parameters:

User Defined Spectrum was chosen as the spectrum model, and the response spectrum obtained from AFAD was uploaded. Strike Slip, active fault type of lateral faulting is observed in Turkey. 360-760 m / s has been chosen as the shear wave velocity range. The reason for this is that the modelled soil is considered as ZC ground class. The time interval in which the earthquake energy is absorbed by 5-95% (significant duration) has been selected for a minimum of 15 seconds and a maximum of 60 seconds. In order to make the response spectrum on the motion and earthquake code spectrum similar, T_A, T_B, T_L values were entered by selecting (Minimized MSE) as the scale method. 0.01-2.2 was chosen as the scaling range.

Earthquake acceleration data obtained from PEER has been uploaded to the SeismoSignal program [12]. With the SeismiSignal program, the appropriateness of the maximum acceleration, duration (significant duration and bracked duration) and spectral acceleration values were checked. Scaling was done according to the acceleration responses Scaling of the Earthquake Records in accordance with the Design Spectrum of TEBR 2018[13].

Table 1. Features of Scaled Earthquake Records

Earthquakes	Station	PGA[a] (g)	Significant Duration	Bracked Duration	Magnitude	Scale Factor
28.06.1992-Landers	Amboy	0,146	25,22	45,00	7,28	3,80
20.06.1990-İran	Abbar	0,515	28,66	45,72	7,37	1,90
21.07.1986-Chanfalt Valley02	Benton	0,209	10,02	16,65	6,19	3,35

[a] *Peak Ground Acceleration.*

IOP Conf. Series: Materials Science and Engineering 1141 (2021) 012030 doi:10.1088/1757-899X/1141/1/012030

3.2. Ground Response Analysis for Single Layer Soil Profiles
In this section, G / G_{max} and damping ratio relationships proposed by Vucetic and Dobry (1991) are used for soil profiles and 0 is chosen as the plasticity index (PI). As the bedrock properties, V_S = 760 m / s, γ= 23 kN / m^3, D (%) = 1 and an elastic solution has been made.

3.2.1 Effect of Soil Layer Thickness
To identify effect of soil layer thickness parametric analysis were perform by keeping the shear wave velocity (Vs) and density (ρ) of the soil layer constant. Analyses were made for 3 different layer thicknesses (Table 2).

Table 2. Soil Profile Properties for Different Layer Thicknesses

Soil Profiles	γ_n (kN/m³)	V_s (m/s)	H (m)
A	20	360	5
B	20	360	15
C	20	360	30

The soil layer thickness is variable 5-15-30 meters, and other parameters are analysed by taking Vs= 360 m / s and γ_n = 20 kN / m3 constant. As the thickness of the soil layer increases, the soil pre-dominant frequency shifts towards smaller frequencies. When this is considered in terms of period, it means that as the thickness of the soil layer on the bedrock increases, the value of the period in which the maximum acceleration is observed increases. When the variation of the peak ground acceleration is examined by looking at the average values, it is seen that the PGA values measured on the surface decrease as the thickness of the soil layer increases. Since the soil pre-dominant frequency is Vs / 4H, as the thickness of the soil layer increases, the frequency value at which the highest soil amplification decreases. The period value in which the maximum spectral acceleration observed is the lowest value on the soil profile A, while it reaches the highest value on the soil profile C. Since there is no change in the impedance ratio between the two layers, it is seen that the maximum amplification values from the Fourier Amplitude Spectrum ratio are around 2.3, and the frequency values at which these values are seen progress towards low frequencies as the layer thickness increases. (Figure 1, Figure 2, Figure 3).

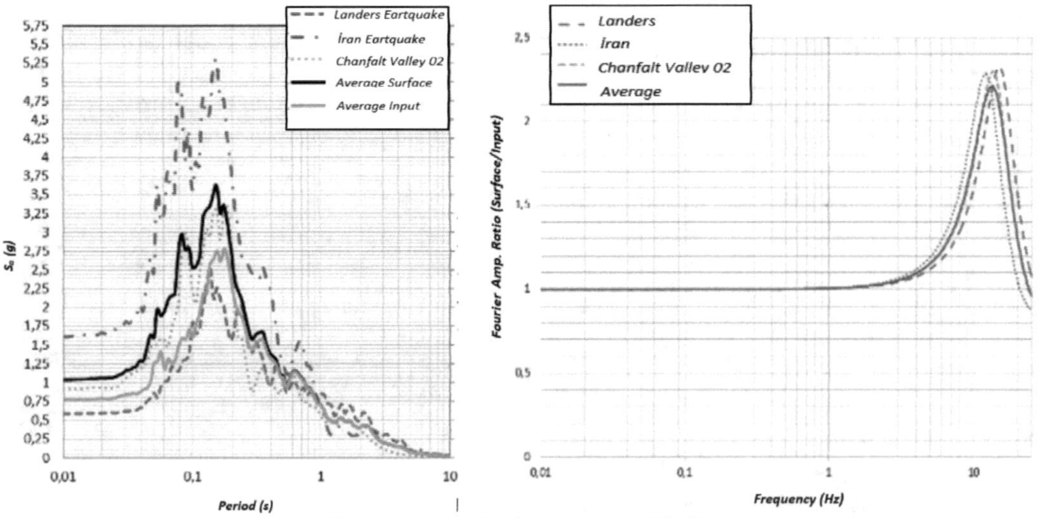

Figure 1. Analysis Results of Soil Layer A

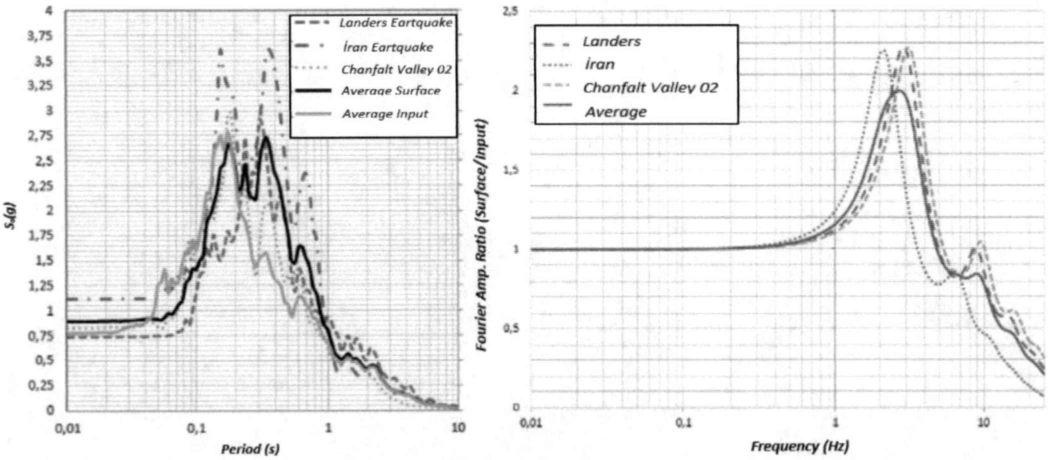

Figure 2. Analysis Results of Soil Layer B

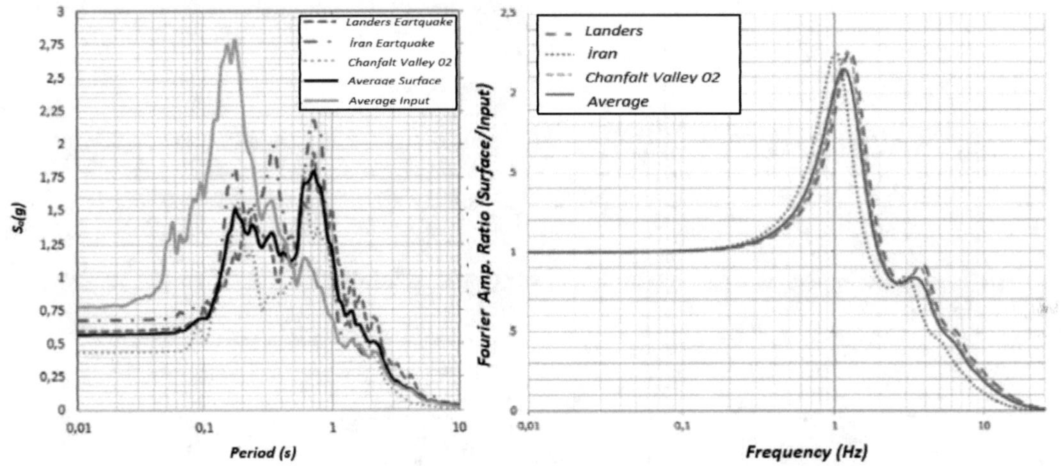

Figure 3. Analysis Results of Soil Layer C

3.2.2. Effect of Soil Layer Shear Wave Velocities

The shear wave velocity (Vs) has been changed by keeping the soil layer density (ρ) and thickness (H) constant. In this section, the variation of different shear wave velocities with layer thickness is also included (Table 3).

Table 3. Soil Profile Properties for Different Shear Wave Velocities H=5 m

Soil Profiles	γ_n (kN/m³)	V_s (m/s)	H (m)
D	20	180	5
E	20	360	5
F	20	760	5

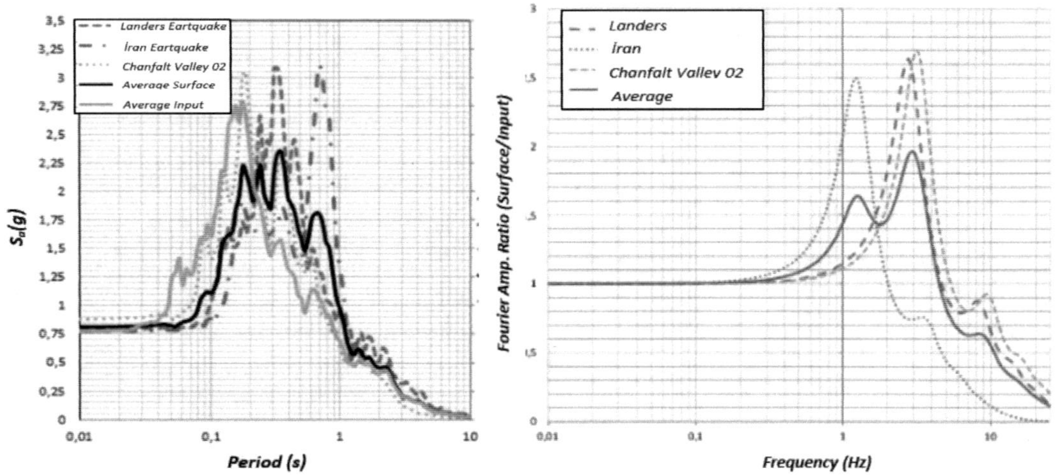

Figure 4. Analysis Results of Soil Layer D

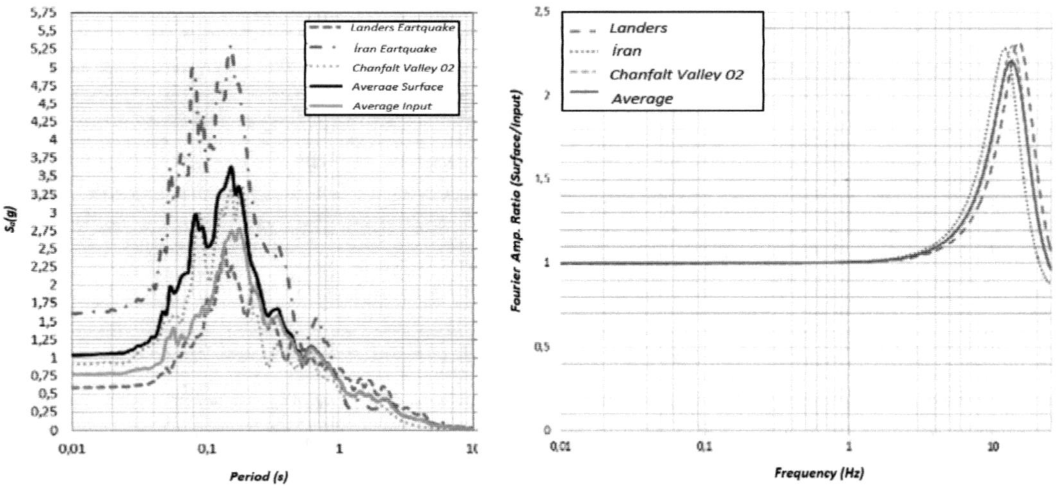

Figure 5. Analysis Results of Soil Layer E

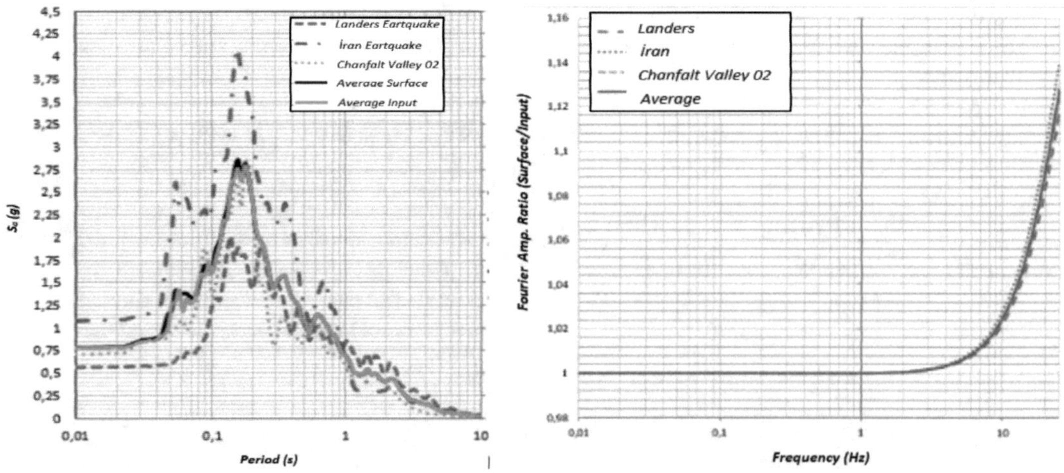

Figure 6. Analysis Results of Soil Layer F

IOP Conf. Series: Materials Science and Engineering 1141 (2021) 012030 doi:10.1088/1757-899X/1141/1/012030

Table 4. Soil Profile Properties for Different Shear Wave Velocities H=15 m

Soil Profiles	γ_n (kN/m^3)	V_s (m/s)	H (m)
G	20	180	15
H	20	360	15
I	20	760	15

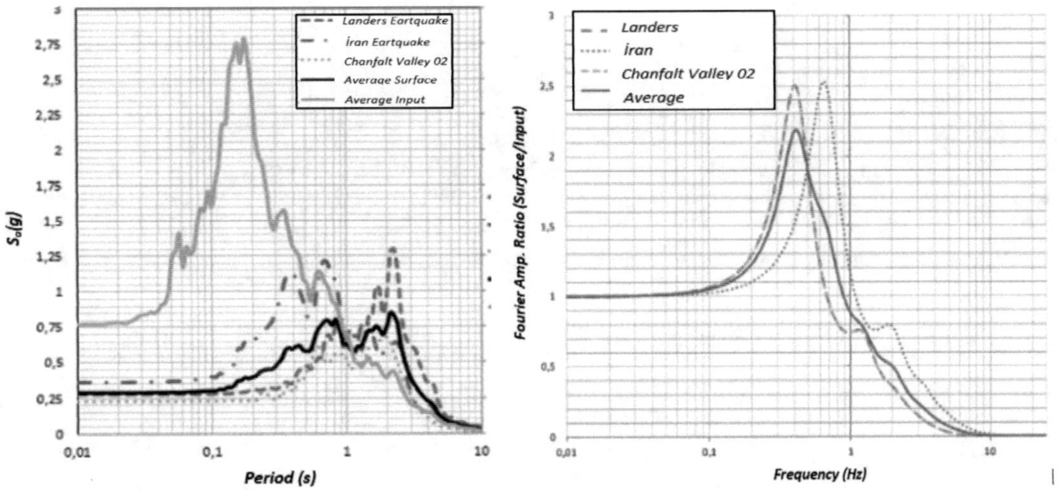

Figure 7. Analysis Results of Soil Layer G

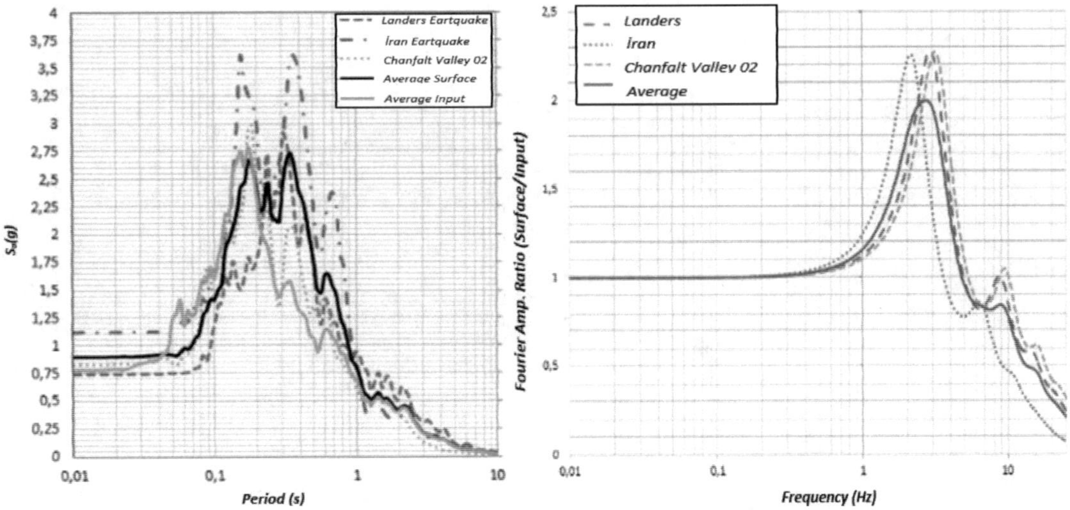

Figure 8. Analysis Results of Soil Layer H

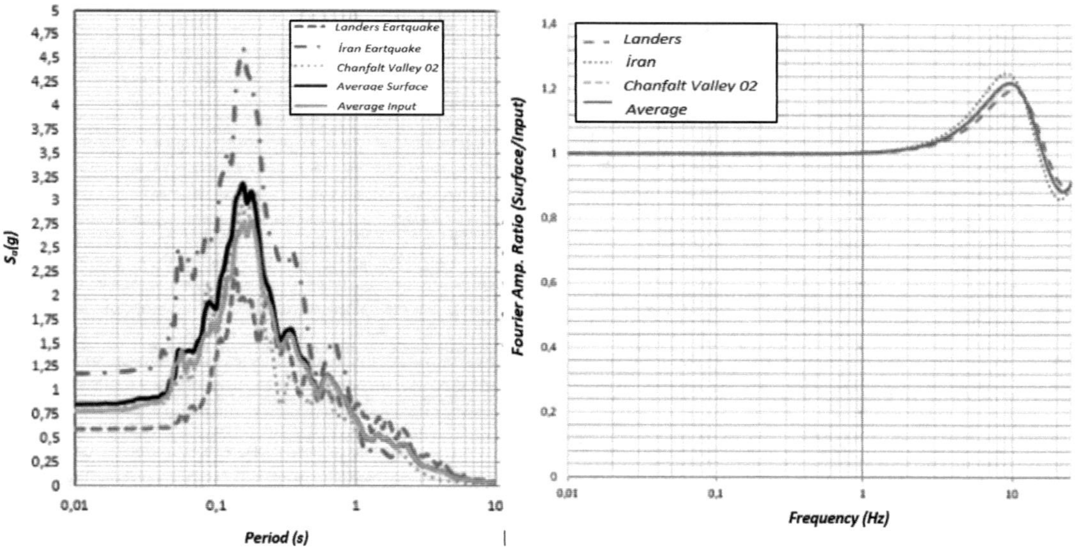

Figure 9. Analysis Results of Soil Layer I

Table 5. Soil Profile Properties for Different Shear Wave Velocities H=30 m

Soil Profiles	γ_n (kN/m³)	V_s (m/s)	H (m)
J	20	180	30
K	20	360	30
L	20	760	30

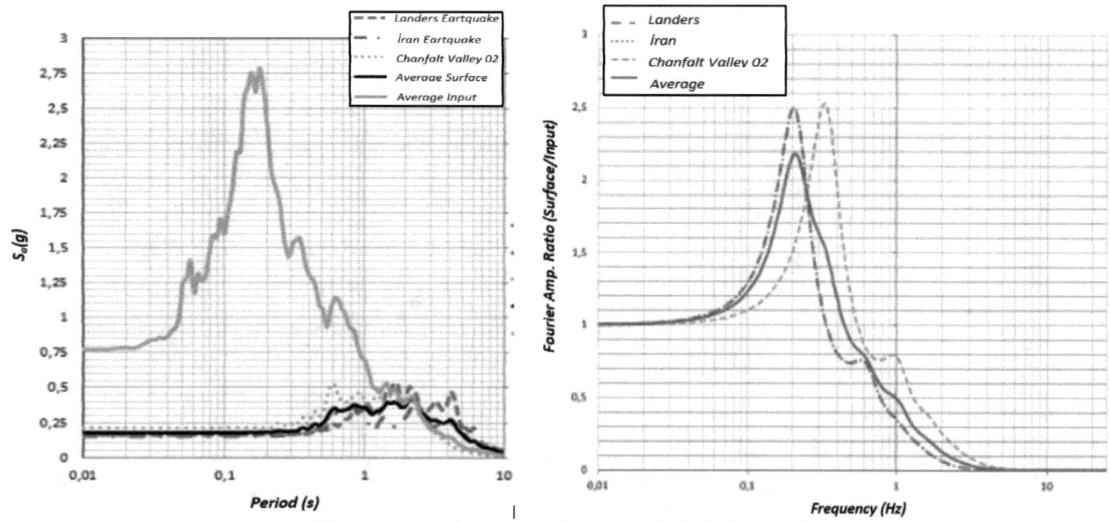

Figure 10. Analysis Results of Soil Layer J

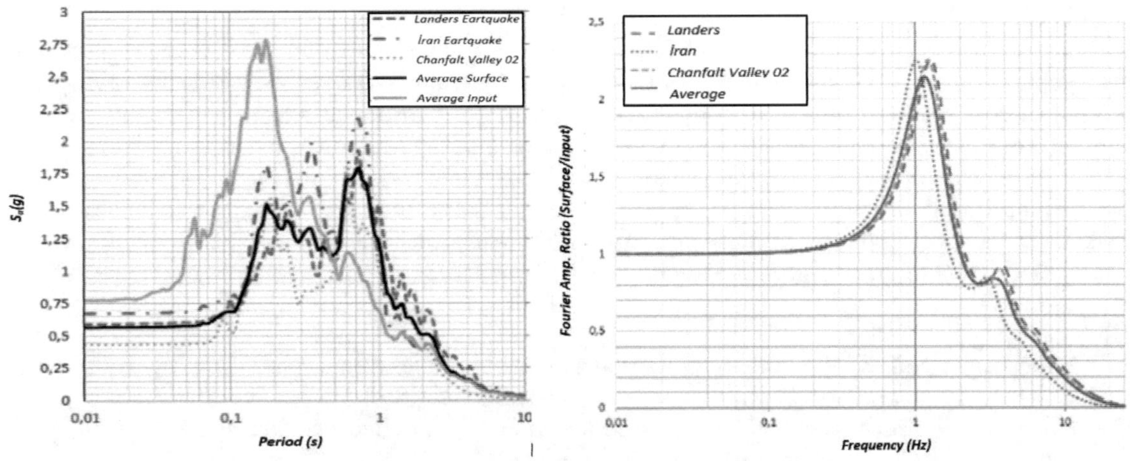

Figure 11. Analysis Results of Soil Layer K

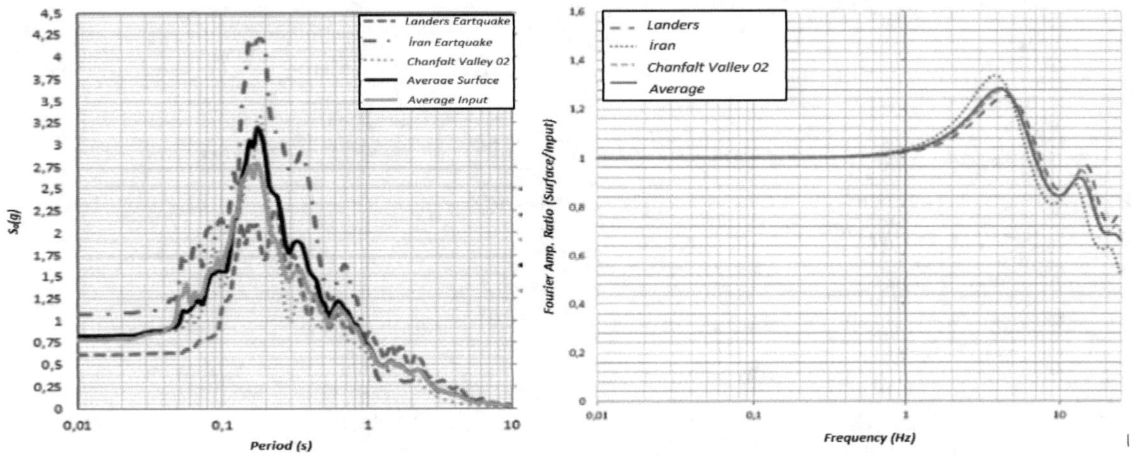

Figure 12. Analysis Results of Soil Layer L

Considering the average values as a result of the analysis made by taking the thickness of the soil layer as 5 meters and the shear wave velocity as 180-360-760 m / s. When the shear wave velocity profile is 760, it is seen that the impedance ratio ρVs^2 will be very low, so the value measured in the bedrock and the PGA value measured on the surface are very close. When the Fourier Amplitude Ratio graphs are examined, it is seen that as the shear wave velocity increases, the frequency with the highest amplification from Frequency = Vs / 4H increases (Figure 6).

In addition, as a result of the analysis made for the soil layer thickness of 5 meters shear wave velocity 180 m / s, the frequency at which the greatest acceleration will occur is different in Iran earthquake compared to other earthquakes, so the maximum acceleration range has extended (Figure 4).

Analyses for 15 meters and 30 meters of soil layer thickness showed that PGA values decreased, spectral acceleration graphs made smaller peaks, but caused peaks in wider periods, and the frequency at which maximum amplification would be observed was at lower frequency values(Figure 7- Figure 12). As a result, as the shear wave velocity decreases from the shear wave velocity in the bedrock, enlargements or attenuations are observed due to the change in impedance ratio. These amplifications are seen at higher frequencies. As the thickness of the soil layer increases, it is seen that the maximum spectral accelerations decrease and occur in wide periods (Figure 5- Figure 12).

IOP Conf. Series: Materials Science and Engineering 1141 (2021) 012030 doi:10.1088/1757-899X/1141/1/012030

4. Conclusions

In this study, properties of the soil transfer function have been investigated for the different parameters, for example depths and S-wave velocities of soil layers, by using 1-D equivalent linear site response analysis. In conclusion, it has been emphasized that depth of bedrock and S-wave velocity are very important for the fundamental soil period and maximum soil amplification. Physical properties of soil layers determine the frequency interval of earthquake motion that will be amplified by soil layers. If the amplification is at high frequencies, this may not be important for most of the engineering structures. However, the amplification is very important for the cases in which the fundamental frequency of soil is very close to that of the engineering structures.

5. References

[1] Midorikawa S 1987 Prediction of isoseismal map in the Kanto plain due to hypothetical earthquake, *Journal of Structural Engineering* **33B** p 43- 48

[2] Kobayashi H Nagahashi S 1977 Response Spectra on Seismic Bedrock During Earthquake, *Proc. Sixth World Conf. Earthg. Eng*, New-Delhi, p 516-623

[3] Şafak E 2001 Local site effects and dynamic soil behaviour *Soil Dynamics and Earthquake Engineering* V. **21**, Issue 5, p 453-458

[4] Yalçınkaya E 2005 *BYTNet (Bursa-Yalova-Turkey on Meter Ag) Investigation of Local Ground Effects at Stations (Dokuz Eylül University) (*Faculty of Engineering Science and Engineering Journal) c.7, p.2, p 75-86

[5] Aki K Richards P G 1980 *Quantitative Seismology*, (W.H. Freeman and Co., San Francisco, California), vol 1 p137

[6] Kramer S L 1996 *Geotechnical Earthquake Engineering*, (Prentice Hall, Upper Saddle River, New Jersey, USA) p 306

[7] Kramer S L 1996 *Geotechnical Earthquake Engineering*, (Prentice Hall, Upper Saddle River, New Jersey, USA) p 261

[8] Tezcan S S, Kaya E, Özdemir Z 2002 Seismic Amplification at Avcılar, Istanbul*, Engineering Structures* v **24**, p 661-667

[9] 2018 Turkish Earthquake Building Regulations

[10] AFAD Turkey Earthquake Hazard Maps Interactive Web Application, (https://tdth.afad.gov.tr/tdth)

[11] PEER Strong Motion Database, Pasific Earthquake Engineering Research Center. (http://ngawest2.berkeley.edu/)

[12] SeismoSoft,SeismoSignal(2018), www.seismosoft.com

[13] 2018 Turkish Earthquake Response Regulations, p 12

Defining an existing prefabrication workflow to identify measures for increased labour time utilization rates

J Cribbs[1], S Root[2]

[1] Department of Construction Management, Wentworth Institute of Technology, 550 Huntington Avenue, Boston, MA 02115; PH (617) 989-4909, USA

[2] School of Sustainable Energy and the Built Environment, Arizona State University, 660 S. College Avenue, Tempe, AZ 85281; PH (480) 965-3589, USA

E-mail: cribbsj@wit.edu

Abstract. This paper analyses the current operations and labour utilization rates for a project team engaged in the prefabrication of exterior wall panels, in an existing manufacturing facility in Tempe, Arizona. A current state process is defined for the BIM-to-install process of each exterior insulation and finish system (EIFS) wall panel. Through analysis of the current state, it is found that only 17% of time was spent directly modelling components meant for prefabrication. The remaining 83% of time engaged in modelling is subject to optimization. Based on these findings, a discussion is presented regarding in place scheduling procedures, model and data management protocols, and field to shop verification/validation measures impacting modelling activities. Finally, a presentation of a series of future workflow interventions intended to improve labour utilization rates for modelling and shop personnel while enabling the internal shop ideal for "mass customization through prefabrication" is provided.

1. Introduction

1.1. Research Question and Definition

The construction industry is undergoing a paradigm shift from traditional, on-site methods of assembly to more of a focus on engagement of technology enabled assembly methods and related offsite construction techniques such as prefabrication, preassembly, modularization and off-site fabrication [1]. Much of the offsite construction techniques currently deployed by industry leaders are driven through the use of Building Information Modelling tools and workflows [2]. These techniques seek to address issues related to skilled labour shortages, schedule compression and demands, jobsite constraints and logistical issues, safety concerns and minimal access to construction work areas [3]. According to surveys undertaken by the BIMForum organization, prefabrication has seen a major growth since 2016 in response to anecdotal project efficiency gains. However, direct quantification of labour time spent on 3d modelling in support of prefabrication efforts is not well understood or defined.

This research sets out to understand the current state workflow of an existing manufacturing facility for prefabricated wall panels. More specifically, the research seeks to quantify labour time utilization rates of the building information modelling team engaged in the prefabrication process. The facility under study is located in Tempe, Arizona and the focus of operations within this facility is to prefabricate exterior insulation and finish system (EIFS) wall panels for just-in-time (JIT) [4], [5] delivery to various construction sites in the southwest region. Currently, the facility engages internal building information modelling (BIM) processes to create shop drawings which are utilized in the manufacturing process. This research seeks to understand the amount of time utilized in creating accurate and usable BIM's for this manufacturing undertaking. The initial driving question for the research team is as follows:

"What is the ideal modelling workflow to enable mass customization through the engagement of offsite prefabrication?"

This question seeks to provide insight into how a modelling process can allow for design freedom when prefabricating, as well as how to close the loop on ever-changing site conditions once the panels have been installed. The remainder of this paper outlines the current state of operations for the location under study and identifies data collection processes deployed to quantify the amount of time spent in the building information modelling phase of prefabrication.

1.2. Current State Panel Manufacturing Process

1.2.1. Physical EIFS (Exterior Insulation and Finish System) Manufacturing Process

The manufacturing facility at the focus of the case study is located in Tempe, Arizona. It contains 105,000 square feet of production space and 5,000 square feet of office space. The manufacturing floor is separated into the following six (6) sections: sub assembly area, production line 1, production line 2, production line 3, material storage area (ie. Grocery Store) and panel storage and truck loading area [6]. The overall facility layout is shown below in Figure 1.

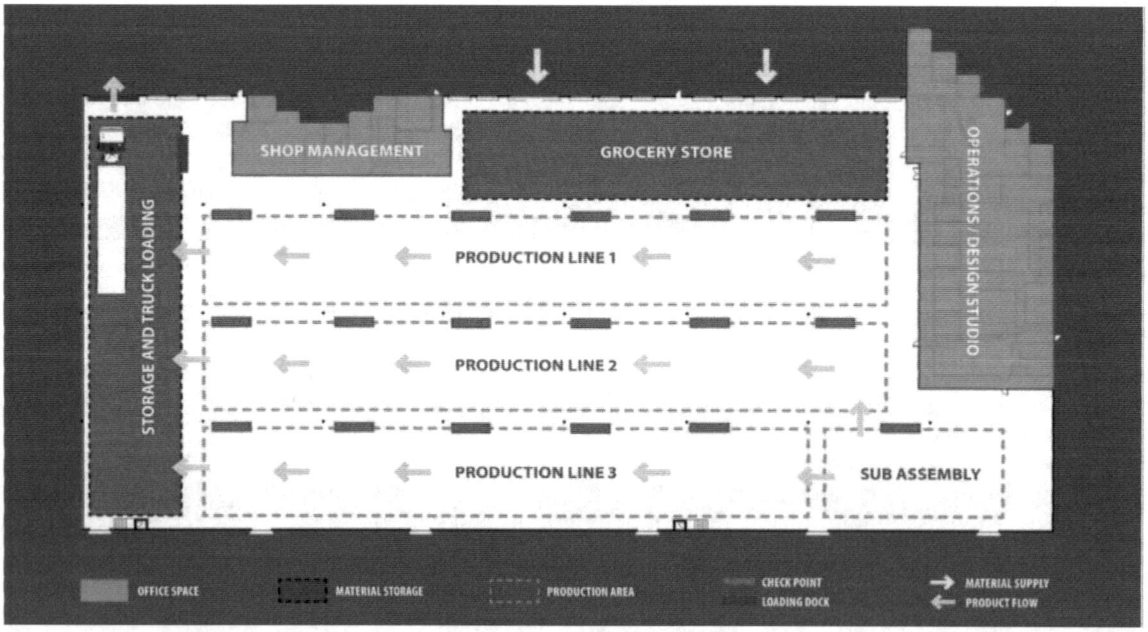

Figure 1. Manufacturing Facility Layout [3] – (Included w/ Authors Permission)

Production line 2 is the main EIFS panel assembly line of focus for this research and it is split into a total of 24 distinct process stations as follows [6]:

Station 1 – Framing
Station 2 – Waterproofing
Stations 3 through 6 – Drying stations for applied waterproofing
Station 7 – EPS foam application
Station 8 – Architectural detailing
Station 9 – Foam rasping and levelling
Station 10 – Basecoat and fiber mesh application
Station 11 through 13 – Drying stations for cementitious basecoat
Station 14 – Sanding and skim coating

Station 15 through 17 – Drying stations for skim coat
Stations 18 through 20 – Finishing
Stations 21 through 22 – Drying stations for finish coat
Station 23 – Stacking and staging of finished panels
Station 24 – Loading of finished panels for delivery

1.2.2. Current State Digital Manufacturing Preparation Process

The manufacturing process for EIFS panels begins with preparation of panel shop drawings using a Building Information Modelling (BIM) interface. The digital process engages parametric attributes of model authoring software (ie. Revit) to generate geometry and an external software application plug-in (Structsoft Metal Wood Framing) to read the model data and populate sheet sets with standardized framing sizes, spacing and connection details. While this plug-in allowed for initial automation of manufacturing standards to be overlayed onto the model, the final detailing process to generate fabrication sheets requires heavy manual coordination between the modelling team and structural engineer [6]. The physical result of the BIM process is a dual sided shop drawing that fits the constraints of a standard 11" x 17" printout sheet. The front side of the sheet contains panel elevations which are structurally dimensioned for framing fabrication requirements. The backside of the sheet contains EIFS elevations and focusses on laying out finish panel dimensions and architectural finish details such as reveals, reliefs and colour schemes.

2. Defining Current State labour Time Utilization Regarding BIM efforts

It is important to understand the current state labour time utilization rate for modelers engaged in the shop drawing process. The purpose of measuring this rate is threefold. First, to gain insight into the allocation of time dedicated to coordination between stakeholders to identify any areas where process changes could enhance the initial modelling process and streamline the panel/shop drawing model output as a result. Second, is to identify areas where manual processes are heavy time constraints in upfront modelling, in order to streamline, engineer out or automate these activities to increase overall productivity within the panel delivery system. Lastly, to begin understanding where time is spent in making an accurate, constructible model for panel fabrication, delivery and installation. From this final piece of insight, it will be possible to identify how best to close the data loop against changing site/field conditions.

This research was embedded in a case-study environment and utilized a mixed methodology for data collection and analysis. The research team engaged in participant observation to gain insight into daily operations of project team members and modelers. A series of unstructured interviews and informal discussions were also engaged to provide more focus to the areas of observation. The structure of these participant observations was borrowed from techniques deployed in "Workflow Management Using Building Information Modelling (BIM) for Prefabrication in a Construction Retrofit Environment" [7]. These unstructured interviews and informal discussions provided researchers with an understanding of how team members perceived the prefabrication process and their role within daily operations. With this knowledge, the research team could identify any deviations in working processes being practiced and those which were formally documented. Ultimately, these findings are out of the scope of this paper, as they are more closely related to fabrication line construction and not building information modelling efforts.

Following participant observation, a series of direct observations took place with the building information modelling team. These direct observations were specifically focused on tasks undertaken to model components which would ultimately be prefabricated within the facility. Again, techniques presented by Cribbs [7] were utilized for data collection throughout the direct observation period. The basis of the applied techniques was focused on the utilization of a pre-designed, categorical observation matrix, which separates physically observed modelling processes into three specific categories: direct modelling work (value added time spent modelling), support modelling work (necessary non-value-added time spent modelling) and any delays (non-value-added time spent

IOP Conf. Series: Materials Science and Engineering 1141 (2021) 012031 doi:10.1088/1757-899X/1141/1/012031

modelling or waste). These categories are further broken down into 15 sub-categories to define typical activities engaged by modelers. All typical activities were pre-defined by the research team through an understanding of required tasks undertaken by modelers and these tasks were verified by modelers during an initial observation period which was not counted towards data collection efforts. Ultimately, the final observation matrix was utilized by researchers when physically sitting next to modelers and watching their workflow, over a specific randomized period of time. Every minute that was observed was categorized and tallied within the observation matrix as a single data point. Overall, the research team collected 367 data points in aggregate. Each data point is related to a 1-minute direct observation period which is summarized, notated and placed into a specific category within the observation matrix. Table 1 below depicts the adapted observation matrix used during observations.

Table 1. Table 1: Direct Observation Matrix (adapted from [7]).

Direct Modelling	Framing
	EIFS panel tickets
	Dimensioning
	Sheathing, finishes, details
Support	Preparatory Work and (Drawing) / Model Set-Up
	Design Package Review / Specification Review
	Existing Conditions Coordinate Verification / Coordination
	Field Walk Verification / Field Modelling
	Internal / External Trade Coordination
	File Searching / Model Load/Download / File Sharing
Delays	Field Re-Verification / Shop Revisions
	Re-Setup / Re-Drawing / Re-Modelling / Redlines
	Inaccurate Existing Conditions
	Waiting / Correspondence Needed
	Personal

3. Measuring BIM use for panelization

3.1. Results of Data Analysis

3.1.1. Aggregate Modelling Time Utilization Rates

In analysing the initial data set, as expected, a substantial amount of time is spent engaged in support work for the modelling process. Figure 2 breaks out the aggregate time utilization rates across the highest tiers of categorization within the observation matrix: direct modelling, support modelling and delays (waste) in the process. Overall, 17% of total modelling time is spent directly modelling a component of the panel which will result in prefabrication. Delays were seen across the random sampling data collection process at a rate of 19% in aggregate. Support work becomes the largest component at a total of 64%-time utilization rate.

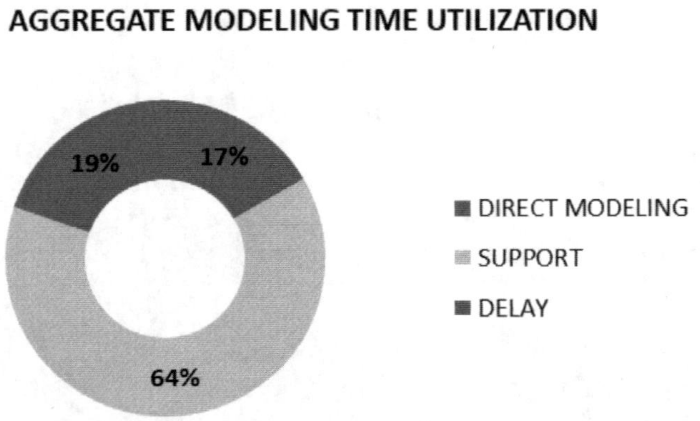

Figure 2. Aggregate Modelling Time Utilization

3.1.2. Aggregate Time Utilization Rates for Direct Modelling Activities
Further analysing the direct modelling time highlights that a large portion of tasks (51%) is focused on the generation of sheathing, finishes and final panel details for prefabrication. This large portion of the direct modelling process can be attributed to the highly manual process of inputting the model data. This is juxtaposed against the initially automated process of standardized framing details via the model authoring tool plug-in at 11% overall time. Finally, the generation of the actual EIFS panel ticket which will be sent out for fabrication takes the remaining 38% of overall direct modelling time. This breakdown can be seen in detail in Figure 3.

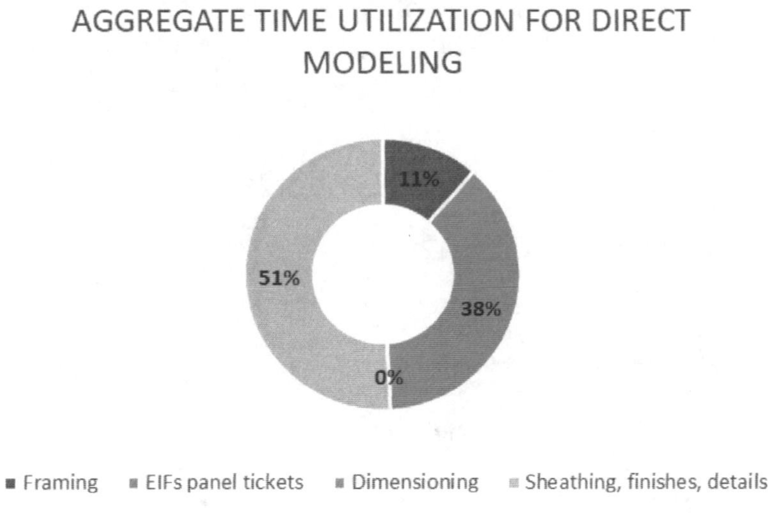

Figure 3. Aggregate Time Utilization for Direct Modelling

The most interesting area to focus on in analysing data is the support modelling category. 63% of total support modelling time is spent in the realm of internal/external coordination to verify model requirements. Again, this is a highly manual process of creating and verifying data from a 2d to 3d format and one which could be the focus for future streamlining of workflows through a model-based, project delivery system. The next area of focus is the verification of existing field conditions throughout the modelling process. In total, verification of existing conditions and then coordinating upcoming models to match these conditions requires the second largest chunk of support work time at 17%. Figure 4 provides a full breakdown of all support work modelling time utilization rates.

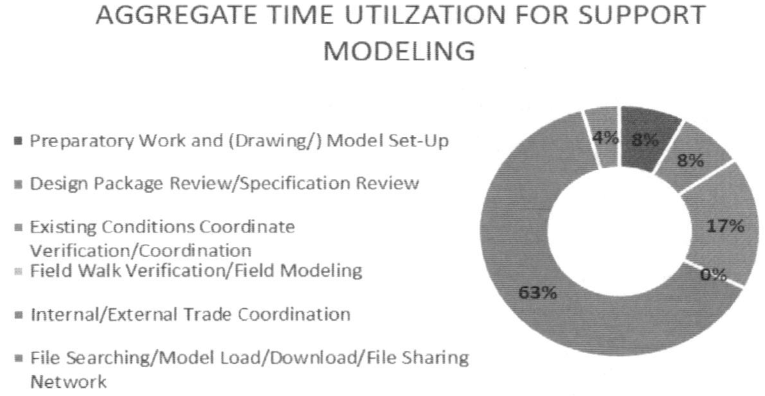

Figure 4. Aggregate Time Utilization for Support Modelling Processes

Finally, a review of delays in the system provides insight into areas of wasted time which should be reallocated to support work and/or direct modelling. Overall, 80% of wasted time is spent either revising shop drawings to match reverified field conditions (after the initial process of modelling to match existing conditions, as stated above in the support work section), or re-modelling due to errors, omissions or issues identified during the manufacturing process. Figure 5 provides an overall breakdown of all delays in modelling.

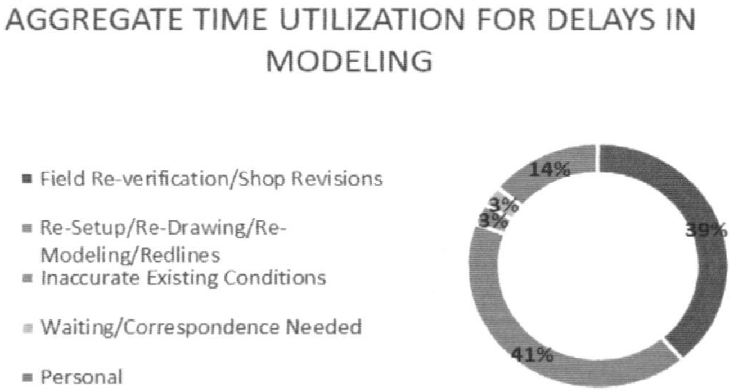

Figure 5. Aggregate Time Utilization for Delays in Modelling

4. A discussion: conclusions & future research direction

Findings from this research show that a significant amount of time is spent in augmenting automated model geometry to meet specific fabrication detail requirements. This signals the need for additional standardized geometric rules to be developed by the plug-in software manufacturer or prefabrication team which could be automatically applied to future models. Additionally, the coordination process related to matching field conditions with design conditions is seen to cause wasted time and effort which can ultimately be streamlined out of the process through revised workflows which close the construction loop. In short, it is recommended that the modelling team utilize a centralized, construction model that is constantly validated for on-site installation accuracy through the engagement of continuous LiDar scans. Upon successful installation of completed prefabrication panels, a LiDar scan should take place to push 3-dimensional, digital data representative of physical site conditions, back to the construction prefabrication team's centralized model for use in continuous panel design and detailing. This process will enable up-front response to changing site conditions rather than downstream reaction via field modifications for completed panels.

While this research is focused on new construction processes, the methodologies for measuring labour time utilization rates are borrowed from a retrofit case study. This can be attributed to the changing conditions on site once the first panel is installed and the project continues in sequence. A retrofit environment is effectively created within this case study environment when attempting to close the modelling data loop and understand the implications of final field install of each panel against the perfect accuracy of an original, digital model. Therefore, the research draws upon the idea that the planning efforts for accurate geometry becomes the critical factor in success when dealing with a BIM for prefabrication workflow in any environment [7].

This research presents the beginning of case study efforts to optimize prefabrication workflows. The focus on upfront modelling seeks to understand where manual activities over-burden the workforce and create bottlenecks for manufacturing personnel downstream. This becomes extremely critical when looking into issues directly related to panel detailing and standardized framing techniques. If these areas require revisions after the initial panel drawing has been sent to the shop floor due to updated site conditions, the amount of time spent revising the model to react to these changes becomes critical in shop flow and output. Ultimately, these areas of modelling can cause a delay in manufacturing which will transpose to the project site.

The next steps in this research are to intervene where the manual modelling activities are heaviest and undertake a second round of observations. The research team intends to engage a similar workflow to that of “prefaBIM” [7] and adapt the overall ideal state to meet the requirements of this manufacturing facility

Following this adaptation and validation process, the direct observations will follow individual panel drawings through the shop floor to begin defining the critical data involved with the manufacturing process. With the help of shop floor cameras utilizing artificial intelligence software, data collection related to individual panel flow along the assembly line floor can be semi-automated to allow for a more robust data set for review, intervention and optimization.

4.1. Final notes: research limitations

This research is limited to the requirements of this specific case study environment and the current personnel involved with the modelling and manufacturing process. Currently, there are two dedicated modelers involved with EIFS panel shop drawings. Both modelers were observed throughout the data collection process and all results were aggregated for an overall look at process requirements. It is noted that the study of a single factory is a limitation, this prefabrication facility is on the cutting-edge of this construction technique and serves as a suitable baseline for future comparison and scalability of best practices. Future research seeks to identify similar prefabrication facilities for identical study and comparison.

5. References

[1] CII 2002b *Preliminary Research on Prefabrication, Preassembly, Modularization, and Off-site Fabrication in Construction*. A Report of Center for Construction Industry Studies, The University of Texas at Austin, Research R(July), 160

[2] Giel B K & Issa, R R A 2013 Return on Investment Analysis of Using Building Information Modelling in Construction. *Journal of Computing in Civil Engineering*, 27(5), 511–521. http://doi.org/10.1061/(ASCE)CP.1943-5487.0000164

[3] Lu Na and Korman T 2014 *Implementation of Building Information Modelling (BIM) in Modular Construction: Benefits and Challenges* Downloaded from ascelibrary.org

[4] Aziz R F, Hafez S M 2013 Applying lean thinking in construction and performance improvement *Alexandria Eng. J*, 52(4), 679–695

[5] Liker J K 2004 T*he Toyota Way - 14 Management Principles from the World's Greatest Manufacturer*. Madison, WI: CWL Publishing Enterprises, Inc.

[6] Root S, Cribbs J, Chasey A D 2019 Case Study: Off-Site Manufacturing of EIFS Panelised Wall Assemblies to Gain Efficiency in Construction Sequencing, *Proceedings of the 2019 Modular and Offsite Construction (MOC) Summit*, Banff, Alberta, Canada
[7] Cribbs J 2016 Workflow Management Using Building Information Modelling (BIM) for Prefabrication in a Construction Retrofit Environment. Doctoral thesis, Arizona State University, https://repository.asu.edu/items/38424

Comparative study on the structural behaviour of a 30m height building with a RC frames structure located in a high intensity seismic area

G Țăranu[1]

[1] Technical University Gheorghe Asachi of Iasi, Faculty of Civil Engineering and Building Services, Department of Structural Mechanics, Iasi, D. Mangeron Street, no. 1, 700050, *Romania*,

E-mail: george.taranu@academic.tuiasi.ro

Abstract. This paper presents an analysis of the Finite Element Model structural system of a building in the Vrancea, Romania seismic area. The studied structure is a part of a set of 3 buildings with a height of 30 m. The height of the levels is 4.5 m and 3.5 m. Due to the vertical irregularities, the distribution of the functions and the location of the building, it is necessary to identify an optimized structural solution for a more balanced behaviour to the seismic action. Thus, various structural models and results are presented regarding the process of establishing the positions and dimensions for structural elements.

1. Introduction

The progress of engineering science in the design of buildings has reached a very high level in the last 100 years. The emergence of computer programs and the development of numerical methods have led to a much faster evolution. The experience and errors encountered in the case of many earthquake-type disasters have led to the identification of vulnerabilities of structural concepts and to the emergence of more efficient design rules. Experimental research programs also facilitate the correct understanding of buildings structural behaviour under seismic actions and capture the potential mechanisms of structural failure.

The shape and design of the resistance structure directly dictate the behaviour of the various actions throughout the life of a building. If the structural elements are correctly positioned in order to be distributed as well as possible and to take over as evenly as possible the loads produced by external actions, the designed construction will have all the chances to withstand all possible actions. This can be achieved if the most important principle is considered: obtaining a balanced dynamic behaviour, namely that the structure is regular both in plane and vertically.

A regular structure both in plane and vertical, is ideal because the evaluation of the seismic structural response by means of simple finite element models is easy to control and verify. At the same time, the structural costs can be optimized in relation to the load's distribution but also because of the lower influence of the inevitable uncertainties related to materials, geometries, eccentricities and other aspects. However, real structures are often neither simple nor regular. In the case of existing old buildings, they were initially designed without any concern about their seismic behaviour. Most of the damage was caused by the effects of torsion which involves a faulty design of the structural elements. [1]

Past earthquakes have frequently shown that structures with an irregular configuration suffer greater damage than regular ones. Some examples are presented in figure 1. The importance of these effects increases with the importance of the first mode of vibration. In literature, multiple articles present different studies and experimental analyses of the seismic response for both symmetric and asymmetric multi-story regular and irregular reinforced concrete structures. [2] [3] [4]

Figure 1 – Collapsed buildings caused by earthquakes [5]

Torsional effects are generally caused by the irregular geometric configuration in plane and/or vertically and the irregular distribution of masses and stiffness in the plane and/or elevation. The evaluation of torsion effects is an important topic in civil engineering design, both for new and existing buildings. Among the basic principles governing the conceptual design of new buildings, some of them are relevant to the configuration and regularity of the building, such as uniformity, symmetry, torsional strength and rigidity.

2. Study case of a RC structural system

The Seismic structural configuration

Seismic design aims to achieve a safe construction in relation to the seismic hazard associated with the site, which meets, in acceptable cost conditions, the fundamental requirements set out in. [6] [7]

The basic conceptual aspects refer to:

- simplicity of the structure
- redundancy of the structure
- the geometry of the structure and of the building, in its entirety, considering the mode of distribution of structural, non-structural elements and masses
- lateral strength and rigidity, in any direction
- making the floors as horizontal diaphragms
- construction of adequate foundations

The seismic structural configuration has three main aspects, as presented in figure 2: 1 - geometry, shape and size of the building, 2 - location and size of structural elements, and 3 - location and size of significant non-structural elements.

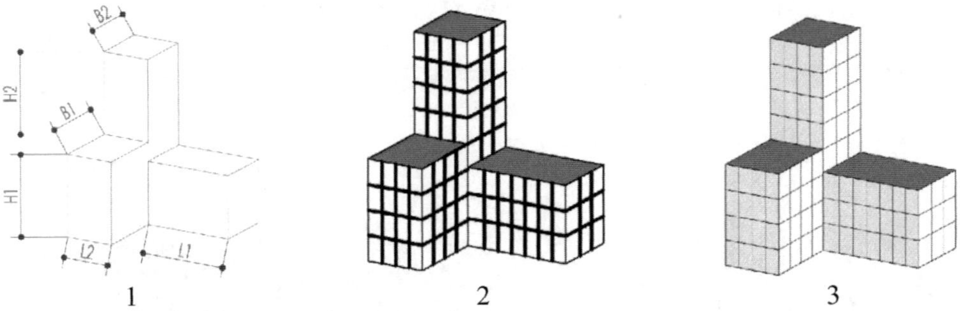

Figure 2. Components of seismic structural configuration: (1) overall geometry, (2) structural elements (e.g., moment resisting frames and structural walls), and (3) significant non-structural elements (e.g., façade glass)

Based on the above, normally built buildings can be classified into two categories: simple and complex. A better structural behaviour is possible when the buildings have rectangular plans and straight elevation, because the seismic forces have a better distribution in structural elements. The complex buildings with setbacks and central openings have geometric constraints to the flow of inertia forces which implies an overloading of some of the structural elements.

In this context, the main purpose of this case - study is to present the results after multiple analyses for the design calculus of the reinforced concrete multi-story irregular structure, in order to reduce or avoid the effects of the overall torsion. The object of the case study is a construction with a height of 30m located in the seismic area of Vrancea (figure 6) characterized by a maximum terrain acceleration of 0.40g and a corner period Tc = 1.0 s [P100]. The building presented in figure 3 has a parallelepiped shape with dimensions in the horizontal plane of 37 by 10 m, which is. The sections of the structural elements were considered differently in 3 scenarios. The views of the structure are presented in figure 3.

The initial structural systems proposed were spatial frames. The first two stories have 4.5 m height each. The next floors are 3.5 m high. At one end of the structure a combined level of 9.0 m height is required, see figure 3.1. Therefore, between the formed frames, parabolic arches are introduced from zero level to + 9.00 m.

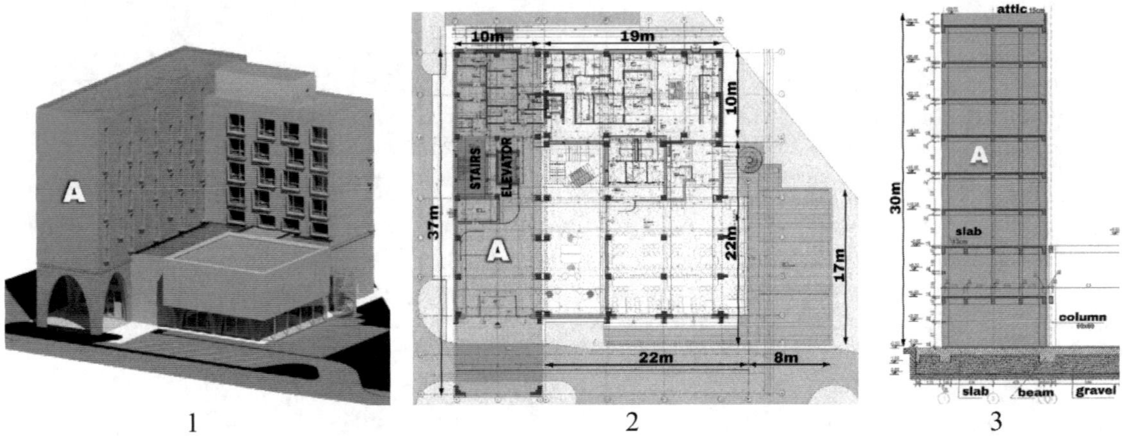

Figure 3. Views of the building: 1 – perspective; 2 – ground floor plane; 3- vertical section

In order to obtain an optimum dynamic behaviour for the building, several comparative structural analyses were performed in order to avoid the first vibration mode as a torsional one. By modifying the section of the columns, beams and introducing additional structural elements like structural walls, the aim was to obtain translation in the first two modes of vibration. The architectural requirements regarding the geometry of the building and the position of the core elevator and stairs presented in figure 4, led in the first stage to an inadequate structural behaviour. [8], [9]

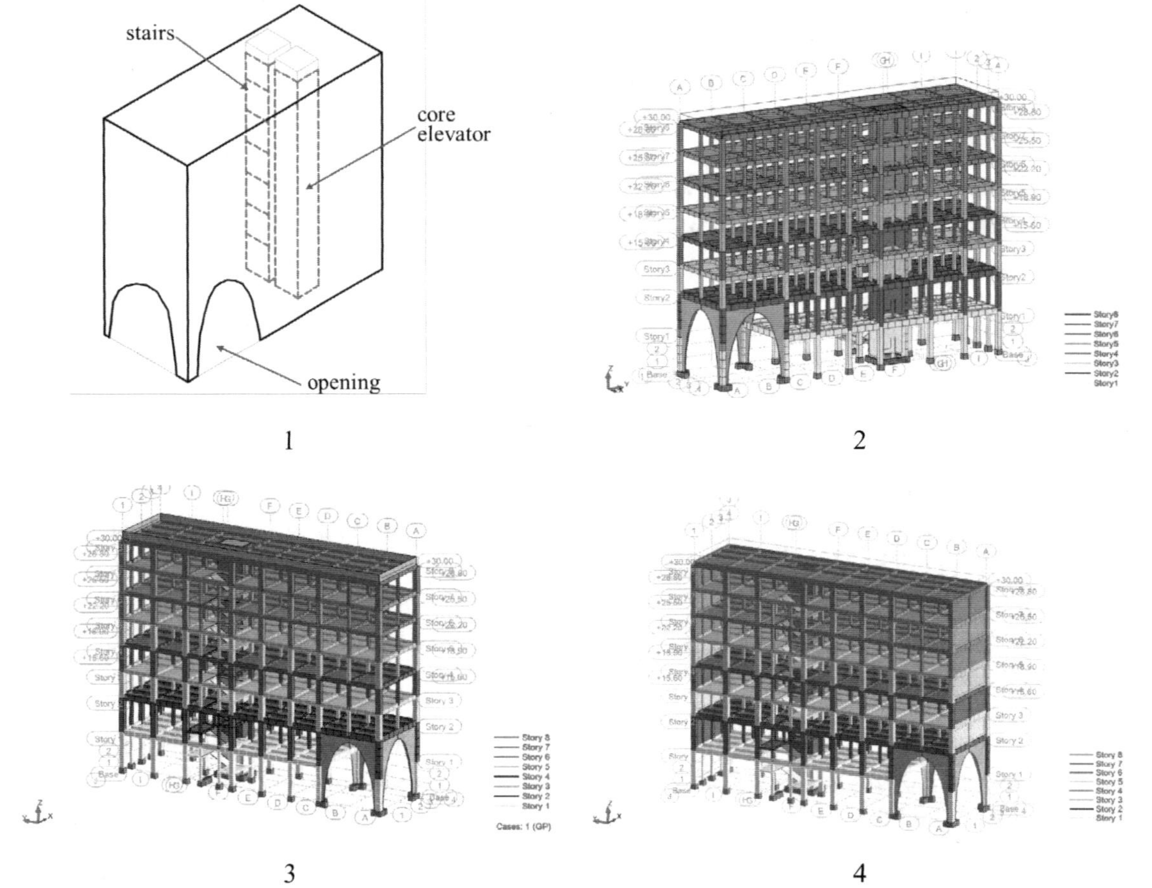

Figure 4. Structural model concept and scenario: 1 – the whole building; 2 – Scenario 1, structure with 50x50 cm columns section; 3 - Scenario 2, structure with 60x60 cm columns section; 4 - Scenario 3 with 60x60 columns section and structural walls at the ends

The materials were chosen respecting recommendation in the design rules considering the strength and elastic modulus. These are presented in table 1. All structural members have been designed considering dead and live loads for civil residential buildings. The structural analysis of reinforced concrete building structure with five stories was carried out by using Robot Structural Analysis 2020 software. [10] In table 2 and 3 are presented loads and combinations taking into consideration. Table 4 presents the section dimension for structural elements and the scenario considered.

Table 1. Materials

Element	Material	E (MPa)	Re (MPa)	Displacement evaluation stiffness reduction factor	Strength evaluation stiffness reduction factor
Columns	C35/45	27200	35	0,5E	0,8E
Walls	C35/45	27200	35	0,5E	0,8E
Beams	C35/45	20400	35	0,5E	0,6E
Slabs	C35/45	20400	35	0,5E	0,6E

Table 2. Loads and combinations

Type	Standard	Value (kN/m^2)	Combination	factor
Self-weight	SR EN 1991-1-1:2004	automated	ULS/ SLS/ ACC	1,35/1,00/1,00
Walls weight	SR EN 1991-1-1:2004		ULS/ SLS/ ACC	1,35/1,00/1,00
Permanent from floors	SR EN 1991-1-1:2004	3,5	ULS/ SLS/ ACC	1,35/1,00/1,00
Imposed loads category A and C	SR EN 1991-1-1:2004	1,5 to 2,5	ULS/ SLS/ ACC	1,5/1,05/0,3
Wind	CR-1-1-3/2012	0,6	ULS/ SLS	1,5/0,6
Snow	CR-1-1-4/2012	2,0	ULS/ SLS/ ACC	1,5/1,00/0,2

Table 3. Seismic design characteristics

Earthquake	Importance class	Importance category	Behaviour factor (q)	Acceleration (ag)	Control period $T_c/T_B/T_D$ (s)	Damping (%)
P100-1/2013	III	C	D.C.H. 4,5	0,40g	1.0/0.20/3.00	5

Table 4. Structural analysis scenarios

Structural elements	Scenarios		
	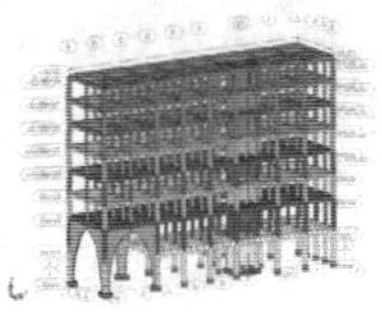 S1	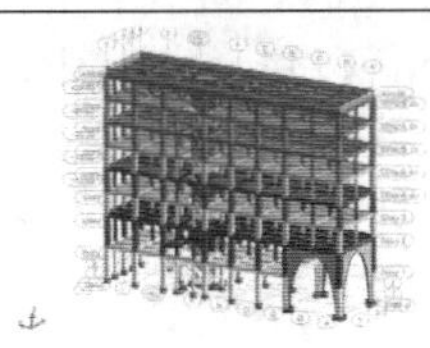 S2	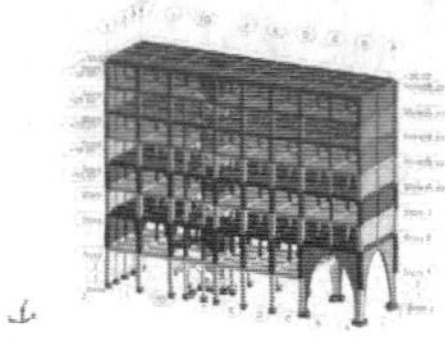 S3
Columns	50x50cm	60x60cm	60x60cm
Beams	30x50cm 30x70cm	30x50cm 30x70cm	30x50cm 30x70cm
Slabs	13cm	13cm	13cm
Shear Walls	NA	NA	20cm

IOP Conf. Series: Materials Science and Engineering 1141 (2021) 012032 doi:10.1088/1757-899X/1141/1/012032

By using the finite element model of asymmetric multi-story reinforced concrete building, the torsional effects are rapidly identified for such type of structure. In seismic areas with high value of peak ground accelerations similar for the present case an optimum vibration of translation in the first two modes of vibration is a demand. In figure 5 and 6 is presented the location of the building site with the peak ground acceleration a_g for design and the control period T_c.

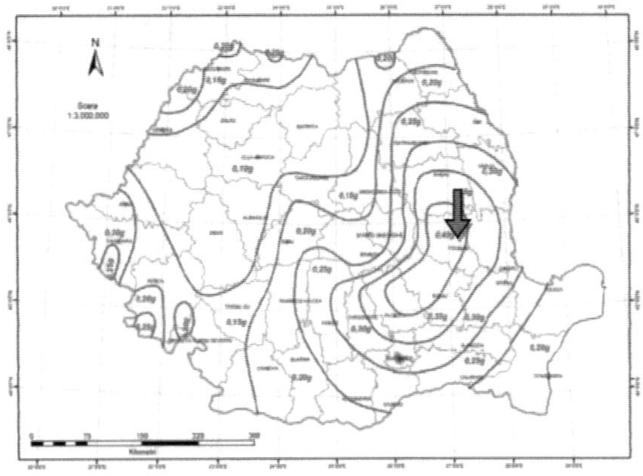

Figure 5. Romania map - Zoning of peak ground acceleration $\mathbf{a_g}$ for design with Mean Recurrence Interval - MRI = 225 years and 20% probability of overtaking in 50 years [6]

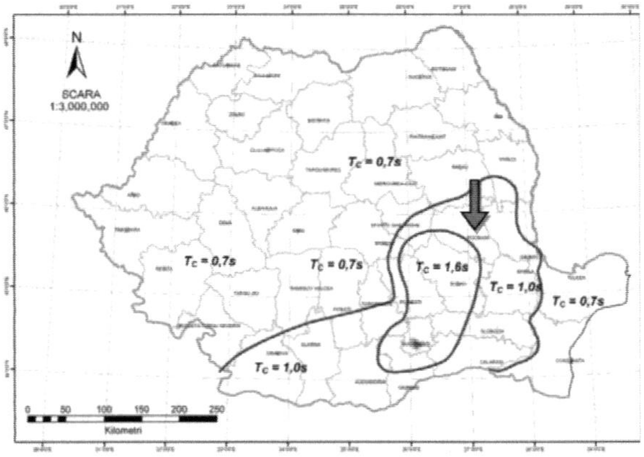

Figure 6. Romania map – distribution in terms of the control period (corner), $\mathbf{T_C}$ of the response spectrum [6]

Obtaining a simple, compact structure, as symmetrical as possible, is the most important objective of the design, because the modelling, calculation, sizing, detailing and execution of simple structures are subject to much lower uncertainties and, therefore, can be imposed on the construction, with a high degree of confidence, the desired seismic behaviour. Since the horizontal action of earthquakes is manifested bidirectionally, the structural elements will be arranged in such geometrical configuration, able to provide sufficient strength and rigidity characteristics in two directions. Structural systems can be different for both directions. In order to fulfil the objectives recommended in the design norms, several scenarios for the distribution of the structural elements were verified. [11], [12], [13]

Open ground story RC frame buildings are common worldwide; they are the dominant set of urban buildings today. The poor performances of such buildings were known from almost a century ago. But there must be compelling reasons (e.g., aesthetics and functionality) other than safety that continues to push the construction of such buildings even today. When glass is used for facades instead of an infill material like brick masonry to the ground story for aesthetics, the building becomes weak in that story. This happens commonly in buildings containing shopping areas and restaurants in their ground story. [14]

3. Results and discussion

In table 5 and 6, the modal mass participating ratios are given for the first six vibrations modes. The deformed shape and modes of vibration of structures analysed structures are shown in figure 7. In the first two scenarios it can be observed that the fundamental period is greater than the control period (1.12s and 1.08s), also in the second mode of vibration the period is near to the control period: this means that is a highly risk possibility for a resonance phenomenon to occur. The reason is that the structure is too flexible. The influence of mass participation and the distribution in the first two modes shows the torsional – translational modes. Also, the deformed shape confirms that.

Table 5. Dynamics characteristics of the structural models considered in the first three modes

Mode of vibration		**Mode 1**			**Mode 2**			**Mode 3**	
Scenario	S1	S2	S3	S1	S2	S3	S1	S2	S3
Frequency (Hz)	0.89	0.93	1.27	1.04	1.09	1.75	1.23	1.29	2.57
Period T (sec)	1.12	1.08	**0.79**	0.96	0.92	0.57	0.81	0.78	0.39
Cur. mas. UX (%)	36.8	38.04	0.01	35.7	34.46	**75.77**	0	0	0.07
Cur. mas. UY (%)	0.2	0.24	**75.52**	0.25	0.26	0.01	74.81	74.92	0.01

Table 6. Dynamics characteristics of the structural models considered in the first three modes

Shape mode		Mode 4			Mode 5			Mode 6	
Scenario	S1	S2	S3	S1	S2	S3	S1	S2	S3
Frequency (Hz)	2.8	3.02	4.22	3.57	3.7	4.99	4.13	4.27	7.33
Period (sec)	0.36	0.33	0.24	0.28	0.27	0.20	0.24	0.23	0.14
Cur. mas. UX (%)	4.29	4.79	0.06	9.6	**9.44**	13.07	0.01	0.01	0
Cur. mas. UY (%)	0.03	0.03	**15.71**	0.04	0.04	0.09	16.19	15.99	0.01

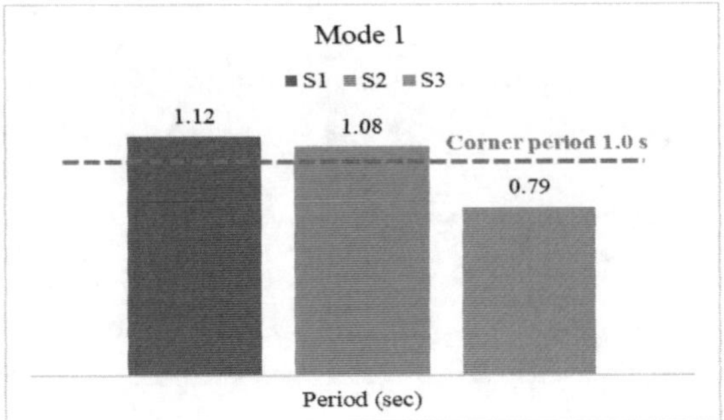

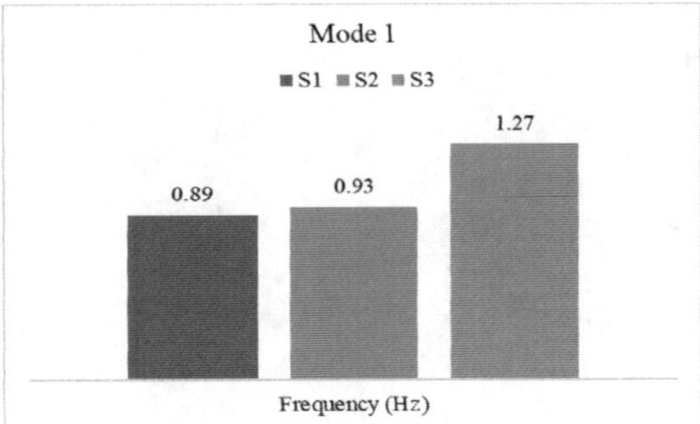

Figure 7. Comparison between frequency and period for the first mode considered

Figure 8. Comparison between mass participation in X direction for the first two modes

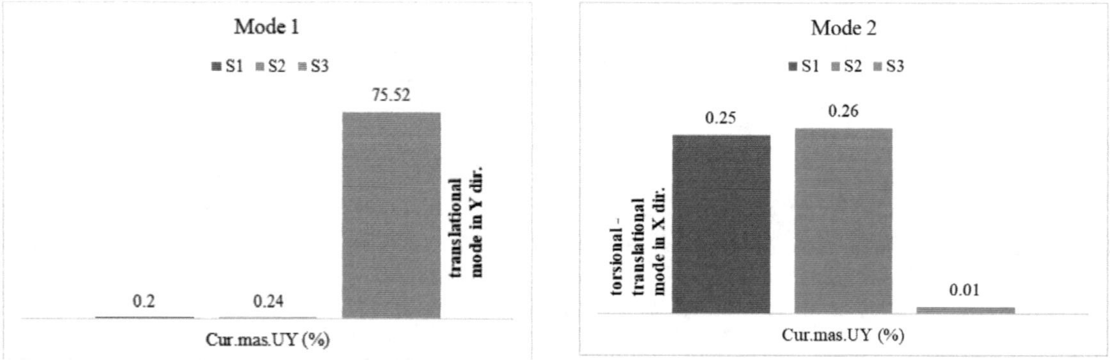

Figure 9. Comparison between mass participation in Y direction for the first two modes

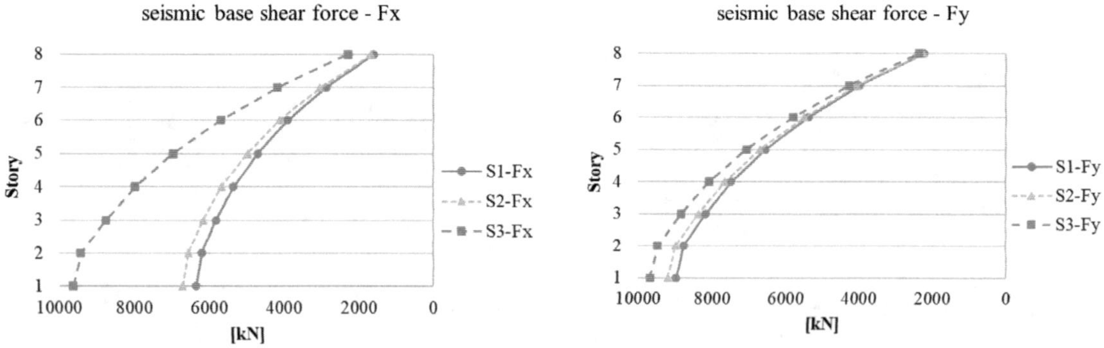

Figure 10. Comparison between seismic base shear distribution on each story

IOP Conf. Series: Materials Science and Engineering 1141 (2021) 012032 doi:10.1088/1757-899X/1141/1/012032

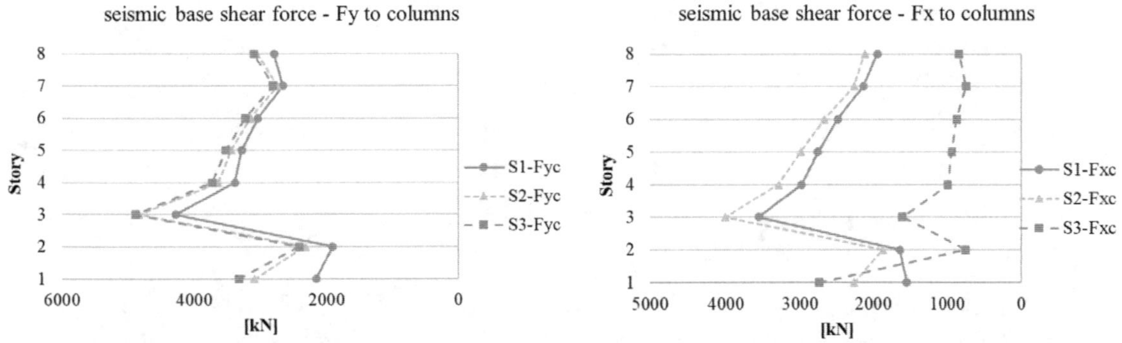

Figure 11. Comparison between seismic base shear distribution to columns on each story

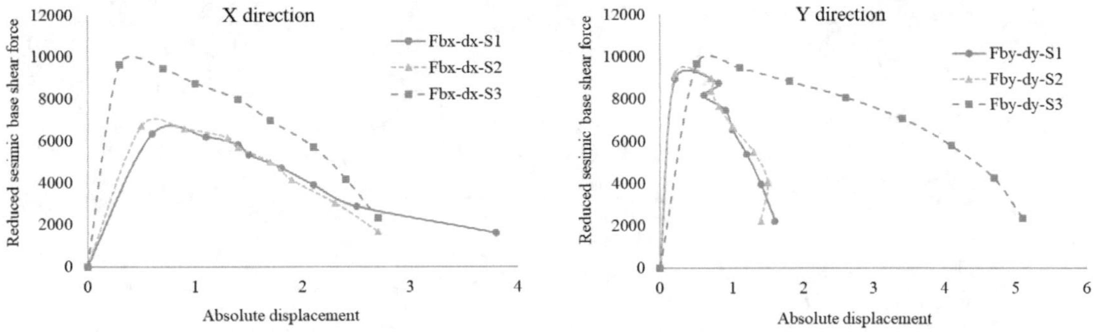

Figure 12. Seismic base shear force – absolute displacement comparison

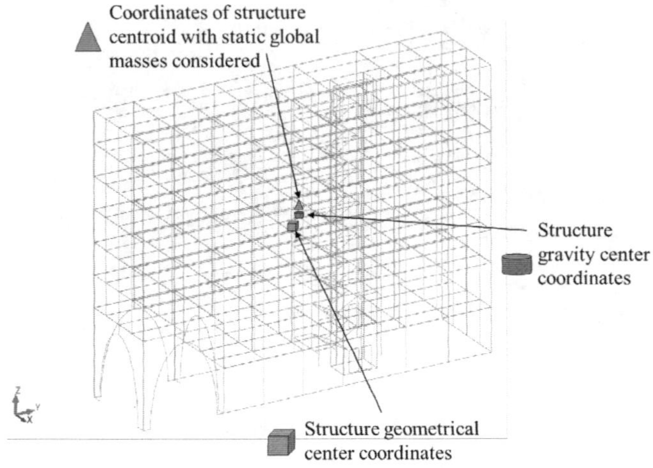

Figure 13. Coordinates of different centres

Table 7. Coordinates of different centres

Case	Center	X (m)	Y (m)	Z (m)
S1		4.50	18.30	15.00
		4.658	18.847	16.081
		4.519	18.839	16.617
S2	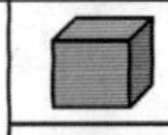	4.50	18.30	15.00
	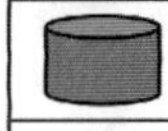	4.65	18.874	15.997
		4.518	18.853	16.561
S3	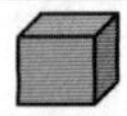	4.50	18.30	15.00
	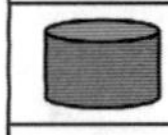	4.638	19.097	16.016
		4.517	18.969	16.547

In the next section figures 14 to 16 presents the deformed shapes of the structure obtained in model analysis.

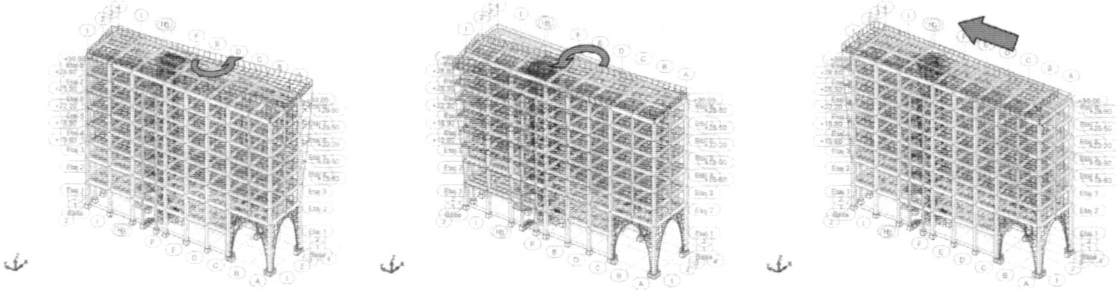

1st mode – torsional – translational X, T=1.12s　　2nd mode – torsional translational X, T=0.96s　　3rd mode – translational Y, T=0.78s

Figure 14. Deformed shape of the structure in case scenario 1

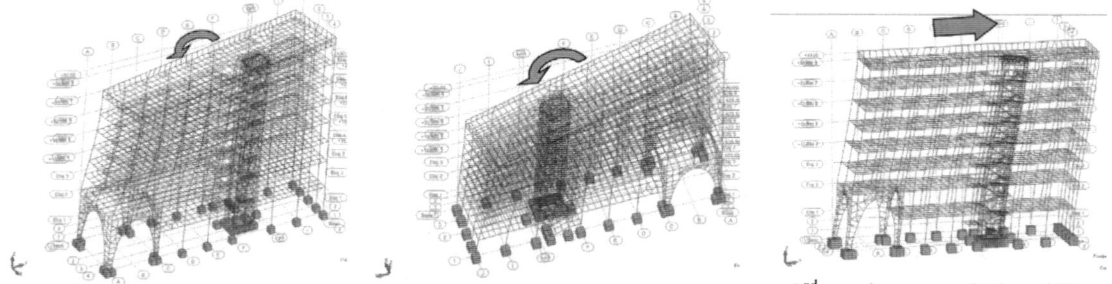

1st mode – torsional – translational X, T=1.08s　　2nd mode – torsional translational X, T=0.92s　　3rd mode – translational Y, T=0.78s

Figure 15. Deformed shape of the structure in case scenario 2

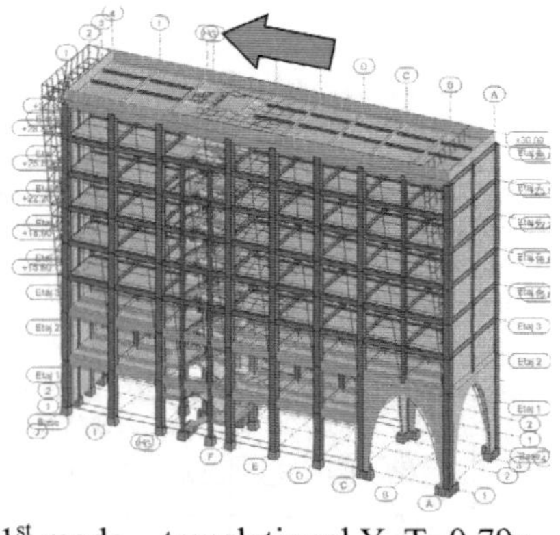

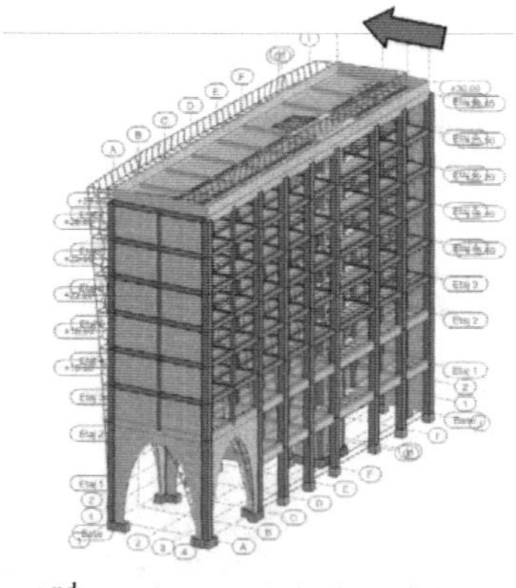

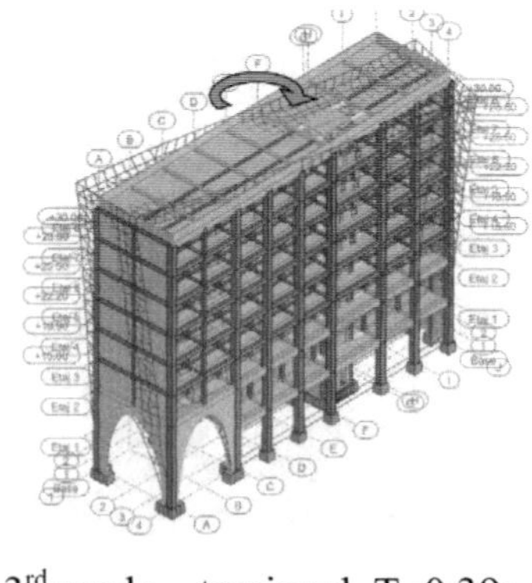

1st mode – translational Y, T=0.79s　　2nd mode –translational X, T=0.57s　　3rd mode – torsional, T=0.39s

Figure 16. Deformed shape of the structure in case scenario 2

Figure 17. Comparison of absolute and relative horizontal displacement obtained in worst loads combination

After comparing the results of the analyses, it can be observed that the considering the elevator core and the stairs, there are substantially influences on the dynamic behaviour of the structure. The appearance of torsional vibration modes in the first two modes is caused by nonsymmetric of the distribution of structural elements in the building plan such as stairs and elevator.

The introduction of structural walls at the ends of the construction balances the structural stiffness and the first two modes are translational with a participation of almost 75% of the modal mass. This confers a greater stiffness to the structure: the structure is stiffer in Scenario 3 by 30% compared to Scenario 1 and by 27% compared to Scenario 2. Also, the distribution of seismic forces depends on the presence of structural walls, almost 70% of the seismic force on each level is taken up by the walls. At the same time, the seismic shear force has an increase of 38% in the X direction (transversal). On the other hand, the configuration in scenario 3 changes the oscillation direction of the structure in longitudinal translation movement, which in the case of the existence of an adjacent building on the transverse direction would influence the size of the seismic joint.

The maximum absolute displacement in X direction (transversal) in Scenario 1 and 2 are bigger with 417% than the structure modelled in Scenario 3. In Y direction (longitudinal), the maximum absolute displacements have close values. In terms of relative displacements in Scenario 3 is an increasing with 25% from 0.6 to 0.8 cm.

4. Conclusions

All buildings are working as vertical cantilevers fixed at the base in terrain. When the earthquakes occur, these structures experience whiplash effects, particularly when the shaking is larger. Buildings designed to be earthquake-resistant have challenging demands. Firstly, buildings become expensive, if they are designed to be indestructible. These must be strong enough to not sustain any structural damage during weak earthquake shaking. Also, they should be stiff enough not to swing too much, even during weak earthquakes. Most important, they should not collapse during the expected strong earthquake shaking to be sustained by them even with significant structural damage. There are two-levels seismic design established explicitly with the two following requirements: [2], [7], [15]

- No-collapse requirement: The structure shall be designed and constructed to withstand the design seismic action without local or global collapse, thus retaining its structural integrity and a residual load bearing capacity after the seismic event.
- Damage limitation requirement: The structure shall be designed and constructed to withstand a seismic action having a larger probability of occurrence than the design seismic action, without the occurrence of damage and the associated limitations of use, the costs of which would be disproportionately high in comparison with the costs of the structure itself.

For real structures a simple configuration is desirable where the first two modes are translational, and the third mode is the torsional mode. When the first mode of vibration is torsional, an unbalanced distribution of seismic forces could occur, which can amplify the stress in certain areas of the structure, especially when subjected to strong earthquakes.

The studied case of a real structure located in the maximum peak ground acceleration area of Romania shows that if there are not taking into consideration correctly the geometry and all the structural components of the building some errors can occur. The modal analysis shows the seismic behaviour so the structure can be rapidly optimized. The asymmetry of the structure and the members distribution led to a problematic behaviour with torsional modes of vibration in the first two modes. Also, the dynamic characteristics of the structure have close values to control period of the terrain which in a probability of an earthquake the resonance phenomenon can occur. These problems can be amended during the design process. To avoid a high level of damage of the structure, there is necessary an optimum design of the structure related to technological demands and financial costs. The solution to introduce new structural elements such as walls on the transversal ends of the building is an acceptable alternative.

References

[1] Ditommaso R, Vona M, Masi A, Mucciarelli M, Ponzo F C 2014 Framed structures: detection of building torsional modes using a simplified experimental approach, *International Journal of Earthquake Engineering*, p 13
[2] Murty C V R, Goswami R, Vijayanarayanan A R, Mehta V V 2012 *Some Concepts in Earthquake Behaviour of Buildings*, Gujarat State Disaster Management Authority
[3] Athanatopoulou A M, Makarios T and Anastassiadis K 2006 Earthquake Analysis of Isotropic Asymmetric Multistory Buildings, *The Structural Design of Tall and Special Buildings*, Wiley Interscience
[4] Shiga T 1963 Torsional Vibrations of Multi-Story Buildings, *Proc. of the Third_World Conference on Earthquake Engineering*, New Zealand
[5] [Online]. https://s01.sgp1.digitaloceanspaces.com/large/803112-article-ipqywhaavs-1454751825.jpeg
[6] P100-1/2013 *Cod de Proiectare Seismica. Partea I — Prevederi de Proiectare Pentru Cladiri. (Seismic Design Code. Part I - Design Provisions for Buildings* 2013 Ministry of Public Works Development and Administration of Romania, MDRAP
[7] Bisch P, Carvalho E, Degee H, Fajfar P, Fardis M, Franchin P, Kreslin M, Pecker A, Pinto P, Plumier A, Somja H and Tsionis G 2012 *Eurocode 8: Seismic Design of Buildings. Worked examples*, EU: Luxembourg: Publications Office of the European Union
[8] Cerbu C, Botis M F 2020 A Method for Reducing of the Overall Torsion for Reinforced Concrete Multi-Story Irregular Structures, *Applied Sciences*, vol **10**, no 5555, p 25
[9] Olteanu I, Canarache R M, Budescu M 2015 Case Study on Vulnerability Increase for a Reinforced Concrete Frame Structure, *Mathematical Modelling in Civil Engineering*, vol **11**, 3, p 38-45
[10] Robot-structural-analysis https://www.autodesk.com/products/robot-structural-analysis/overview?plc=AECCOL&term=1-YEAR&support=ADVANCED&quantity=1. [Accessed 2020].
[11] Păuleț-Crăiniceanu F, Luca S G, Pastia C, Florea V 2019 Improved strategy for optimal control of Civil Engineering structures, *IOP Conference Series: Materials Science and Engineering*, vol **586**
[12] Stratulat M S, Banu O M, Toma A M, Mihai P, Toma I O 2017 Experimental Assessment Techniques for the Dynamic Properties of Concrete, *Advanced Engineering Forum, Trans Tech Publications Ltd*, vol **21**, p 226-233
[13] Toma I O, Petrescu T C, Toma A M, Mihai P, Petcu O A 2020 Evolution of concrete from a traditional material to a next generation sustainable solution, *CPI Concr. Plant Int,* vol **2**, 46-53
[14] Sococol I, Mihai P, Toma I O, VenghiacV M 2020 Influence of concrete strength class on the plastic hinges location for a reinforced concrete moment-resisting frame structure with consideration of the horizontal stiffening effect of the slab, *Bulletin of the Polytechnic Institute of Iasi-Construction & Architecture Section*, vol **70**, 2

[15] Tuna Z 2012 Seismic Performance, Modelling, and Failure Assessment of Reinforced Concrete Shear Wall Buildings, Peer reviewed | Thesis/dissertation, *UCLA Electronic Theses and Dissertations*

FEM analysis of a floor structural system made of thin-walled cold-formed steel profiles

G Ţăranu[1], I-O Toma[2]
[1, 2] Technical University Gheorghe Asachi of Iasi, Faculty of Civil Engineering and Building Services, Department of Structural Mechanics, Iasi, D. Mangeron Street, no. 1, 700050, *Romania*,

E-mail: george.taranu@academic.tuiasi.ro

Abstract. The paper presents the results of a nonlinear FEM analysis of a floor system made of plane truss beams. The steel profiles are made of U-shaped thin walled cold-formed stell. Experimental data were used to model the stiffness of the joints. To observe the structural behaviour of the floor system, nonlinear constraints were defined as a parabolic function considering previous experimental data.

1. Introduction

Cold-formed steel (CFS) profiles as structural elements and systems are widely used in the building construction industry. Truss-made floors, stud walls, modular building panels, purlins and even as a load-bearing structure in low - to mid-rise buildings use CFS profiles. An example of a house with a structure made of cold formed steel profiles is presented in figure 1. Compared to their hot-rolled counterparts, CFS members are often found to be more economical and efficient, due to inherent advantages such as light weight, ease and speed of erection and a greater flexibility in manufacturing cross-sectional profiles and sizes. Many cold-rolling companies can customise different roll sections on demand. With the versatility of the production, the problem of finding the optimum cross-sectional shape is an important objective for the engineers [1], [2].

The behaviour of this type of elements must be correctly understood and future projects well-designed because of the joints, which in general, are hinges. Some experiments on joints and structural elements reveal that the weakest part of the system is the connection between elements [3], [4]. Furthermore, a large number of experimental studies found that the damage of cold-formed thin-walled steel members is mainly caused by the tilting, shearing and pull-out of screws [5], [6]. The connection between cold-formed thin-walled steel members needs to be tested, verified and improved. [7].

An alternative for screws is the self-piercing rivet (SPR) connection process also called semi hollow rivet self-piercing riveting, which is commonly used in the fields of automobile and mechanical engineering [7], [8], [9]. In the case of self-piercing rivet (SPR) connection there are some disadvantages regarding the precision of the hole position of the mounted parts and the amount of time. The connections between parts in floor truss systems based on 2 or 4 screws positioned in an imprint alveolus. The experimental results obtained on a few types of joints were used to calibrate the finite element model. This paper presents the results of an incremental non-linear analysis of a truss floor structural system considering non-linear joints using experimental data results on 90° joint connection tests. The model was analysed with Robot Structural Analysis [10] in order to provide a simple approach for the current design process.

Figure 1. Cold formed steel profile structural system of a house

2. Materials

The structure of the floors is made of cold formed steel profiles type C12/41/89/41/12x1, as is presented in figure 2, produced from DX51D + Z sheet with a thickness of 1mm. The specific elongation of steel is 35.4 %. The mechanical characteristics of the profile and the material properties are presented in table 1 and 2. The cold formed steel profiles are thin-walled profiles with a C-shaped section and are provided with holes and imprinted alveola for screw assembly.

The joints between the profiles for making the subassemblies are made with self-tapping screws Ø4.8x16 (DIN 7981) or GRABBERGARD screws type GFS 4.8x19 R as presented in figure 3. The clamps of the subassemblies are made with screws type Ø5.5x25 (DIN 7504N) or DRIVALL® FP102275LYZ # 10 19 mm. At the top of the beams there is mounted an 18 mm thick OSB board with mechanical fastening at 200 mm distance.

Table 1. Section properties

Section	Section mechanical properties		
	A_x (mm^2)	I_y (mm^4)	I_z (mm^4)
C 89x41x1	189	243100	45600

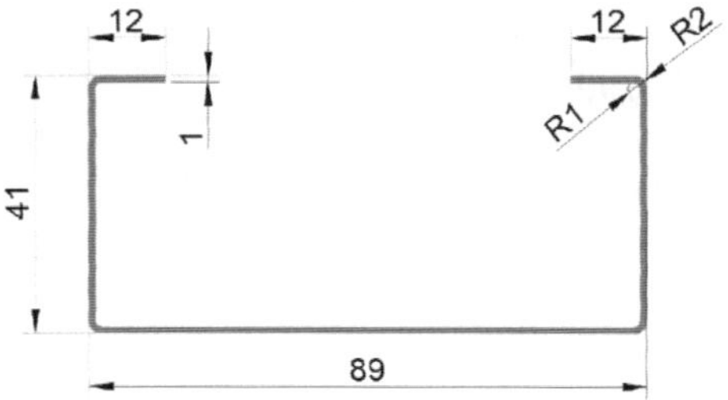

Figure 2. Section dimension (mm)

Table 2. Material properties

Material	Elastic modulus – E (N/mm^2)	Shear modulus – G (N/mm^2)	Tensile strength (N/mm^2)
steel DX51D+Z	150000	45000	140

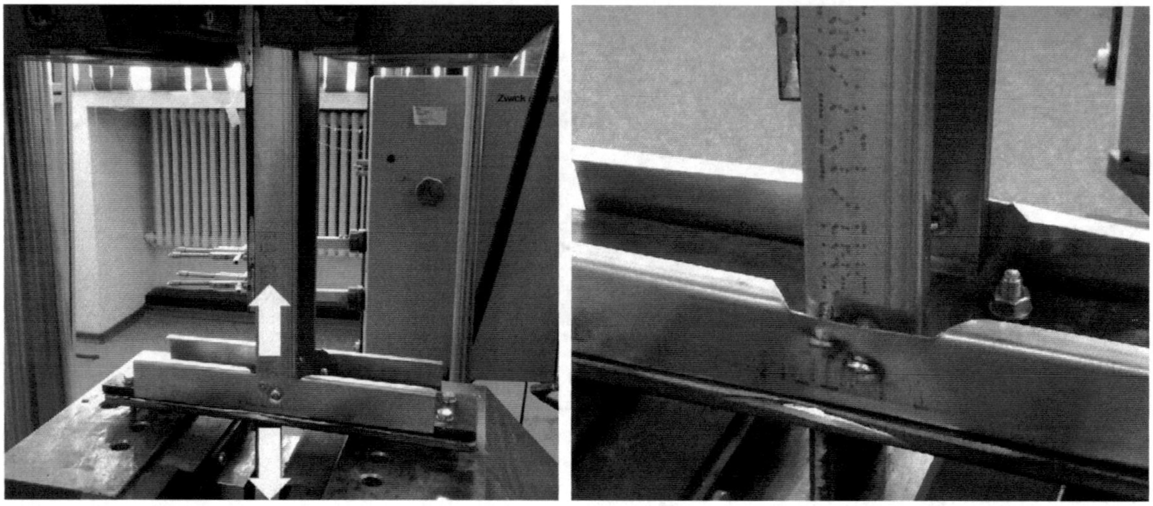

Figure 3. Experimental testing of 90º joint connection with two self-tapping screws on lateral sides

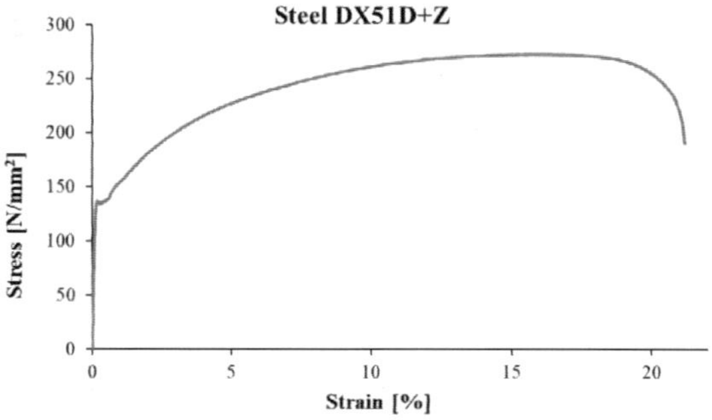

Figure 4. Tensile stress strain experimental test curve for steel DX51D+Z

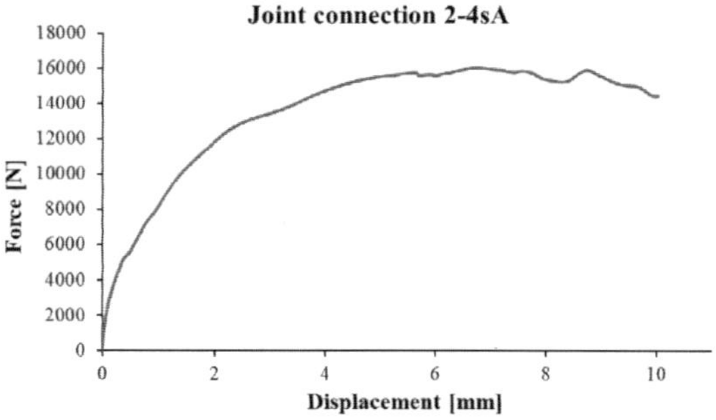

Figure 5. Force displacement joint connection experimental test curve

In order to calibrate the numerical model for joints with a relation of the stiffness of a bolted joint between cold-formed elements, experimental studies were performed [11].

3. FEM analyses

Numerical modelling and, in particular, the non-linear analysis of cold-formed steel profiles using finite elements as bars or shells has been well studied. [12], [13]. The non-linear analysis (including the effects of residual stresses and imperfections) capturing the local, distortional and overall buckling behaviours (and their interactions), as well as the modelling of the local distortions at the anchorages and connections of the CFS members under axial and lateral loading are well established [3], [14].

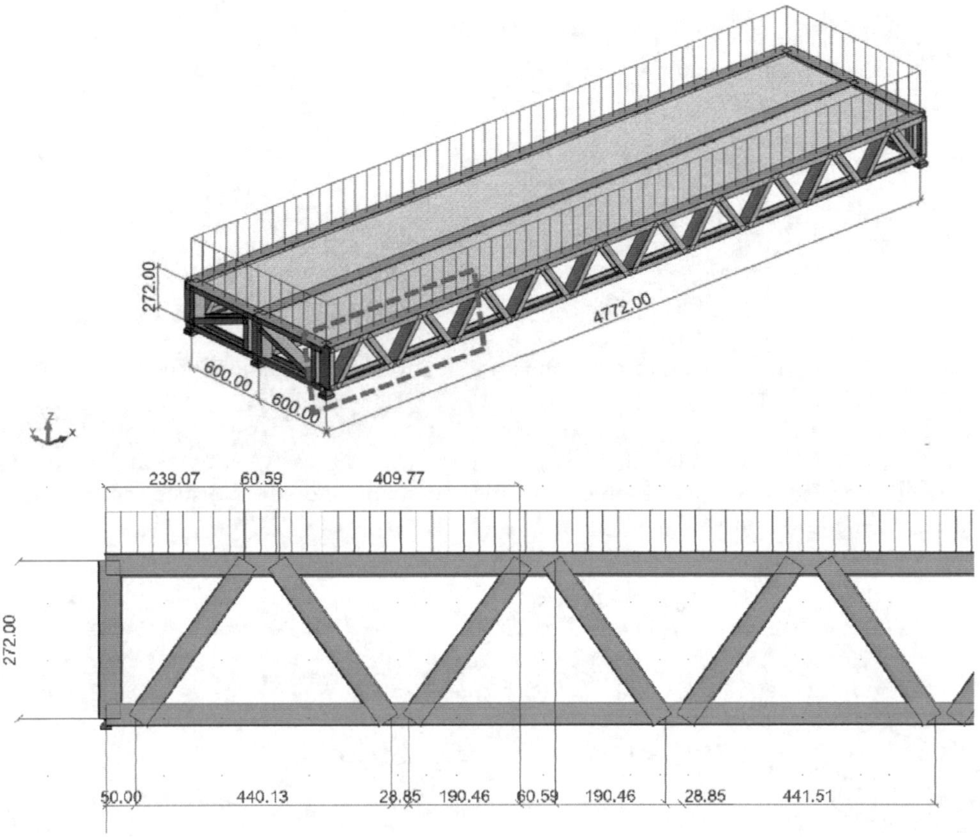

Figure 6. Model geometry.

Robot Structural Analysis finite element software was used to analyse the structural model. Finite element bars are 236 and 312 nodes. The material is considered linear elastic based on the modulus of elasticity value presented in table 2. Figure 6 presents the model geometry and the load distribution in figure 7. The load was uniformly distributed on claddings on top of the trusses.

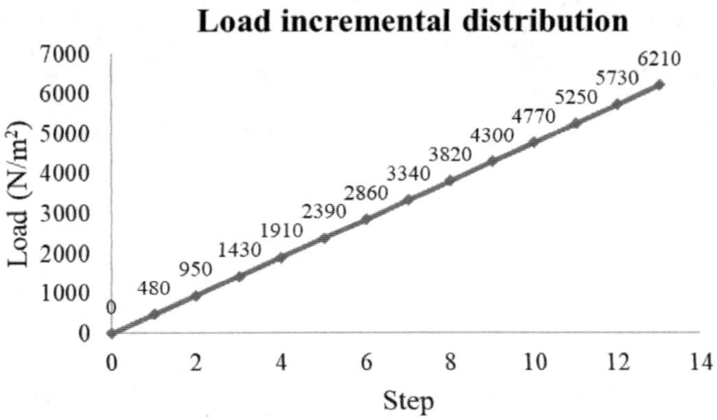

Figure 7. Load distribution on the structural model

In order to obtain a behaviour close to the real one in the FEM software, the nonlinear constraints were defined considering the previous experimental data. A parabolic function was defined considering the force – displacement curve presented in figure 5 and 8.

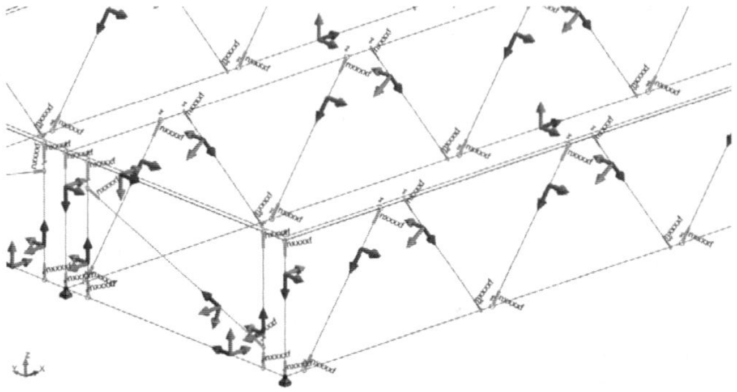

Figure 8. Definition of the joint connection with axial stiffness

4. Results and discussion

The results consist of maximum tensile stresses, maximum compressive stresses and maximum vertical displacements are also presented in table 3. Deformed shape and stress distribution on the structural model are presented in figure 6.

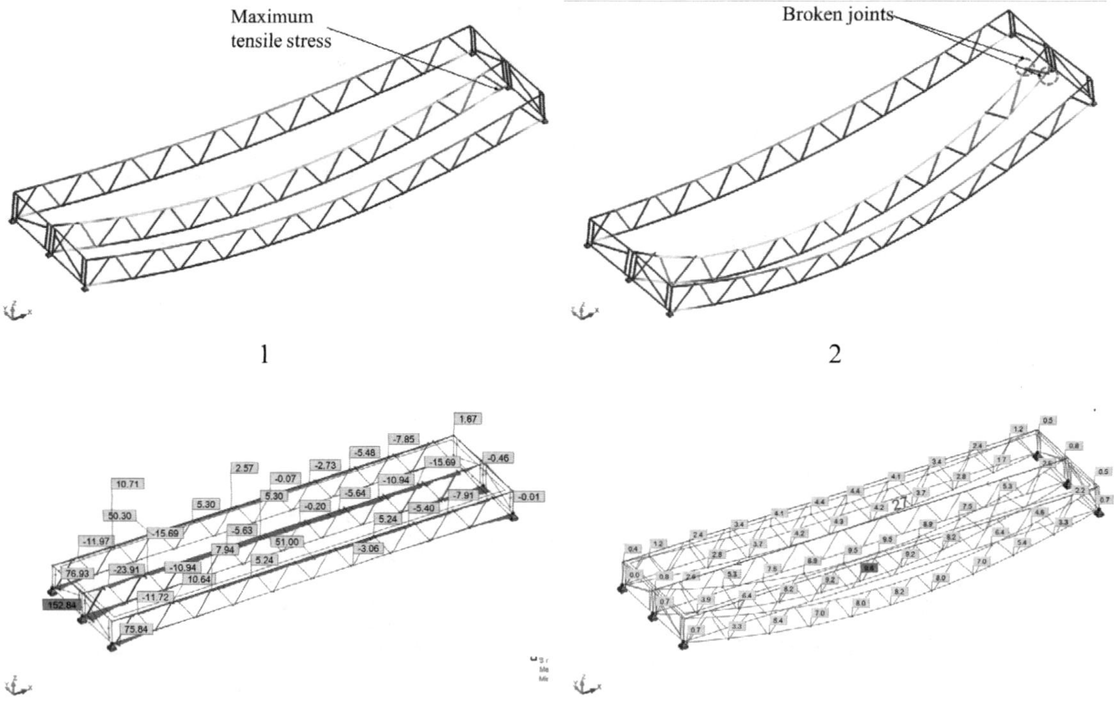

Figure 9. Results of the static analysis: 1 – maximum tensile stresses in case 5; 2 – the joint connection that yields first; 3 – Tensile normal maximum stress diagram; 4 – deformation diagrams in case 5

Comparing the results obtained from the analyses that are presented in table 6 and expressing them graphically in the figure 10, it is observed that at a maximum value of the vertical displacement of 9.6 mm, the maximum stress that appears is 152.84 N / mm^2. At the same time, the diagonal of the truss next to the support is out of the strength capacity range.

Table 3. Load distribution and static analysis results

Load step	Load Value (N/m^2)	Maximum tensile stress (N/mm^2)	Maximum compressive stress (N/mm^2)	Maximum vertical displacement (mm)
1	480	31.04	15.4	1.6
2	950	61.09	29.95	3.3
3	1430	91.55	45.09	5.2
4	1910	122.48	60.22	7.3
5	2390	152.84	75.39	9.6
6	2860	180.4	90.18	12
7	3340	208.3	105.29	16
8	3820	237.82	120.46	20.3
9	4300	266.9	135.57	25.1
10	4770	294.85	150.35	30.4
11	5250	323.35	165.44	36.1
12	5730	347.56	180.61	42.6
13	6210	368.54	195.72	52.6

As it can be seen in the graphical presentation of the results in figure 10, the linear elastic behaviour is marked. After the breaking point of the joint connection caused by the shear stresses, the displacements increase more compared to the rate of increase of force.

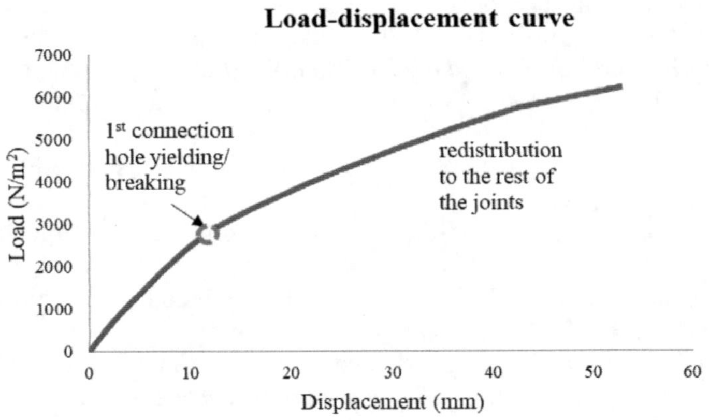

Figure 10. Load - displacement curve obtained from the numerical model

5. Conclusions

Structural systems made of U-shaped thin walled cold-formed profiles are used a lot lately due to their advantages such as low cost, low own weight, fast assembly through a system of prefabricated elements and accessible technological conditions. The structural floor systems are made of trusses. The profiles are obtained in workshops by bending and cold rolling sheet metal strips. In the forming process, the component elements of the beams are cut, stamped and drilled. Self-tapping screws are used in the holes with the stamped socket for mounting. Although the tensile strength of the material is usually up to 275 N/mm2, failure occurs most often by shearing into the screw holes.

The analysis presented in this article aims to identify a nonlinear calculation approximation model for truss structural floor systems. Modelling a nonlinear behaviour of joints between finite elements is a reasonable approximation to be used in the current process of designing these types of structures. The definition of an axial stiffness with a parabolic function type proved to obtain good results. These results must be validated by more experimental analysis of truss-made floors. The analysis shows the deformations of the model and the maximum tensile stresses that can appear near to the ends of the trusses, close to the supports. However, there is a disadvantage in this approach, that is the phenomenon of loss of stability of the compressed bars being unavailable. A superior calculation model in terms of quality must be achieved by considering a curve of material curve behaviour and its calibration with experimental results at scale.

6. References

[1] Ye J, Hajirasouliha I, Becque J, Eslami A 2016 Optimum design of cold-formed steel beams using Particle Swarm Optimisation method, *Journal of Constructional Steel Research*, vol **122**, pp 80-93

[2] Vatin N I, Havula J, Martikainen L, Sinelnikov A, Orlova A, Salamakhin S 2014 Thin-walled cross-sections and their joints: tests and FEM-modelling, *Advanced Materials Research*, pp 945-949

[3] Nithyadharan M, Kalyanaraman V 2021 A new screw connection model and FEA of CFS shear wall panels, *Journal of Constructional Steel Research*, vol **176**

[4] Cao K 2017 Research on Joint and Structural Behavior of Column-Supported Modular Steel Buildings, Doctoral dissertation, Tongji University

[5] Lacey A W, Chen W, Hao H, Bi K 2019 New interlocking inter-module connection for modular steel buildings: Experimental and numerical studies, *Engineering Structures*, vol **198**

[6] Henriques J, Rosa N, Gervasio H, Santos P, Simões da Silva L 2017 Structural performance of light steel framing panels using screw connections subjected to lateral loading, *Thin-Walled Structures*, vol **121**, pp 67-88

[7] Song L, Yan W, Yu C, Xie Z, Tan Q 2019 Flexural behavior investigation of the CFS truss beams with self-piercing, *Journal of Constructional Steel Research*, vol **156**, pp 28-45

[8] Tusnin A 2016, Finite Element for Calculation of Structures Made of Thin-Walled Open Profile Rods, *Procedia Engineering*, vol **150**, pp 1673-1679

[9] Eurocode 3: Design of steel structures - part 1–8: design of joints 2005 *European Committee for Standardization*, Brussels

[10] [online] Available: https://www.autodesk.com/products/robot-structural-analysis/overview [Accessed 2021]

[11] Zaharia R and Dubina D 2006 Stiffness of joints in bolted connected cold-formed steel, *Journal of Constructional Steel Research*, vol **62**, pp 240-249

[12] Avery P, Mahendran M 2000 Distributed plasticity analysis of steel frame structures comprising non-compact sections, *Engineering Structures*, vol **22**

[13] Usefi N, Sharafi P, Ronagh H 2019 Numerical models for lateral behaviour analysis of cold-formed steel framed walls: state of the art, evaluation and challenges, *Thin Walled Structures*, vol **138**, pp 252-285

[14] Vieira Jr L C M, Schafer B W 2016 Fastener demands for sheathing-braced cold-formed steel stud gravity walls, *Lat. Am. j. solids struct.*, vol **13**, no 6, pp 1167-1185

Performance analysis for an improved strategy in optimal control of civil engineering structures

F Păuleț-Crăiniceanu, V Florea, C Pastia, S G Luca and **O V Roșca**
"Gheorghe Asachi" Technical University of Iași, 1 Mangeron Blvd., *Romania*

E-mail: fideliu.paulet-crainiceanu@academic.tuiasi.ro

Abstract. Civil Engineering structures are already an important part in the Smart Cities concept. In connection to this, integration of new technologies is under way in the field of Structural Engineering as in all Civil and Buildings/Bridges Engineering. One of the main goals is to protect structures from possible harm from earthquakes. In order to achieve this goal, Optimal Active Control has been proposed and used with a high degree of success. However, the application strategies are continuously improving. A strategy consisting in simplifying the Optimal Active control for full states has been proposed by the authors and relatively recently modified for an increased efficiency. That was a theoretical research taking into account the possibility to reduce the order of magnitude of matrices involved in computations. The present paper is assuming the previous theoretical result and simulates its application to the structural model of a ground floor plus ten floors building specially designed for this purpose. The reinforced concrete frame building's behaviour is analysed under seismic actions. Efficiency and the control strategy are observed in different scenarios. The structural response is revealing that the new strategy is working faster in limiting the effects of earthquake input. In addition, the reduced computational time leads to a faster control and smaller phase-shift in practical use.

1. Introduction

The actual human society is relying more and more on technological advances that must lead to an improved assurance of community safety and functionality. In the case of the Civil Engineering domain, the population is expecting to use buildings and bridges with the highest standards of strength and comfort even during and after very strong natural or anthropic actions.

For the earthquake effects mitigation, a solution is to provide flexible constructions with mechanical and computational means that prevent damages and losses while permitting the functionality to be uninterrupted. In this line, the use of passive and active methods has been intensively studied and, more or less experimentally, applied during the last decades [1, 2].

Though the cost of using structural control is still high, the last years have shown that the price of many electronics components – as sensors, communication devices, processors – and advanced mechanical components – as Tuned Mass Dampers (TMDs) or Active Mass Dampers (AMDs) – has dropped dramatically while their quality has been increased. This obvious trend is leading to keep researches on the topic [3, 4].

One of the problems in the case of active structural control applied to structures under strong seismic input is to get better methodologies leading to better performances, especially for large structures with many devices, e.g. [5, 6], where computational time is critical.

In a relatively recent work, [7], the active structural method developed in [5] and shown in detail in [6] was theoretically revised and improved. The improvement is based on considering the particular case of proposed control matrices and the calculus of the gain matrix that contain some zero submatrices. That was leading to the conclusion of solving two Riccati matrix equations (one for symmetrical and one for non-symmetrical solutions) instead of a double-dimensional Riccati symmetrical matrix equation.

This scale down of the main problem is now debated in terms of a numerical study taking into focus a ground floor plus ten stories flexible building, designed for this study.

2. Theoretical background

To implement active structural control in Civil Engineering using Active Mass Dampers (AMDs), the full state Linear Quadratic Regulator method was used [5] with some simplifications [6].

In this method, the central point from computational point of view is to solve the *2n*-dimensional Riccati matrix equation

$$\mathbf{PA} - \mathbf{PBL}_1\mathbf{R}^{-1}\mathbf{L}_1'\mathbf{B}'\mathbf{P} + \mathbf{A}'\mathbf{P} + \mathbf{Q} = \mathbf{0} \quad (1)$$

where $\mathbf{P}$ is the unknown $2n \times 2n$ matrix and where n is the number of dynamic degrees of freedom involved by the structure in the classical Structural Dynamics description.

In equation (1), the matrices $\mathbf{A}$ and $\mathbf{B}$ are

$$\mathbf{A} = \begin{bmatrix} \mathbf{0} & \mathbf{I} \\ -\mathbf{M}_s^{-1}\mathbf{K}_s & -\mathbf{M}_s^{-1}\mathbf{C}_s \end{bmatrix}, \quad \mathbf{B} = \begin{bmatrix} \mathbf{0} \\ \mathbf{M}_s^{-1} \end{bmatrix} \quad (2)$$

where $\mathbf{M}_S$ is the $n \times n$ structural mass matrix, $\mathbf{C}_S$ is the $n \times n$ structural damping matrix and $\mathbf{K}_S$ is the $n \times n$ structural stiffness matrix.

In addition, in equation (1) the $2n \times 2n$ weighting matrix $\mathbf{Q}$ is imposed as

$$\mathbf{Q} = \begin{bmatrix} \mathbf{K}_S & 0 \\ 0 & \mathbf{M}_S \end{bmatrix} \quad (3)$$

and the $m \times m$ weighting matrix $\mathbf{R}$ as

$$\mathbf{R} = diag\{r_1 \ \cdots \ r_m\} \quad (4)$$

where m is the number of active devices, AMDs in this application. For a simpler implementation, the matrix $\mathbf{R}$ is iteratively obtained through trials on only one scalar r on its diagonal

$$\mathbf{R} = r\mathbf{I} \quad (5)$$

and $\mathbf{I}$ is the identity matrix having $m \times m$ dimension.

To finish the description of equation (1), it should be noted that $\mathbf{L}_1$ is an $n \times m$ matrix showing the distribution of active forces on structural degrees of freedom.

In this paper, the full states control low is implemented as

$$\mathbf{u} = \mathbf{L}_1\mathbf{Kx} \quad (6)$$

where $\mathbf{u}$ is the $n \times 1$ control forces vector, time dependent, and $\mathbf{K}$ is the $m \times 2n$ control (gain) matrix, with $\mathbf{K} = \mathbf{R}^{-1}\mathbf{L}_1'\mathbf{B}'\mathbf{P}$.

As shown in [7], solving the *2n*-dimensional Riccati equation (1) for a symmetric solution $\mathbf{P}$ might be replaced with solving two n-dimensional Riccati equations:

$$\mathbf{P}_{12}\mathbf{A}_{21} - \mathbf{P}_{12}\mathbf{B}_2\mathbf{R}^{-1}\mathbf{B}_2'\mathbf{P}_{21} + \mathbf{A'}_{21}\mathbf{P}_{21} + \mathbf{Q}_1 = \mathbf{0} \quad (7)$$

and

$$\mathbf{P}_{21} + \mathbf{P}_{22}\mathbf{A}_{22} - \mathbf{P}_{22}\mathbf{B}_2\mathbf{R}^{-1}\mathbf{B}_2'\mathbf{P}_{22} + \mathbf{P}_{12} + \mathbf{A'}_{22}\mathbf{P}_{22} + \mathbf{Q}_2 = \mathbf{0} \quad (8)$$

where notations are from decomposition of the $2n \times 2n$ matrix $\mathbf{P}$, the symmetric solution of equation (1), into four $n \times n$ submatrices

$$\mathbf{P} = \begin{bmatrix} \mathbf{P}_{11} & \mathbf{P}_{12} \\ \mathbf{P}_{21} & \mathbf{P}_{22} \end{bmatrix} \tag{9}$$

also from decomposing $2n \times 2n$ matrix $\mathbf{Q}$ in halves

$$\mathbf{Q} = \begin{bmatrix} \mathbf{Q}_1 & 0 \\ 0 & \mathbf{Q}_2 \end{bmatrix} \tag{10}$$

and, similarly, decomposing

$$\mathbf{A} = \begin{bmatrix} \mathbf{0} & \mathbf{I} \\ \mathbf{A}_{21} & \mathbf{A}_{22} \end{bmatrix}, \quad \mathbf{BL}_1 = \begin{bmatrix} \mathbf{0} \\ \mathbf{M}_s^{-1}\mathbf{L}_1 \end{bmatrix} = \begin{bmatrix} \mathbf{0} \\ \mathbf{B}_2 \end{bmatrix} \tag{11}$$

Knowing that $\mathbf{P}_{12} = \mathbf{P}'_{21}$, the first equation of the two equations above, (7) and (8), is solved for $\mathbf{P}_{12}$ and the result is used in equation (8). While equation (7) has a non-symmetric solution, equation (8)'s solution, $\mathbf{P}_{22}$, is symmetric.

In this paper, the theoretical approach showing that the gain matrix, $\mathbf{K}$, can be obtained using only n-dimensional equations and is validated through a numerical case study in the next section. However, it must be noted that aspects of the efficiency in computation are rising news challenges for future studies.

3. Numerical simulation and performance analysis

In order to numerically validate the above theoretical strategy, the goal of simulating the active control is eleven stories building designed for this case. As computational tool, a personal computer running Matlab [8] was employed. Results are shown the validity and the interesting aspects concerning the proposed strategy.

The seismic action used for this application is the north-south component of the acceleration recorded during the 1977 Vrancea, Romania, earthquake, shown in Figure 1.

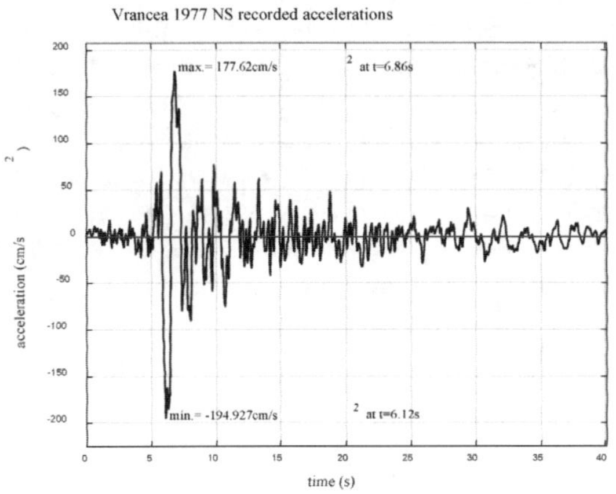

Figure 1. The seismic input used by numerical simulation.

3.1. *The structure involved in simulation*

For achieving the goals of this study (numerical validation and the analysis for computational performance of the strategy proposed in [7]), the structural matrices from a finite element method

(FEM) model shown in Figure 2 were imposed. This structure has a ground floor plus ten floors, the last one being a smaller, technical one.

The structure is composed from 206 nodes, first 20 nodes being the connection to foundations. Therefore, the number of equations generated for this model is 1116. A number of 472 beams are also part of the model. The number of lumped masses is 372, equal to the number of dynamic degrees of freedom that constitute also the dimension of the classic structural dynamic's matrix problem. Adding the dynamics of the eight TMDs / AMDs will increase the size of the problem to 380.

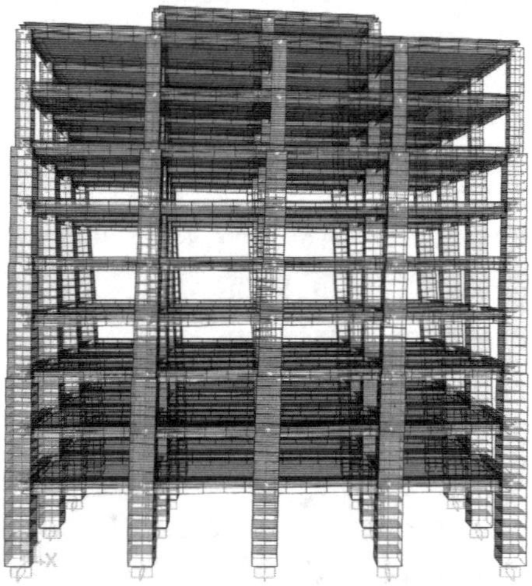

Figure 2. Structural model used in numerical simulation, during vibration.

In order to serve the present work, the structure is a flexible one with the first period of vibration equal to T=1.301s. The proposed control system is strengthening the structure.

3.2. TMDs and AMDs characteristics

The numerical simulation employed a number of eight identically TMDs. They are located in a number of four at levels 7 and 9 over the ground floor. Each of these TMDs are characterised by the 4-t mass, 100 kN/m stiffness and 10% damping, resulting in a tuned period of vibration 1.257 s, closer to 1.301 s, [8], the first structural natural period of vibration. Masses of the TMDs total 0.38% of the building's mass. The action of these TMDs is along the X axis direction of the building model. The case of using TMDs is denoted as the "Passive control" case of computer runs in this paper.

Later, the TMDs are transformed in AMDs, involving actuators that can move the masses conforming to the control law calculated by equation (1) and referred as "Control 1". Then, the computations were under the control law implying the solution proposed in [7], and shown by equations (7) and (8), situation labelled as "Control 2" in this study. The TMDs and corresponding AMDs are numbered from 1 to 8, first four being located at the level 7 and the others at the level 9 above the ground floor.

3.3. Programming the problem

In order to simulate all the planned situations ("No control", "Passive control", "Control 1" and "Control 2"), Matlab computer software was run [9].

The "No control" program was run in order to study the structural characteristics of the structure and to obtain the corresponding frequency response and seismic time response to Vrancea 1977 NS component earthquake. In addition, comparisons of the two types of responses have been done. Comparisons for each level frequency response were also performed. Responses were calculated in terms of displacements, velocities and accelerations.

A second computer program was targeting the "Passive control" scenario, when the 8 tuned masses were placed on the structure, acting on x direction. The very same time of responses as in the case of "No control" scenario had been issued. Comparisons "No control" versus "Passive control" were also done.

The third Matlab program (referring to the "Control 1" scenario) is concerned with the Linear Quadratic Regulator involving the solution of the equation described in equation (1), i.e. using *2n*-dimensional matrices. In this case, two ways to calculate the solution of the Riccati equation (1) had been employed:

- using the *lqr* Matlab function (with calculus of the gain matrix) that is based on a general solution method and implying a longer time in computation
- using the *Riccati* function for a symmetric *2n*-dimensional solution obtained from *MATLAB Central File Exchange* [10], shown in the Appendix.

For the "Control 2" scenario, a Matlab program was created for solving the two equations, (7) and (8). These equations were solved through:

- using *Riccati* function for symmetrical solution (n-dimensional) [10], for equation (7)
- using *Solve_BQMEs* function, shown in Appendix, for a non-symmetrical solution (*n*-dimensional) of equation (8). This function is adapted from the original one from *MATLAB Central File Exchange*, [11].

3.4. Computational means

The computer used in this study was a personal computer with the next main computational characteristics: Intel I5-7400 processor clocked at 3 GHz with 4 cores; 32 GB RAM; 1TB SSD; Radeon RX580 video card with 8GB GDDR5 RAM.

For obtaining the results, the runs were done under the "same" (as possible) conditions, i.e. without the interference of other programs or internet connection.

3.5. Results and comments

From the runs, a series of selected results show the quality of the control strategy (with better responses compared to passive and non-controlled cases) together with the comparisons between the "Control 1" and "Control 2" scenarios.

The responses in the two control situations are practically identical in all cases (time responses, responses to harmonic loads and maximum responses).

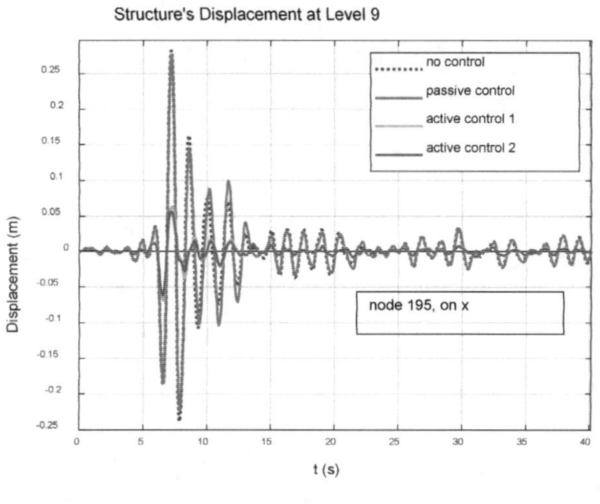

Figure 3. Level 9's displacement on the four scenarios.

Figures 3 and 4 prove the above statement. There, the seismic time response under Vrancea NS 1977 earthquake in terms of displacement and acceleration for the level 9 of the building is shown in the four scenarios taken into views of this study. In addition, the cases without devices ("No control") and with TMDs ("Passive case") are presented.

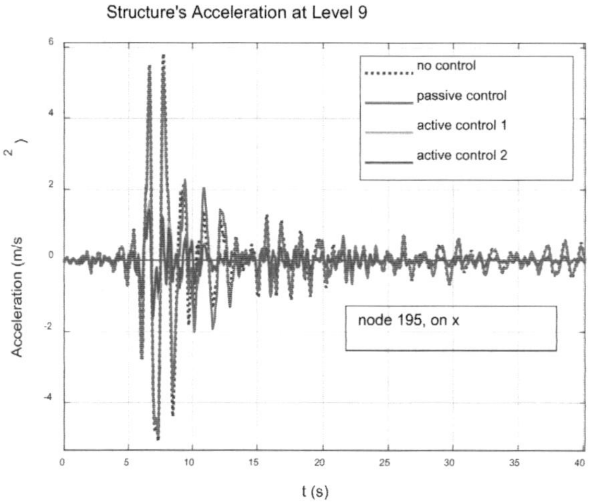

Figure 4. Level 9's acceleration on the four scenarios.

In addition, it can be seen that the passive case is not too effective. This is because the number of devices is not enough to lead to better response. Adding more devices would have increased the total additional mass, creating an unfavourable situation. It should be mentioned that the role of this study is to focus on active control. The passive case is used mainly as a reference for comparisons with no-control case. Figures 3 and 4 are showing the almost identical responses in the two control cases, proving that the two computational versions presented in section 2 have the same performance level.

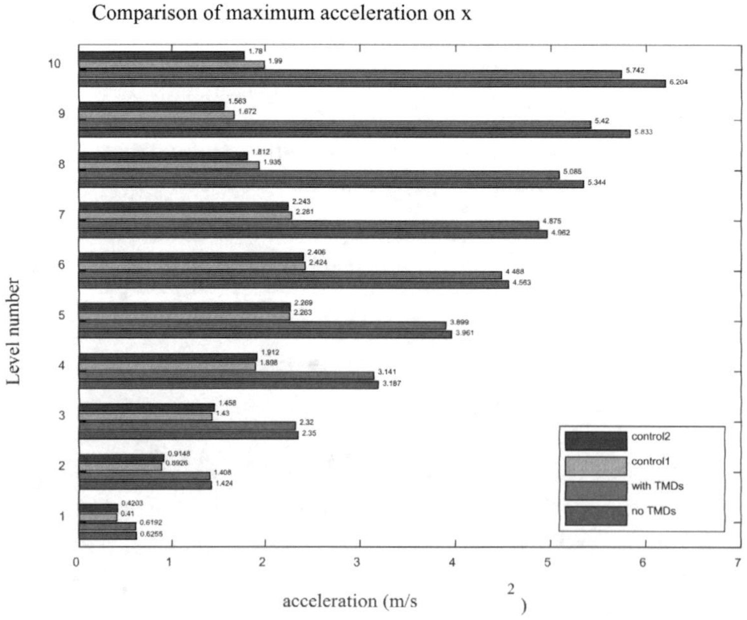

Figure 5. Maximum acceleration on level in the four cases.

Figures 5 and 6 prove the same. The performance of the control strategies is shown in terms of maximum accelerations during the earthquake action at level 9 above ground floor. The first of the two figures refer to absolute floor displacement, while in the second figure the storey drift is the element of comparison.

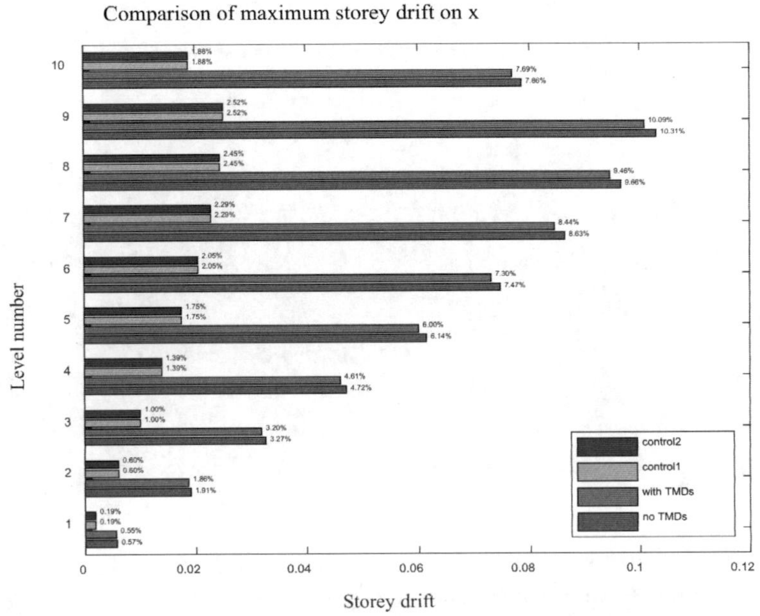

Figure 6. Maximum storey drift on level in each of the four scenarios.

In Figures 7 and 8, the frequency response is in use for proving the correctness of the theoretical proposals in Section 2. Figure 7 shows that the control strategies are very close in performance and much better than the other two results from the non-control and passive cases. If Figure 7 is making comparisons in terms of accelerations, Figure 8 is employing the same kind of comparisons in terms of velocity of the ninth floor of the structure used in this study.

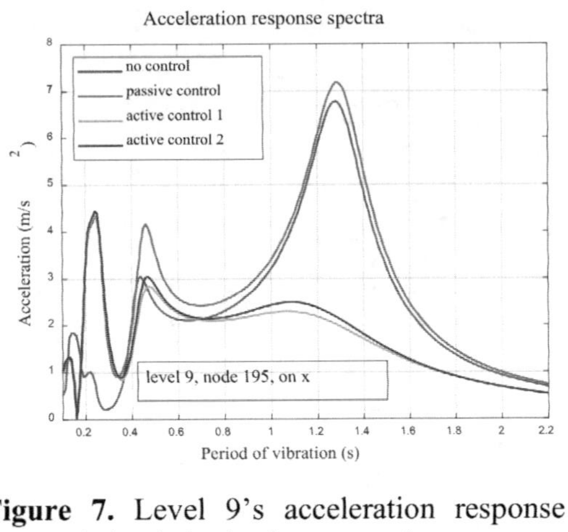

Figure 7. Level 9's acceleration response to harmonic loads in the four scenarios.

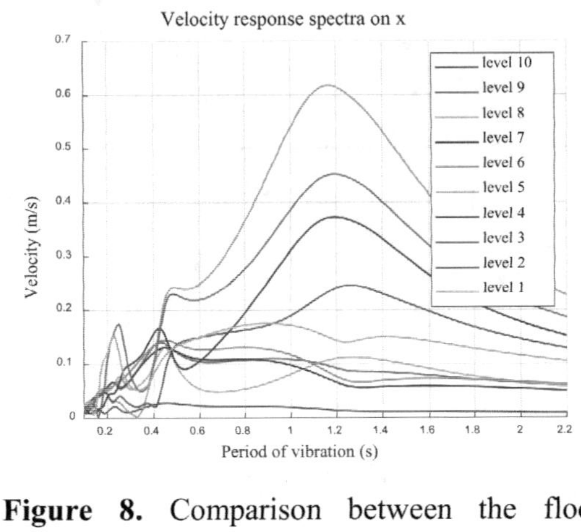

Figure 8. Comparison between the floors' response to harmonic loads in velocity terms.

Figure 9 adds insights of the control strategies requirements in terms of actuating need of forces (showing again the almost no differences between the "Control 1" and "Control 2" cases). This figure is stressing once more on the validity of the proposed strategy because the control 1 and 2 versions are showing very close effectiveness for the upper floor responses.

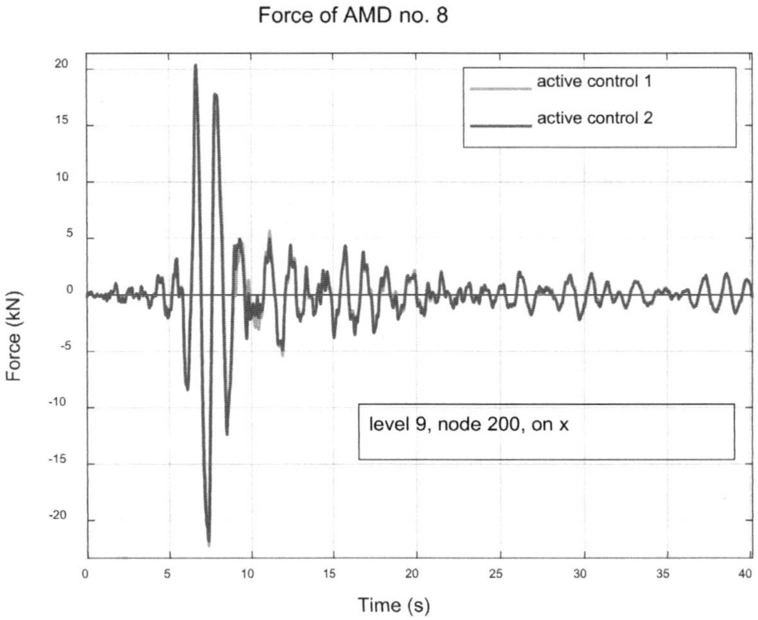

Figure 9. A typical actuator requirement in terms of forces.

Figure 10 is a comparison in terms of displacements of TMS versus AMD case (in order to observe the need of large displacements required by active control compared to the passive control). This figure helps understanding the mechanism of control imposed by the control law. Also, as in the other figures, the almost no difference between the two control strategies is adding more confirmation to the methodology.

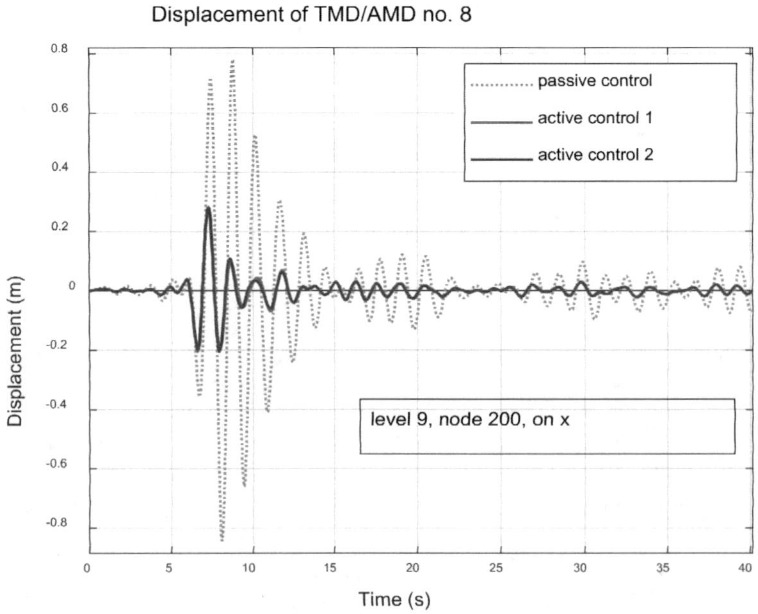

Figure 10. Comparison between control 1 and 2 with AMDs and TMDs movement.

In conclusion, the numerical validation of the "Control 2" scenario is strongly established. Regarding the time efficiency of the "Control 2" scenario over the "Control 1" scenario, the discussion must be based on numbers presented in Table 1.

Table 1. Running times for the control strategies, in seconds.

Run	Control 1[a]			Control 2[b]			Performance[c]	
no.	lqr[d]	Riccati-s[e]	Bome-n[f]	Riccati-s[g]	Gain[h]	total[i]	absolute[j]	%[k]
1	57.5205	1.87137	1.55026	0.33857	0.00157	1.89040	0.019026	1.01667
2	57.5070	1.88718	1.51286	0.32639	0.00167	1.84092	-0.04626	-2.45138
3	57.9074	1.88794	1.48044	0.32316	0.00167	1.80527	-0.08267	-4.37878
4	57.4805	1.87917	1.47524	0.32438	0.00163	1.80125	-0.07792	-4.14669
5	57.5284	1.88980	1.46890	0.32920	0.00203	1.80023	-0.08957	-4.73959
6	57.4184	1.88706	1.47544	0.32077	0.00150	1.79771	-0.08935	-4.73486
7	57.9965	1.89199	1.47898	0.32121	0.00147	1.80166	-0.09033	-4.77416
8	57.4732	1.86806	1.47985	0.31979	0.00154	1.80118	-0.06689	-3.5806
9	57.3530	1.87716	1.48543	0.32526	0.00150	1.81220	-0.06497	-3.46092
10	58.4764	1.92876	1.46947	0.34506	0.00150	1.81603	-0.11274	-5.84495
Mean	57.6662	1.88685	1.48770	0.32738	0.00161	1.81669	-0.07017	-3.70953

[a - k] See explanations in text, paragraph 3.5.

Table 1 shows groups of 10 simulations time needed for running scenarios for validating the control proposed strategy. Therefore, the next items had been taken into account:

a) First control strategy, as described in [5, 6]. The involved matrices are *2n*-dimensional (i.e. 760 x760 elements for the solution matrix).
b) Second control strategy, as described in [7]. The involved matrices are *n*-dimensional (380x380 elements for solution matrices).
c) Performance evaluated as in differences in values in "Control 2" minus values in "Control 1".
d) Run time, in seconds, using *lqr* Matlab function.
e) Run time, in seconds, using *Riccati* function for symmetrical solution (*2n*-dimensional) [10].
f) Run time, in seconds, using *Solve_BQME* function for non-symmetrical solution (n-dimensional) [11].
g) Run time, in seconds, using *Riccati* function for symmetrical solution (*n*-dimensional) [10].
h) Run time, in seconds, for calculus of the gain matrix in the case of "Control 2" strategy.
i) Total run time for "Control 2" (column *f* + column *g* + column *h*).
j) Performance evaluation in absolute running time, in seconds (column *i* – column *e*).
k) Performance evaluation in relative running time, in percent, i.e. (column *i* – column *e*)/column *i**100.

From Table 1, it can be withdrawn that the time for running the *lqr* function (for the 760x760 size solution) is very long, more than 57 seconds on average (column *d*). More efficient is to use in this case the *Riccati* function [10] for the same size of symmetric solution matrix. An average of 1.88685s was measured (column *e*).

From the same Table 1, it can be observed that dividing the problem in two smaller Riccati-like problems leads to a small efficiency in time (an average of only 3.7%, column *k*) compared to the initial problem, column *e*. However, the use of *Riccati* function for the half problem (380x380 size symmetric solution, column *g*) is showing an average of 0.32738s (column *g*) compared to 1.88685s (column *e*) for the 760x760 size solution. In other words, the efficiency in time is around 1/6.

A conclusion from above aspects in Table 1 is that for non-symmetric solution, column *f*, the function *Solve_BQMEs* (shown in the Appendix) should get some further refinement or should be replaced by a more efficient function. This is the point to focus on in the near researches.

4. Conclusions

In a previous work, [7], a proposal for improving the algorithm of optimal active control in Civil Engineering structures has been shown. The proposed approach is asking for solutions of two n-dimensional Riccati matrix equations, one with a symmetrical solution and another with a non-symmetrical solution. This is done in comparison to classical $2n$-dimensional problem with symmetrical solution.

The numerical simulation of this proposed approach was applied to a ground level plus ten stories building. The numerical results, using the classical method and the proposed one, presented practical no differences, this way validating the method.

This validation is proved by the presented figures. Seismic time responses of the structure, together with the spectral responses or maximum story displacements and storey drifts are proving that the two approaches have almost the same performance.

From the computational point of view, a very good improvement in time computation is observed in comparison to *lqr* Matlab function applied to the original $2n$-dimensional problem. However, in the case of using other functions (as *Riccati.m* calling Schur approach for symmetrical solutions together with *Solve_BQME.m* - an iterative adapted function for non-symmetrical solutions, shown in the Appendix) the performances are closer but still better (lower time) with an average of around -3.7% in favour of the proposed methodology.

As a consequence, the method being numerical validated, the future researches will focus on obtaining faster computational means for matrical Riccati equations applied to non-symmetrical solution cases. In addition, based on previous work, e.g. [12], more practical features might be added to the procedure.

Acknowledgments

Authors wish to acknowledge and be grateful for the kind support from The MathWorks, Inc. that provided full Matlab and Simulink software.

5. References

[1] Ghaedi K Ibrahim Z Adeli H and Javanmardi A 2017 Invited Review: Recent developments in vibration control of building and bridge structures *Journal of Vibroengineering* **19** 5 p 3564-3580

[2] Li H and Huo L 2010 Advances in Structural Control in Civil Engineering in China *Mathematical Problems in Engineering* **2010** 936081 pp 1-23

[3] Preumont A Seto K 2008 *Active Control of Structures* (Chichester: JohnWiley & Sons)

[4] Xu Y L and He J 2017 *Smart civil structures* (Boca Raton: Taylor & Francis Group)

[5] Paulet-Crainiceanu F 1997 *Active Control Approach for Long Span Bridge Responses to Strong Earthquakes* Ph.D. Thesis (Yokohama: Yokohama National University)

[6] Paulet-Crainiceanu F 1999 Seismic Response Control of Long Cable-stayed Bridges *Proc. Of the 2nd World Conf. on Structural Control (Kyoto)* (Chichester: John Willey & Sons) vol 2 p 959-964

[7] Paulet-Crainiceanu F, Luca S-G, Pastia C and Florea V 2019 Improved strategy for optimal control of Civil Engineering structures *15th Int. Conf. on Computational Civil Engineering (Iasi)* IOP Conference Series: Materials Science and Engineering Vol 586 pp 1-6 (IOP Publishing Ltd)

[8] Pastia C Luca S G Paulet-Crainiceanu F and Florea V 2016 Passive, Active and Semi-Active Vibration Control of a Classical Mass-Spring-Dashpot System *16th Int*

Multidisciplinary Scientific GeoConference SGEM2016 (Sofia) (Sofia: Stef92 Technology Ltd.) book1 vol. 3 p 575
[9] The MathWorks Inc *Matlab R2021a*, www.mathworks.com
[10] Emmett 2021 Algebraic Riccati Equation Solver *MATLAB Central File Exchange* (https://www.mathworks.com/matlabcentral/fileexchange/36263-algebraic-riccati-equation-solver) Retrieved March 20, 2021
[11] Sadek L 2021 Nonsymmetric differential Riccati matrix equations *MATLAB Central File Exchange* (https://www.mathworks.com/matlabcentral/fileexchange/72895-nonsymmetric-differential-riccati-matrix-equations) Retrieved March 20, 2021
[12] Bakule L Paulet-Crainiceanu F Rodellar J and Rossell J M 2005 Overlapping Reliable Control for a Cable-Stayed Bridge Benchmark *IEEE Trans. C.S.T.* **13** 4 663–9

Appendix. Some functions used in simulations

The modified `Solve_BQME.m`, [11], used under the name `Solve_BQMEs.m` used together with the also modified `EvalBiQuad.m` function:

```
function X=Solve_BQMEs(A,B,C,D)
% This Function is adapted to solve the equation
% of the form A'X'+XB+XCX'+D = 0 for non-symmetric X
[p1,p2]=size(A);
[p3,p4]=size(B);
X=zeros(p2,p3);
iters=0;
Omega=EvalBiQuad(A,B,C,D,X);
while (iters<100*sum(size(X))) && (norm(Omega)>1e-9)
Delta = sylvester(A'+X*C,B+C*X',-Omega);
if norm(Delta) > 1e10
error('Cannot Find Real Valued Solution')
end
iters=iters+1;
X=X+Delta;
Omega=EvalBiQuad(A,B,C,D,X);
end
end
function Omega=EvalBiQuad(A,B,C,D,X)
Omega=A'*X'+X*B+X*C*X'+D;
end
```

The next function [10], `Riccati.m`, solves the classical matrical Riccati equation with symmetrical solution using the Schur method:

```
function X=Riccati(A,G,Q)
%RICCATI  Solves an algebraic Riccati equation
%   X = Riccati(A,G,Q) solves the algebraic Riccati equation of the form:
%       A'*X + X*A' - X*G*X + Q = 0, where X is symmetric.
n=size(A,1);
Z=[A -G
    -Q -A'];
[U1,S1]=schur(Z);
[U,S]=ordschur(U1,S1,'lhp');
X=U(n+1:end,1:n)*U(1:n,1:n)^-1;
end
```

IOP Conf. Series: Materials Science and Engineering 1141 (2021) 012035 doi:10.1088/1757-899X/1141/1/012035

Numerical studies on steel-concrete composite structures

G Urian[1], A Haupt-Karp[2],
[1] Technical University of Cluj-Napoca, Faculty of Civil Engineering, Structures Department, 15 Constantin Daicoviciu street, Cluj-Napoca, *Romania*,
[2] S.C. Romsoft Comimpex S.R.L., 6 Portile de Fier street, Cluj-Napoca, *Romania*

E-mail: gabrielurian@yahoo.com

Abstract. The paper focuses on the seismic performances of steel-concrete composite structures made with fully encased steel-concrete composite columns and steel beams. An important objective is to study the influence of the structural steel ratio on the behaviour of composite columns. To meet the proposed objectives, a numerical study was developed on composite structures. Some comments, design recommendations and an economical study conclude the paper.

1. Introduction

The paper concentrates on the seismic performances of steel-concrete composite frames, with emphasis on the behaviour of composite columns. Concrete encased hot-rolled steel sections are still interesting for researchers from all over the world, being studied intensively for more than three decades. In order to improve design recommendations of current norms researchers are still developing extensive experimental tests, the latest in Singapore and China [1], especially on columns made with high strength materials [2]. Based on experimental tests, numerical models are implemented in different types of calculus programs (commercial or specially developed) in order to perform a large variation of sections, types of embedded profiles and quality of materials.

To achieve the proposed objectives a numerical study on composite structures was developed. The structures were designed with fully encased composite columns and steel beams. The analysis included five types of similar structures regarding the floor plan, but different heights. The studied structures had twelve, ten, eight, six and two levels. The columns were designed using three structural steel ratios: low, medium and high. The loads taken into consideration were identical for all types of structures. To investigate the seismic performances two types of analysis were performed: pushover and dynamic time history, based on the numerical model developed [3,4]. Beside the seismic performances of the structures, an economic study was conducted on the chosen frames.

2. Numerical model

The numerical model used was developed in 2013 in FineLg, a finite element program implemented at Liège University, ArGenCo department. The calibration and validation of the numerical model [3,4] was made using six experimental tests taken from the international literature on fully embedded steel-concrete composite columns.

2.1. Numerical model

The finite element used was a classic beam element (Bernoulli) for concrete plane frames with steel reinforcement and embedded beams with three nodes, as shown in figure 1, without taking into account the shear force effect. The total number of degrees of freedom corresponds to: one relative translational degree of freedom for the node situated at the mid-length of the beam element and one rotational and two translational degrees of freedom for each two nodes located at beam element ends and, as shown in figure 1. Nodes 1 and 3 have three degrees of freedom (u, v, θ), and node 2 has a single degree of freedom u, which allows to consider a possible relative displacement between steel and concrete.

This type of element does not allow the involvement of the local buckling phenomenon of the section. Because this analysis is two-dimensional, the bending phenomenon applied outside the section plane, such as torsional bending, is not taken into account in this analysis. It is considered a perfect connection between steel and concrete.

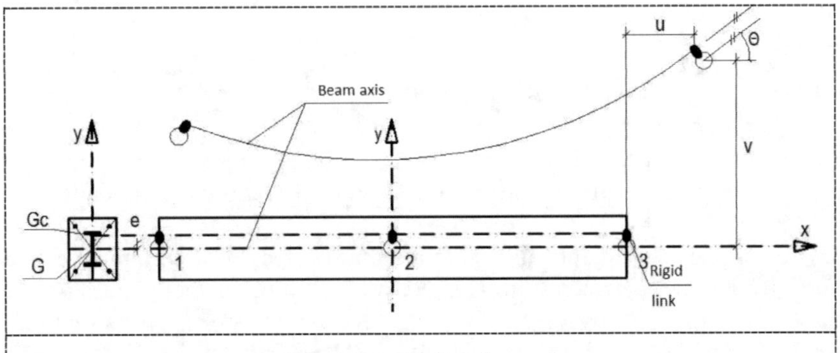

Figure 1. Finite element

The materials behaviour model used are presented in figure 2 for concrete and figure 3 for steel.

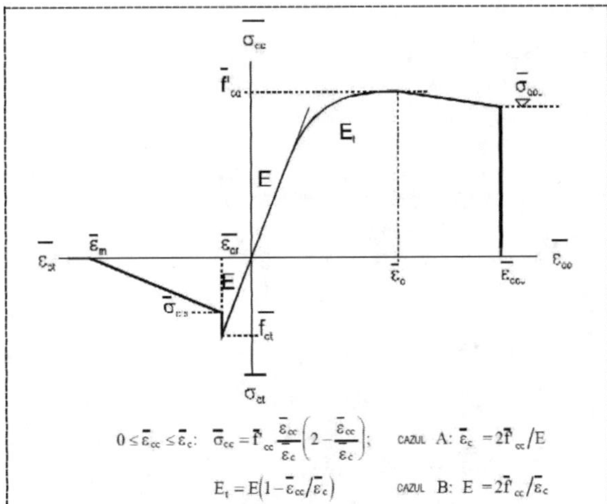

Figure 2.
Material behaviour model for concrete

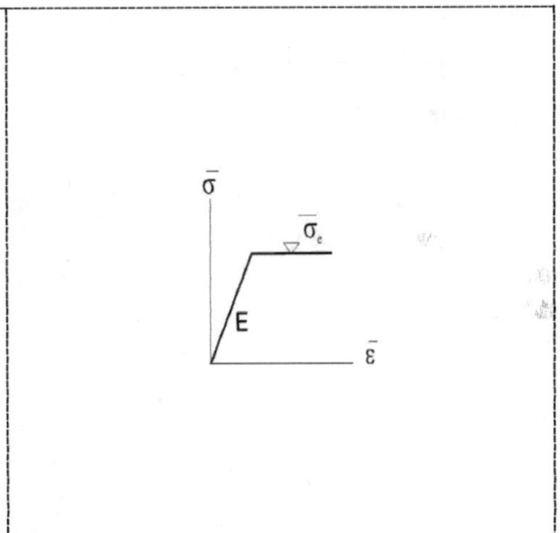

Figure 3.
Material behaviour model for steel

2.2. Calibration and certification of the numerical model

The calibration and certification of the numerical model was made using six experimental tests taken from the international literature on fully encased steel-concrete composite columns. The experimentally tested columns had different types of concrete or steel, different structural steel or reinforcing steel ratios. The columns were tested (both monotonic and cyclic) to constant axial force and lateral forces. The subject of the paper is not the detailed presentation of the experimental tests used for the calibration and certification of the numerical model, but on the study of the seismic performances of the structures with composite columns, based on the case study developed with the numerical model validated using these experimental tests. Figure 4 presents the types of cross sections of the columns used for calibration and certification of the numerical model. The experimental programs used for calibration and certification of the numerical model were developed at Technical University of Cluj-Napoca, in Romania [5,6], at NCU, Chung-Li, in Taiwan [7], at UC, San Diego, California, in USA [8,9], at CTU, Hsinchu, in China [10], and at Lakehead University, Thunder Bay, Ontario, in Canada [11].

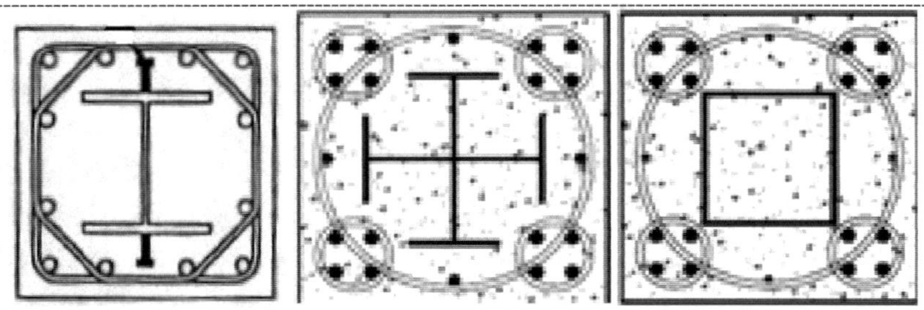

Figure 4. Type of cross sections used for calibration and validation

The main characteristics of test specimens are as follows: length of columns between: 0.80 ÷ 4.00 m, cross section of columns: between 160x180 ÷ 600x600 mm, characteristic compressive cylinder strength of concrete at 28 days: between 22.50 ÷ 92.30 MPa, yield strength of structural steel: between 293.40 ÷ 442.00 MPa, yield strength of longitudinal reinforcement: between 330 ÷ 634 MPa. The main parameters of test specimens are: structural steel ratio between: 0.16 ÷ 0.471 and slenderness ratio between: 0.32 ÷ 1.16.
The validation of the numerical model was performed by comparison of experimental force-displacement curves with numerical ones, as showed in figure 5 and figure 6.

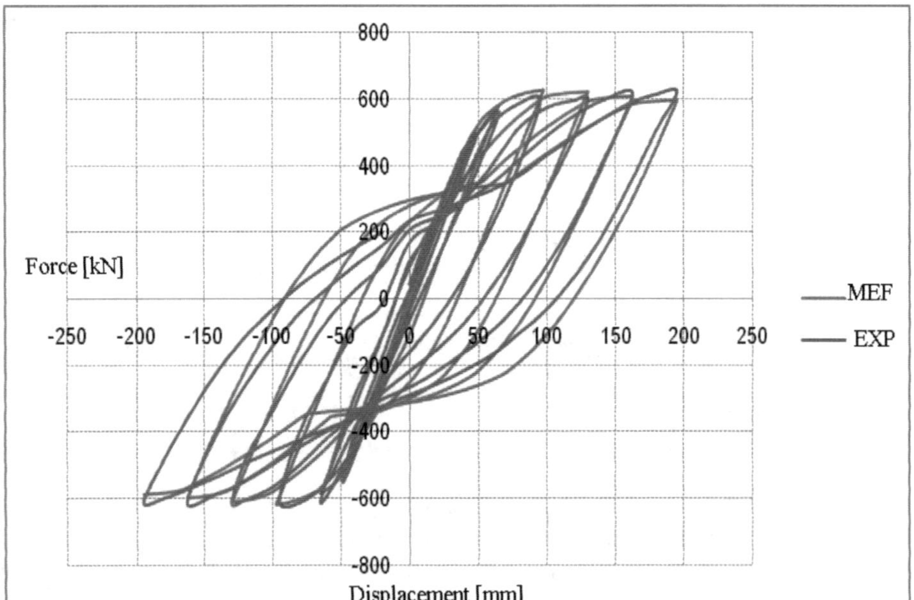

Figure 5. Validation of numerical model [4]

With red colour are presented the experimental curves and with blue colour the numerically obtained ones. The difference between experimental and numerical values was between 0÷15 percent with a mean value of 5% [3,4]. For exemplification, in figure 5 is presented the comparison with a cyclically tested column in China [10] and in figure 6 with a monotonically tested column in Canada [11].

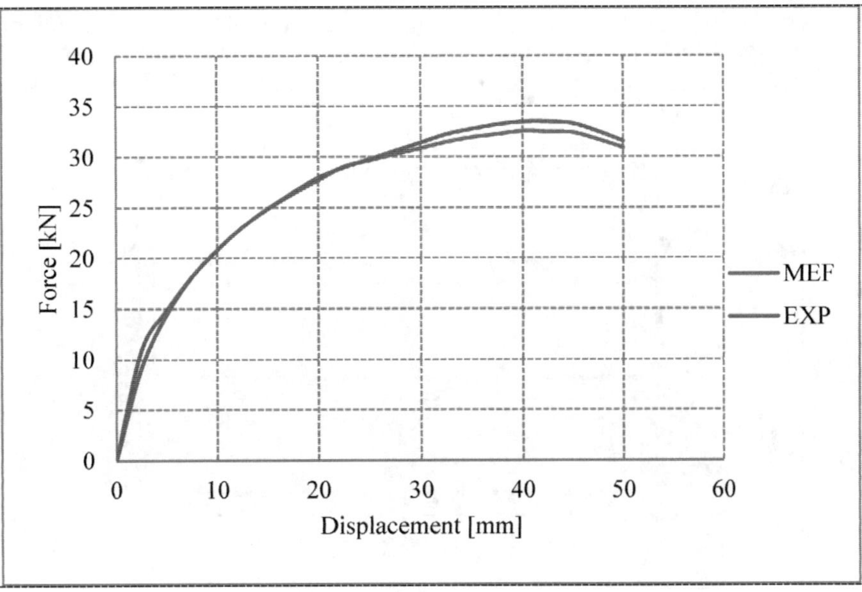

Figure 6. Validation of numerical model [3]

3. Case study

The developed case study [3] included five similar types of composite frames, with the same floor plan as showed in figure 7 and with twelve, ten, eight, six and two levels, as showed in figure 8 for the six levels structure. The composite frames had five openings of 6 meters in longitudinal direction and two openings of 7 meters in transversal direction.

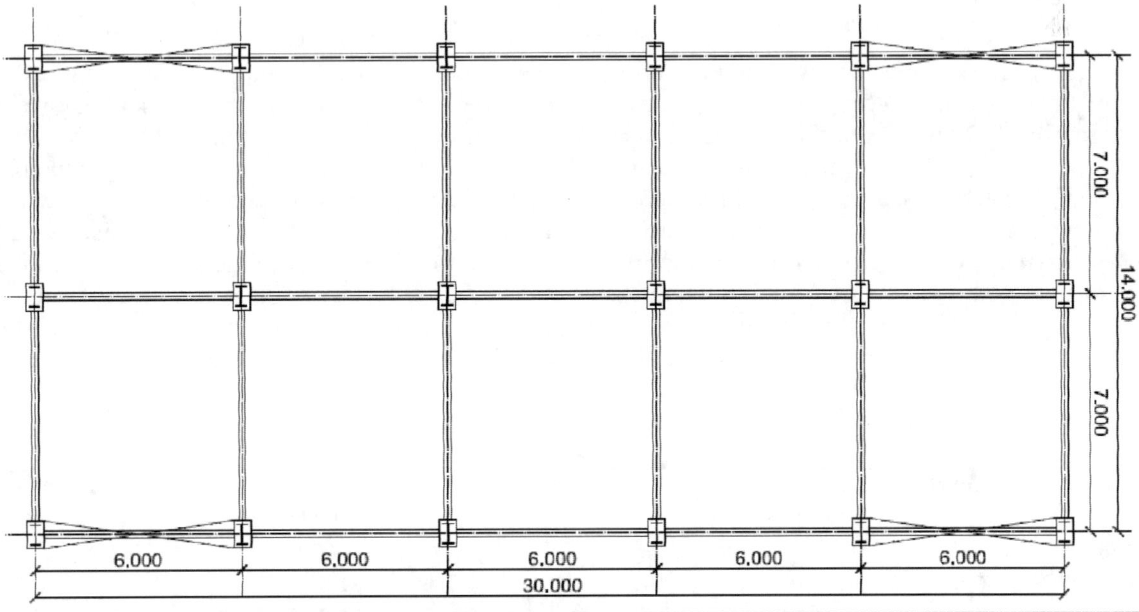

Figure 7. Current floor plan

The level height was identical for all levels, 3.20 m. The considered loads were identical for all levels: permanent load 6.50 kN/m^2 and live load 3.00 kN/m^2.

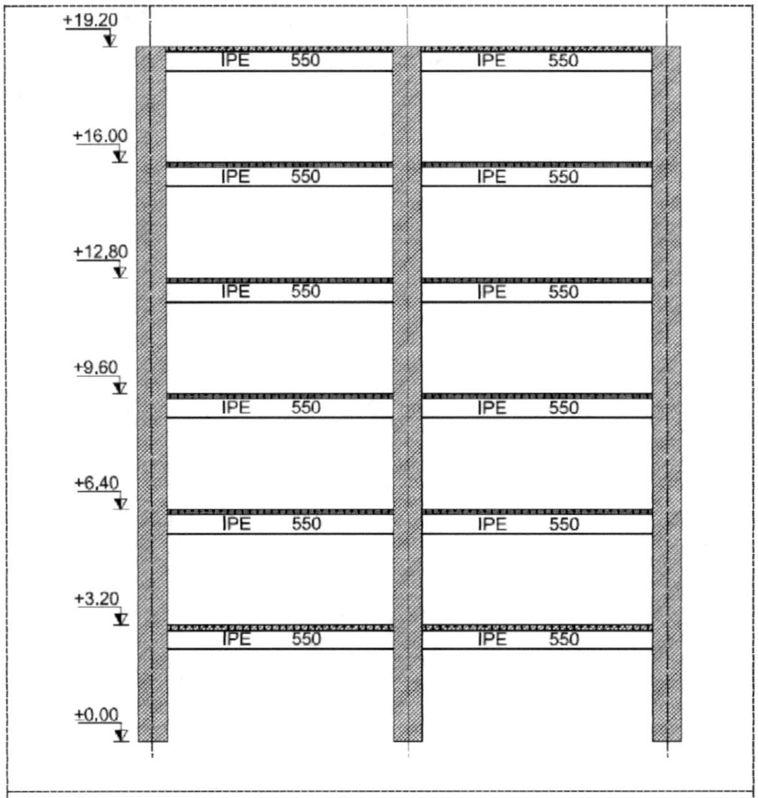

Figure 8. Transversal frame for six levels structure

The chosen seismic zone was the one corresponding to a peak ground acceleration of 0.40 g and a corner period of 1.6 s. After a preliminary design analysis performed with a commercial software were chosen IPE 550 profiles for the steel beams and the following materials for columns: C40/50 concrete, S355 steel for the structural steel profiles embedded in concrete and S500 steel for the longitudinal reinforcement. For each type of structure were chosen three types of composite columns. The difference between the columns was the structural steel ratio: low, medium and high. In table 1 are presented all the characteristics of the composite columns used in the analysis. The names of the structures have the following meaning: for example, 6Fb: 6 represents the number of levels, F comes from floors and the last number (b) represents the structural steel ratio chosen (a for low, b for medium and c for high). The columns of the twelve, ten and eight level structures were also varied in height, as showed in table 1.

Table 1. Characteristics of case study structures

Structure	Level	Column section (mm x mm)	Embedded profile	Longitudinal reinforcement	Structural steel ratio (δ)
2Fa	all	390 x 400	HE 200 A	16Ø16	0.288
2Fb	all	350 x 360	HE 140 M	16Ø14	0.439
2Fc	all	350 x 360	160 x 150 x 18 x 28	16Ø14	0.506
6Fa	all	500 x 590	HE 400 A	14Ø22	0.320
6Fb	all	490 x 510	HE 260 M	14Ø20	0.543
6Fc	all	450 x 460	260 x 250 x 25 x 40	14Ø18	0.610

8Fa	1÷4	520 x 900	HE 500 AA	20Ø25	0.253
	5÷8	520 x 670	HE 400 AA	20Ø20	0.291
8Fb	1÷4	520 x 770	HE 450 A	20Ø22	0.349
	5÷8	520 x 570	HE 360 A	20Ø20	0.368
8Fc	1÷4	510 x 580	HE 340 M	16Ø20	0.582
	5÷8	470 x 490	HE 260 M	16Ø18	0.550
10Fa	1÷5	500 x 980	HE 700 AA	20Ø25	0.315
	6÷10	500 x 670	HE 400 AA	20Ø20	0.297
10Fb	1÷5	500 x 840	HE 650 A	20Ø22	0.415
	6-10	500 x 550	HE 350 A	20Ø18	0.389
10Fc	1÷5	510 x 680	HE 450 M	16Ø22	0.553
	6÷10	470 x 490	HE 260 M	16Ø18	0.550
12Fa	1÷4	600 x 2000	HE 1000 AA	30Ø32	0.215
	5÷8	500 x 1600	HE 700 AA	30Ø28	0.209
	9÷12	500 x 670	HE 400 AA	20Ø20	0.297
12Fb	1÷4	520 x 1650	HE 1000 B	26Ø28	0.361
	5÷8	520 x 1200	HE 700 B	26Ø25	0.370
	9÷12	500 x 550	HE 360 A	20Ø18	0.389
12Fc	1÷4	520 x 1150	HE 900 x 466	22Ø25	0.559
	5÷8	520 x 850	HE 600 x 399	22Ø22	0.595
	9÷12	470 x 490	HE 260 M	16Ø18	0.550

In the preliminary design stage, all recommendations from P100-1/2013 [12] were considered. The ductility class chosen was DCH, with a q behaviour factor equal to 6.5 for the transversal frames and 4 for the longitudinal ones.

After the preliminary design stage, using the FineLg program, based on the numerical model presented at chapter 2.1., to study the seismic performances of the chosen frames two types of analysis were performed: pushover and dynamic time history. In the pushover analysis, the seismic forces had a triangular distribution. The time history analysis was performed using three artificial accelerograms, in accordance with P100-1/2013 [12] and one real (Vrancea 1977). The followed parameters were: the evolution of inter-storey drift at all levels, the global pushover curve and rotation capacity. In addition, the q behaviour factor was evaluated for all analysed frames.

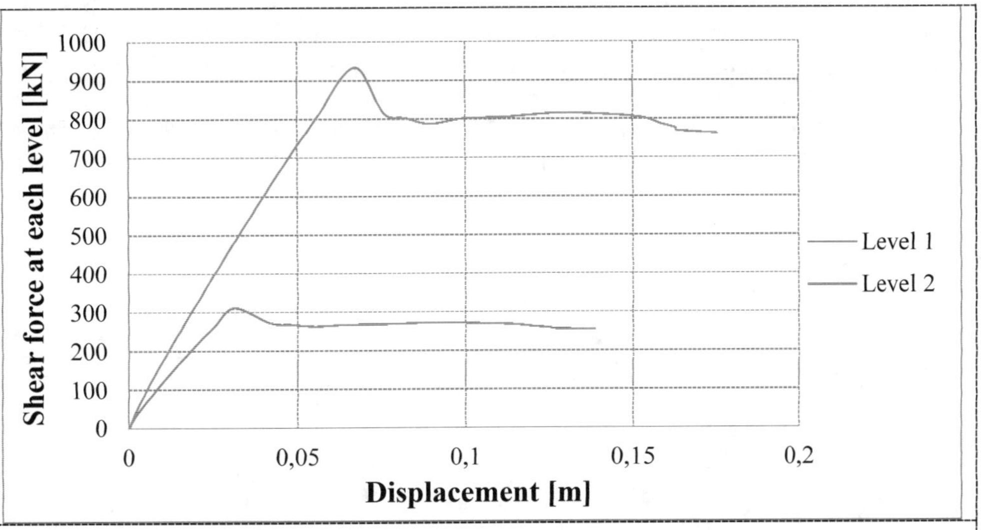

Figure 9. Pushover curves for 2Fa structure [3]

For exemplification, in figure 9 are presented the pushover curves for 2Fa structure and in figure 10 the evolution of inter-storey drift at all levels for the same structure.

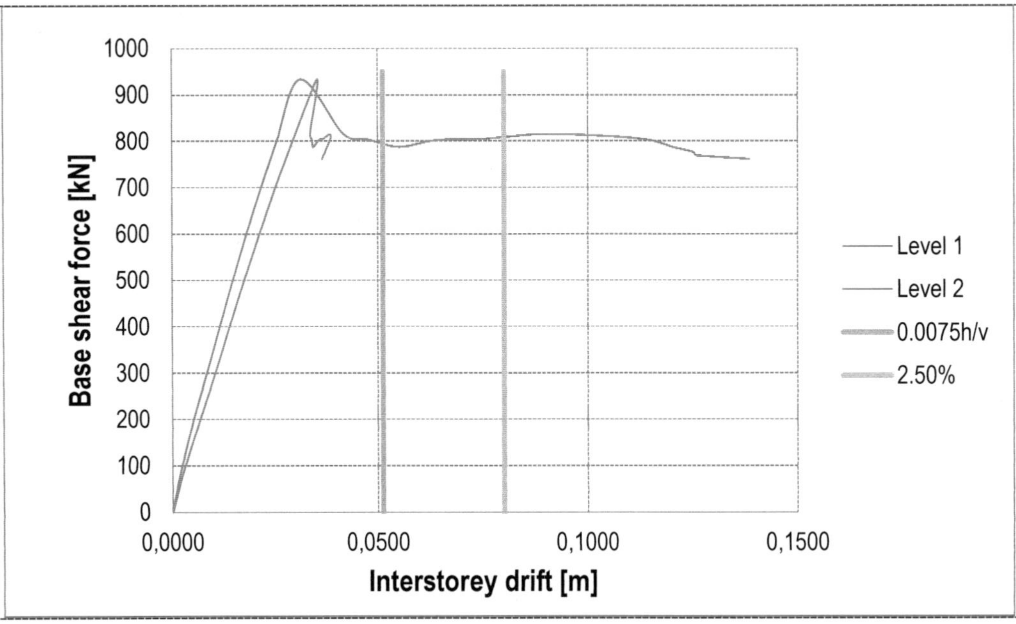

Figure 10. Evolution of inter-storey drift for 2Fa structure [13]

With a green vertical line is marked the inter-storey drift limitation of 0.0075h/ν, where ν represents the reduction factor that takes into account the lower return period of the seismic action associated with the damage limitation requirement and h is the height on structure. The 0.0075 value corresponds to buildings having non-structural elements with high dissipation capacity, attached to the structure, according to the seismic norm P100/1-2013 [12]. The yellow line represents the inter-storey drift limitation of 2.5%, FEMA 356-2000 [13] criteria for Life Safety.

The pushover analysis results for all studied composite structures are presented centralized in table 2. Table 2 presents the displacement (d_c) and corresponding force (F_b) for 0.0075h criteria [12], 2.5% drift limitation according to FEMA 356-2000 [13] and the values at concrete failure, when ε_{cu2} reaches 3.5‰ value. The last column presents the corresponding force when θ_p reaches 35mrad value, where θ_p represents the rotation capacity of the plastic hinge region. The six and two floor structures did not achieve a minimum rotation capacity of the plastic hinge region of 35 mrad, necessary to design the structure in DCH ductility class, as can be seen in table 2. From the eight floor, analysed frames reached a superior rotation capacity of the plastic hinge region, 37 mrad for 8Fa structure to 69 mrad for 12Fc structure.

Table 2. Pushover analysis results

Structure	0.0075h/ ν		2.50%		Concrete failure		35 mrad corresponding force (kN)
	F_b (kN)	d_c (m)	F_b (kN)	d_c (m)	F_b (kN)	d_c (m)	
2Fa	676	0.046	804	0.112	934	0.067	-
2Fb	610	0.049	804	0.121	858	0.082	-
2Fc	616	0.050	891	0.131	891	0.113	-
6Fa	744	0.112	1443	0.307	1480	0.369	-
6Fb	733	0.115	1267	0.314	1351	0.374	-

IOP Conf. Series: Materials Science and Engineering 1141 (2021) 012035 doi:10.1088/1757-899X/1141/1/012035

6Fc	660	0.116	1193	0.324	1244	0.377	-
8Fa	859	0.151	1531	0.375	1754	0.555	1677
8Fb	773	0.155	1329	0.385	1516	0.630	1488
8Fc	739	0.158	1252	0.411	1417	0.674	1375
10Fa	724	0.155	1355	0.464	1644	0.911	1547
10Fb	711	0.175	1330	0.482	1580	0.975	1393
10Fc	753	0.195	1281	0.549	1496	1.078	1348
12Fa	678	0.198	1400	0.541	2068	1.501	1764
12Fb	672	0.213	1284	0.578	1867	1.567	1565
12Fc	653	0.219	1221	0.623	1707	1.700	1372

Table 3 presents the q behaviour factor obtained in the pushover analysis and dynamic one, using artificial accelerograms, according to P100/1-2013 [12] and real ones (Vrancea 1977).

Table 3. q behaviour factor results

Structure	q_{max} Pushover	q_{max} P100-1-2013	q_{max} Vrancea
2Fa	4.00	4.40	4.20
2Fb	4.20	4.50	4.30
2Fc	4.50	4.75	4.35
6Fa	5.00	5.25	5.10
6Fb	6.10	6.10	5.98
6Fc	6.30	6.30	6.20
8Fa	5.30	5.30	5.20
8Fb	6.60	6.40	6.30
8Fc	7.20	7.40	7.10
10Fa	5.80	6.10	5.95
10Fb	6.40	6.45	6.35
10Fc	7.00	7.20	7.00
12Fa	5.60	5.45	5.20
12Fb	6.50	6.60	6.40
12Fc	7.20	7.10	6.95

Based on the performed analysis it is recommended that low level structures (with one up to six-seven levels) to be designed in medium ductility class. Structures with more than eight levels can be designed in medium or high ductility class, depending on the architectural and/or structural restrictions. Higher structural steel ratio leads to an important increase of structure ductility, more pronounced from low to medium than from medium to high. Given the financial importance, the structural analysis was completed with an economical study of the frames for an optimal choice of structural steel ratio. In 'Table 4' is presented the cost of each type of designed column per 1 linear meter of element. The final price was obtained by summing the costs of all materials (structural steel, concrete and reinforcement), formwork and labour. The prices were calculated based on average offers received from local suppliers. The most economical solution is offered by choosing a low structural steel ratio. Up to eight floors, the price difference between columns with a low structural steel ratio and medium steel ratio is about 15%. This difference decreases substantially at higher structures up to about 5%. In comparison with low structural steel ratio, a medium one offers an important increase of structural ductility and slimmer cross-sections, so the 5% cost difference is considered acceptable. The values of the structural steel ratio for the chosen structures were for low: between 0.209÷0.32, with a mean value of 0.276, for medium: between 0.349÷0.543, with a mean value of 0.403 and for high: between 0.506÷0.610, with a mean value of 0.562.

Table 4. Costs of case study columns – per 1 linear meter of column

Structure	Level	Concrete price/1m (euro)	Formwork price/1m (euro)	Reinforcement price/1m (euro)	Steel price/1m (euro)	Column price/1m (euro)
2Fa	all	15.5	12.6	27.5	57.6	113
2Fb	all	12.5	11.3	21.1	86.0	131
2Fc	all	12.5	11.3	21.1	109.8	155
6Fa	all	29.3	17.3	45.5	170.2	262
6Lb	all	24.8	15.9	37.6	234.2	312
6Fc	all	20.5	14.5	30.5	261.8	327
8Fa	1÷4	46.4	22.6	83.9	145.7	299
	5÷8	34.6	18.9	53.7	125.8	233
8Fb	1÷4	39.7	20.5	65.0	190.6	316
	5÷8	29.4	17.3	53.7	152.5	253
8Fc	1÷4	29.4	17.3	43.0	337.6	427
	5÷8	22.9	15.3	34.8	234.2	307
10Fa	1÷5	48.6	23.5	83.9	204.2	360
	6÷10	33.2	18.6	53.7	125.8	231
10Fb	1÷5	41.7	21.3	65.0	258.7	387
	6÷10	27.3	16.7	43.5	152.5	240
10Fc	1÷5	34.4	18.9	52.0	358.0	463
	6÷10	22.9	15.3	34.8	234.2	307
12Fa	1÷4	119.1	41.3	206.3	302.2	669
	5÷8	79.4	33.4	157.9	204.2	475
	9÷12	33.2	18.6	53.7	125.8	231
12Fb	1÷4	85.2	34.5	136.9	427.5	684
	5÷8	61.9	27.3	109.1	328.1	526
	9-12	27.3	16.7	43.5	152.5	240
12Fc	1÷4	59.4	26.5	92.3	634.4	813
	5÷8	43.9	21.8	71.5	543.2	680
	9÷12	22.9	15.3	34.8	234.2	307

4. Conclusions
Composite frames made with steel beams and fully encased steel-concrete composite columns can be an efficient solution for buildings situated in medium and high seismicity zones. From the case study developed some notable conclusions can be drawn: small structures (up to 6-7 levels) are recommended to be designed in medium ductility class; for higher structures a medium or high ductility class can be adopted, the solution chosen being optimized from different points of view: cross-section dimensions, necessary rotation capacity, costs, etc. When considering only the economical point of view the structures with low steel ratio offered the best results, but considering the 5% (for tall buildings) difference in using low or medium steel ratio it is recommended a medium one when designing a composite column.

5. References
[1] Binglin Lai B, Richard Liew J Y and Mingxiang X 2019 Experimental study on high strength concrete encased steel composite short columns, *Construction and Building Materials,* Vol **228**

[2] Binglin Lai, Richard Liew J Y, Akshay V, Shan Li and Mingxiang X 2020 Assessment of high-strength concrete encased steel composite columns subject to axial compression *Journal of Constructional Steel Research,* Vol **164**

[3] Urian G 2015 Steel-concrete composite structures. Frame behaviour analysis *PhD. Thesis* (Cluj-Napoca)
[4] Vermesan (mar. Haupt-Karp) A 2013 Behavior of fully encased steel-concrete composite columns *PhD. Thesis* (Cluj-Napoca)
[5] Campian C 2010 Contribution a l'etude du comportement et au calcul de poteaux mixtes acier-beton (sous des charges transversales de variation monotone ou cyclique alternee *PhD. Thesis* (Cluj-Napoca)
[6] Sav V 2011 Steel-concrete composite columns with high strength concrete *PhD. Thesis* (Cluj-Napoca)
[7] Hsu H L, Jan F and Juang J L 2009 Performance of composite members subjected to axial an bi-axial bending *Journal of Constructional Research 65,* 869-878
[8] Ricles J M and Paboojian S 1994 Seismic performance of steel-encased composite columns *Journal of Structural Engineering,* Vol **120**, No 8
[9] Ricles J M and Paboojian S 1992 Behavior of composite columns under seismic conditions *Tenth World Conference,* Balkerna, Rotterdam ISBN 90 54100 605
[10] Weng C C and Yin Y L 2008 Seismic cyclic loading test of SRC columns confined with 5-spirals *Science in China Series E: Technological Sciences,* Vol. 51, No. 5, pp. 529-555
[11] Mirza S A, Hyttinen V and Hyttinen E 1996 Physical tests and analyses of composite steel-concrete beam-columns *Journal of Structural Engineering,* 122(11), pp. 1317–1326
[12] P100-1/2013 Cod de proiectare seismica. Partea I – Prevederi de proiectare seismică pentru clădiri (Seismic design – Part I – Seismic recommendations for buildings)
[13] FEMA 356-2000 Prestandard and commentary for the seismic rehabilitation of buildings

Vibration analysis of the simply supported R/C beam

I Stoynova

USEA "Lyuben Karavelov" Bulgaria, Faculty of Construction, Department "Mechanics and Mathematics", 1373 Sofia, 175 Suhodolska str.

E-mail: stoynova@vsu.bg

Abstract. A mechanical vibration can be of different origins – caused by machines, due to a shock impact, by motors, etc. In this paper the behaviour of a reinforced concrete beam with a static scheme of a simply supported beam under the action of vibrations has been studied. The paper presents differences between static and dynamic analysis of the simply supported reinforced concrete beam. Concrete and the reinforcement bars are modelled with their material characteristics by applying linear-elastic material models. The load area is in the middle of the beam. A numerical model by the Finite Element Method was developed. To perform the vibration analysis, a time-history analysis has been employed. Dynamic analysis of the numerical model was performed using a software product based on the Finite Element Method. The dynamic time-history analysis was adopted to determine the dynamic response of the simply supported R/C beam under the action of time variables.

1. Introduction

A mechanical vibration can be of different origins – caused by machines, due to a shock impact, by motors, cranes etc. Transient or time-history dynamic analysis is a technique used to determine the dynamic response of a structure under the action of time-dependent vibrations. This type of analysis may be utilized to determine the time-varying displacements, strains, stresses, and forces in a structure, as it responds to any combination of static, transient and harmonic loads [1].

Harmonic vibrations

The equation describing the harmonic damping vibrations of a system with a final number of degrees of freedom, and taking into account the resistance forces:

$$[m]\{\ddot{u}\} + [c]\{\dot{u}\} + [k]\{u\} = \{F(t)\},$$

and without taking into account the resistance forces

$$[m]\{\ddot{u}\} + [k]\{u\} = \{F(t)\}$$

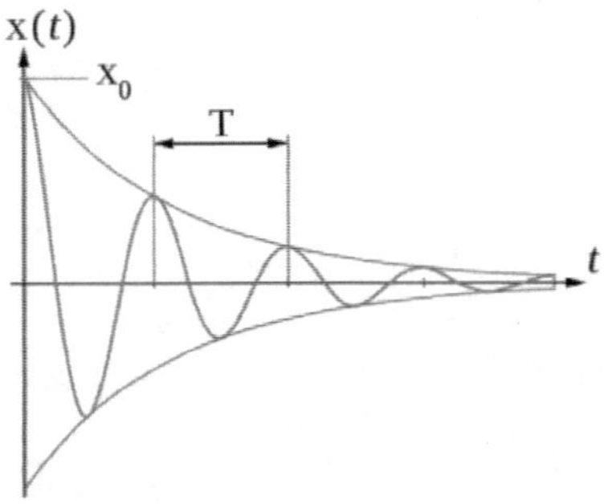

Figure 1. Harmonic damping vibrations

2. Numerical models

The concrete and reinforcement material properties is according to Eurocode 2 [4]. More about calculating anchoring length can be found in [8].

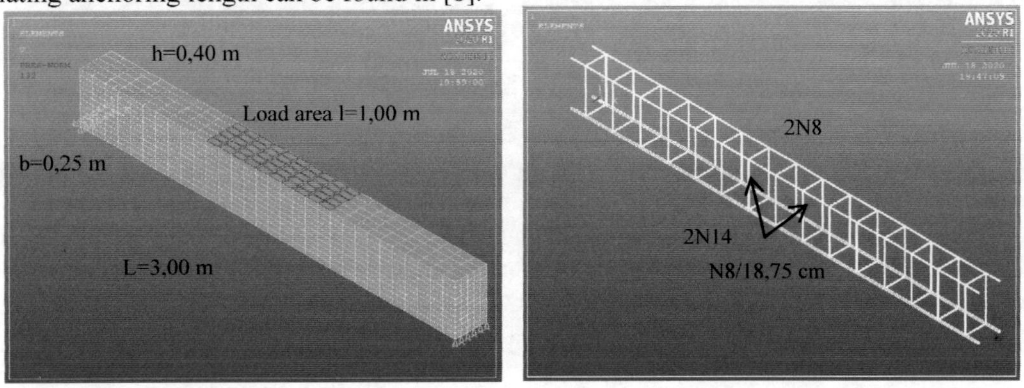

Figure 2. a) Scheme and loading area b) Position of reinforced bars

The concrete properties are: C20/25, f_{cm}=28 MPa, E_{cm}=30 GPa, Poison`s ratio is 0,2.
Reinforcement bars' diameters and position is shown on Figure 2b).
Cross section of the simply supported beam – b=0,25 m, h=0,40 m, L=3,00 m.

2.1. Static analysis. Linear elastic concrete model.

First numerical model (M1) represents modelling by ANSYS. The load stamp size is 0,25/1,00 m. A static load of 132 kN/m^2 (R_q=33 kN) on the load stamp is applied. The concrete is modelled by SOLID 65 (CONCRET 65) finite elements as linear elastic concrete material model – Table1. Reinforcement bars are modelled via BEAM188 finite elements – Table 2.

Table 1. Concrete properties - linear isotropic

Concrete material model for C20/25
Young's modulus E_c=3x10^7 kN/ m^2
Poison`s ratio ν=0,2

Table 2. Reinforcement properties - linear isotropic

Reinforcement steel B500 model
Young's modulus E_s=2x10^8 kN/ m^2
Poison's ratio ν=0,3

2.2. Static analysis. Nonlinear inelastic concrete model.

Second model (M2) is modelling with material properties according to Table 3, are presented. The boundary conditions adopted are restraints applied on the left side of the simply supported beam - displacements by axes X and Y. The load stamp size is 0,25/1,00 m. The applied force starts at 40 kN/m^2, by a step of 40 kN/m^2. The beam is modelled by SOLID 65 (CONCRET 65) finite elements with nonlinear plasticity material model. Different variables for applying the Willam and Warnke's material model [5] in ANSYS are required (Table 3). Reinforcement bars are modelled with BEAM188 finite elements – Table 2. In Figure 3, the distribution of the normal stresses and displacement u_y in the simply supported beam are shown. Figure 4 shows the formed cracks at model M2.

Tensile strength of concrete under bending - the average tensile strength in the bending moment of reinforced concrete elements depends on the average tensile strength and the height of the cross section. [9]

Table 3. Concrete material nonlinear inelastic model in tension zone

1. Shear transfer coefficient for an open crack (β_t)	0,2
2. Shear transfer coefficient for an close crack (β_c)	0,9
3. Uniaxial cracking stress (f_r)	2800 kN/m^2
4. Uniaxial crushing stress (f_c)	-1

2.3. Dynamic analysis. Harmonic vibrations.

The numerical model of a simply supported beam is a process of modelling run in ANSYS. Applied load at the end of the beam on the load stamp (Figure 2a). The load stamp size is 0,25/1,00 m. The stamp load case is presented. The concrete is modelled by SOLID 65 (CONCRET 65) finite element as linear elastic concrete material model – Table 1. Reinforcement bars are modelled with BEAM188 finite elements – Table 2. The most important point in the study of building structures of dynamic impacts is the choice of a suitable dynamic model. More information about the dynamic study in structural mechanics can be found in [1], [3] [6] and [7].

The behaviour of the simply supported beam subjected to dynamic vibrations, which become changeable under a harmonious law, has been investigated. For the purpose of the study, on the numerical model, excitations defined with the following dependencies were applied:

$$q(t) = q.\sin(\theta_i.t),$$

$$\theta_i = 2.\pi.\widetilde{f}_1, \text{ as } \widetilde{f}_1\text{=5,10,20,25,100 s}^{-1}.$$

Amplitude decay factor: $\gamma = 0{,}005$,

and

$$q(t) = q.e^{-0{,}05.\theta_i.t}.\sin(\theta_i.t)\ ,$$

$$\theta_i = 2.\pi.\widetilde{f}_1, \text{ as } \widetilde{f}_1\text{=5,10,20,25,100 s}^{-1}.$$

Amplitude decay factor: $\gamma = 0{,}005$.

3. Numerical results and discussion

3.1. Static analysis

In Figure 3a) displacement u_y is presented and Figure 3b) shows the distribution of σ_x under a load of 132 kN/m^2. The results from static analysis are presented in Table 4.

3.1.1. Linear elastic material models

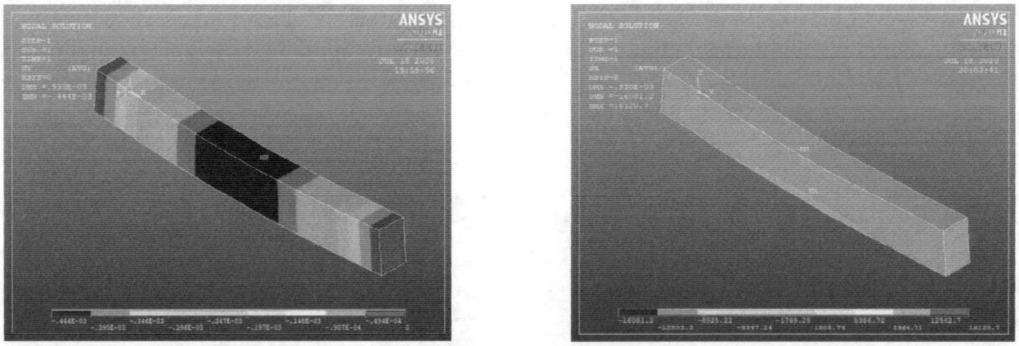

a) b)

Figure 3. a) Displacement u_y, b) Distribution of normal stress σ_x at the simply supported beam, q=132 kN/m^2

3.1.2. Nonlinear material model

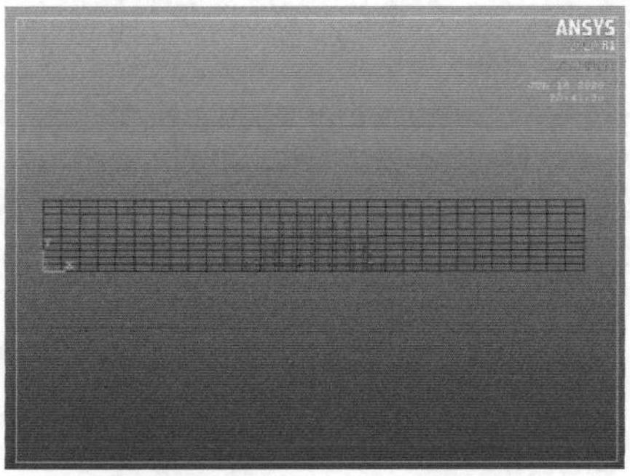

Figure 4. Nonlinear concrete material model. Crack initiation at applied force q=132 kN/m^2

Table 4. Results from static analysis max u_y, q=132 kN/m^2.

Parameter	Linear elastic material model (M1)	Plasticity material model q=132 kN/m2 (M2)	Plasticity material model q>132 kN/m2 (M2)
max $\|u_y\|$	0,000441525 m	0,00441943 m	0,00160113

3.2. Dynamic analysis – Harmonic vibrations.

Table 5. Results from dynamic analysis – harmonic vibrations displacement max u_y.

Load	max u_y, m $q(t)=q.\sin(\theta_i.t)$, $\gamma=0{,}005$	max u_y, m $q(t)=q.e^{-0{,}05.\theta i.t}.\sin(\theta_i.t)$, $\gamma=0{,}005$
Static	0,000441525	0,000441525
Dynamic-harmonic		
q(t)=132.sin(2.π.5.t)	0,000575430	0,000519005
q(t)=132.sin(2.π.10.t)	0,000840289	0,000700378
q(t)=132.sin(2.π.20.t)	0,016037000	0,003785650
q(t)=132.sin(2.π.25.t)	0,001941520	0,001379130
q(t)=132.sin(2.π.50.t)	0,000292689	0,000261438

In Table 5 are presented the results from the harmonic analysis – max u_y upon the applied load: $\boldsymbol{q(t)=q.\sin(\theta_i.t)}$, with the amplitude decay factor $\boldsymbol{\gamma=0{,}005}$ and $\mathbf{q(t)=q.e^{-0{,}05.\theta i.t}.\sin(\theta_i.t)}$, with the amplitude decay factor $\boldsymbol{\gamma=0{,}005}$

In figure 5, the "time-displacement" relation due to harmonic analysis $\boldsymbol{q(t)=q.\sin(\theta_i.t)}$, $\boldsymbol{\gamma=0{,}005}$ is shown.

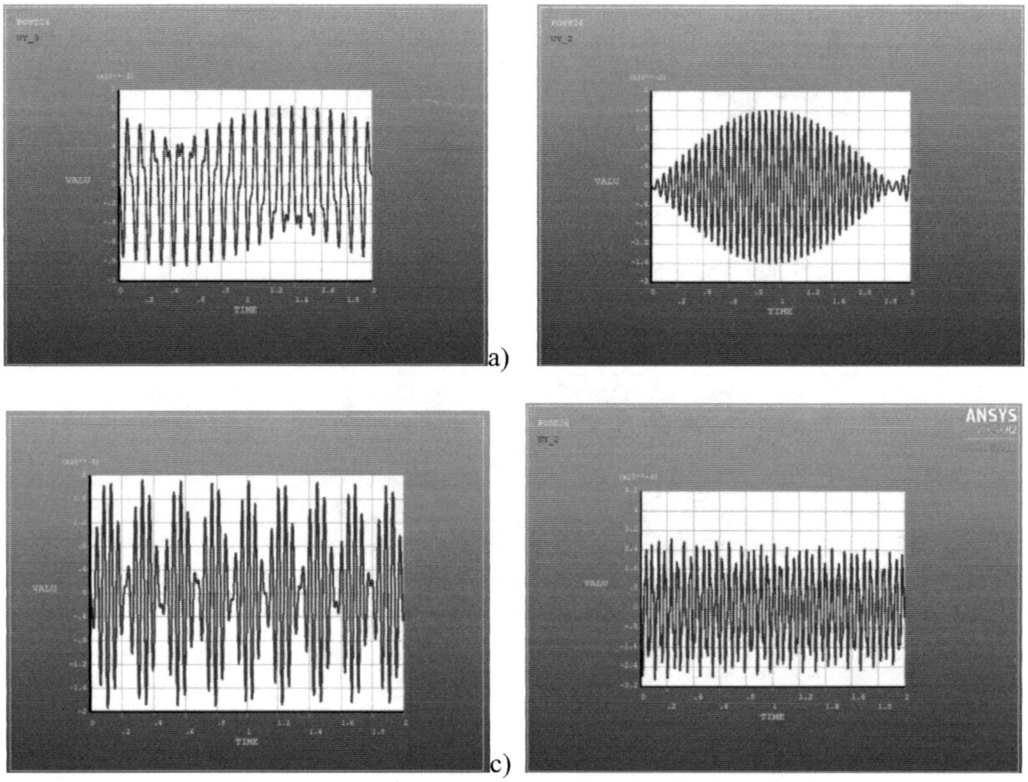

Figure 5. Displacements max u_y from harmonic analysis
$\mathbf{q(t)=q.\sin(\theta_i.t),\ \gamma=0{,}005}$
a) $\widetilde{f}_1=10\ s^{-1}$; b) $\widetilde{f}_1=20\ s^{-1}$; c) $\widetilde{f}_1=25\ s^{-1}$; d) $\widetilde{f}_1=50\ s^{-1}$

In figure 6, the "time-displacement" relation due to harmonic analysis $\boldsymbol{q(t)=q.e^{-0{,}05.\theta i.t}.\sin(\theta_i.t)}$, $\boldsymbol{\gamma=0{,}005}$ is shown.

In Figures 7 and 8 "q(t)- u_y" relations from harmonic vibrations are presented.

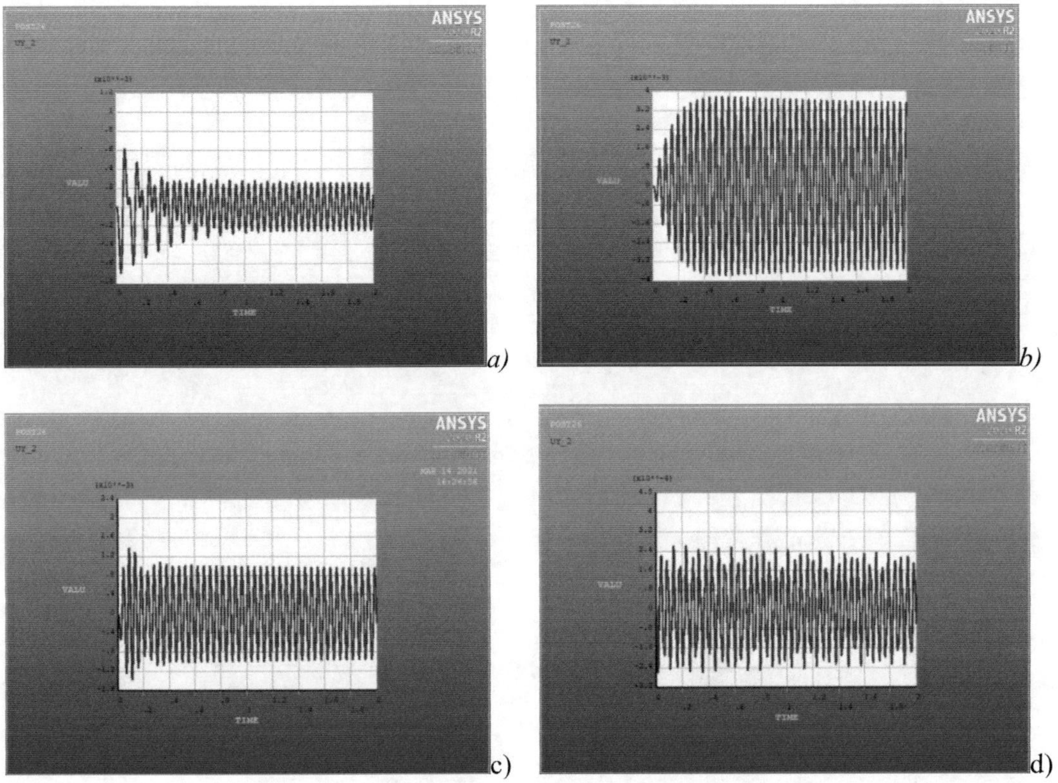

Figure 6. Displacements max u_y from harmonic analysis
$q(t)=q.e^{-0,05.\theta i.t}.\sin(\theta_i.t)$, $\gamma=0,005$
a) $\tilde{f}_1=10\ s^{-1}$; b) $\tilde{f}_1=20\ s^{-1}$; c) $\tilde{f}_1=25\ s^{-1}$; d) $\tilde{f}_1=50\ s^{-1}$

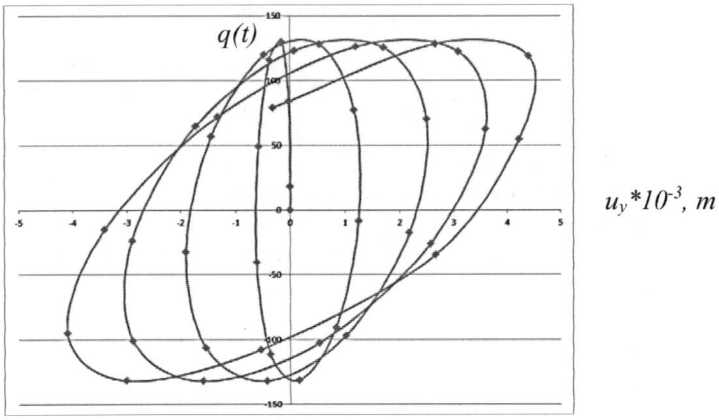

Figure 7. "q(t)- u_y" relation from harmonic analysis
$q(t)=q.\sin(\theta_i.t)$, $\gamma=0,005$, $\tilde{f}i=20\ s^{-1}$.

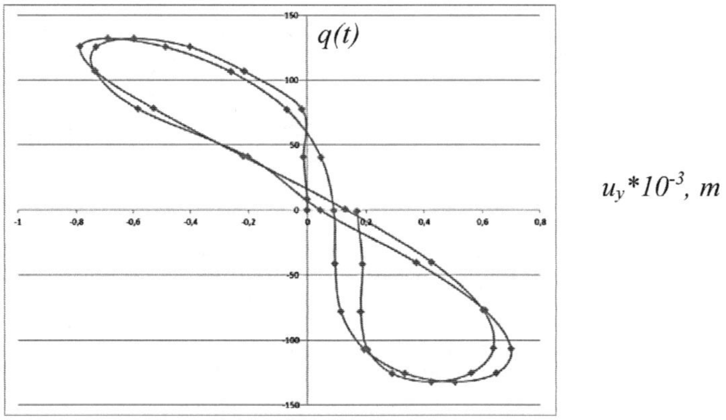

Figure 8. "q(t)- u_y" relation from harmonic analysis
$q(t)=q.e^{-0,05.\theta i.t}.sin(\theta_i.t)$, γ=0,005, $\tilde{f}i$=20 s^{-1}.

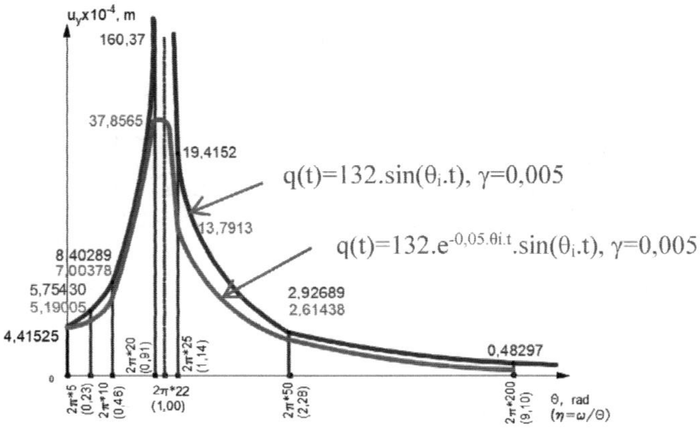

Figure 9. Harmonic vibrations. Relation " u_y- η".

4. Conclusions

The FEM provides opportunities for further studying a stress-and-strain state of constructive members subjected to local concentrated static load and dynamic vibrations, taking into account the material and geometric nonlinearity. The results obtained by these numerical models are much closer to the actual behaviour of the investigated systems and their constructive members.

At static analysis, loads are constant in size and location and are applied slowly to the structure. The accelerations and inertial forces that develop during deformation and cause additional load are very small. In most cases, the loads on the structures and elements have a dynamic character. They cause deformations in elastic systems with high accelerations, so that the additional inertial load cannot be ignored. Dynamic analysis clarifies both quantitatively and qualitatively the actual phenomena related to stresses and strains caused from dynamic loads. At harmonic vibrations of increasing frequency likely to reach frequencies close to resonance – Figure 9, the vertical displacements increase, reaching values from 2 to over 40 times higher than the values obtained in static analysis. For small values of the ratio η, i.e. when the force changes slowly, the amplitude of the vertical displacements is close to the static load displacements. The curves of the dependence u_y-η for the different values of the amplitude decay factor are approaching. When the displacements and deformations continue to increase, at some point in time, destruction occurs if it is made of a brittle material, and if it is made of ductile material, yielding will occur. At values of η> 1,20 or η <1,20 again approaching curves u_y–η are observed.

Dynamic response at small values of η - when the force is changing slowly, is approximately the same as a static load and is predetermined by the stiffness. Dynamic response at low values - when the force is changing very quickly and controlled by the mass. Increasing the η ratio values to reach the resonant frequency $\theta=\omega$ is expressed via a sharp increase in vertical displacements, respectively strains and deformations. After passing the resonant frequency, the vertical displacement values begin to decrease sharply. In damped systems, when the $\eta\approx 1$ system is very sensitive to damping. The displacements are significantly higher than the values of u_y at static loading. It is necessary to protect the structure from frequencies close to the resonant one.

Acknowledgements
This paper is financially supported by USEA (VSU) "Lyuben Karavelov", Sofia, project 08/2019.

5. References

[1] Bankov B 2004 *Theory of elasticity, stability and dynamic of elastic systems*, UACEG, Sofia
[2] *ANSYS Academic Reference Manual,* www.ansys.com
[3] Chopra A 1995 *Dynamic of structures*, Prentice Hall Inc. ISBN 0-13-855214-2
[4] *European standard Eurocode 2* – BDS EN 1992-1-1
[5] Willam K J and Warnke E P 1975 Constitutive models for the triaxial behavior of concrete. *Proc. Int. Assoc. for Bridge and Structural Engineering*, vol **19**, pp 1- 30
[6] Handruleva An, Matuski V, Kazakov K, Bankov B 2011 *Practical examples on Theory of elasticity and dynamics of structures*, VSU Publishing house, Sofia, 2011, ISBN 978-954-331-033-3, p.121
[7] Handruleva An 2020 *Mechanisms of destruction of spatial frame structures*, Sofia 2020, ISBN: 978-954-331-114-9
[8] Cvetkov St 2013 Anchoring length of the bearing reinforcement at reinforced concrete beams, In *Proc.:XIII International Scientific Conference VSU2013*, ISSN 1314-071X, pp. II-108-II-111
[9] Cvetkov St, *Constructors for civil engineers - the prefabricated one-storey buildings made of reinforced concrete elements*, VSU Publishing house, ISBN: 978-954-331-103-3

Seismic vulnerability of pre-code reinforced concrete shear wall buildings in Romania

M Barnaure[1]
[1] Faculty of Civil, Industrial and Agricultural Buildings, Technical University of Civil Engineering, Bucharest, 122 – 124, Lacul Tei Blv, 020396, Bucharest, *Romania*

E-mail: mircea.barnaure@utcb.ro

Abstract. Prior to 1963, when the first earthquake design code was implemented in Romania, mid height (over 8 storeys) shear wall buildings were already being built in some of the major cities. These buildings have important structural weaknesses when assessed using the current regulations. An 11 storeys residential building in Bucharest built in 1962 is analysed. Results show that the building is extremely vulnerable on the longitudinal direction, where the lateral resisting system is composed by only two shear walls. If bending capacity is lacking mainly at the base of the building, shear strength deficiencies are present over the entire height of the building, due to a lack of shear reinforcement in the walls. An analysis of several possible strengthening solutions is made, and conclusions are drawn regarding the feasibility of these solutions.

1. Introduction

A large part of Romania, including some of the major cities, are in earthquake prone areas [1]. While lateral loads were sometimes taken into account for the design of buildings even prior to the introduction of the first official provision in 1963 [2], the strong earthquake from March 4, 1977 [3] showed the structural vulnerability of previously constructed buildings. The Romanian earthquake design codes were amended several times based on both local and international expertise, with the latest version being adopted in 2013: P100-1:2013 [4]. The vulnerability of existing buildings is assessed by comparing their strength with that of a code-compliant building. The current Romanian code, P100-3:2019 [5] is in many aspects similar to the European norm EN 1998-3:2005 [6].

Of particular interest when analysing the seismic vulnerability of existing buildings are the mid-rise (over 8 storeys) buildings constructed prior to the introduction of the earthquake codes. Such buildings, with a reinforced concrete structural system, were built in the 1960's in some of the major Romanian cities, including Bucharest. These buildings are unsafe and require structural strengthening. But a reasonable intervention solution might be difficult to choose, as the implications of the possible structural engineering solutions are multiple: economic, legal, functional, aesthetic, and so on. Based on the analysis of an actual building, some intervention possibilities are discussed, and some general conclusions are made.

2. Description of the analysed building

A basement, ground floor and 10 storeys apartment building built in 1962 in Bucharest was analysed. It consists of two very similar (from a structural point of view) sections, each with a rectangular shape in plan, 25.85 m by 11.55 m, separated over the full height by a seismic joint. These sections are called b1 and a5. They were named as such in the original project, as the analysed building is one of many similar buildings constructed in that period, and these buildings were made by adjoining several "a" or "b" section types. The height of the ground floor and of the stories is 2.75 m, while the basement is a technical space of 1.90 m height. The ground around the building is at level -0.75 m (measured from the level of the ground floor). The level at the top of the slab above the 10th storey is +30.17. Throughout the paper the references will only be made to results obtained for section b1, as results obtained for section a5 are very similar qualitatively and quantitatively.

2.1. The structural system

The structural system is with reinforced concrete shear walls, this constructive solution being typical for the design of similar structures from that period (before or immediately after the appearance of the first normative document of seismic design, P13-63 [2]).

The lateral resisting system (needed to withstand the earthquake forces) is dominated in the longitudinal direction by two "T"-shaped walls (in the longitudinal direction these walls take over practically the entirety of the seismic forces). On the longitudinal facades, there are reinforced concrete frames, but their stiffness is very low when compared to the shear walls from the axis of the building. In the transverse direction, there are a series of lamellar walls. The thickness of the walls is generally 15 cm, but they sometimes have flanges or small columns at their ends. Reinforced concrete floors are 11 cm thick, except for the balconies (8 cm) and the staircase (13 cm) – figure 1.

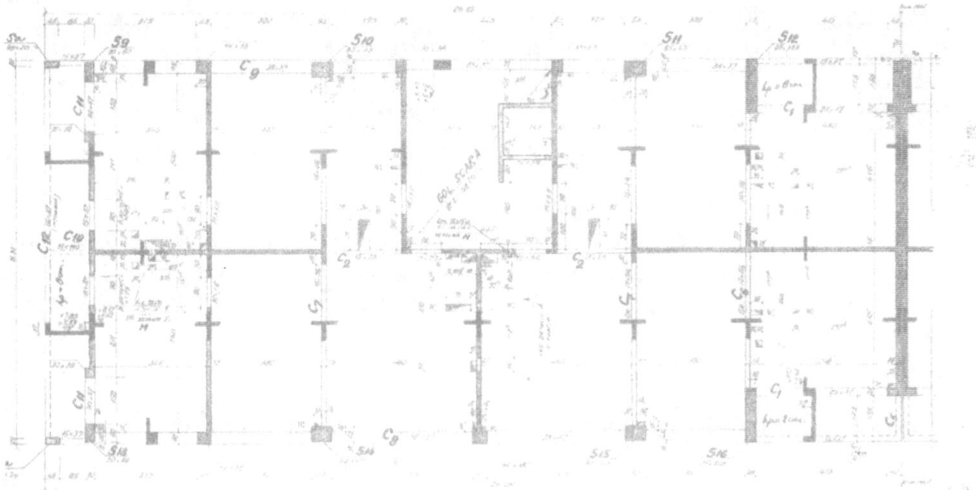

Figure 1. Structural configuration of the stories – section b1 (extract from the original project).

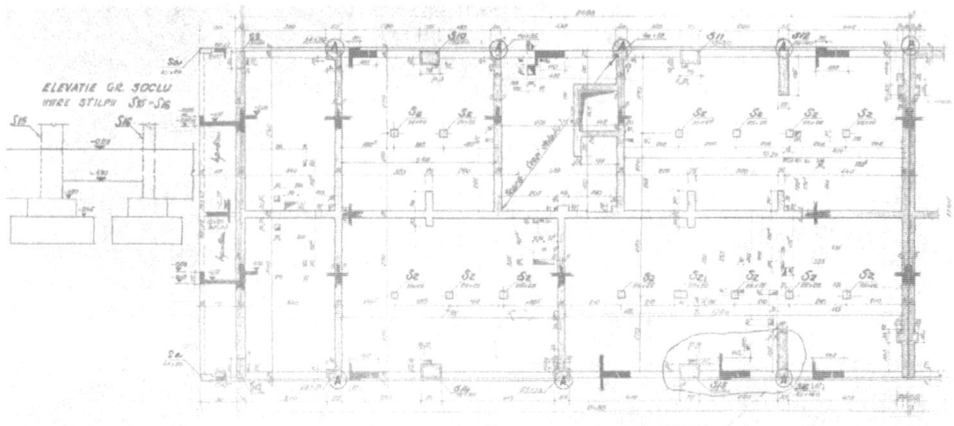

Figure 2. Structural configuration of the basement – section b1 (extract from the original project).

The vertical structural elements in the superstructure are continuous in the basement. The thickness of the structural walls in the basement is higher than in the superstructure (25 cm) and a few additional columns 25x25 cm or 25x50 cm act as supports in the central area of the slab above the basement – figure 2.

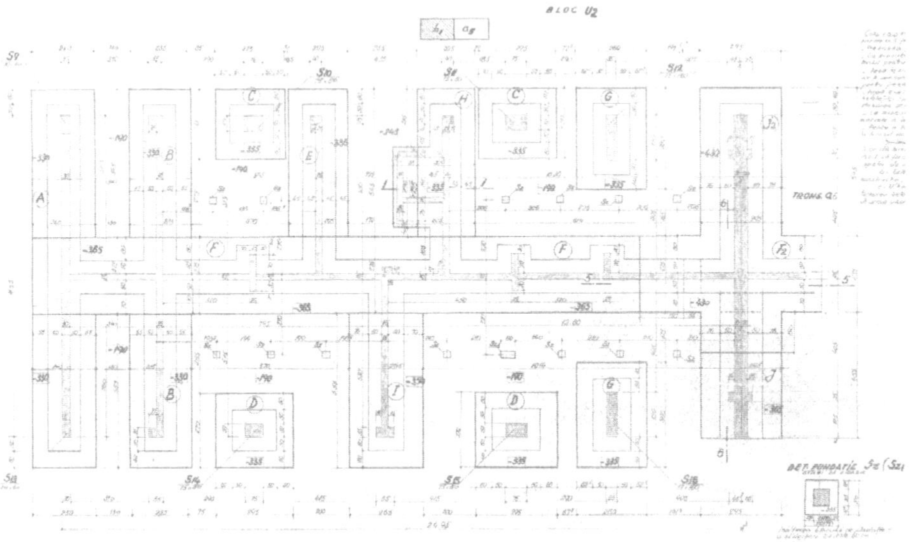

Figure 3. Structural configuration of the foundations – section b1 (extract from the original project).

The shallow foundations (figure 3) are continuous (with an average width of approximatively 250 cm) under the main shear walls, while isolated foundations exist under the columns. The connection between the foundations is made on the longitudinal facades by high (182 or 142 cm) beams beneath the slab above the basement. The base of the foundation is at a level varying from -3.35 m to -5.25 m.

2.2. Materials

The strength classes for the concrete mentioned in the original project are:

- B50 (C 2.8/3.5) in lean concrete foundation blocks
- B110 (C 6/7.5) in reinforced foundation beams
- B200 (C 12/15) in the structural elements for ground floor up to the 5th storey
- B170 (intermediate class between C8/10 and C12/15) from the 6th story upwards

For the typical concrete in the superstructure, the f_{cm} = 20 N/mm^2 was considered within the study.

The rebars (not deformed) are made of mild steel OL38 type. The f_{ym} = 284 N/mm^2 value was considered within the study.

2.3. The reinforcing detailing for the shear walls

The reinforcement of the elements is specific for the period of the 1960s. Thus, the columns and the intersections of the walls are reinforced (at the base of the building) longitudinally with Φ14 - Φ20 (less often Φ25) rebars and stirrups Φ8 / 30 cm (less often Φ8 / 25 cm). The diameters of the longitudinal bars and of the stirrups are lower in the upper floors. The stirrups are generally more densely spaced on the overlapping zones of the longitudinal reinforcements.

In figure 4 the reinforcing details of one of the main shear walls are shown. It is to be noticed that even in the lower storeys there is very little shear reinforcement in the web (Φ8 / 30 cm in basement and ground floor, Φ8 / 25 cm in storeys 1 and 2) while in the upper storeys there is no continuous reinforcement in the web, but only in the intersections between web and flanges.

IOP Conf. Series: Materials Science and Engineering　　1141 (2021) 012037　　doi:10.1088/1757-899X/1141/1/012037

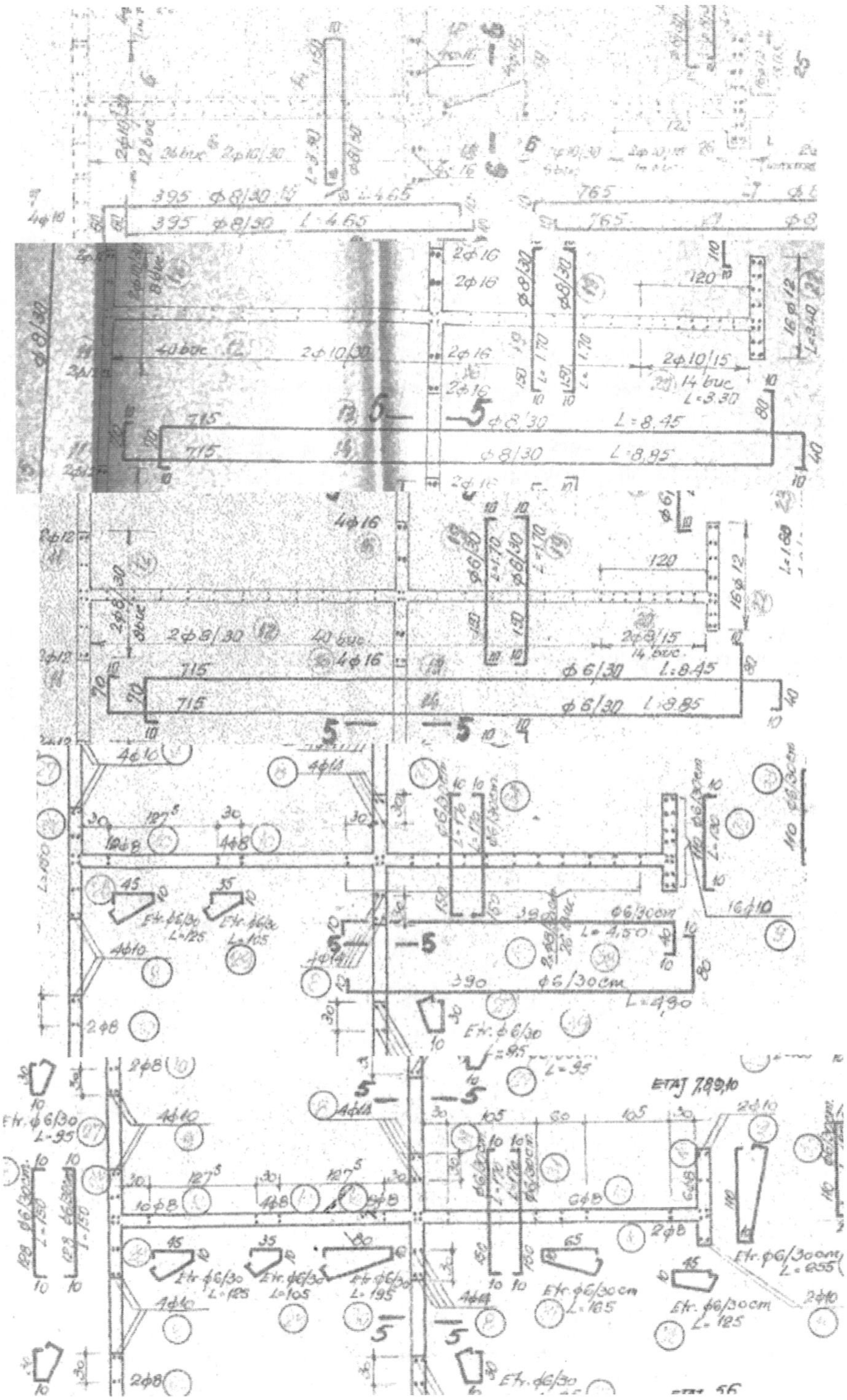

Figure 4. Reinforcing details for one of the main shear walls in the longitudinal direction – section b1 (extract from the original project).

2.4. Previous damages

In the approximately 60 years of existence, there were several earthquakes of Vrancea origin (the 1977 one having the largest magnitude). No reliable information was found about damages caused by these earthquakes. Still, an investigation report done on the building in 1993 mentions that damage was observed (cracks in the walls, cracks in the beams and lintels) and also that the building was damaged both during the 1977 earthquake (requiring repairs), as well as after the 1986 and 1990 earthquakes. The site investigation performed in 2020 identified no such damages. The only visible damage with possible seismic causes were cracks in the slabs in the staircase area (developed in the transverse direction of the building).

During the discussions with the tenants of the building, some of them living in the building in 1977, with one exception, all replied that they did not know that the building had suffered any damage to the structural elements or that structural interventions had been required and made. The mentioned damages were vertical cracks (cracks in plaster) in the areas where the reinforced concrete structural walls were continued with masonry (non-structural) partition walls. Also, the tenants who had recently carried out refurbishment works in the apartments said that they did not notice any cracks (even repaired) or other previous interventions when they removed the old finishes. Only one tenant living in the building in 1977 described damage to one of the walls (deep cracks, which required resin injection and wall cladding), but the investigations carried out on that wall (radar survey) did not validate the existence of these cracks.

However, as the 1993 report mentioned that: "the building was strengthened after the 1977 earthquake, when local repairs were made consisting in injections and local plastering of the shear walls with epoxy resin and fiberglass fabric", the existence of previous damage can not be excluded. In fact, as discussed further below, the area where the tenant mentions structural damage during the 1997 earthquake (see figure 5) corresponds to one of the most vulnerable areas of the structure.

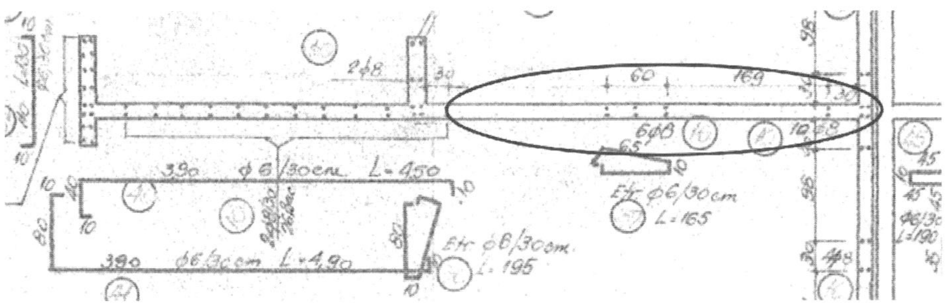

Figure 5. Reinforcing details for one of the main shear walls in the longitudinal direction – section b1 in story 3 (extract from the original project). The highlighted area is the one where the owner of one of the apartments mentioned damages (inclined cracks) appearing during the 1977 earthquake.

3. Numerical modelling

Several numerical models were developed using the ETABS software [7]. The shear walls were modelled using shell elements, the beams and columns were modelled using frame elements, while slabs were modelled as membrane elements. Some numerical models only include the superstructure of the building, while others also include the substructure. In the models that include the substructure, the foundations were modelled as point or surface elements, while the restraints were defined as linear springs, with characteristics in accordance with the results of the geotechnical survey. The analysis for the seismic combination was made considering static loads. No significant differences in results were observed between the equivalent lateral load and the response spectrum methods. Linear analyses were performed, considering the ductility factor of the building q=2.5.

Figure 6. Numerical 3D model including the substructure.

The vertical structural elements over the height of one storey for the b1 section, as defined in the numerical models, are shown in figure 7.

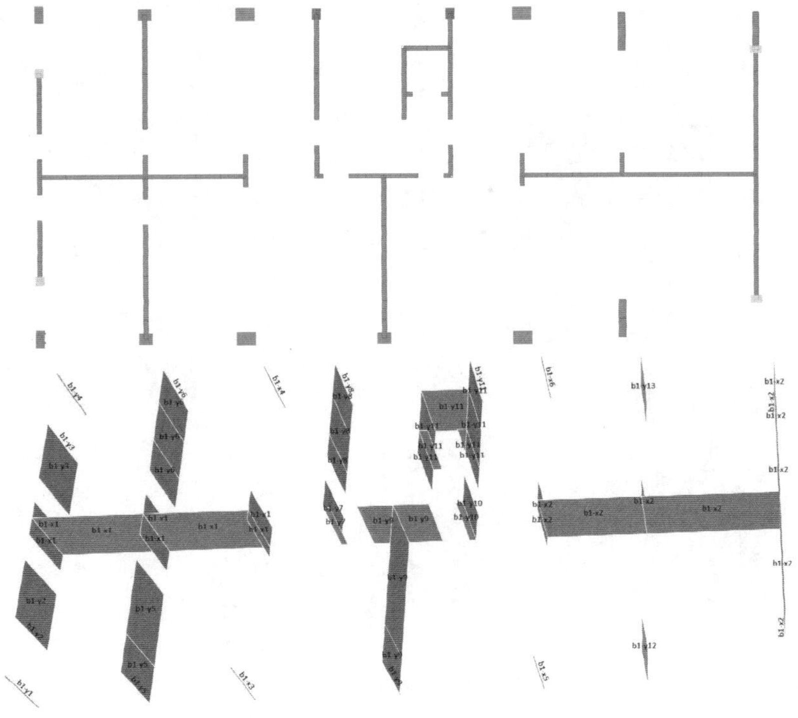

Figure 7. Vertical structural elements over the height of a storey for the b1 section of the building the numerical 3D (plan view and 3D view respectively).

For the b1 building considered fixed at its base, the natural periods of vibration are 1.2 s (transversal translation), 0.96 s (torsion) and 0.93 s (longitudinal translation). The modal participation mass ratios are over 70% for each of the main 3 modes. For the a5 building, values are slightly lower (periods 1.1 s, 0.85 s and 0.77 s for transversal translation, longitudinal translation, and torsion respectively.

4. Results

4.1. Preliminary qualitative assessment

The qualitative assessment, based on the analysis of the distribution and values of stresses in the shear walls – figure 8 -, indicated a significant susceptibility to damage under earthquake loads for the walls in the longitudinal direction and in particular to the wall b1-x2 (longitudinal wall from body b1 adjacent to the joint to section a5). This is the wall where potential previous damage during the 1977 earthquake had been mentioned by one of the occupants of the building.

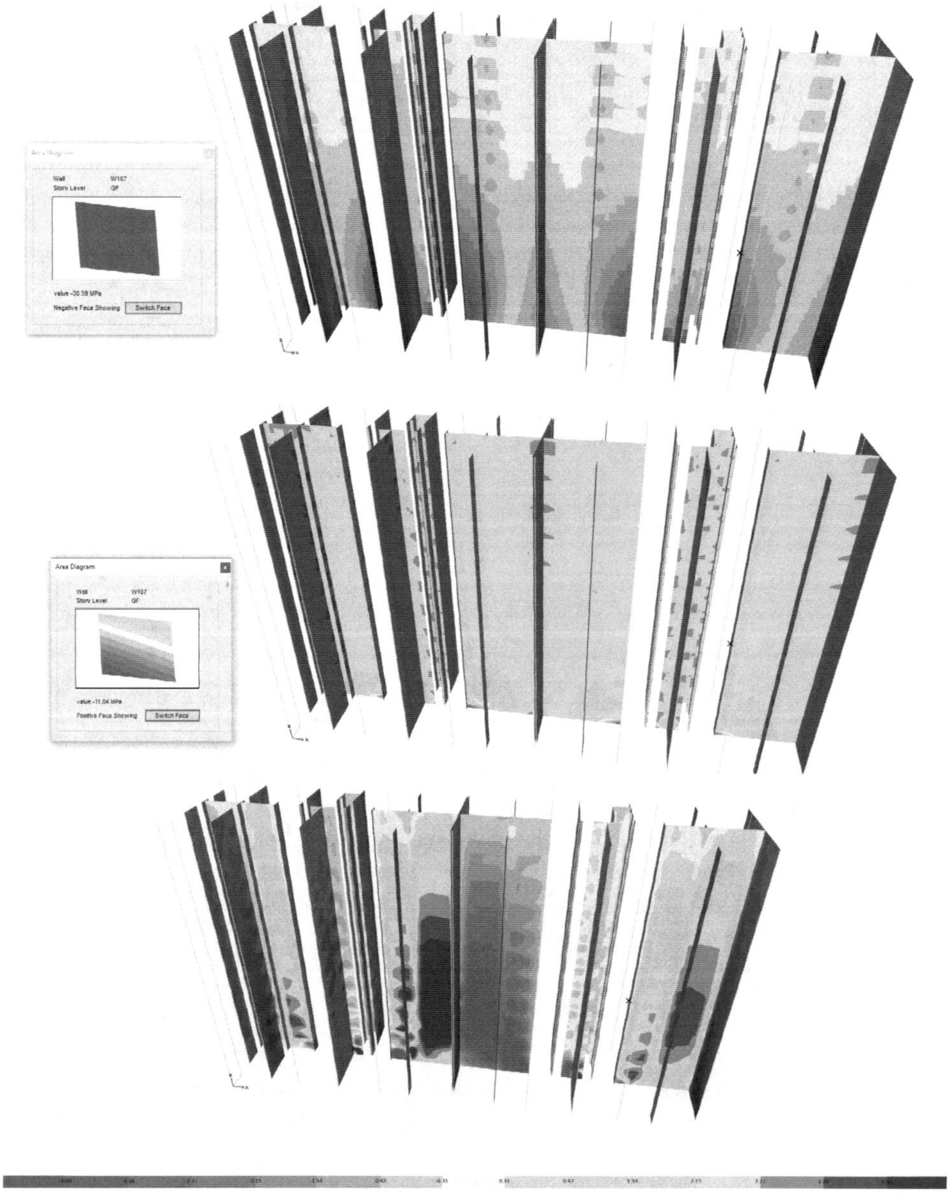

S_{22} stresses in shear walls – envelope for earthquake load combinations

S_{11} stresses in shear walls – envelope for earthquake load combinations

S_{12} stresses in shear walls – envelope for earthquake load combinations: magenta and blue areas correspond to values higher than 4 N/mm^2

Figure 8. Stresses in shear walls of the buildings b1 and a5 under earthquake load combinations

In terms of story drifts, the computed values for ULS are within the acceptable limit (2.5%), but for SLS the maximum computed value is twice the acceptable (0.5%) limit, which shows a risk related to damage of the non-structural elements.

4.2. Bending capacity check

The capacity calculation for each wall was determined using CSiCOL software [8]. The software determines the stress distribution in a given element, based on the geometry of the element, the properties of the materials (concrete and rebars) and the loads. An example of results provided by the software is shown in figure 9. Based on the capacity of the element and the actual level of stresses, the software computes the capacity ratio in bending.

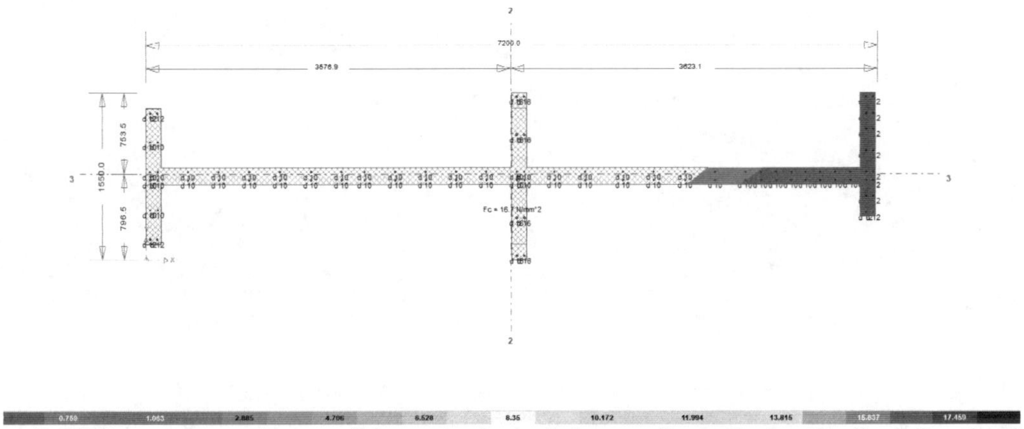

Figure 9. Ultimate stresses in one of the two principal shear walls on the longitudinal direction in the b1 section of the building under earthquake loads in the direction of the wall.

4.3. Shear capacity check

Shear capacity check was made following the rules in EN 1998-3:2005 [6], using μ_Δ=4, as stipulated in the P100-3 code [5].

By analysing the capacity ratios for bending and shear, shown in table 1, several observations can be made. The first one is that the building is much more vulnerable in the longitudinal direction, with capacity ratios as low as 19%, while capacity ratios in the transverse direction are higher than 0.55. The second observation is that, while bending capacity is low mostly in the levels at the base of the building, an important lack of shear strength is observed over the entire height of the building, due to the lack of shear reinforcement in the walls.

It is to be noted that in table 1, the values for the capacity in bending are average values (considered more relevant than individual values for each wall), as the ductile failure mechanism would lead to a redistribution of loads between the walls in a given direction. On the contrary, for shear stresses, the minimal value for the walls is shown. There is no real possibility of shear stress redistribution between the walls, mainly in the longitudinal direction. The brittle failure of one of the main walls in the longitudinal direction will cause the transfer of shear forces to the second wall and a rapid domino effect, which can lead to local or even global collapse of the building.

Table 1. Capacity check for the building under code earthquake loads

Story	Capacity ratio in bending		Capacity ratio in shear	
	Longitudinal direction	Transverse direction	Longitudinal direction	Transverse direction
GF	0.25	0.55	0.22	0.68
1F	0.26	0.62	0.19	0.67
3F	0.31	0.73	0.20	0.73
7F	0.88	1.0	0.43	0.90

4.4. Substructure check

The maximum pressure on the foundation ground in the fundamental load combination is lower than the 390 kPa allowable pressure determined in the geotechnical survey. Under earthquake loads along the longitudinal direction, the pressure levels remain moderate, with a maximum value of 310 kPa, but important shear forces develop in the foundation beam in the longitudinal axis (between the long shear walls) – figure 10.

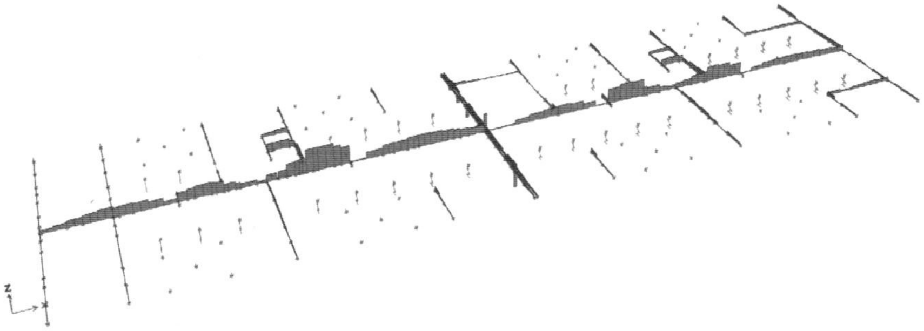

Figure 10. Qualitative representation of shear forces in the foundation beams under earthquake loads. The maximum values develop in the area between the two main shear walls in section b1.

In the case of earthquake load in the transverse direction, the global tilting effect is important, with the maximum allowable pressure under the external foundations corresponding to a value of the earthquake load equal to 55% of the code load.

4.5. General earthquake safety of the building

The analysis of the building under earthquake loads showed that the overall capacity of the building is only 20% of the required one, the most unfavourable failure pattern being shear failure of the two main walls along the longitudinal direction, either at the ground level or 1st story (where shear forces are the highest, but continuous shear reinforcement in the webs of the walls is scarce) or at 3rd and 4th level, where shear stresses are somewhat lower but there is no continuous web reinforcement.

This value corresponds to classifying this building in the Rs I (worst) earthquake risk class as defined by the P100-3 [5] code. As explained in the code, this category of buildings is susceptible to local or total collapse when submitted to the design earthquake corresponding to the Ultimate Limit State. Retrofit measures are mandatory for such buildings, with the minimal strength level to be attained being 65% (this value is the lower value for a building to be included in the Rs III class).

4.6. Discussion on the results

It is interesting to analyse a few of the reasons that make such a building which did not suffer extensive damage during the 1977 earthquake be considered today in the highest risk category. Even more as the analysed building is definitely not a singular case. In fact, even in the P100-3 [5] code, an example assessment analyses a different building from the 1960's, and that building also resulted to correspond to the highest risk class.

- The risk assessment is made for a reference earthquake (estimated return period of 225 years), which corresponds to peak ground acceleration values much higher than the values actually recorded in Bucharest in 1977;
- As the building is higher than 28 meters, it falls into the 2nd category of importance [4], which corresponds to a coefficient $\gamma_{I,e}$ = 1.20 (20% higher design earthquake loads);
- The evaluation methodology takes into account the uncertainties regarding the structure of the building through the confidence factor, CF=1.20, which translates in a reduction of the strength of elements taken into account for the assessment;
- In the case of brittle failures, the current codes (both national [5] and European [6]) impose

considering the safety factors for different materials when establishing the strength of the elements;
- The building is particularly vulnerable in the longitudinal direction, while the 1977 earthquake might have had a different principal direction;
- There are various possible mechanisms (rocking at the base, over-strength of rebars, energy dissipation due to degradation of floors, beams, non-structural elements, etc.) that may have contributed to reducing the level of damage to structural walls in 1977, but these mechanisms are not controllable and cannot be considered in the structural assessment;
- Although the level of damage suffered by the building in 1977 is uncertain (in 2020 only cracks in the floors were identified, although a previous report also mentions cracks in walls and beams that required local retrofit), a building that has already been damaged in a previous seismic event has lost some of its post-elastic deformation capacity and is likely to be much more severely damaged in the event of a new similar earthquake;
- Results indicate a significant risk of brittle wall failure; thus, until the strength of the structural walls is reached, the building can resist, without damage or with minor damage, but when one of the two longitudinal walls fails, it causes an immediate transfer of the entire lateral load to the second wall, with a sudden domino effect that might lead to local or even global collapse.

5. Possible retrofit solutions

As previously mentioned, the minimal acceptable capacity ratio for a strengthened building in Romania is 65%. Based on the results in the structural assessment, for the analysed building this goal would require interventions in the substructure (foundations and basement elements) as well as interventions in all the superstructure storeys. Under these conditions, the economic implications may be much greater than those relating exclusively to the strengthening of the structure, as they involve for example costs related to refurbishment of finishes damaged by the structural interventions or costs related to providing accommodation for the current building tenants during the intervention. In fact, studies analysing the decision-making processes after the L'Aquila [9] or Christchurch [10] earthquakes concluded that the solution for intervention (either repair or demolish and rebuild) is most often based on financial considerations. Still, some other variables, including among others the perception of risks and government decisions influence the final decision, which is rarely solely based on the structural requirements.

Several possible solutions are further discussed. It is to be noted that these solutions exclusively cover the structural safety aspects. Yet, an intervention project should be based on a complex and comprehensive study that analyses all the implications, not only technical, but also financial, legal, functional, aesthetic, and so on.

5.1. Solution type 1

The first retrofit solution consists in RC jacketing for all the vertical structural elements (walls and columns) in all the stories where a superior strength is required. The jacket could be 7 cm thick when applied on both sides of an element, or 15 cm thick for one-side intervention. This solution allows to:
- Favour ductile failure in bending and avoid brittle shear failure;
- Limit the compression stress levels in the vertical elements;
- Develop the compressed area at the free ends of the wall section;
- Direct the potentially plastic areas of the walls at their base.

In order to increase the capacity of foundations, two types of interventions are needed:
- Extending the surface of the foundations, especially on the longitudinal facades, by creating new foundation blocks between the existing foundation blocks;
- Jacketing the areas where high stresses are transmitted by the walls (mainly along the central axis of the building), and thus creating a continuous reinforced concrete foundation beam.

The advantages of this solution are:

- Interventions are performed differently from one element to another and from one floor to another, depending on the required increase in strength;
- The ground surface occupied by the building is not modified (or just to a very small extent).

As a sidenote, such a solution is also proposed in the case of the mentioned example building in the P100-3 [5] code.

At the same time, the solution has a significant disadvantage:

- As it is necessary to retrofit most of the elements (at least on the first floors), and as the longitudinal shear walls need to be jacketed over the whole height of the building, the implementation of the solution will inevitably involve extensive damage to finishes in all the apartments (with associated repair / replacement costs), as well as the need to relocate the tenants during the works.

Another disadvantage is related to the fact that, by jacketing the vertical elements, the usable area of the apartments is reduced, while some free-passage areas (corridors, staircases, etc.) become smaller (with possible functionality problems). Given the disadvantages of this solution, the authors consider that, as far as possible, intervention solutions 2 or 3, discussed below, should be favoured.

5.2. Solution type 2

The second intervention solution (figure 11) consists in adding reinforced concrete cores and reinforced concrete frames on the outside of the building. These newly introduced structural elements would withstand most of the lateral forces, especially in the longitudinal (deficient) direction, so that existing elements in the superstructure would no longer need interventions inside the building (or only small interventions).

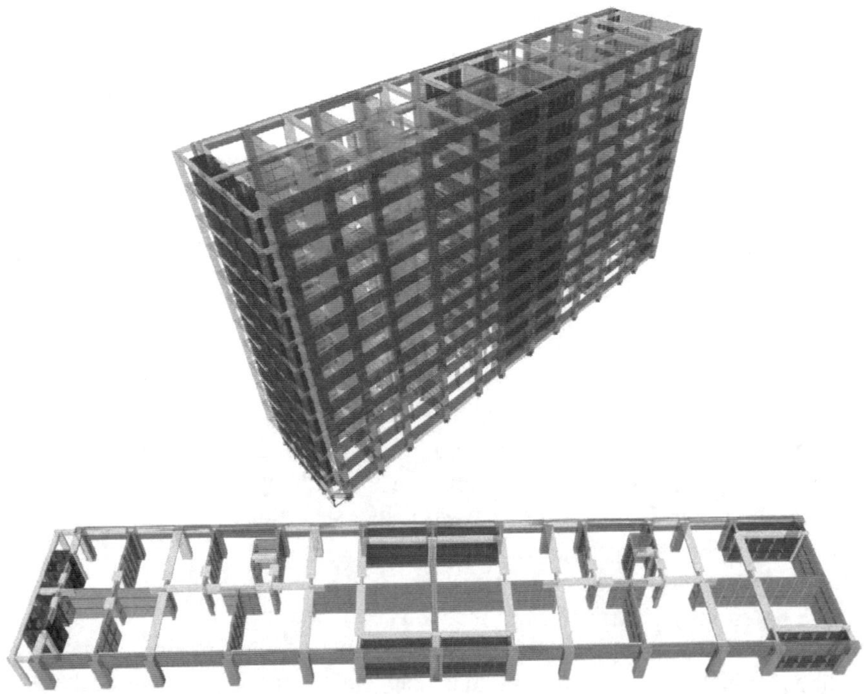

Figure 11. Type 2 strengthening solution (3D and current level view).

The newly introduced cores could be positioned in the balconies area and possibly in the staircase area. The main advantage of this solution is that most of the interventions could be carried out without the need to relocate the occupants of the apartments for long periods of time, and the interventions would not involve extensive degradation of interior finishes.

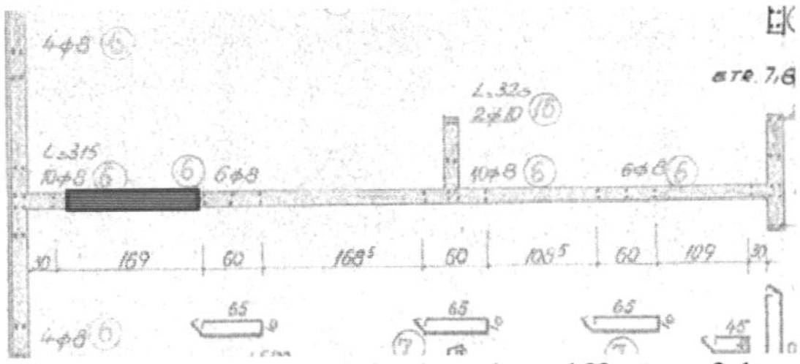

Figure 12. Possible intervention solution for reducing the stiffness of the existing longitudinal walls.

However, the solution also raises certain problems:

- The ground footprint of the building and its exterior appearance change;
- It is difficult to connect the new exterior frames to the existing elements on the facade as to ensure the transfer of seismic forces to the cores;
- As the new elements should be very stiff compared to the existing structure (in order to limit the forces transmitted to the existing elements), the new RC cores should have important in-plane dimensions as well as limited door / windows openings;
    - In order to reduce the stiffness of the existing longitudinal walls in the axis of the building (and thus the lateral loads on them), voids could be created (by cutting with a diamond disc) – as shown in figure 12. This type of intervention, even if it is done inside the apartments, is much less invasive than wall jacketing.
- The foundations of the newly introduced cores would most likely need to be deep foundations (piles), as the new elements will be very stressed in bending, but with a low level of vertical loads; this implies high costs for the foundations.

5.3. Solution type 3

The third intervention solution is a mixed intervention, which has some of the advantages, but also of the disadvantages of the previously discussed solutions. The solution consists in adding reinforced concrete frames on the outside of the building, on the longitudinal facades, combined with additionally jacketing the elements with insufficient strength.

Figure 13. Type 2 strengthening solution (3D view of the superstructure).

The main advantage of this intervention solution is that it allows making intervention in two steps. Such a possibility is allowed by the P100-3 code [5].

Thus, in the first stage:

- Interventions are made at the foundations on the main facades and in the central axis;

- A system of reinforced concrete frames is created, on the outside of the building, on its entire height; the frames are connected to the existing structure using chemical anchors fixed in the existing beams and pillars on the façade;
- The intervention on the facades can be completed with thermal rehabilitation works for the building; the facade repair works are useful to avoid further damage to the building until the funds are available to carry out the remaining major repair works;
- Following this first stage, the building can pass from the current Rs I seismic risk class into Rs II class (capacity ratio higher than 35%), limiting the risk of collapse in case of a severe earthquake.

In the second stage:

- Jacketing is conducted inside the building, but, as frames were introduced on the façade in the previous stage, the intervention extent for the elements inside the apartments is reduced;
- Following this second stage of intervention, strength of the building will correspond to the Rs III seismic risk class (capacity ratio higher than 65%).

6. Conclusions

The analysis of a reinforced concrete shear wall structure built prior to the introduction of the seismic design code provisions in Romania led to the conclusion that the existing building is highly vulnerable and is susceptible to local or total collapse when submitted to the design earthquake corresponding to the Ultimate Limit State. The main deficiency is the very low shear strength along the longitudinal direction, which is observed not only at the base of the building, but also in the upper storeys. Extensive interventions are needed for the structural strengthening of the building. This situation poses difficulties in choosing the most appropriate solution. Several intervention possibilities were analysed, and their advantages and disadvantages discussed. As the vulnerabilities of studied building are actually representative for many similar buildings from the 1960's, further analysis should be carried out in the direction of identifying the optimal retrofit solution for this building type. Such analysis should not only involve structural engineers, as the possible technical solutions have different architectural, economic and legal implications that should be taken into account.

7. References

[1] Oncescu M C, Marza V I, Rizescu M and Popa M 1999 The Romanian earthquake catalogue between 984–1997 *In Vrancea earthquakes: tectonics, hazard and risk mitigation* Springer, Dordrecht

[2] P13-63 *Conditioned Standards for Civil and Industrial Constructions Design in Seismic Regions (in Romanian)*

[3] BERG G V 1980 *Earthquake in Romania, March 4, 1977: an engineering report.* National Academy Press.

[4] P100-1:2013 *Seismic Design Code. Part 1: Design Provisions for Buildings (in Romanian)*

[5] P100-3:2019 *Seismic Design Code. Part 3: Seismic assessment of existing buildings (in Romanian)*

[6] EN 1998-3:2005 *Eurocode 8: Design of structures for earthquake resistance – Part 3: Assessment and retrofitting of buildings*

[7] CSI ETABS 18.1.1 2018 Computers & Structures, Inc.

[8] CSI Column 9.0.1 2014 Computers & Structures, Inc.

[9] Polese M, Di Ludovico M and Prota A 2018 Post-earthquake reconstruction: A study on the factors influencing demolition decisions after 2009 L'Aquila earthquake *Soil dynamics and earthquake engineering* **105** 139

[10] Marquis F Kim J J Elwood K J and Chang S E 2017 Understanding post-earthquake decisions on multi-storey concrete buildings in Christchurch, New Zealand *Bulletin of Earthquake Engineering* **15(2)** 731

The effects of parameter uncertainties on the numerical plastic analysis of non-prismatic reinforced concrete beams

S Khaleel Ibrahim[1], M Movahedi Rad[1] and M A Habashneh[1]
[1] Department of Structural and Geotechnical Engineering, Széchenyi István University, H9026 Győr, Hungary

E-mail: Sarahibrahim8391@gmail.com

Abstract. Structural models and parameters are usually considered deterministic in most studies. However, test results revealed that uncertainties exist in most of these cases due to natural variability or lack of awareness. Therefore, uncertainty in structures is taking huge attention as it has an important role in the structures analysis and the calibration of designing code factors where it enhances the prediction of the models' behavior. Despite this significance, existing knowledge concerning uncertainties and their parameters seems to suffer from a lack of knowledge especially in unique structures like non-prismatic reinforced concrete beams which are considered a special case in structural engineering for having variable depth across beam section. The present paper aims to examine some uncertainties facing structural engineers and their role in the non-prismatic reinforced concrete beam behavior considering different variables and parameters. Concrete Damage Plasticity (CDP) constitutive model is used to calibrate the numerical model according to the data obtained experimentally. Then different uncertainty parameters were studied by applying several simulations to find out their effect on models' behavior. The results seem to provide a better understanding of the uncertainties and their effect on the non-prismatic reinforced concrete beams. Lastly, finite element simulations are applied using ABAQUS.

1. Introduction

The lack of experimental researches studying the behavior of reinforced concrete (RC) non-prismatic beams leads to challenges in both deterministic and probabilistic analyses. Non-prismatic beams are slender structures with cross-section parameters (like shape, dimensions and centroid position) that could vary along the beam length. Researchers are interested in this type of structures as it is possible to optimize structures shapes according to the design conditions. For instance, if the size of the beam cross-section changes proportionally to the internal stress amount, the beam gains the needed structural strength with a minimum material amount. Consequently, non-prismatic structures are broadly used in several engineering areas, such as buildings and bridges design [1].

The thought of nonlinearity behavior of concrete is more difficult than reinforcement steel behavior in finite element models. Several constitutive models exist to represent the nonlinear response of concrete. However, damage mechanics and plasticity, or the mixture of them together is commonly utilized as the principal concepts of modeling concrete material. Still, the strain is divided into elastic and plastic sections describing the unloading and path dependency well, besides the decrease of the unloading stiffness can't be described in plasticity models. In opposite, damage mechanics models are built on the concept of a progressive loss of elastic stiffness whilst they are usually incapable of describing permanent deformations. The mixture of plasticity and isotropic damage to enhance their abilities leads to the concrete damage plasticity [2].

Recently, Finite Element (FE) analyses participated in providing remarkably detailed descriptions. Where modern constructions need further critical and complicated designs, the demand for accurate methods to estimate uncertainties in computer models, loadings, material and geometry properties has progressed importantly. Generally, in problems where randomness is comparatively small, it is preferred to use a deterministic model more than a stochastic model. Yet, as the effect of uncertainty is essential, stochastic methods are required for both analysis and design. Regularly, both load and strength are variables, the values of which are distributed about their mean values and the factor of safety usually is too conservative, causing a higher costly design. As opposed to the deterministic approach which is built on safety factors, the stochastic method enhances design reliability where it provides several benefits to engineers [3].

Kappos et al., [4] studied the strength and ductility of RC beams and columns cross-sections considering different geometries and levels of restriction where material characteristics are considered as random variables and their effect on section behavior is evaluated. Mainly, the ductility of the section is in several states decided to be essentially reliant on the ultimate concrete strain. Also, the used Monte Carlo simulations reveal that a notable quantity of variability exists in both the strength and ductility of restricted concrete sections because of the changes in the properties of the material.

Numerical investigations have become an important common approach for analyzing different engineering systems. Simulation is usually described as the means of replacing the actual world based on a combination of theories and imagined models of reality. Monte Carlo simulation was developed to be an experimentally probabilistic approach to determine complex deterministic difficulties as computers simply simulate a huge number of experimental tests which hold random results [5].

Nicolis [6] showed that in nature, the method of determination, on which the detective interacts with a physical system, is bounded with finite accuracy. As a consequence, the 'state' of a system needs to be recognized in reality, not like a point in phase space but preferably like a small area that the finite accuracy of the measuring apparatus is reflected by size. So, the applicability of Monte Carlo simulation in the uncertainty evaluation of various phases of a system appears to give a further realistic method. Remarkable benefits of Monte Carlo simulation have been reported by Basil and Jamieson [7] where the method can manage both small and large uncertainties in the input quantities.

Herrador and González [8] introduced the principal boundaries of the Guide to the expression of Uncertainty Measurement (GUM) method for assessing the uncertainty measurement. The benefits of adopting Monte Carlo simulation upon the GUM approach are described and explained and the main propagation of distributions is illustrated.

Papadopoulos and Yeung [9] Confirmed that the Monte Carlo simulation approach is completely agreeable with the conventional uncertainty evaluation techniques for linear systems and systems with small uncertainties. Where Monte Carlo simulation owns the capacity to take into account partially correlated estimation input uncertainties. It likewise checks the uncertainties of the results of some essential manipulations that may or may not be connected.

Kioumarsi et al., [10] investigated the influence of an interference of corrosion pits on nearby reinforcement steel on the probability of bending failure of a rusted RC beam. Where the probability of failure is evaluated by applying Monte Carlo simulation. Moreover, uncertainties in material properties, geometry, loading, and corrosion damage are taken into consideration.

This study aims to investigate the effect of probability failure on the reinforced concrete non-prismatic beams using deterministic and probabilistic finite element analysis. Firstly, validation of the numerical model was made applying the concrete damage plasticity (CDP) constitutive model after getting the required data from the experimental tests. Then, a MATLAB code was developed to achieve the desired aim considering the probability of failure as a bound while the compressive strength (f_c') and the modulus of elasticity (E_c) of concrete are considered variables with a 5% standard deviation. The mean values of the variables are therefore specified by ($f_c' = 35$ N/mm^2) and ($E_c = 26420$ N/mm^2). Besides, an advanced Monte Carlo sampling technique is used to create the samples of the material properties depending on the statistical parameters of the material where the final step includes the calculation of the corresponding load and displacement values.

This paper structure is detailed as the following: Section 1 defines the introduction while Section 2 develops the concrete damage plasticity model. Furthermore, Section 3 introduces the probability analysis; meanwhile, the numerical model validation is illustrated in section 4. Finally, Sections 5 and 6 conclude the discussion of the results and conclusions respectively.

2. Concrete damage plasticity model

To investigate the behaviour of concrete in an inelastic form, the isotropic damaged elasticity, the isotropic tensile and compressive plasticity are exhibited in the concrete damaged plasticity model. The overall strain amount ε is consisted of the elastic division ε^{el} and the plastic division ε^{pl} as simplified below:

$$\varepsilon = \varepsilon^{el} + \varepsilon^{pl} \quad (1)$$

$$\sigma = D^{el}:(\varepsilon - \varepsilon^{pl}) \quad (2)$$

$$\bar{\sigma} = D_0^{el}:(\varepsilon - \varepsilon^{pl}) \quad (3)$$

$$D^{el} = (1-d)D_0^{el} \quad (4)$$

Hence D^{el} refers to the elastic degraded stiffness and D_0^{el} is the initial (undamaged) elastic stiffness of the material, while d is the variable of the scalar stiffness degradation. Wherever equation (2) can be updated considering the nominal stress with the decreased elastic tensor given in equation (4) providing the resulting equation:

$$\sigma = (1-d)D_0^{el}:(\varepsilon - \varepsilon^{pl}) \quad (5)$$

Consequently, the Cauchy stress σ is related to the effective stress $\bar{\sigma}$ by the scalar reduction relation:

$$\sigma = (1-d)\,.\bar{\sigma} \rightarrow \sigma = (1-d_t)\bar{\sigma}_t + (1-d_c)\bar{\sigma}_c \quad (6)$$

Hence the variables of the scalar damage, d_t and d_c, are varying from 0 expressing the undamaged state until 1 expressing a completely damaged state. Generally, the damage model applied on the concrete got into account the failure manner of tensile cracking and compressive crushing. The uniaxial tension and compression response of concrete are deemed to be affected by plasticity damage as revealed in figure 1. The uniaxial tension and compression response of concrete in accordance to the concrete damage plasticity model affected by compression and tension loads are given by:

$$\sigma_t = (1-d_t)E_0(\varepsilon_t - \varepsilon_t^{pl,h}) \quad (7)$$

$$\sigma_c = (1-d_c)E_0(\varepsilon_c - \varepsilon_c^{pl,h}) \quad (8)$$

While $\varepsilon_t^{pl,h}$ and $\varepsilon_c^{pl,h}$ are the equivalent plastic strains in tension and compression, respectively. Therefore, the effective uniaxial compression and tension stresses $\bar{\sigma}_t$ and $\bar{\sigma}_c$ are concluded as follows:

$$\bar{\sigma}_t = \frac{\sigma_t}{(1-d_t)} = E_0(\varepsilon_t - \varepsilon_t^{pl,h}) \quad (9)$$

$$\bar{\sigma}_c = \frac{\sigma_C}{(1-d_c)} = E_0(\varepsilon_C - \varepsilon_c^{pl,h}) \quad (10)$$

As tensile strain ε_t equals to $\varepsilon_t^{pl,h} + \varepsilon_t^{el}$, and compressive strain ε_c equals to $\varepsilon_c^{pl,h} + \varepsilon_c^{el}$.

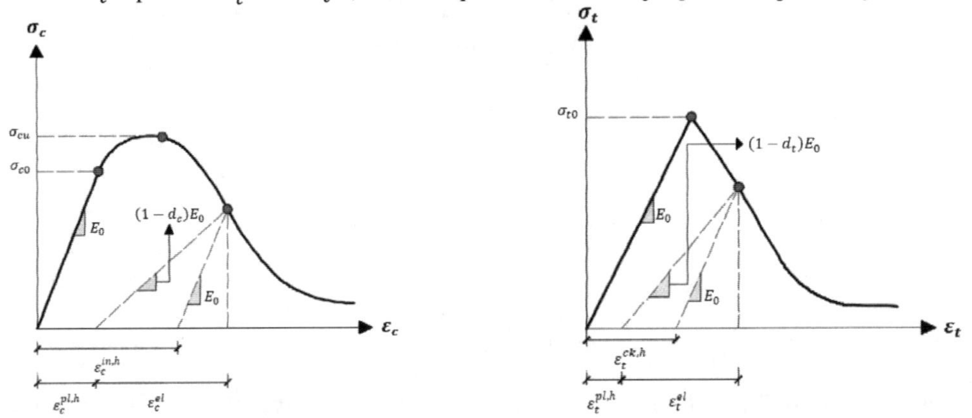

(a) Compression (b) Tension

Figure 1. Concrete response to uniaxial loading condition

2.1. Probability analysis

By presenting the main concepts of probability analysis and by presuming that X_R expresses the nonnegative limit for X_S, the failure problem could be determined by $X_R \leq X_S$. Let X_R and X_S be the independent random variables accompanied by the functions of density probability $f_R\,(X_R)$ and $f_S\,(X_S)$. Then the failure probability can be calculated from the following:

$$P_f = P[X_R \leq X_S] = \iint_{X_R \leq X_S} f_R\,(X_R)\, f_S\,(X_S) dX_R dX_S \tag{11}$$

An optional description of the earlier problem is given in terms of the so-called boundary state function identified by:

$$\mathrm{g}(X_R, X_S) = X_R - X_S \tag{12}$$

Seeing that $g \leq 0$ designates the domain of failure D_f. Consequently, the failure probability P_f is denoted by:

$$P_f = F_g(0) \tag{13}$$

Additionally, P_f can be estimated as:

$$P_f = \int_{\mathrm{g}(X_R, X_S) \leq 0} f(X)dX = \int_{D_f} f(X)dX \tag{14}$$

Monte Carlo (MC) technique includes creating realizations x of the random vector X of their collective probability density function $f_X(x)$ and verifying if a delivered realization drives to failure or not. Points number in the domain of failure concerning the entire number of formed points is an estimator of failure probability. This concept may be signified by proposing the next indicator function of the failure domain D_f:

$$\chi_{D_f}(x) = \begin{Bmatrix} 1 & \text{if } x \in D_f \\ 0 & if\ x \notin D_f \end{Bmatrix} \tag{15}$$

Therefore, the probability of failure in equation (14) can be rewritten as:

$$P_f = \int_{-\infty}^{+\infty} \cdots \int_{-\infty}^{+\infty} \chi_{D_f}(x) f_X(x) dx \tag{16}$$

Then, the function $\chi_{D_f}(X)$ is accordingly considered as a random variable with two points distribution.

$$\mathbb{P}\left[\chi_{D_f}(X) = 1\right] = P_f \tag{17}$$

$$\mathbb{P}\left[\chi_{D_f}(X) = 0\right] = 1 - P_f \tag{18}$$

As $P_f = \mathbb{P}[\, X \in D_f\,]$, so the variance and the mean value of $\chi_{D_f}(X)$ are provided by:

$$\mathbb{E}\left[\chi_{D_f}(X)\right] = 1 \cdot P_f + 0 \cdot (1 - P_f) = P_f \tag{19}$$

$$\mathbb{V}ar\left[\chi_{D_f}(X)\right] = \mathbb{E}\left[\chi_{D_f}^2(X)\right] - (\mathbb{E}\left[\chi_{D_f}(X)\right])^2 = P_f - P_f^2 = P_f(1 - P_f) \tag{20}$$

Furthermore, the following mean value estimator is applied to estimate the failure probability in the MC sampling:

$$\hat{\mathbb{E}}\left[\chi_{D_f}(X)\right] = \frac{1}{K}\sum_{k=1}^{K}\chi_{D_f}(X^{(k)}) = \hat{P}_f \tag{21}$$

As $X^{(k)}$ are considered independent random vectors (where $k = 1, \dots, K$) with functions of probability of density presented by $f_X(x)$. Moreover, the fraction of volume is described as a random variable and for determination of effortlessness, it supports the Gaussian distribution with the delivered mean $\mathbb{E}$ and the variance by $\mathbb{V}ar$. Then, the mean and the variance of the estimator can be simply estimated as follows:

$$\mathbb{E}[\hat{P}_f] = \frac{1}{K}\sum_{k=1}^{K}\mathbb{E}\left[\chi_{D_f}(X^{(k)})\right] = \frac{1}{K}KP_f = P_f \tag{22}$$

$$\mathbb{V}ar[\hat{P}_f] = \frac{1}{K^2}\sum_{k=1}^{K}\mathbb{V}ar\left[\chi_{D_f}(X^{(k)})\right] = \frac{1}{K^2}KP_f(1-P_f) = \frac{1}{K}P_f(1-P_f) \tag{23}$$

2.2. Numerical model validation

In this study, the experimental results considered are presented by Ibrahim and Rad [11]. According to that, reinforced concrete non-prismatic beams were experimentally built with an overall length of 2000 mm, depth at supports (h_s) of 250 mm while the width was about 150 mm, and 9° haunched angle (α) value which creates a mid-span depth (h_m) of 150 mm, figure 2 displays beams geometry and testing conditions as beams were tested under monotonic loading up to failure. Furthermore, in order to obtain shear failure, beams have been provided with steel reinforcement bars that have different diameters as illustrated in figure 3 while the properties of materials obtained experimentally are presented in table 1.

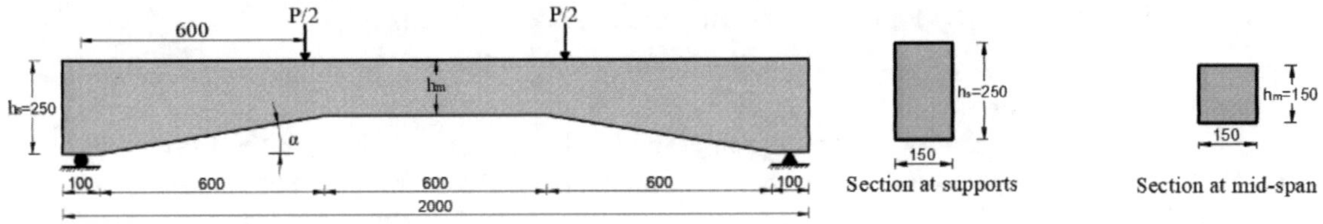

Figure 2. Beams geometry and loading condition.

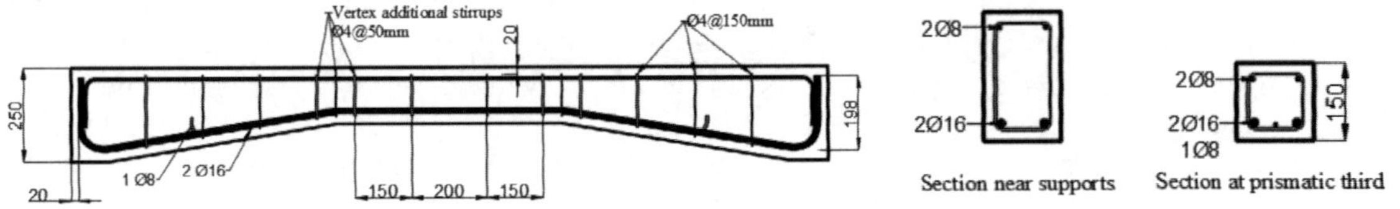

Figure 3. Reinforcement details.

Table 1. Materials properties.

Material	Specifications	Yield strength (N/mm²)	Compressive strength, f_c' (N/mm²)	Ultimate tensile strength (N/mm²)	Flexural strength (N/mm²)	Elastic modulus, E_c (N/mm²)
Steel	ϕ = 16 mm	490		558.58		210000
	ϕ = 8 mm	556		632		210000
	ϕ = 4 mm	548		629		210000
Concrete			35	3.27	3.6	26420

Further, Finite element (FE) analysis was conducted to model the nonlinear behaviour of the non-prismatic reinforced concrete beams considering three different hunched angles (α = 11.7°, α = 9°, 6.18°) applying damage plasticity model. Additionally, the boundary conditions were chosen to resemble experimental conditions as a pin is set at the first support to generate rotation while a roller is set at the other support to generate rotation and horizontal movement. Figure 4 shows the meshed numerical model while figure 5 illustrates the element types used to build up the model.

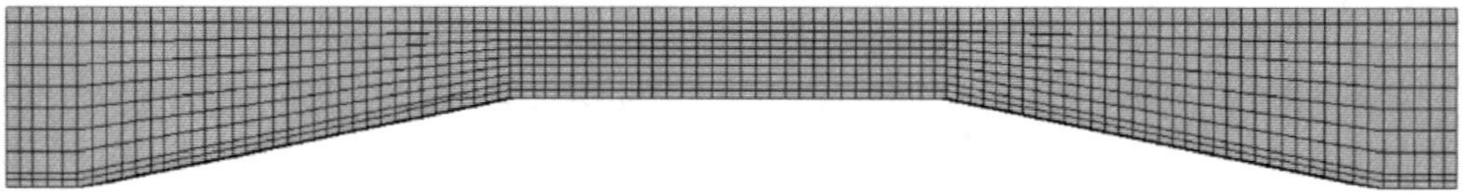

Figure 4. The numerical model.

Figure 5. The numerical model elements.

3. Results and discussions

Basically, ABAQUS is used to validate the numerical model using the data collected from experimental tests. Then, a MATLAB program is written and connected to ABAQUS to start an analysis considering the probability of failure as bound and concrete properties (f_c' and E_c) as variables with a 5% standard deviation. The mean values for theses variables are then specified by ($f_c' = 35$ N/mm^2) and ($E_c =$ 26420 N/mm^2). Monte Carlo simulation method is used to analyse the samples under different material properties to study the effect of the probability of failure on the behaviour of the beams considering a specific number of sample point ($K = 1.5 \times 10^3$).

For illustration, tables 2,3 and 4 show some iterations results for three different hunched angles (α= 11.7°, α =9°, 6.18°) values. Clearly, at a specific failure probability value (P_f), displacement value (U) is changing according to the geometry of the beams whereas α value decreases, the displacement values are decreasing too demonstrating that beams are acting less elastically for smaller α values. Also, it can be observed that slimmer beams (with larger α values) are exposed to fail under lower loads. Besides, the effect of P_f is obvious on the results as it works as a bound alongside the varied material properties to give the corresponding loads (F) and displacements (U).

Sensibly, adopting higher P_f values will produce higher loads, thus higher displacements, so a high P_f value will require more load to be achieved. However, having variable material properties with a variance will cause the operation to provide random material properties for each iteration and this is how the uncertainty part is taken into consideration in this study, and as a result, unexpected lower loads are produced for higher P_f values when the randomly chosen concrete properties are small and cause the beam to fail earlier due to its weakness.

Table 2. α =11.7° beam analysis results.

P_f	f_c' (N/mm^2)	E_c (N/mm^2)	U (mm)	F (N)
0.00066	34.26720889	27122.79498	20.83770457	23900
0.00330	34.21558567	26263.07422	22.23618428	24400
0.00530	34.15450458	27140.87577	20.63089823	24100

Table 3. α =9° beam analysis results.

P_f	f_c' (N/mm^2)	E_c (N/mm^2)	U (mm)	F (N)
0.00066	35.43147989	26495.14933	14.50823663	37600
0.00330	32.27323598	27915.95509	14.93224562	37900
0.00530	33.10731313	25562.37959	15.39884225	37800

Table 4. α =6.18° beam analysis results.

P_f	f_c' (N/mm^2)	E_c (N/mm^2)	U (mm)	F (N)
0.00066	34.22	26263	13.13	45900
0.00330	35.84	26663	13.73	46300
0.00530	34.71	26257	13.25	46100

Moreover, the variation of f_c' and E_c, illustrated in tables (2-4), proves the performance of 5% standard deviation applied on these two values and the results are varied accordance where in all the tables above the material properties are affecting directly the displacement and load values cooperating with the inserted P_f values.

On the other hand, another comparison is made to clarify the damage pattern of different models under different conditions as shown in table 5. As three different models having different α values were considered, taking into consideration two cases for each, the first case shows probabilistic results and damage patterns of models at P_f = 0.002 while the second case shows the deterministic results considering ultimate load (F_u) and the corresponding displacement (U). Obviously, damaged areas represented by red colour are more intensive in the deterministic cases compared to the probabilistic cases. In this way, the P_f is working as a limit for a safe design controlling the damage in the models, where the intensity of the damaged areas is graded from maximum red to minimum blue as seen in table 5.

Table 5. Damage pattern of different models under different conditions.

α Value	Properties	Damage pattern
α =11.7°	F = 24000 N, U = 20.85 mm, P_f = 0.002	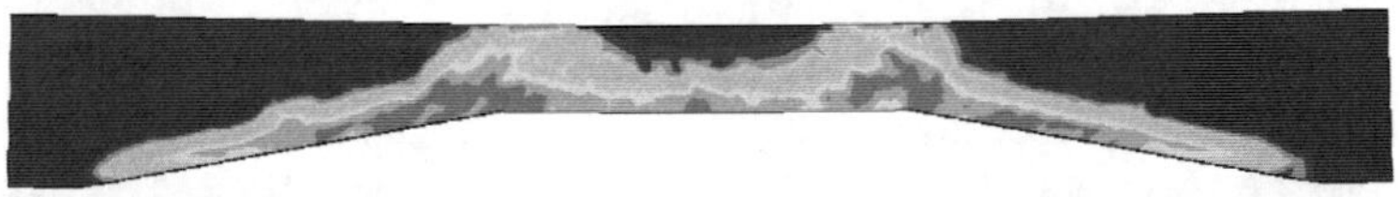
	F_u = 26210 N, U = 23.666mm	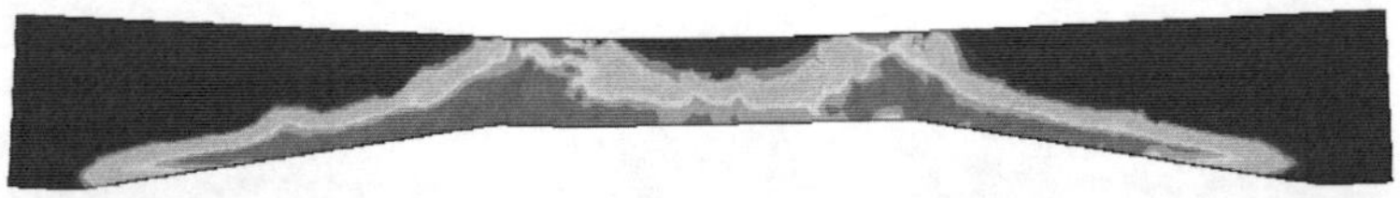
α =9°	F = 38300 N, U = 13.67mm, P_f = 0.002	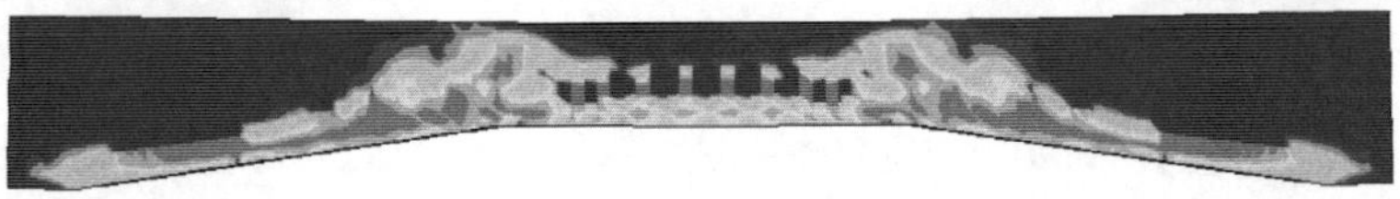

IOP Conf. Series: Materials Science and Engineering 1141 (2021) 012038 doi:10.1088/1757-899X/1141/1/012038

	$F_u = 43130$ N, $U = 16.72$ mm	
α =6.18°	$F = 46400$ N, $U = 11.12$ mm, $P_f = 0.002$	
	$F_u = 48830$ N, $U = 15.13$ mm	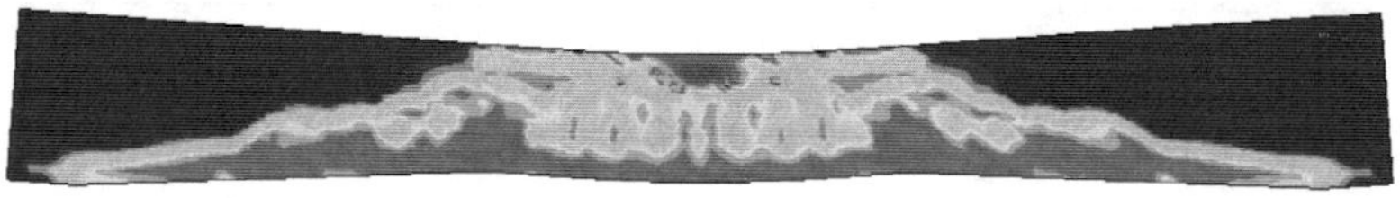

4. Conclusions

In this research, the effect of probability failure on the reinforced concrete non-prismatic beams is investigated using deterministic and probabilistic finite element analysis. Validation of the numerical model was done using the concrete damage plasticity (CDP) constitutive model after collecting the required data from the experimental tests. Then, a MATLAB code was developed to achieve the desired purpose considering the probability failure as a bound while the compressive strength and the modulus of elasticity of concrete are considered variables with a 5% standard deviation. The mean values for the variables are then specified by ($f_c^{'} = 35$ N/mm^2) and ($E_c = 26420$ N/mm^2). Therefore, based on the above, the following conclusions can be drawn:

- The effect of P_f is obvious on the results where it is working as a bound alongside the varied material properties to give the corresponding loads (F) and displacements (U).
- At a specific probability of failure value (P_f), displacement value (U) is changing according to the geometry of the beams whereas α value decreases, the displacement values are decreasing too demonstrating that beams are acting less elastically for smaller α values.
- Slimmer beams (with larger α values) are exposed to failure under lower loads.
- Choosing higher P_f values will produce higher loads, thus higher displacements. However, having variable material properties with a specific variance will cause the operation to provide random material properties for each iteration, and as a result, unexpected lower loads are formed for higher P_f values that the randomly chosen concrete properties are small and cause the beam to fail earlier because of its weakness.
- The 5% standard deviation of f_c' and E_c values causes random varied results where the material properties are affecting directly the displacement and load values cooperating with the inserted P_f values.
- Damaged areas are more intensive in the deterministic cases compared to the probabilistic cases, thus the P_f is working as a limit parameter controlling the damage in the models.

5. References

[1] Beltempo A, Balduzzi G, Alfano G and Auricchio F 2015 Analytical derivation of a general 2D non-prismatic beam model based on the Hellinger–Reissner principle *Engineering Structures* **101** pp. 88-98.

[2] Grassl P, Xenos D, Nyström U, Rempling R and Gylltoft K 2013 CDPM2: A damage-plasticity approach to modelling the failure of concrete *International Journal of Solids and Structures*

50(24) pp. 3805-3816.

[3] Choi S K, Grandhi R and Canfield R A 2006 Reliability-based structural design *Springer Science & Business Media.*

[4] Kappos A J, Chryssanthopoulos M K and Dymiotis C 1999 Uncertainty analysis of strength and ductility of confined reinforced concrete members *Engineering structures* **21**(3) pp. 195-208.

[5] Kottegoda N T and Rosso R 2008 *Applied statistics for civil and environmental engineers* Malden, MA: Blackwell, Chapter 8 pp. 487.

[6] Nicolis G 1995 *Introduction to nonlinear science.* Cambridge university press Cambridge (UK) [ISBN 0 521 46228 2].

[7] Basil M and Jamieson AW 1998 *Uncertainty of complex systems using Monte Carlo techniques In North Sea Flow Measurement Workshop* **32**(1) pp. 16-20.

[8] Herrador M Á and González A G 2004. Evaluation of measurement uncertainty in analytical assays by means of Monte-Carlo simulation *Talanta,* **64**(4) pp. 415-422.

[9] Papadopoulos C E and Yeung H 2001 Uncertainty estimation and Monte Carlo simulation method *Flow Measurement and Instrumentation* **12**(4) pp. 291-298.

[10] Kioumarsi M M, Hendriks M A, Kohler J and Geiker M R 2016 The effect of interference of corrosion pits on the failure probability of a reinforced concrete beam *Engineering Structures* **114** pp.113-121.

[11] Ibrahim S K and Rad M M 2020 Numerical Plastic Analysis of Non-Prismatic Reinforced Concrete Beams Strengthened by Carbon Fiber Reinforced Polymers *Proc. of the 13th fib International PhD Symposium in Civil Engineering* Marne-la-Vallée Paris France pp. 208-215.

IOP Conf. Series: Materials Science and Engineering 1141 (2021) 012039 doi:10.1088/1757-899X/1141/1/012039

Modeling of rectangular plates by the elastic scheme method

I Stoynova, A Handruleva and A Dimitrova
USEA "Lyuben Karavelov" Bulgaria, Faculty of Construction, Department "Mechanics and Mathematics", 1373 Sofia, 175 Suhodolska str.

E-mail: stoynova@vsu.bg

Abstract. This study presents an approach, based on correction coefficients, by which bending moments in rectangular plates when obtained by the elastic scheme method, can be calculated more accurately. At first, the well-known generalized coefficients, used in the elastic scheme method, and also taken from different literature sources are juxtaposed. The correction coefficients are obtained by the FEM according to the thin elastic plate theory. On the next stage, by looking into the shear deformability of the plate cross- section, the values of these coefficients are provided in a more accurate manner. The research has been done for a quadratic plate, but the aforenamed approach can be used to investigate any general rectangular plate, as well.

1. Introduction

The article presents a methodology based on correction coefficients, as it specifies the expressions for obtaining the bending moments in rectangular plates, calculated by an elastic scheme. Initially, a comparison of the generalized coefficients found from various literature sources was made. Correction coefficients are obtained by the Finite Element Method (FEM) as they refer to a thin elastic plates model. Additionally, their value-expressions have been compared upon calculating the angular deformability of the cross section, i.e. a model of a moderately thick plate. The study was conducted for the geometry of a square plate, but the method can be applied to a rectangular shape, as well.

Modelling is required taking into account the actual behaviour of the structure, including taking into account the rheological properties – more information in [5].

2. Theoretical framework of the method of calculation of slabs while observing the elastic scheme

Bending moments across cross-reinforced slabs have been determined by the principles of elasticity theory or the force-distribution method.

When calculating the slabs under the Elastic Frame Method, most commonly utilized appear to be the tables compiled by Marcus to determine the forces depending on the boundary conditions of the slab and the relationship between the sides of the field, according to (1).

$$M_x = \frac{1}{m_{xi}} q l_x^2, \quad M_y = \frac{1}{m_{yi}} q l_y^2 \tag{1}$$

where m_{xi} and m_{yi} (for being required to determine the bending moments in the field of the slab in both directions) are calculated by replacing each slab with two mutually perpendicular stripes of 1 m

width. By using these coefficients, also known as generalized ones, the interaction between the individual stripes, the boundary conditions, as well as the behaviour of load-transferring towards the supports (supporting structure) may be calculated. The total load q acting upon the slab becomes distributed between the two stripes of q_{xi} and q_{yi}, so that at the point of intersection, the two stripes may achieve equal sagging moments. In this way, a continuity of the stripe configuration is simulated.

The hypotheses on which this method is based determine the scope of its applicability. It is well known that the method can only be applied to model rectangular smooth elastic plates loaded with uniformly distributed continuous load, and also supported but continuously on the contour sides. Moreover, the results obtained relate to specific and well-defined points in the scheme.

3. Numerical models

The purpose of this study is to initially find the correction coefficients by the Finite Element Method (FEM) of a model of a thin elastic plate, and then only to compare the results with those obtained from modelling a moderately thick plate, whereupon the angular deformability of the cross section has been reported, as well.

Two types of computational models have been considered as objects of the study of elastic plates:

- type I – thin elastic plate. Conditionally, the computational models have been denoted accordingly, with P1-P10.
- type II – a moderately thick plate (the angular deformability of the cross section is also taken into account). Conditionally, the computational models have been denoted accordingly, with P11- P20.

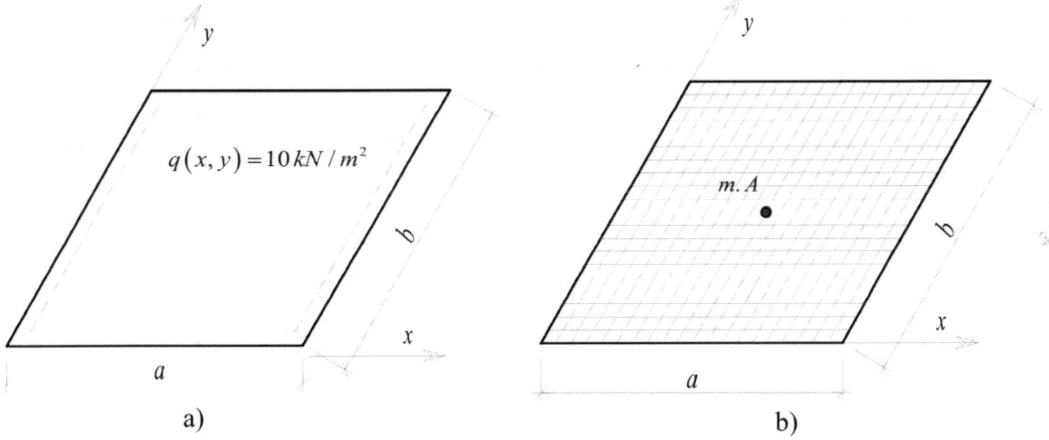

Figure 1. Geometric and discretization model

Of the two types of computational models, respectively, ten types of square elastic plates of the following fixed parameters have been studied, Figure 1a:

- plate thickness $h = 0.1\,m$;
- boundary conditions – free surface support along the contour
- load - evenly distributed load of intensity $q(x, y) = 10\,kN/m^2$.

The study is a parametric one with alterations of the dimensions of the plate $a\,(b)$, in both directions within an interval of $(3-12)\,m$ with a step of 1 m.

The Plane Finite-Element Method, shell type, with 12 degrees of freedom was used in order to build and simulate the computational models.

The finite elements dimensionality, having been used for discretization, is 0.1x0.1m.

4. Numerical solutions

As a main indicator for the comparison of the computational models of elastic plates, the bending moment Mx (respectively My) at node A has been identified, as shown on Figure 1b. SAP2000 is used for FEA analysis.

Throughout Tables 1-6, the obtained numerical results for the selected computational models are seeded in a consecutive order.

Table 1. Numerical results for P1 - P10 models

Type	M_x $[kNm/m]$	ΔM_x $[kNm/m]$	$\Delta^2 M_x$ $[kNm/m]$
P1	3.981		
P2	7.075	3.094	
P3	11.053	3.978	
P4	15.916	4.862	
P5	21.662	5.746	0.884
P6	28.292	6.630	
P7	35.807	7.514	
P8	44.206	8.398	
P9	53.488	9.28	
P10	63.654	10.167	

Furthermore, for the calculation models of thin elastic plates, manual calculation was done under the method of the elastic scheme on different values of the m_x coefficient (18.00 and 27.43) [1,2,3,4]. The aim is to obtain the bending moment value of the investigated node A, as well as to make the comparison of its value with the results obtained by the Finite Element Method. The models worked out through the method of the elastic scheme are conditionally denoted as P1 '- P10'.

Table 2. Numerical results for models P1' – P10' when $m_x = 18.00$

Type	M_x $[kNm/m]$	ΔM_x $[kNm/m]$	$\Delta^2 M_x$ $[kNm/m]$
P1'	5.00		
P2'	8.89	3.89	
P3'	13.89	5.00	
P4'	20.00	6.11	
P5'	27.22	7.22	1.11
P6'	35.56	8.34	
P7'	45.00	9.44	
P8'	55.56	10.56	
P9'	67.22	11.66	
P10'	80.00	12.78	

Table 3. Numerical results for m_{xi} and m_{yi} coefficients of the P1' – P10' models

M_x $[kNm/m]$	$\Delta^2 M_x$ $[kNm/m]$
3.981	
7.075	
11.053	
15.916	
21.662	22.62
28.292	
35.807	
44.206	
53.488	
63.654	

Throughout Figure 2 to Figure 4, the following numerical results ($M_x, \Delta M_x, \Delta^2 M_x$) for the considered computational models are illustrated graphically

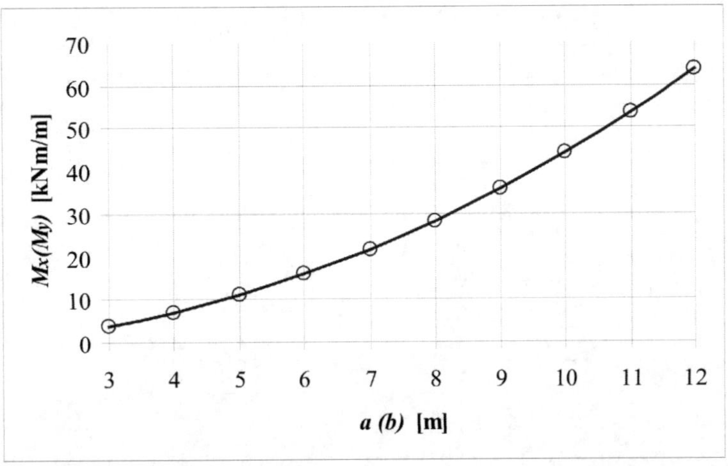

Figure 2. Modification of M_x (M_y) for different computational models

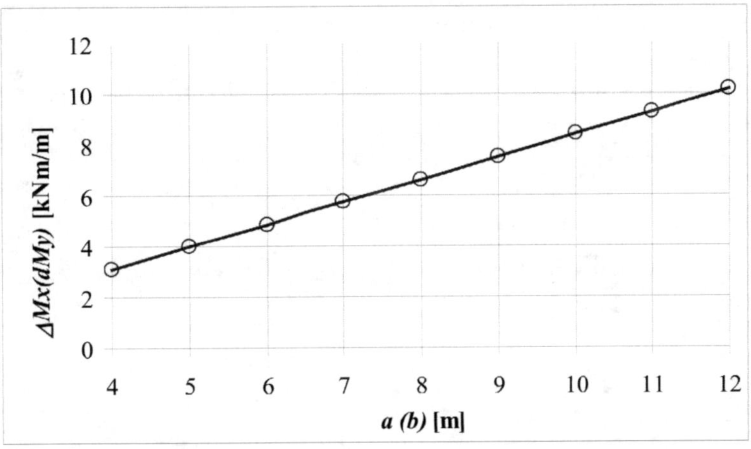

Figure 3. Modification of ΔM_x (ΔM_y) for different computational models

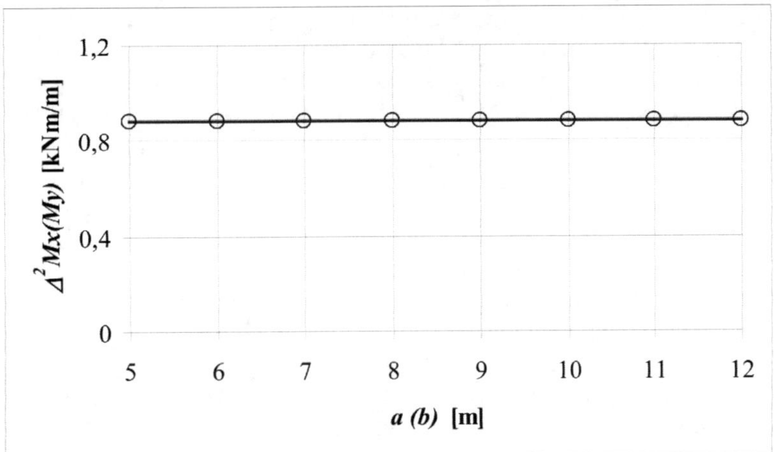

Figure 4. Modification of $\Delta^2 M_x$ (ΔM_y) for different computational models

Table 4. Numerical results for P11 – P18 models

Type	M_x $[kNm/m]$	ΔM_x $[kNm/m]$	$\Delta^2 M_x$ $[kNm/m]$
P11	3.989		
P12	7.083	3.094	
P13	11.061	3.978	
P14	15.924	4.862	
P15	21.670	5.746	0.884
P16	28.300	6.630	
P17	35.815	7.514	
P18	44.214	8.398	

Table 5. Numerical results for models P11' – P20' when $m_x = 18.00$

Type	M_x $[kNm/m]$	ΔM_x $[kNm/m]$	$\Delta^2 M_x$ $[kNm/m]$
P11'	5.00		
P12'	8.89	3.89	
P13'	13.89	5.00	
P14'	20.00	6.11	
P15'	27.22	7.22	1.11
P16'	35.56	8.34	
P17'	45.00	9.44	
P18'	55.56	10.56	
P19'	67.22	11.66	
P20'	80.00	12.78	

Table 6. Numerical results for m_{xi} and m_{yi} coefficients of the P11 – P20 models

M_x $[kNm/m]$	$\Delta^2 M_x$ $[kNm/m]$
3.989	
7.083	
11.061	
15.924	
21.670	2.61
28.300	
35.815	
44.214	
53.497	
63.663	

5. Conclusions

The article demonstrates an applied methodology for obtaining bending moments across rectangular smooth elastic plates, calculated upon observing an elastic scheme. It is based on generalized corrective coefficients derived under the Finite Element Method, with and without taking into account the angular deformation of the cross section. It can be summarized that the proposed methodology is easily applicable for practical purposes, as it may also facilitate a preliminary assessment of a stress state of a slab.

6. References

[1] Dimitrov B 1993 *Reinforced concrete structures*, Technika Publishing House

[2] Manoilov L 2002 *Reinforced Concrete*, Technika Publishing House

[3] Cvetkov St, Georgiev G 2012 *Methodological guide for development of course project reinforced concrete according to Eurocode 2*, VSU Publishing house, Sofia, 2012, ISBN: 978-954-331-115-6, p.45

[4] Georgiev G, Tsvetkov St 2013 *Methodical guide for development of a course project in Reinforced concrete structures according to Eurocode*, part One, Chapter One: Flat slabs and stairs, VSU Publishing house, Sofia

[5] Ivanov Ya, Stoyanov V, Kotsilkova R 2000 Rheological Estimation of the Methods for Concrete Design, *In: Proc. XIII International Congress on Rheology, Cambridge*, UK, August 20-25, 2000, (eds. D.M. Binding, N.E. Hudson, J. Mewis, J.-M. Piau, C.J.S. Petrie, P. Townsend,

M.H. Wagner, K. Walters), *British Society of Rheology*, Vol. **4**, 2000, pp.205-207. ISBN: 978-095-389-040-8 (set), 978-095-389-046-0 (pbk.)

Dynamic research of dome structures

A Handruleva

USEA "Lyuben Karavelov" Bulgaria, Faculty of Construction, Department "Mechanics and Mathematics", 1373 Sofia, 175 Suhodolska str.

E-mail: rector@vsu.bg

Abstract. In this paper the dynamic responses of discrete single-layer domes are discussed. As objects of the study, computational models of domes have been analysed. These models differ in the type of their boundary conditions, the grid configuration, and in the height of the spheres. The analysis was conducted in two stages: a modal and a spectral analysis. In the modal analysis, a coefficient of the effective modal mass and the type of natural forms of vibration have been observed. Comparison has been made between CQC and SRSS being two different methods for combining modal contributions, as well as between two other methods of combining modal responses in the direction of seismic effects, namely: SRSS method and the „Rule of 30%". The purpose of the study is to work out the dynamic response of the discrete single-layer spherical domes.

1. Introduction

Thus far, the combination of coordinate directions following our norms is carried out by applying the SRSS method (square-root-of-the-sum-of-squares), which practically in two directions, it appears to be the Pythagorean theorem. It is written as follows:

$$\mathrm{E_E} = \sqrt{\mathrm{E}^2_{\mathrm{E},x} + \mathrm{E}^2_{\mathrm{E},y}} \tag{1}$$

where $\mathrm{E_E}$ is the force exerted in section E, in turn used for dimensional purposes;

$\mathrm{E}^2_{\mathrm{E},x}$ is a force exerted in section E, generated from an activity of seismic forces along axis x, i.e. having resulted from the application of a spectrum in this direction;

$\mathrm{E}^2_{\mathrm{E},y}$ is a force exercised in section E, generated from the action of seismic forces along axis y, i.e. having resulted from the application of a spectrum in this direction;

Other ways of combining, called methods of coordinate combination, can be used, as well. One of them is based on expressions/equations (2) and (3) of Eurocode 8, Part I. [1]. It is a combination of the full effects obtained from the impact in one coordinate direction with those from the other, but multiplied by ± 0.3, i.e. 30% of their full dimension(value). The equation is as it follows:

$$\begin{aligned} \mathrm{E_E} &= \pm \mathrm{E_{E,x}} \pm 0.3\mathrm{E_{E,y}} \\ \mathrm{E_E} &= \pm 0.3\mathrm{E_{E,x}} \pm \mathrm{E_{E,y}} \end{aligned} \tag{2}$$

when calculating the horizontal forces only.

There is also a number of national norms for which the coefficient to be combined with is ± 0.4. It is most common within the range from ± 0.3 to ± 0.5. [3] Thence in two directions, the general expression of this method of coordinate combination can be written as

$$\begin{aligned} E_E &= \pm E_{E,x} \pm \alpha E_{E,y} \\ E_E &= \pm \alpha E_{E,x} \pm E_{E,y} \end{aligned} \tag{3}$$

when $\alpha \in [0.3 \div 0.5]$.

The most common method of combined effects in coordinate directions is expressed in the following equation:

$$E_E = \sqrt{\left(E_{E,x}\right)^2 + 2\beta E_{E,x}E_{E,y} + \left(E_{E,y}\right)^2} \tag{4}$$

where β is a coefficient and its common boundaries are $\beta \in [-1 \div 1]$. Most often, however, it is accepted within the following allowance of $\beta \in [0 \div 0.3]$. It can be likened or related to the CQC method when combining the modal maxima. It is quite obvious that if we accept β as the lower boundary (end-boundary), i.e. $\beta = 0$, thence this method gets transmuted into SRSS method, i.e. as a particular, or rather as its special single case.

Viewing it all from a mathematical perspective, the CQC method is a generic and a common one. The SRSS method is considered as a single special case, whereas the method based on α coefficients appears to be an initial (a first step) approximation.

It is assumed that in general the SRSS method of combined effects in coordinate directions results in rather safe computations, i.e. presents better results than those using coefficient $\alpha = \pm 0.3$. However, that is not always the case. It is valid when the effects in both directions are near in value. For example, if they are the same, SRSS corresponds to a coefficient $\alpha = 0.44$ or SRSS gives about 10% higher results. It can be recognized or taken as the Pythagorean theorem of an isosceles right triangle. However, when the mentioned combined effects differ significantly (one effect is less than 83% of that in the orthogonal direction), the opposite is observed. This can be easily proven. If we denote by ζ the relationship of the two effects, i.e. $\zeta = E_{E,y}/E_{E,x}$ and if we assume that $E_{E,y} < E_{E,x}$, then $E_E = \sqrt{(1+\zeta^2)\left(E_{E,x}\right)^2}$.

This means that if $\sqrt{(1+\zeta^2)} < 1.3$ the method based on coefficient $\alpha = \pm 0.3$ yields higher values than the combined one. This is the case when $\zeta < 0.83$.

The method with coefficient is recommended when rotating eigenforms dominate. However, in case of strong irregularity across plane, it is appropriate to compare the results by adopting other combination techniques for analysis. This is substantiated (validated) in detail in [4].

2. Description of the selected computing models

Eleven representative elements were considered as objects of research for studying the dome-like rod structures. Correspondingly, the domes are identified and marked as K1, K2, K3, K4, K5, K6, K7, K8, K9, K10 and K11, respectively, shown in Figure 1a **(K1)** and Figure 1.b **(K11)**. The study is a parametric one, as diameter is subjected to change(s) in direction **y** within a range of 30 ÷ 15m. SAP 2000 was used to build and study the computational models.

The following fixed parameters have been adopted for the selected models of **ribbed domes**:

- diameter of base in x direction Dx=30m;
- height of dome h = 7m

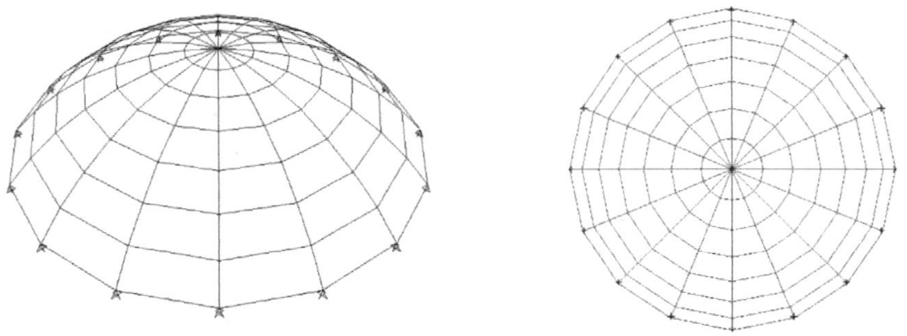

Dome K1 (30x30) a)

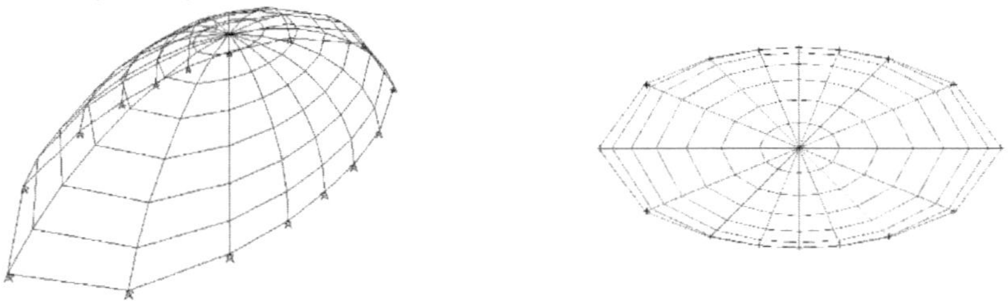

Dome K11 (30x15) b)

Figure 1. Selected rod domes

In generating computational models, the following:

- Boundary support conditions: supporting ring on radially movable supports
- Rigid connection of the elements with the support ring
- Rigid connection between the elements of the ribs and the inner rings of the lattice.

Tubular cross-sections and standard hot-rolled profiles were adopted for the structural elements. Table 1 shows the overall geometric characteristics of the selected domes.

Table 1. Geometric characteristics of the computational models

	K1	K2	K3	K4	K5	K6	K7	K8	K9	K10	K11
height of dome, h [m]	7	7	7	7	7	7	7	7	7	7	7
diameter of dome D_x [m]	30	30	30	30	30	30	30	30	30	30	30
diameter of dome D_y [m]	30	30	29	29	28	28	25	23	20	18	15
cover/span area [m^2]	707	695	683	671	660	648	589	530	471	412	353

For the representative elements of the seven domes along the meridian and the rings, the same geometrical characteristics of the cross sections given in Table 2 were adopted.

Table 2. Geometric characteristics of the cross-sections of the elements

K1, K2, K3, K4, K5, K6, K7, K8, K9, K10, K11			
	supporting ring	meridian ribs	tension-and-compression ribs
cross-section	IPE 300	Ø88.9x4	Ø88,9x4
cross-section surface [cm^2]	53.8	10.7	10.7
moment of inertia [cm^4]	8356	96.3	96.3
modulus of elasticity - E=206 000 MPa			

The frame type beam-finite-element method of 6 degrees of freedom was used to build and simulate the computational models. [2] Based on the accepted parameters and the subsequent discretization, the following characteristics of the eleven computational models were obtained: number of elements - 224, number of nodes - 113, number of internal rings (without the support ring) - 6.

3. Comparative analysis of the results

In tables below from table 3 to table 5, in a sequential order the following are provided:
periods of oscillation and seismic forces acting in a selected element; results on the different method of combined effects in coordinate directions; absolute difference between the maximum forces obtained with different methods of combination in the coordinate direction.

Table 3. Periods of oscillation and seismic forces applied in a selected element

	T_1 [s]	T_2 [s]	ΔT [%]	E_x*1000 [kN]	E_y*1000 [kN]	E_y/E_x
K1 (30x30)	0.988	0.988	0.00	19.04	19.04	1.00
K2 (30x29.5)	0.979	0.971	0.89	18.57	20.41	1.10
K3 (30x29)	0.970	0.953	1.81	17.56	22.59	1.29
K4 (30x28.5)	0.961	0.935	2.78	16.60	24.94	1.50
K5 (30x28)	0.953	0.916	3.99	12.28	23.14	1.88
K6 (30x27.5)	0.943	0.899	4,82	15.39	31.40	2.04
K7 (30x25)	0.901	0.810	10.64	17.97	48.44	2.70
K8 (30x22.5)	0.862	0.722	17.68	25.60	68.39	2.67
K9 (30x20)	0.830	0.637	26.29	35.66	88.83	2.49
K10 (30x17.5)	0.810	0.557	36.92	43.70	106.72	2.44
K11 (30x15)	0.836	0.506	49.10	43.90	114.76	2.61

Table 4. Results upon the various methods of combined effects in coordinate directions

	Methods of combined effects in coordinate directions								
	SRSS	E_x+0.3E_y	E_x-0.3E_y	-E_x+0.3E_y	-E_x-0.3E_y	0.3E_x+E_y	0.3E_x-E_y	-0.3E_x+E_y	-0.3E_x-E_y
K1	26.92	24.75	13.33	-13.33	-24.75	24.75	-13.33	13.33	-24.75
K2	27.59	24.70	12.45	-12.45	-24.70	25.98	-14.83	14.83	-25.98
K3	28.61	24.34	10.79	-10.79	-24.34	27.85	-17.32	17.32	-27.85
K4	29.96	24.08	9.12	-9.12	-24.08	29.92	-19.96	19.96	-29.92
K5	26.20	19.22	5.34	-5.34	-19.22	26.83	-19.46	19.46	-26.83
K6	34.97	24.81	5.97	-5.97	-24.81	36.02	-26.78	26.78	-36.02
K7	51.67	32.50	3.44	-3.44	-32.50	53.83	-43.05	43.05	-53.83
K8	73.02	46.12	5.08	-5.08	-46.12	76.07	-60.71	60.71	-76.07
K9	95.72	62.31	9.01	-9.01	-62.31	99.53	-78.13	78.13	-99.53
K10	115.32	75.72	11.69	-11.69	-75.72	119.83	-93.61	93.61	-119.83
K11	122.87	78.33	9.47	-9.47	-78.33	127.93	-101.59	101.59	-127.93

Table 5. Absolute difference between the maximum forces as obtained upon different methods of combined effects in coordinate directions

	SRSS	$0.3E_x+E_y$	$\lvert\Delta\rvert$ [kN]
K1	26.92	24.75	2.17
K2	27.59	25.98	1.62
K3	28.61	27.85	0.76
K4	29.96	29.92	0.04
K5	26.20	26.83	0.63
K6	34.97	36.02	1.05
K7	51.67	53.83	2.17
K8	73.02	76.07	3.05
K9	95.72	99.53	3.81
K10	115.32	119.83	4.51
K11	122.87	127.93	5.06

From Figure 2 to Figure 8, accordingly are shown the following: periods of oscillation; ΔT chart; seismic forces E_x and E_y; the E_y/E_x ratio; comparison between the methods of combined effects in coordinate directions; alteration(s) of $\lvert\Delta\rvert$; $\lvert\Delta\rvert$ as a function of E_y/E_x.

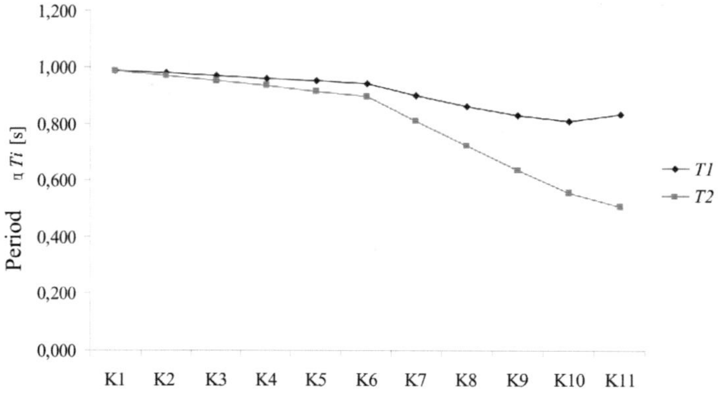

Figure 2. Periods of oscillation

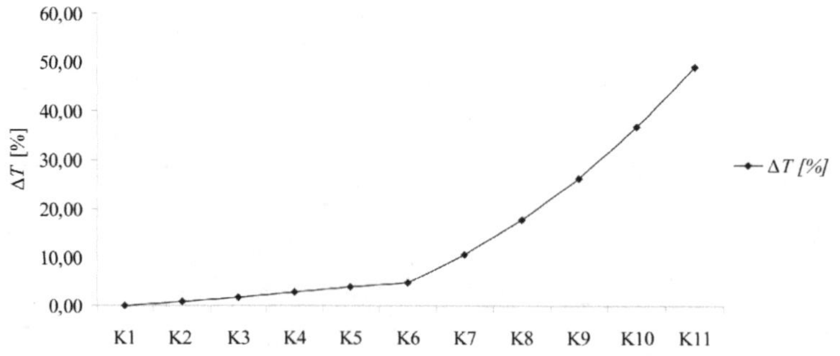

Figure 3. ΔT Graph

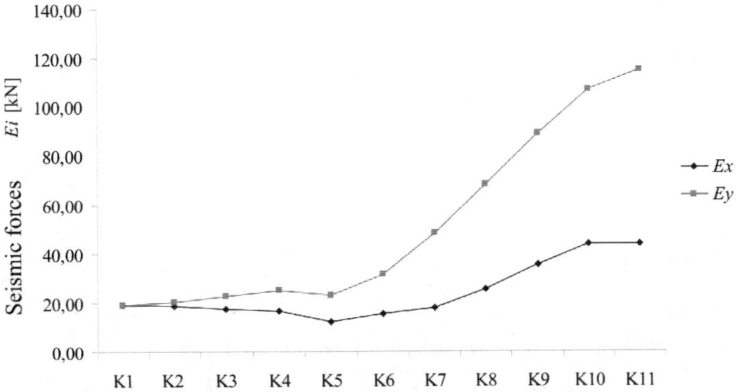

Figure 4. Seismic forces E_x and E_y

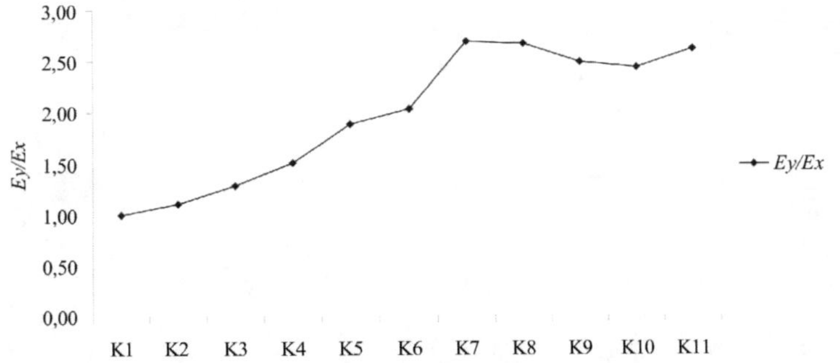

Figure 5. The E_y/E_x Ratio

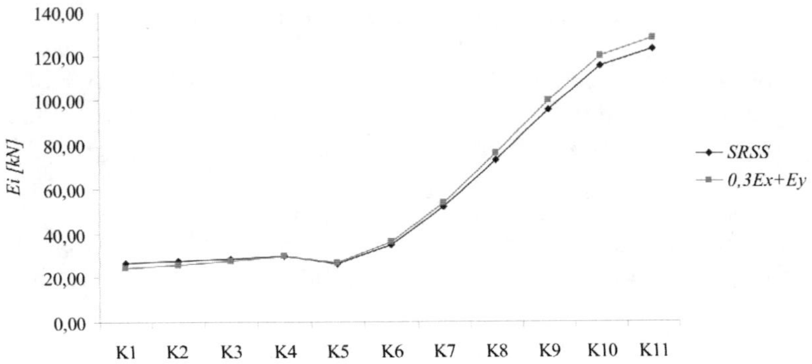

Figure 6. Comparison between the methods of combined effects in coordinate directions

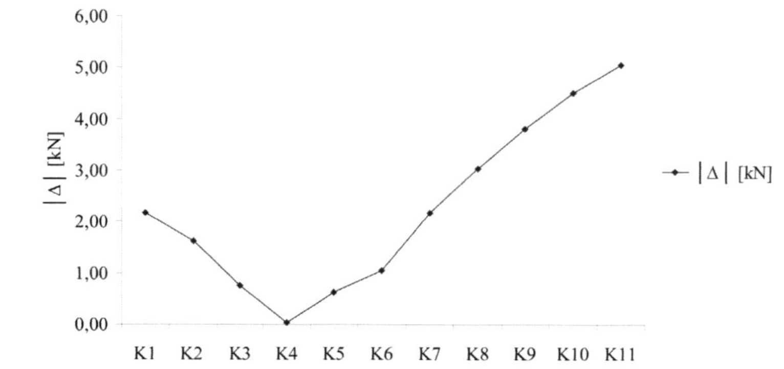

Figure 7. Alteration(s) of |Δ|

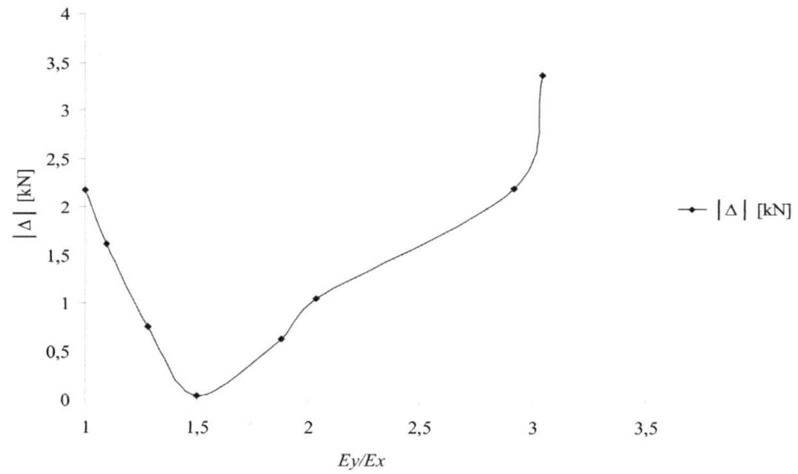

Figure 8. |Δ| as a function of Ey / Ex

The consideration of the real interaction of the structure and the ground base are of special importance both for achieving optimality and economy of the design solution, as well as for the management and minimization of the generated waste. More information can be found in [5], [6] and [7].

4. Conclusions

In the present study, three methods of combined effects in coordinate directions, as well as the acting forces in a particular single case obtained upon independent seismic impact in these same directions have been theoretically compared. These are the three possible technologies provided by the Unified European Building Standards concerning the design of structures in seismic areas, i.e. Eurocode 8. Based on a numerical experiment, two of these methods were practically compared, namely SRSS and the Rule of 30%. This was performed through the computational models of rod domes.

From the obtained results, the following more important conclusions may be drawn:

- for effects relatively close in value in/along the two horizontal coordinate directions, the SRSS method gives higher (more optimal) results, i.e. resulting in higher consistency;
- in cases where the effects differ significantly (by more than 80% for the specific models), the Rule of 30% yields higher (optimal) end-values;
- the CQC combination method, i.e. method based on corelation (4), it can be considered as the most general one and when an appropriate choice of coefficient β is made, it will always develop results in the direction of the safety of the structure.

It is rather advisable, especially for geometrically complex space-frame systems, that separate comparisons of the results attained by the SRSS and the Rule of 30 methods be made. This is especially true for some particular areas or certain elements that are rather fundamental for the load-bearing capacity of the structure, i.e. elements that mainly predetermine its capacity. Studies of a similar nature provide further clarity on dispositive regulations of Eurocode 8. It is absolutely desirable that studies of such nature may continue to be conducted further, while using different structural configurations, as it is quite vital to look for a reference value of coefficient β that leads to a higher overall safety of the structure.

5. References

[1] European standard Eurocode 8 – BDS EN 1998-1-1;
[2] Stoynova I 2012 *Dynamic response of reinforced concrete members and connections*, VSU Publishing house, Sofia, ISBN : 978-954-331-116-3, p.32-33;
[3] Georgiev and St Tsvetkov, Guide for Design of special Reinforced Concrete Structures According to Eurocodes, Part 1, Chapter 1, General Information for Analysis and Construction of Special Reinforced Concrete Structures, USEA (VSU), Sofia, Bulgaria;
[4] Tsvetkov St 2019 Design of steel arched roofs and implementation of the connection "steel-concrete (reinforced concrete)" at the foundations, In *Proc. IX International Scientific Conference on Architecture and Construction ArCivE 2019*, section II, paper № 9 ISSN: 2535-078;
[5] Stoyanov V A, Petkova V, Andonova V 2020 Analysis of the construction site waste management plans in Bulgaria, adopted in the period 2015-2017, *IOP Conf. Ser.: Mater. Sci. Eng.*, **951**, (2020), 012007, http://doi.org/10.1088/1757- 899X/951/1/012027 SJR=0.2 for 2019 г.
[6] Frangov G, Zayakova H, Frangov S 2017 Destructive influence of technogenic factors and precipitations on the landslide support structure. *Proceedings of the 4thWLF, Ljubjana*, Slovenia, published in *Advancing Culture of Living with Landslides*, Volume **3**: Advances in Landslide Technology, ISBN 978-3-319-53487-9, pp. 529-536.
[7] Hamova M, Zayakova H, Frangov G 2015 Passive grouted anchors for emergency reinforcement of retaining structures", 6th ACES Conference GEOTECHNICS IN CIVIL ENGINEERING, 2-6 November 2015, Vrsac, Serbia.

IOP Conf. Series: Materials Science and Engineering 1141 (2021) 012041 doi:10.1088/1757-899X/1141/1/012041

FEM computational models of shear walls with openings

A Handruleva and A Dimitrova
USEA "Lyuben Karavelov" Bulgaria, Faculty of Construction, Department "Mechanics and Mathematics", 1373 Sofia, 175 Suhodolska str.

E-mail: rector@vsu.bg

Abstract. This study is devoted to the comparison of shear wall computational models and the juxtaposition of displacement and stress fields. Two types of numerical models are compared. The first type concerns discretization of the geometry by using the beam finite elements model, and the second type – with plane finite elements. The displacements and the internal forces are used as the key indicators for the comparison of the mechanical behaviour. Some conclusions have been drawn, as well as recommendations have been made on the application of the juxtaposed type computational models and their practical application in real life.

1. Introduction

When designing buildings subjected to seismic actions, vertical bearing elements (earthquake washers) are widely utilized to absorb the seismic forces in the coordinate directions – more information in [6], [7], [10], [11]. Such cases determine the need to know their real mechanical behaviour in order to more adequately model them. In the studied models, it is accepted that the wall is fixed into the foundation, but in the next stage of the research, the soil – structure interaction will be taken into account more precisely-more information in [8], [9].

2. Computational models

Two types of computational models of vertical load-bearing elements are considered as objects for study:

-type I – one-storey earthquake washer, discretized by beam finite elements approach. Conditionally viewed, the computational models are denoted throughout M1-M4 accordingly, as shown in Figure 1a. The following characteristics have been adopted and confirmed for the different types of models:

- model M1 - elements s_1 and s_2 are modeled upon using real elastic modulus ($E=3e^7$ kN/m^2), without taking into account the shear stress. Element b is accepted as a non-finite rigid beam ($E=3e^{20}$ kN/m^2);
- model M2 - elements s_1 and s_2 are modeled upon using real elastic modulus ($E=3e^7$ kN/m^2) and taking into account the shear stress. Element b is accepted as a non-finite rigid beam ($E=3e^{20}$ kN/m^2);
- model M3 - elements s_1 and s_2 are modeled upon using real elastic modulus ($E=3e^7$ kN/m^2), without taking into account the shear stress. Element b is accepted with its real rigidity (E=3e7 kN/m^2);
- model M4 - elements s_1 and s_2 are modeled upon using real elastic modulus ($E=3e^7$ kN/m^2) and taking into account the shear stress. Element b is accepted with its real rigidity ($E=3e^7$ kN/m^2);

- type II – one-storey and three-storey earthquake washer, discretized by plane-finite-elements method. Conditionally viewed, the computational models are denoted throughout M5- M17 and M18 – M28 respectively, as shown in Figure 1a

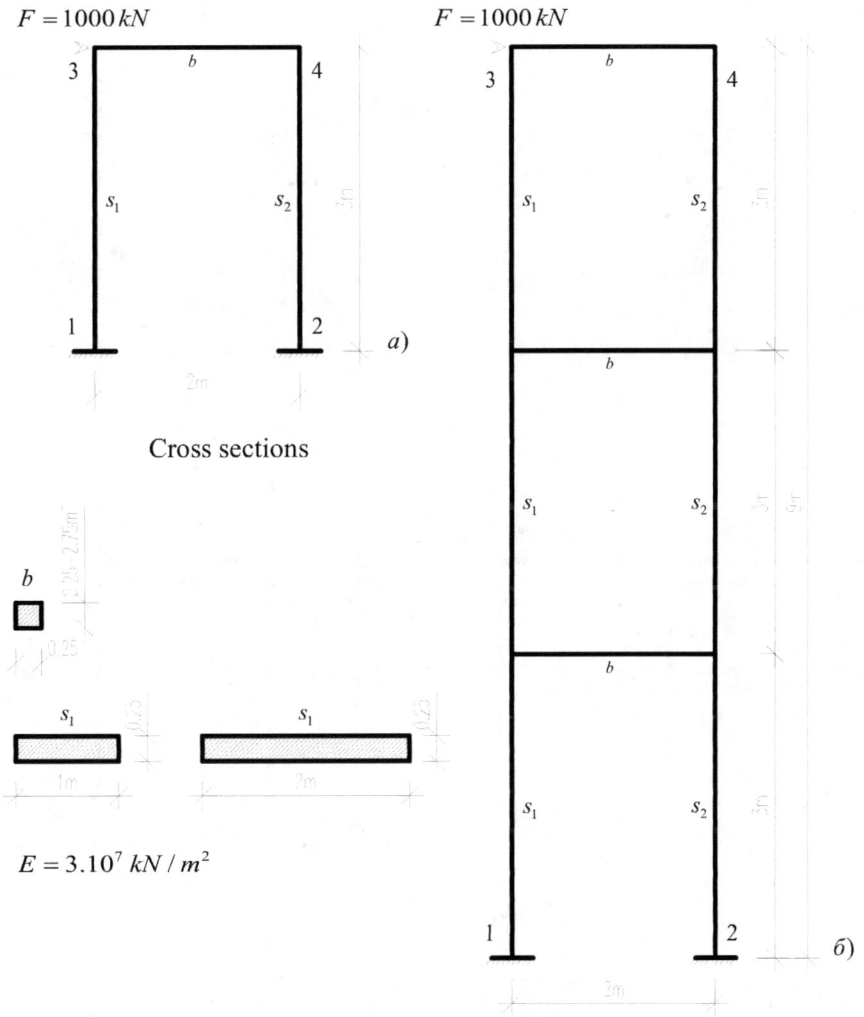

Figure 1. Computational models

The following fixed parameters have been adopted for the selected models:

- cross sections of the elements s_1 and s_2, shown in Figure 1;
- floor elevation 3m;
- horizontal force F=1000 kN.

Boundary conditions, as follows:

- rigid connection (for firm-fixing) of the ground level;
- rigid connection of the vertical elements s_1 and s_2 with the horizontal beam (b);

The study is parametrically based on alteration in the geometric (height) and modification of the material characteristics (elastic modulus) of the horizontal beam (b).

Frame type beam finite elements of 6 degrees of freedom, as well as shell type planar finite elements with 12 degrees of freedom [4] were used to build and simulate the computational models. More information about FEM modelling can be found in [3], [4], [5]. Based on the parameters accepted and upon a subsequent discretization, discretized models have been obtained, shown in Figure 2.

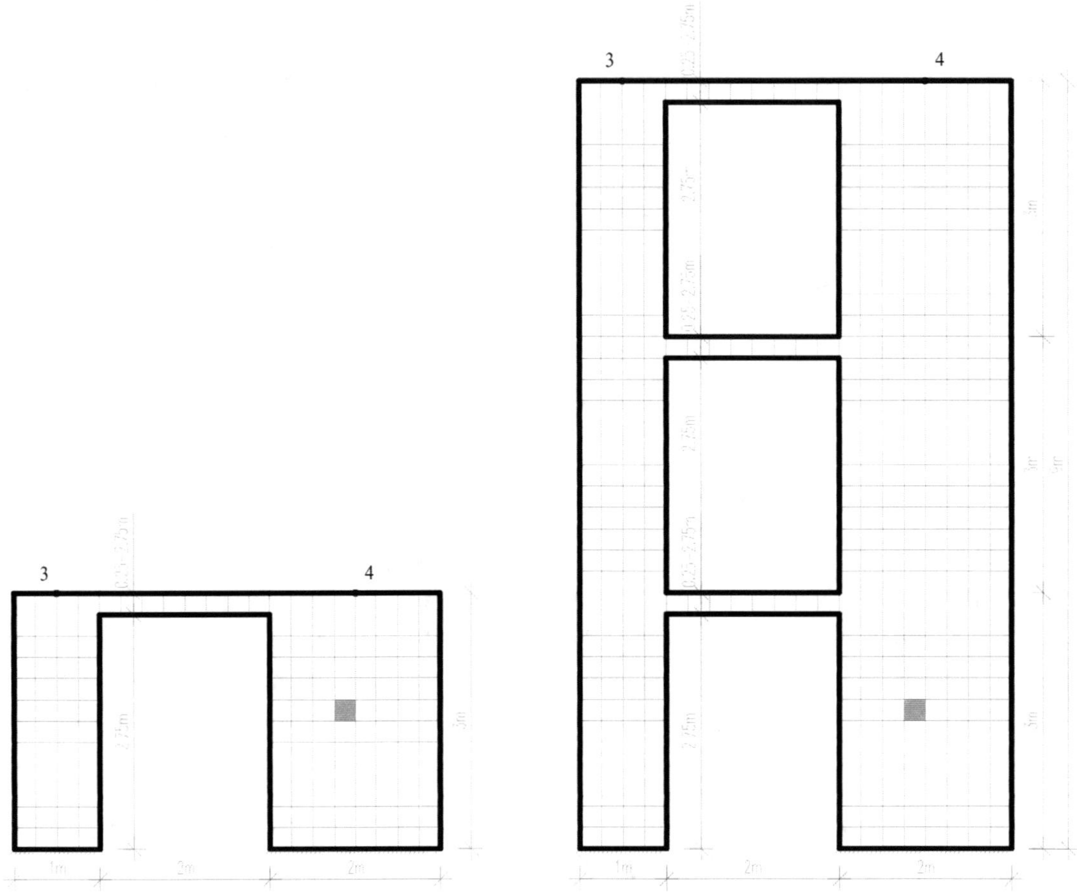

Figure 2. Discretized models

In generating computational models accepted were the following:

- Boundary support conditions: supporting ring on radially movable supports;
- Rigid connection of the elements with the support ring;
- Rigid connection between the elements of the ribs and the inner rings of the lattice.

Tubular cross-sections and standard hot-rolled profiles were adopted for the structural elements.

3. Results

As main indicators for the comparison of the selected computational models of a vertical bearing element, displacements at nodes 3 and 4, as well as shear forces at nodes 1 and 2 have been used, Figure 1.

In tables 1-3 the numerical results (displacements and shear forces) are presented for the considered computational models.

Table 1. Numerical results for models from M1 to M4

	$u_{x,3}$ [e^{-3}m]	$u_{x,4}$ [e^{-3}m]	Q_1 [kN]	Q_2 [kN]	M_1 [kN.m]	M_2 [kN.m]
M1	0.40	0.40	111.23	888.93	173.33	1385.33
M2	0.90	0.90	181.94	818.13	274.40	1283.51
M3	2.68	1.22	220.92	779.08	616.51	2225.36
M4	3.06	1.60	231.62	768.38	646.60	2194.32

Table 2. Numerical results for M5 to M17 models

	$u_{x,3}$ [e^{-3}m]	$u_{x,4}$ [e^{-3}m]	Q_1 [kN]	Q_2 [kN]
M5	2.20	1.78	169.93	830.09
M8	1.65	1.47	203.10	796.92
M9	1.25	1.19	239.00	761.02
M10	1.00	1.00	264.18	735.84
M11	0.80	0.82	281.91	718.11
M12	0.70	0.70	296.92	703.10
M13	0.58	0.61	312.40	685.62
M14	0.52	0.53	329.86	670.16
M15	0.47	0.47	349.50	650.52
M16	0.43	0.43	370.38	629.64
M17	0.41	0.40	390.95	609.07

Table 3. Numerical results for M18 to M28 models

	$u_{x,3}$ [e^{-3}m]	$u_{x,4}$ [e^{-3}m]	Q_1 [kN]	Q_2 [kN]
M18	33.07	32.82	169.25	830.77
M19	17.19	17.19	228.67	771.35
M20	11.08	11.08	267.50	732.52
M21	8.43	8.43	295.01	705.01
M22	7.03	7.03	317.78	682.24
M23	6.16	6.16	339.57	660.45
M24	5.48	5.57	362.73	637.28
M25	5.08	5.14	388.31	611.71
M26	4.77	4.81	416.22	583.80
M27	4.52	4.55	445.35	554.67
M28	4.32	4.33	474.08	525.94

In the following figures, from Figure3 to Figure 5, the Q_2/Q_1 ratio for the various types of models is correspondingly presented.

IOP Conf. Series: Materials Science and Engineering 1141 (2021) 012041 doi:10.1088/1757-899X/1141/1/012041

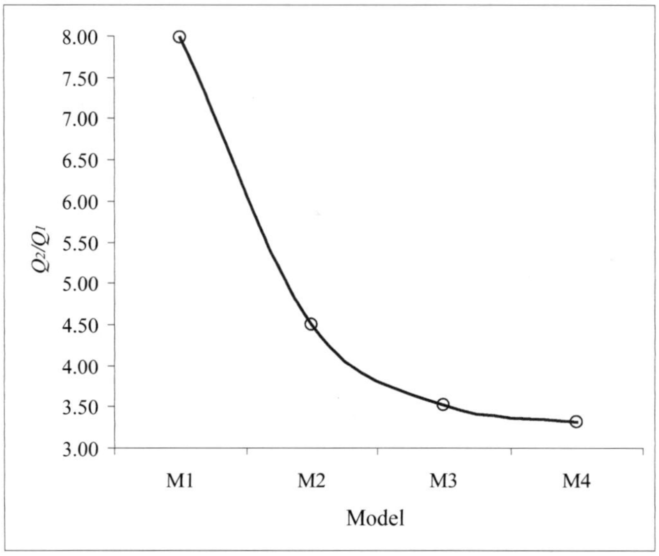

Figure 3. The Q_2/Q_1 ratio for M1-M4 models

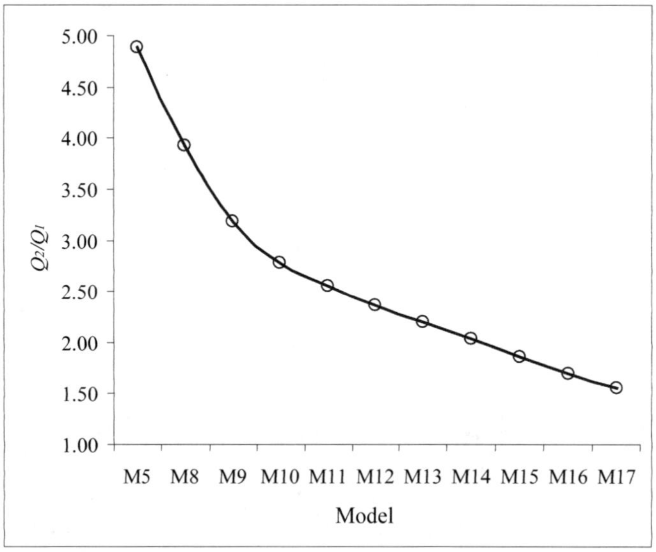

Figure 4. The Q_2/Q_1 ratio for M5-M17 models

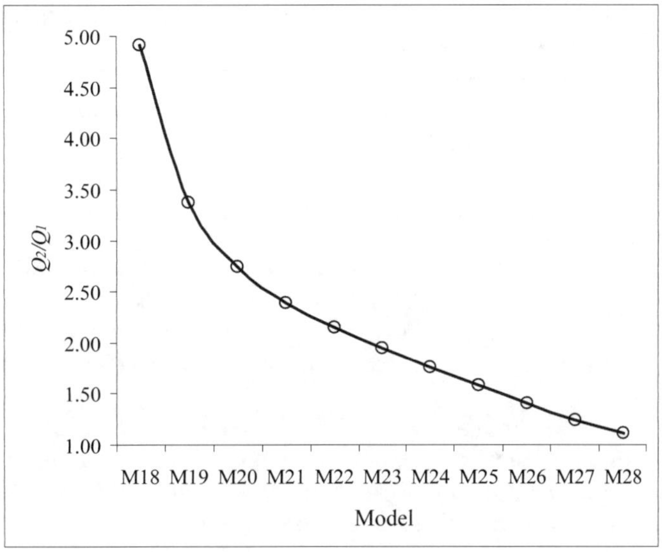

Figure 5. The Q_2/Q_1 ratio for M18-M28 models

4. Conclusions

Upon the results presented in the previous section, the following more significant conclusions could be made:

1. A manual calculation based on the displacement method gives an approximate solution, which is known to be based on the well-known work-hypotheses of Construction Statics, Part II. Therefore, the results for the M1 model approximate the manual calculations mentioned, herein afore;

2. The axial deformability of the beam section exercises a strong influence on the distribution of forces in-between the two vertical sections. This is clearly seen via the comparison of models M1 through M4 when developing the modelling by using beam elements, as it is clearly represented by M5-M17 when modelling with KE planar elements;

3. As the height of the beam increases (leading to an increase in its axial stiffness), the displacements at nodes 3 and 4 draw nearer in value, whereas in the simulation of the boundary case of a finite rigid beam (M1) they are the same, just like the behaviour of the M12 - M17 models;

4. As for the ratio Q_2/Q_1, there is a decreasing trend in value ranging from 8.00 to 3.32 for models from M1 to M4 (type I), and ranging from 4.88 to 1.56 for models from M5 to M17 (type II), Table 1 and Table 2.; The same is observed with regard to the bending moments at p. 1 and p. 2, table 1;

5. The higher beam corresponds to an increased value of an equivalent rotation stiffness at node 3 and p.4, which can be expected to lead to an increment in the bending moments at node 3 and p.2. / for all types of computational models /;

6. The higher beam, corresponding to the expansion of the overall geometry of the computational model, leads to a redistribution of the energy of deformation, which in turn reduces the displacements and the ratios (in) between the shear forces.

5. References

[1] Dimitrov B 1993 *Reinforced concrete structures*, Technika Publishing House,

[2] Manoilov L 2002 *Reinforced Concrete*, Technika Publishing House

[3] Stoynova I, Hristov T 2019 Comparative analysis between by classic solutions methods of flat slab and numerical solutions. Verification of finite elements, *18-th International Symposium MASE 2019*, 2-5 October 2019, Ohrid, North Macedonia, ISBN 978-608-4510-35-2, on USB, pp 1020-1029

[4] Stoynova I, T Christov 2018 Response of a reinforced concrete cantilever beam subject to impulse impact loads, *IOP Conf. Series: Materials Science and Engineering, vol.* **951** 012015 IOP Publishing, DOI:10.1088/1757-899X/951/1/012015

[5] Stoynova I 2020 *Dynamic response of reinforced concrete members and connections*, VSU Publishing house, Sofia, ISBN: 978-954-331-116-3, pp.32-33;

[6] Cvetkov St, Georgiev G 2012 *Methodological guide for development of course project reinforced concrete according to Eurocode 2*, VSU Publishing house, Sofia, ISBN: 978-954-331-115-6, pp.45

[7] Cvetkov St, Georgiev G 2020 *Reinforced concrete structural shear walls - antiseismic "discs"*, VSU Publishing house, Sofia, ISBN: 978-954-331-037-1, pp.78;

[8] Hamova, M, Zayakova H, Frangov G 2015 Passive grouted anchors for emergency reinforcement of retaining structures, *6th ACES Conference Geotechnics in Civil Engineering*, 2-6 November 2015, Vrsac, Serbia;

[9] Frangov G, Zayakova H, Frangov S 2017 Destructive influence of technogenic factors and precipitations on the landslide support structure, *Proceedings of the 4thWLF, Ljubjana, Slovenia, published in Advancing Culture of Living with Landslides*, Volume **3**: Advances in Landslide Technology, ISBN 978-3-319-53487-9, pp. 529-536;

[10] Ivanov Ya, Stoyanov V, Kotsilkova R 2000 Rheological Estimation of the Methods for Concrete Design, *In: Proc. XIII International Congress on Rheology, Cambridge*, UK, August 20-25, 2000, (eds. D.M. Binding, N.E. Hudson, J. Mewis, J.-M. Piau, C.J.S. Petrie, P. Townsend, M.H. Wagner, K. Walters), British Society of Rheology, Vol. **4**, 2000, pp.205-207. ISBN: 978-095-389-040-8 (set), 978-095-389-046-0 (pbk.);

[11] Ivanova M, Ivanov Ya, Stoyanov V 2013 Concrete and Requirements for Sustainable Development, *In: Education, Science, Innovations (Proc. III Int. Conf. of the European Polytechnical University, June 9-10, 2013, Pernik, Bulgaria)*, European Polytechnical University, pp.109-116., ISSN: 1314-5711 (print).

IOP Conf. Series: Materials Science and Engineering 1141 (2021) 012042 doi:10.1088/1757-899X/1141/1/012042

Numerical study of random material properties within a soil specimen

M Amad[1], R P Ray[1]

[1] Széchenyi István University, Egyetem tér 1, Győr, 9026, Hungary

E-mail: ahmad.majd@hallgato.sze.hu

Abstract. Material properties derived from laboratory soil tests often assume that the property is uniform throughout the specimen. For some exceptional soils, this may hold true, but for many others it is obviously false. We have been performing cyclic and irregular torsional simple shear (TOSS) tests on hollow cylinder samples for decades and were intrigued by the idea of how to model inherently non-uniform specimens. As an added corollary, we wanted to understand the influence of imperfections (voids, inclusions) on the measured stress-strain behaviour in these tests. This paper examines two general classes of problems: (a) uniform specimens with inclusions of voids or imperfections, and (b) non-uniform specimens with random distributions of material properties within the specimen. Finite element modelling was performed on a TOSS specimen (ID=4cm, OD=6cm, L=14cm) using a set of over 500 different elastoplastic material properties within the specimen. Various distributions (Normal, Log-normal, Bimodal) of stiffness and strength properties were examined. The results were examined as torque vs. twist curves since those values are typically measured in the TOSS laboratory test before being converted (with assumptions of uniformity) to shear stress-shear strain hysteresis. The Bimodal distributions were used to represent soils with distinct hard and soft zones. Additionally, distributions with some degree of spatial correlation were also examined.

1. Introduction

In a paper written in 1926 [1], Masing discussed the behaviour of a brass specimen undergoing loading and unloading. His underlying premise about modelling the stress-strain response of the specimen was that it behaved as a collection of elasto-plastic elements that were loaded to their yield point at slightly different times during the test. This resulted in the load-deformation curve in figure 1a where location ***l*** is the onset of yielding. He believed that the specimen did not fail uniformly because the stresses along the specimen cross-section were not uniform and therefore, some failed while others were still undergoing elastic loading. The progress of gradual transition from fully elastic condition to (nearly) fully plastic could be seen by observing the normalized tangent modulus versus strain shown in red (figure 1b). Presently, this is a common way to represent soil behaviour under small strain conditions as a modulus reduction curve where G/G_{max} would replace the parameter ***a*** on the y-axis.

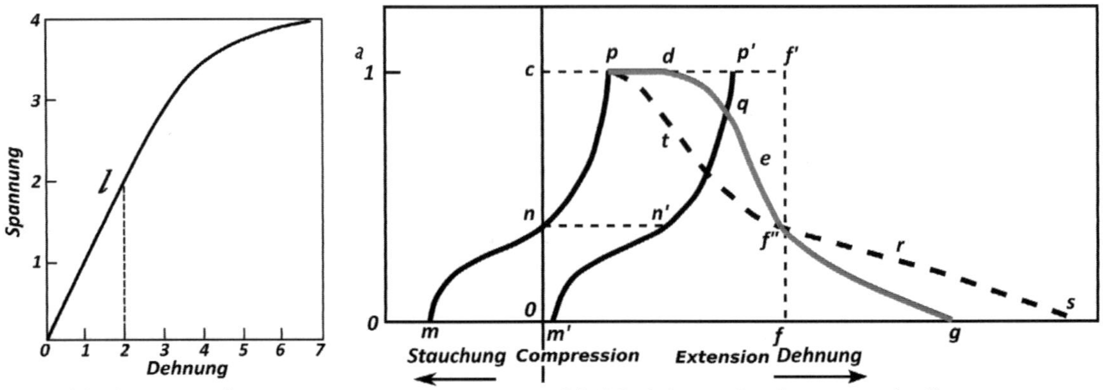

(a) Stress-strain curve (b) Modulus reduction curve (red)

Figure 1. Nonlinear stress-strain curve and tangent modulus reduction curve redrawn from Masing (1926)

2. Concept

Masing's idea prompted the start of a numerical study where we assigned various simple elasto-plastic material properties throughout a specimen to generate nonlinear stress-strain behaviour. The purpose was to better understand the impacts of varied simple material properties on the collective behaviour of a soil specimen. This presentation is part of a broader investigation on modelling the torsional simple shear (TOSS) and resonant column (RC) tests by finite element methods. Some results have been published while other publications are still in progress. Here we present the results of modelling our test specimen by random distributions of simple elasto-plastic material behaviour.

3. Tresca Model and Implementation in Midas GTS

The TOSS-RC test specimen is a hollow cylinder with inner and outer diameters of 4 and 6 cm. Specimen height is approximately 14cm but may vary slightly due to sample preparation methods. Variations of this test are described in Szilvagyi [2], Szilvagyi and Ray [3], Woods and Ray [4], Ray [5]. The advantage of a hollow cylinder is to create better shear strain uniformity across the specimen cross section. This is helpful when shear strains (γ) reach levels where significant softening occurs (γ~10^{-3} to 10^{-2} mm/mm) in the realm of small strain behaviour. At these levels, shear modulus ($\boldsymbol{G}$) may reduce to 20-30% of its original, low-strain value ($\boldsymbol{G_{max}}$).

The stress strain model that we have relied on has been the Ramberg-Osgood model. While it was originally developed for non-linear behaviour of aluminium alloys [6], the model was adapted for soils and used extensively by Richart and Woods [7]. Another, similar model is the hyperbolic model adopted and modified by Hardin and Drnevich [8] and later by Stokoe and Darendeli [9],[10] and others. The Ramberg-Osgood model is used here only as a reference for comparison. The simple elasto-plastic model used extensively in this study is the one developed by Tresca. It has an elastic modulus, Poisson's ratio and a yield value (1D), yield line (2D), or more generally, yield surface (3D). The yield surface is shaped like a tube with a hexagonal cross section centred along the hydrostatic axis. The behaviour in torsional shear would follow a path shown in Figure 2a. The conceptual load and stress conditions for an element in the model is shown in figure 2b.

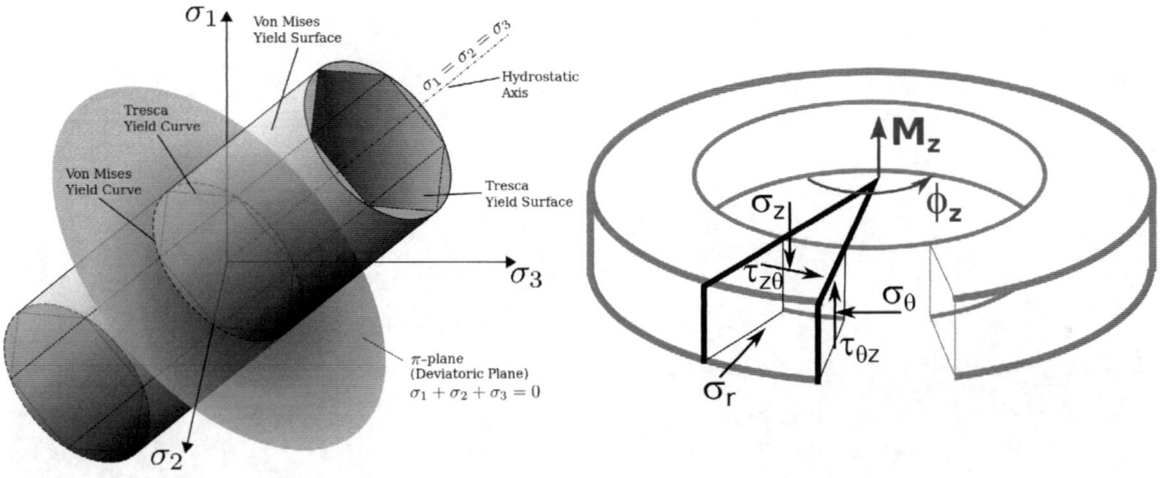

Figure 2. (a) Tresca and Von Mises failure surfaces [11]. (b) Applied twist, reaction moment and stress conditions in hollow cylinder element.

The finite element software used for this study was MidasGTS-NX. We have had a great deal of experience with the software, and it allows for 3-D modelling using cylindrical coordinates which is very helpful for torsional problems. The mesh consisted of 4033 quadratic hexahedral elements and 20,689 nodes (figure 3a). To create random properties, we specified 512 different elasto-plastic materials in the original model. The Midas program could export the mesh and materials to a text file (.fpn) that was identical to a (.csv) format file and easily imported into Excel. The material properties could then be randomized and inserted using a VBA program within Excel. Once the material was randomized, the file could be imported back into Midas. A work-flow chart and example of randomized elements from Excel is shown in figure 3b. One should note that step (4) could generate, say, a normal distribution of yield stresses and moduli based on prescribed mean and standard deviations for that material. There could also be up to ten different distributions of Tresca materials, each with their own mean and standard deviation values for yield and modulus. The only requirement was that the total number of possible property combinations (yield stress and modulus) had to be 512. At this stage, log-normal, and random (flat histogram) distributions could be created as well. For step (5) the 512 properties created from step (4) were distributed throughout the specimen with even randomness, thereby creating a specimen with a normal distribution of properties. The property map shown (step 5) is the inner "cylinder" of the finite element mesh laid out flat so that we could visually inspect the distribution of properties. Each cell has a property number from 1 to 512 and a corresponding colour. The middle and outer cylinders were also mapped in a similar fashion.

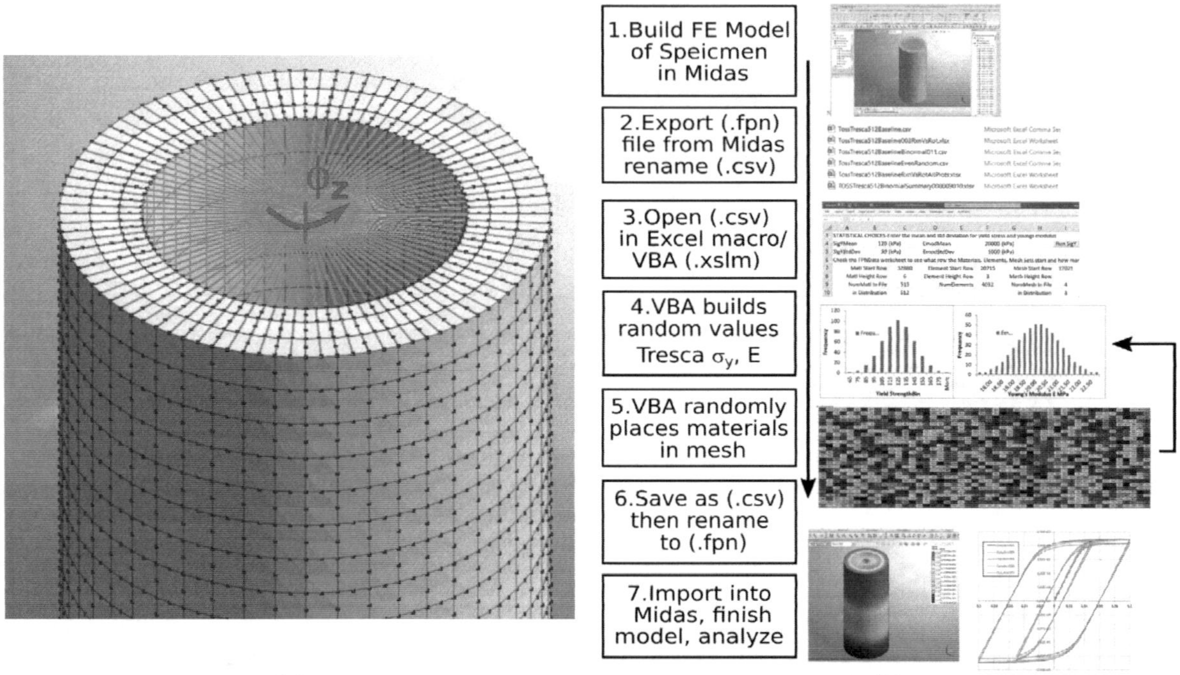

Figure 3. (a) Mesh with rigid links attached and applied twist (**ϕ_z**). (b) workflow for analysis.

Once the randomized mesh was imported into Midas, the remaining analysis parameters were input. Boundary conditions were added at the base as pinned nodes that represented the fixed base of the experimental apparatus. The top of the specimen was connected to a series of rigid links (figure 3a light blue lines). These links crossed the top of the mesh along every diameter, connecting nodes on opposite sides; a total of 529 nodes. The links were all connected to a single node at the centre where a single moment ($\boldsymbol{M_z}$) or prescribed rotation ($\boldsymbol{\phi_z}$) was applied. This arrangement avoided any difficulty with having a uniform rotation along the top of the specimen throughout the test. Generally, a rotation was prescribed since it was more likely to remain stable as the specimen approached full yield.

Since Tresca material strength is not influenced by confining stress, there was no need to simulate confinement during the analysis. The prescribed rotation was entered as a static displacement; however, the final load condition was a time-varying static load. The time function was often a sine wave that had a duration of 1.25 cycles, simulating initial (one-way) loading then one full cycle to produce a hysteresis loop. The time function was set to a slow speed to avoid any inertial effects, however the analysis type was set to non-linear time history, since it was the most convenient way to model the loading process.

Analysis options were set to seek convergence via normalized load, deflection, and work parameters to be below 1.0e-03, 1.0e-03, and 1.0e-06, respectively. The solution method used a full Newton Raphson stiffness update scheme, allowing for line search. A maximum number of iterations for a single loading step was set to 50, and the loading history itself was set to 1200 steps for the 1.25 cycle excursion.

4. Analysis Cases and Results

The initial analyses were performed to verify that the Tresca material model was performing in a way we understood. As a verification, three realizations with different Tresca $\boldsymbol{\sigma_y}$ and $\boldsymbol{E}$ values were analysed and centrally located element stress-strain data was extracted. The results (figure 4) showed what we expected: in pure rotational shear: a) the stiffness was equal to the shear modulus G, where

$$G = \frac{E}{2(1+\nu)} \quad (1)$$

and b) The yield stress in pure shear was equal to one-half the Tresca yield. Sinc $\upsilon = 0.3$ for all tests, $\boldsymbol{G}$ and $\boldsymbol{E}$ were related by a factor of 2.6. This is shown in the linear portions of the results in figure 4a.

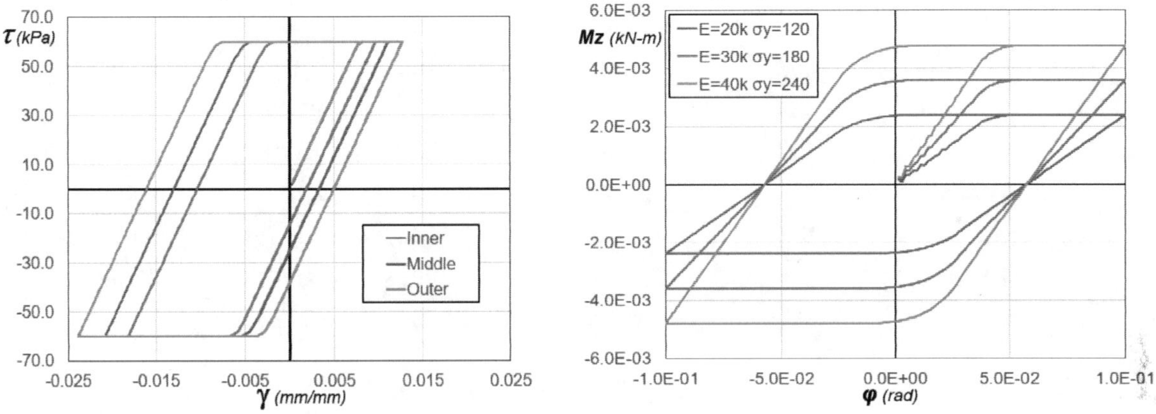

(a) (b)

Figure 4. a) Stress-strain curves for $\boldsymbol{E}$=20000 kPa $\boldsymbol{\sigma_y}$=120, b) Moment rotation curves for three speciemens with unform properties

An introductory set of analyses were performed to investigate the behaviour of the test specimen when random distributions of properties were input. The series of configurations is presented in table 1. The analysis number indicates normal distribution (N), log-normal distributions (LN), bimodal normal even split (BE), and bimodal normal offset (BO). All distributions started with population of 512. Normal distributions were described by mean and standard deviation values for both $\boldsymbol{E}$ and $\boldsymbol{\sigma_y}$. The log-normal distributions had much higher values for standard deviation. The reader should note the drastic differences in standard deviation between the two sets. This is expected since, the whole purpose for using log-normal distributions is to account for the extensive degree of variability in geotechnical properties one may find in nature. The even bi-modal normal distributions were simply two distributions of 256 values with two mean values and two standard deviation values. Both distributions were randomly spread over the element mesh as there were no soil layers or zones separating them. The offset bi-modal normal distributions were two distributions with different populations that added to 512. Additionally, the overall mean remained the same ($\boldsymbol{\sigma_y}$ = 120 kPa, $\boldsymbol{E}$=20000 kPa). The placement of all elements in this set was also entirely random.

Table 1. Material conditions, random distributions, loading conditions.

Analysis Number	Number of Materials	Yield Stress, $\boldsymbol{\sigma_y}$ (kPa)		Tresca Modulus $\boldsymbol{E}$ (kPa)	
		Mean	Std Dev	Mean	Std Dev
N-01	512	120	30	20000	1000
N-02	512	120	20	20000	1000
N-03	512	120	20	20000	5000
LN-01	512	120	60	20000	10000
LN-02	512	120	120	20000	20000
LN-03	512	120	240	20000	40000

BE-01	256	80	20	10000	2000
	256	160	20	30000	5000
BE-02	256	80	20	16000	2000
	256	160	20	24000	2000
BE-03	256	60	20	14000	2000
	256	180	20	26000	2000
BO-01	384	80	20	14000	2000
	128	240	20	38000	2000
BO-02	435	80	20	14000	2000
	77	346	20	53900	2000
BO-03	486	80	20	14000	2000
	26	868	20	132000	2000

Reaction moment vs rotation plots are shown for each group in figures 5a,b and 6a,b. Figure 5 shows the reaction moment vs. rotation for specimens with normal and log normal distributions of ***E*** and $\boldsymbol{\sigma}_y$. Test numbers correspond to the conditions listed in table 1.The first characteristic is the "wobbly" path of the initial loading curve. This is not a numerical oscillation, but rather demonstrates the effect of different elements reaching yield, and perhaps unloading/reloading over short intervals as the main body of the specimen continues to rotate. The unloading/reloading bursts were verified by viewing the plastic states of all elements during the entire loading history. We recorded animations of each test and they proved to be quite remarkable. Since every test was a prescribed rotation between 0.1 and -0.1 radians, the plots all match along the horizontal axis.

Figure 5 presents a set of: (a) normally distributed properties, and (b) log-normally distributed properties. The effect is obvious in that the paths are more smoothly curved and mimic more closely the typical behaviour of soils in our torsional shear tests.

A similar effect occurs in figure 6a where the pairs of mean values are separated greater and greater distances. Analysis BE-03 has the widest differences and shows the smoothest curvature. In figure 6b the bimodal concept was changed so that the stiffer, stronger component had a reduced population. However, the overall average ***E*** and $\boldsymbol{\sigma}_y$ remained constant. The progression is similar to the previous groups but less pronounced.

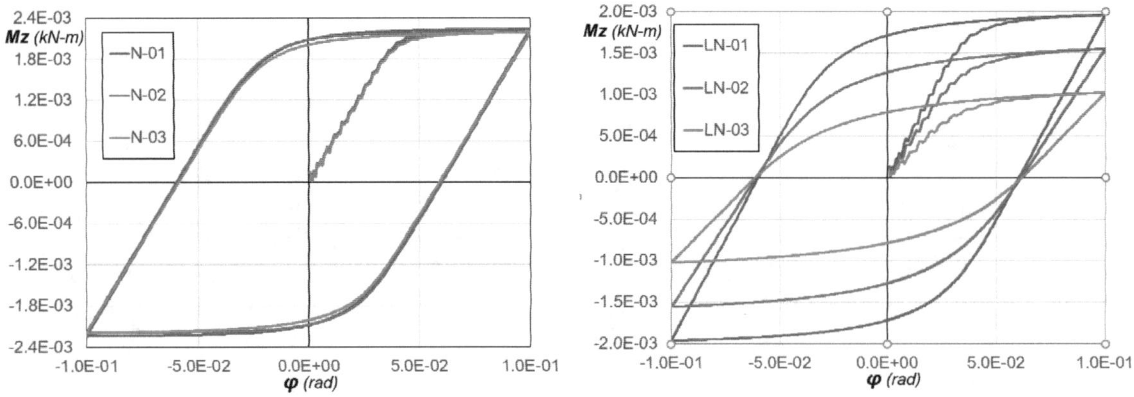

(a) (b)

Figure 5. Reaction moment vs rotation for a) Normally distributed, b) Log-Normally distributed values of ***E*** and $\boldsymbol{\sigma}_y$

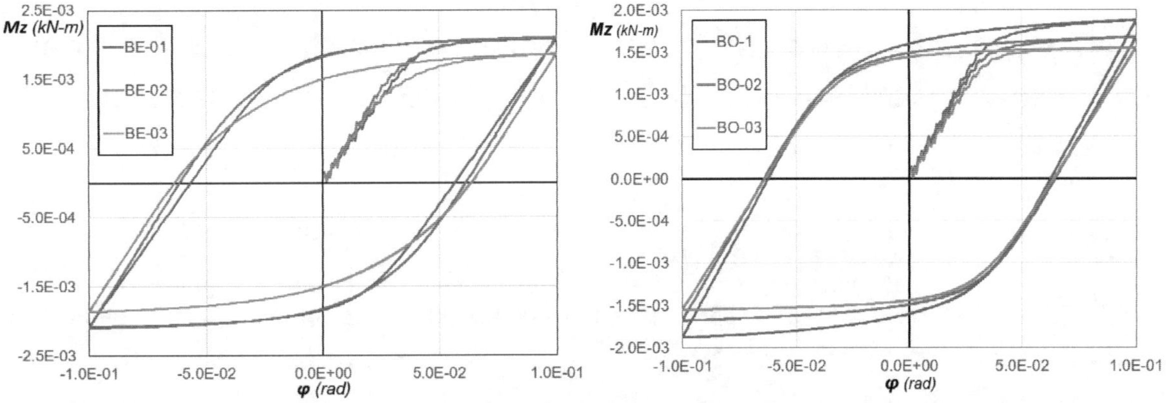

(a) (b)

Figure 6. Reaction moment vs. rotation a) even bimodal normal distribution and b) offset bimodal normal distribution of values of $\boldsymbol{E}$ and $\boldsymbol{\sigma_y}$

5. Conclusions and Further Research

The observations presented by Masing were investigated by applying a simple elasto-plastic model to a carefully analysed test specimen. His observation that such a simple material model can produce complex behaviour is also demonstrated here. Our main conclusion is that material non-linearity, as a continuous process, becomes more apparent as the material properties vary over a wider range, become more variable. This supports the intuitive notion of particulate mechanics where the particle-particle contact behaviour may be simple, but the aggregate behaviour becomes quite complex.

We anticipate our research will examine a broader range of material models and more sophisticated approaches to representing their variability in a laboratory specimen.

6. References

[1] Masing G 1926 Self stretching and hardening for brass, (in German) in *Proc. 2nd Int. Congr. Appl. Mech., Zurich*, pp 332–335

[2] Szilvágyi Z 2017 *Dynamic soil properties of Danube sands*, PhD Thesis, Győr, Hungary: Széchenyi István University

[3] Szilvágyi Z and Ray R P 2018 Verification of the Ramberg-Osgood material model in Midas GTS NX with the modeling of torsional simple shear tests, *Periodica Polytechnica, Civil Engineering,* vol **62**, no 3, pp 629–635

[4] Ray R and Woods R 1988 Modulus and damping due to uniform and variable cyclic loading, *Journal of Geotechnical Engineering*, vol **114**, no 8, pp 861–876

[5] Ray R P 1983 *Changes in shear modulus and damping in cohesionless soils due to repeated loading*, PhD Thesis, University of Michigan

[6] Ramberg W and Osgood W R 1943 *Description of stress strain curves by three parameters*, Technical Notes, National Advisory Committee for Aeronautics, vol. 902

[7] Richart F, Hall J and Woods R 1970 *Vibrations of soils and foundations*, Prentice-Hall, Englewood Cliffs, N.J., 1970.

[8] Hardin B and Drnevich V 1972 Shear modulus and damping in soils: Design equations and curves, *Journal of the Soil Mechanics and Foundations Division*, vol **98**, pp 667–692

[9] Darendeli M B 2001 *Development of a new family of normalized modulus reduction and material damping curves*, PhD Thesis, University of Texas at Austin, USA

[10] Stokoe K H II and Darendeli M B 1998 *Laboratory Evaluation of the Dynamic Properties of Intact Soil Specimens: Garner Valley, California,* Geotechnical Engineering Report GR98-3, Civil Engineering Department, The University of Texas at Austin, Austin, TX, March

[11] SCIA engineers help page Retrieved from http://kc.scia.net/Help/kik/functionalities/nonlinearities/general_plasticity/general_plasticity.htm

Self-stressed steel-reinforced concrete floor slab stress-strain state numerical analysis taking into account the concreting stages

O V Semko and A V Hasenko
Educational and Scientific Institute of Architecture and Construction, National University "Yuri Kondratyuk Poltava Polytechnic", Pershotravnevyj Ave. 24, Poltava (36011), *Ukraine*,

E-mail: gasentk@gmail.com

Abstract. In steel-reinforced concrete floor slabs, structurally ensuring the joint operation of monolithic reinforced concrete slab and steel crossbars, a successful combination of its components positive qualities is realized at a high level: compression forces are perceived by monolithic slab, and tensile forces by steel beams. Reinforced concrete slabs in such structures have a relatively equal weight with a payload on interfloor slabs for public buildings. Having effectively developed the sequence of concreting the slab through the span in two stages, it is possible to create pre-stressing in the floor steel crossbars, located in the spans of the second concreting stage, and in monolithic reinforced concrete slab of the first concreting stage. In this sense, the term "self-stressed" is used in this work. By creating pre-stresses in the steel crossbars of the floor, it is possible to reduce their cross-section or to increase their span and to achieve savings in metal costs on the building frame. Pre-stressing of the monolithic reinforced concrete slab of the first concreting stage in the opposite to operational bending form, will increase crack resistance of the specified sites of a monolithic overlapping and will give the chance to reduce its reinforcement.

1. Introduction

Born as a building structure about a century ago, steel-concrete structure has recently become more widely used in various fields of construction, proof of which is the development of relevant regulations. Steel-concrete structures are made by a successful combination of steel, mainly rod elements and monolithic concrete. In some cases, to increase the overall load-bearing capacity of the combined cross section, flexible or rigid reinforcement rods are inserted into monolithic concrete, especially in the case of monolithic reinforced concrete beyond rigid reinforcement (for example, in steel-reinforced concrete floor slab). The number and cross section of such reinforcement is selected by a separate calculation only for the reinforced concrete floor slab [19].

In steel-reinforced concrete floor slabs (see figure 1), structurally ensuring the joint operation of monolithic reinforced concrete slab and steel crossbars, a successful combination of its components positive qualities is realized at a high level: compression efforts are perceived by monolithic reinforced concrete, and tensile efforts - by steel beams [8]. The high level of reliability, durability and resource saving during the construction and operation of such floor slabs is confirmed by the significant volume of its installation for both public and industrial buildings [13]. Quite often, to increase the manufacturability of such floor slab installation, a monolithic reinforced concrete slab is arranged on a fixed formwork from profiled flooring [6].

Monolithic reinforced concrete floor slabs in steel-reinforced concrete floor slabs have a relatively equal weight with a payload on interfloor slabs for public and civil buildings [15]. Having effectively developed the sequence of concreting the slab through the span in two stages, it is possible to create prestressing in the floor steel crossbars, located in the spans of the second concreting stage, and in monolithic reinforced concrete slab of the first concreting stage. In this sense, the term "self-stressed" is used in this work.

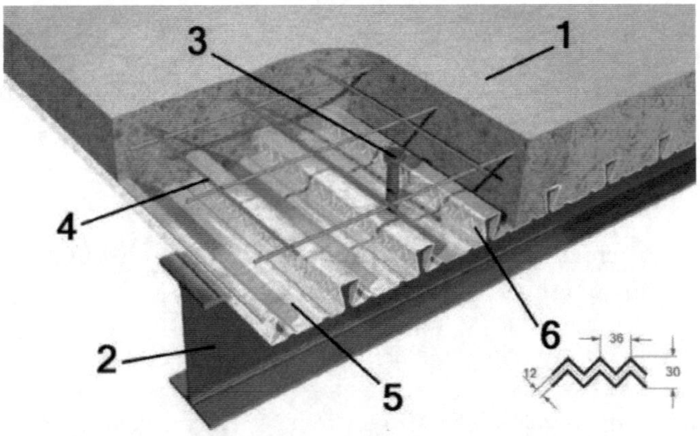

Figure 1. Construction of effective monolithic steel-reinforced concrete floor slab with "lock" profiled flooring:
1 – concrete monolithic slab;
2 – steel beams;
3 – anchor bolts;
4 – reinforcing mesh;
5 – "lock" construction profiled flooring;
6 – profiled flooring corrugated surface to increase the anchoring properties

2. Analysis of recent research and publications

Based on the results of research in recent decades, today steel-reinforced concrete elements that work on the bend are calculated using an analytical apparatus for calculating similar reinforced concrete elements [3; 7; 9]. This analytical apparatus is based on the deformation method taking into account complete nonlinear real diagrams of material deformation [2]. The essence of the adaptation of reinforced concrete structures working on the bend calculation method, to the calculation of similar reinforced concrete elements is as follows [16]. The steel part of the calculated cross section is divided into separate "strips". Within the height of these "strips" stresses are considered constant, and the deformations distribution in the height of the steel element - according to a linear law. This approach allows you to depict a steel-reinforced concrete beam as a reinforced concrete with multi-row external reinforcement. Comparison of experimental and theoretical values of steel-concrete beams bearing capacity allows us to conclude that the considered analytical apparatus for calculating bending steel-reinforced concrete elements, fairly accurately assesses their stress-strain state and bearing capacity.

However, in most experimental and theoretical studies in this direction, the definition of parameters that characterize the initial (before inclusion in the joint work of the combined section) state - the initial deformations and stresses of both steel beams and monolithic concrete in steel- concrete floor slab, was not given due attention [1]. Therefore, the analysis of their influence on the overall stress-strain state of the structure is difficult. It is assumed that ignoring the initial conditions provides a margin of 10%. The calculated analytical apparatus, the essence of which is given above, take into account the initial conditions, such as pre-stresses and deformations of steel reinforcement [16].

A linear approach to taking into account the initial state of the elements of the composite floor slab can give errors that can lead to the destruction of the structure or, conversely, the design of structures with a large margin of safety [17]. Therefore, the application of such a linear approach, despite its simplicity and availability in terms of a relatively small amount of knowledge and skills to use it, is unjustified. An accurate study of the above problem requires nonlinear analysis [4]. This analysis takes into account nonlinear effects, such as nonlinear material properties [5], contacts, large deformations [10], loading history [18], etc.

For the preliminary analysis of the real stress-strain state taking into account the stage loading of the steel-reinforced concrete floor slab, was used the software package Femap 2020.2 with NX Nastran Trial Version Siemens Digital Industries Software with personal activation code 2827301401535961. This finite element analysis program has proven its reliability, accuracy, and speed many times over the course of about fifty years [14]. Due to a widely developed set of output parameters, this program is used by engineers in various fields of technology, in particular in mechanical engineering, construction, electro mechanics, thermodynamics and others. To solve the most complex nonlinear problems, including if you need to take into account the history of loading the structure, the Femap software package has a module of multi-step nonlinear solutions Multistep Nonlinear (SOL401 / SOL402) [11].

3. Formulating of the article's goals

The aim of the article is to consider the algorithms of finite element modelling using computer software for the stages of two different materials inclusion in the work during the analysis of the stress-strain state of self-stressed steel-reinforced concrete floor slab at the stage of taking into account the stages of concreting monolithic reinforced concrete slab.

4. Description of the research problem

The object of research is a steel-reinforced concrete floor slab, arranged according to a three-span continuous scheme (see figure 2), the construction of which is shown in figure 1.

The subject of research is a distribution of stresses and strains along the height of the combined cross section of steel-reinforced concrete floor slab during two-stage concreting of a monolithic slab and, accordingly, its two-stage inclusion in the work.

As indicated in the introduction to this work, the own weight of steel-reinforced concrete floor slab is approximately equal to the payload on floor slabs in residential and public buildings. Therefore, during the construction of the building before its exploitation in the cross sections of steel-reinforced concrete floor slabs from its own weight will be quite significant stresses. Their level can reach 30…50% of the maximum design stresses. The authors propose to use this effect to create favourable pre-stressing components of steel-reinforced concrete floor slabs.

Creation of preliminary stresses is offered to realize by the following sequence of slab concreting in two stages. At the first stage concreting through span is carried out. That is, for the considered three span scheme of floor slab – two extreme spans (see figure 2, a). At this stage, the steel beams of the extreme spans (spans of the first stage of concreting) are bent down from the weight of the freshly laid concrete mixture. In this position there is a setting (transformation) of the concrete mixture into a monolithic concrete stone. That is, the monolithic concrete floor slab in these spans has an initial negative bending, which repeats the bending of steel beams in these spans. The stress in the concrete at this level is zero. *Steel beams in spans without concrete mix* (for the considered three span scheme – average span) due to the inseparable scheme of work beams *are bent upwards, then the preliminary self-stressed is created in them.*

After setting the design strength by the concrete of the first stage of concreting, the second stage of concreting for other spans is performed. For the considered three-span scheme of floor slab – the average span (see figure 2, b). At this stage, the steel beam of the middle span (span of the second stage of concreting) bends down from the weight of the freshly laid concrete mixture. However, due to the inseparable scheme of beams and inclusion in the work the concrete shelves extreme spans the first stage of concreting, the beam deflection in the middle span is less than the deflections of the beams in the extreme spans. Thus, the effect of steel beams self-tension begins to be shown. Beams of the extreme spans (spans of the first stage of concreting) in the second stage of concreting begin to bend upwards towards the initial position. Thus, *the concrete in these areas is stretched, which creates a preliminary self-stress.*

By creating pre-stresses in the steel crossbars of the floor slab in the areas of the second stage of concreting, it is possible to reduce the cross-section of the latter and achieve savings in metal costs on the building frame. Another option for using the increased bearing capacity of pre-stressed crossbars is to increase their span. That is, to design the location of columns with a variable step through the span. Then the cost of metal on the building frame is saved by reducing the number of columns. Pre-stressing of the monolithic reinforced concrete slab of the first stage of concreting in the form of opposite to operational bending, will increase crack resistance of a monolithic floor slab specified sites and will give the chance to reduce its reinforcement.

Figure 2, c shows a deformed view of the steel-reinforced concrete floor slab in the operating position. Also, in figure 2 for clarity of the above statements about the creation of pre-tension of the steel-reinforced concrete floor slab elements, shows the expected plots of stress distribution in the corresponding sections at each stage of concreting.

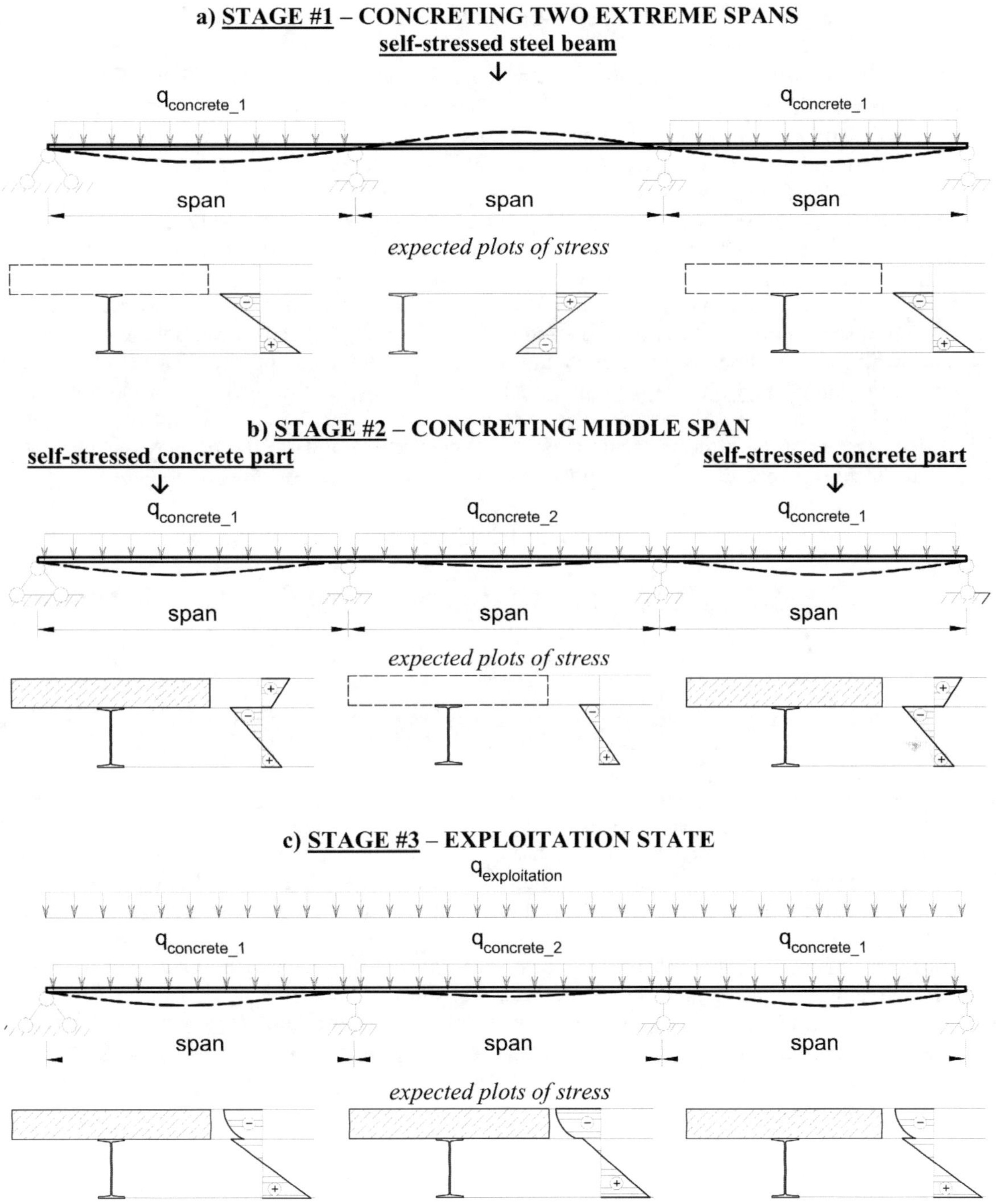

Figure 2. The stage of the work steel-reinforced concrete floor slab

5. Ways to solve the problem

The presence of pre-operational stresses in the components of the composite reinforced concrete floor with the simultaneous absence of stresses in other components during the finite element analysis in the software package Femap 2020.2 with NX Nastran may be considered in different ways. These include:

- initial tension (stress) of the steel floor beam (similar to the previous tension of bolted joints – Bolt Preload);

- loading of the composite reinforced concrete structure with the initial data (Output Load) of the preliminary calculations of the steel floor beams;
- installation of a preliminary gap between steel and concrete parts (Initial Gap);
- temperature effects on monolithic reinforced concrete in order to reduce to zero stresses in it at the time of concrete setting (Temperature Load);
- taking into account the stages of two materials (steel and concrete) inclusion in the work with the module of multi-step nonlinear solutions (Multistep Nonlinear).

The following is a brief overview, advantages and disadvantages of the above ways to solve the problem, some of which are quite cumbersome when creating a finite element model and may give calculation errors.

During bending of steel beams of floor from weight of a monolithic reinforced concrete plate, distribution of stresses on height of their section is uneven (see figure 2). For example, when the beam is bent down, the lower fibers are stretched and the upper ones are compressed. Therefore, to set the correct pre-tension of the Bolt Preload type, it is necessary to divide the steel cross-section into strips within which the stresses will be considered constant (see figure 3). This procedure leads to an unjustified increase in the time to create these bands, determine and set the pre-stresses in them, as well as to a significant increase in the number of finite elements of the steel beam model.

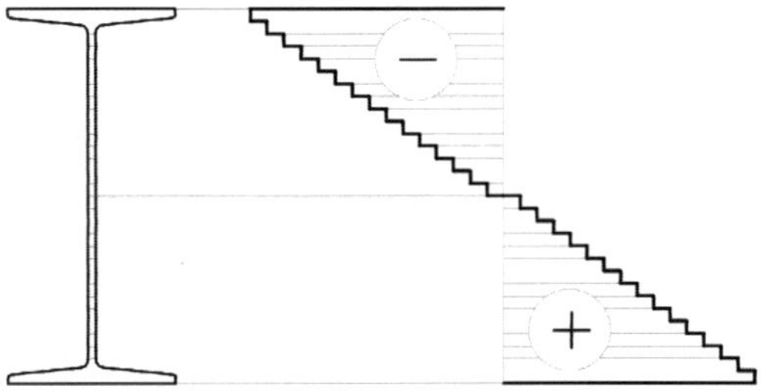

Figure 3. The strips on the steel cross-section

When loading the composite reinforced concrete structure with the initial data of the preliminary calculations of the steel floor beams, the concrete part of the structure is immediately included in the work (see figure 4). This process is not true, as in the concrete part of the section at the time of concrete set design strength until the next stage of concreting or other external load, no stress. This disadvantage is also present for the first of the above ways to solve the problem (pre-stress of the steel beam).

Since during bending of steel beams under the action of evenly distributed load from freshly laid concrete mixture their maximum deflection occurs in the middle of the span, and on the supports vertical displacements are zero (see figure 5), the establishment of an initial gap between concrete and steel parts of the composite structure is also time consuming. In this case, it is necessary to divide the beam along the length of the span into an infinite number of sections within which the deflections will be considered constant.

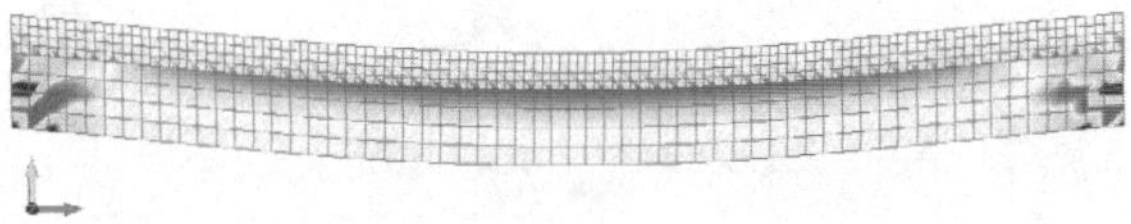

Figure 4. The concrete part is immediately included in the work

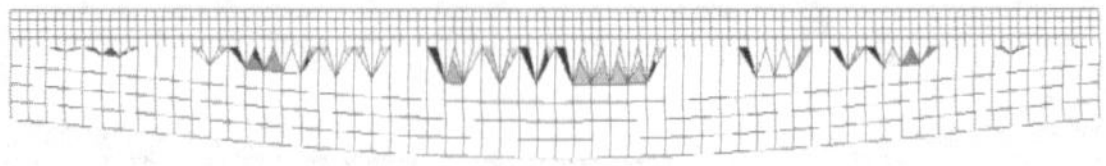

Figure 5. An initial gap between concrete and steel parts

Setting different temperatures along the length of the concrete shelf in order to obtain temperature stresses in it, which would reset the stress at the time of the concrete mixture setting, is also time consuming. This is due to the different values of the bending moment and, consequently, the different level of stresses along the length of the beam axis. Therefore, it is necessary to divide the beam along the length of the span into an infinite number of sections within which the bending moment will be considered constant. With the help of the last two ways to solve this problem, it is possible to achieve zero stresses in the concrete shelf with a stress-strain steel part of the composite structure.

The module of multistep nonlinear solutions Multistep Nonlinear gives the chance to create subcases of calculation by means of which materials of a composite design will be included in work at various necessary for the user stages of the investigated construction work.

6. Conclusion

Using the proposed two-stage technology of concreting a monolithic slab of steel-reinforced concrete floor structure by creating self-pre-stressed in the steel crossbars of the steel-reinforced concrete floor, it is possible to reduce their cross-section or to increase their span and to achieve savings in metal costs on the building frame. Pre-stressing of the monolithic reinforced concrete slab of the first concreting stage in the opposite to operational bending form, will increase crack resistance of the specified sites of a monolithic overlapping and will give the chance to reduce its reinforcement.

It is recommended to use the module of multi-step nonlinear solutions Multistep Nonlinear in Femap 2020.2 with NX Nastran to model by the method of finite elements of the stage of a monolithic concrete shelf inclusion in the work of steel-reinforced concrete floor slab, as in this case it is necessary to spend the least resources.

7. References

[1] Ametov U G 2014 Some aspects of the calculation of combined joints of steel-concrete composite structures *Academic journal. Industrial Machine Building, Civil Engineering* (Ukraine, Poltava: PoltNTU) №3 (42) book 1

[2] Azizov T, Kochkarev D and Galinska T 2020 Reinforced concrete rod elements stiffness considering concrete nonlinear properties *Lecture Notes in Civil Engineering* (London: Chapman and Hall) volume 47 pp 1–6 https://doi.org/10.1007/978-3-030-27011-7_1

[3] Bachinsky V Ya, Bambura A N, Vatagin S S and Zhuravlyova N V 1985 *Methodical recommendations for determining the parameters of the diagram "σ-ε" of concrete during short-term compression* (Ukraine, Kiev) p. 15

[4] *Basic Nonlinear Analysis User's Guide* 2014 (Siemens Industry Software) p. 144

[5] Bolotov A U and Karpenko A A 2017 Calculation of buildings of social housing by a deformation method based on time factor with using transformed diagram of concrete deformation *Construction, materials science, mechanical engineering* (Ukraine, Dnipro: PSACEA) No 99 pp 35–43

[6] Cherednikov V, Voskobiinyk O and Cherednikova O 2017 Evaluation of the warping model for analysis of polystyrene concrete slabs with profiled steel sheeting *Periodica Polytechnica Civil Engineering* (Hungary: Budapest University of Technology and Economics) volume 61(3) pp 483–490

[7] Gasii G, Hasii O and Klimenko V 2020 Testing of the combined structural elements of support of a mine opening *In E3S Web of Conferences* (France, Les Ulis Cedex A) volume 168 **8** 00028

[8] Hudz S, Storozhenko L, Gasii G and Hasii O 2020 Features of operation and design of steel sloping roof purlin *Proceedings of the 2nd International Conference on Building Innovations. ICBI 2019. Lecture Notes in Civil Engineering* (Springer, Cham) volume 73 pp 65–73 https://doi.org/10.1007/978-3-030-42939-3_8

[9] Lazariev D, Avramenko Y, Zyma O and Pasichnyk P 2020 Experimental researches of concrete ultimate characteristics and strength of compressed and bended reinforced concrete elements *Lecture Notes in Civil Engineering* (Springer, Cham) volume 73 pp 133–141

[10] Liang K, Ruess M and Abdalla M 2016 An eigenanalysis-based bifurcation indicator proposed in the framework of a reduced-order modeling technique for non-linear structural analysis *International Journal of Non-Linear Mechanics* (Elsevier) volume 81 pp 129–138 doi:

10.1016/j.ijnonlinmec.2016.01.013
[11] *Multi-Step Nonlinear User's Guide (SOL 401 and SOL 402)* 2018 (Siemens Industry Software) p. 292
[12] Pavlikov A M Mykytenko S M and Hasenko A V 2018 Effective structural system for the construction of affordable housing *International Journal of Engineering & Technology: Publisher of International Academic Journals: Science Publishing Corporation, RAK Free Trade Zone* 7 (UAE, Al Mamourah Area) No 3.2 pp 291–298 doi:10.14419/ijet.v7i3.2.14422
[13] Pavlikov A, Harkava O, Prykhodko Y and Baryliak B 2019 Highly constructed precast flat slab frame structural system of buildings and research of its slabs *Proceedings of the International fib Symposium on Conceptual Design of Structures* (Torroja Institute, Madrid, Spain) volume pp 493–499
[14] Rudakov K N 2011 *Femap 10.2.0. Geometric and finite element modeling of structures* (Ukraine, Kiev: Igor Sikorsky Kyiv Polytechnic Institute) p. 317
[15] Semko O V, Hasenko A V, Kyrychenko V A and Sirobaba V O 2020 The rational parameters of the civil building steel frame with struts *Proceedings of the 2nd International Conference on Building Innovations – Part of the Lecture Notes in Civil Engineering book series* (LNCE) volume 73 pp 235–243
[16] Storozhenko L I, Semko O V and Pents V F 2005 *Steel-reinforced concrete constrictions (Tutorial)* (Poltava: PoltNTU) p 181
[17] Su Y Y and Gao X L 2014 Analytical model for adhesively bonded composite panel-flange joints based on the Timoshenko beam theory *Composite Structures* (Elsevier) vol 107 pp 112–118
[18] Titarenko F 2020 Simple about nonlinear finite element analysis *Nanosoft blog (Electronic Materials)*
[19] Vatulia G L, Lobiak O V, Deryzemlia S V, Verevicheva M A and Orel Y F 2019 Rationalization of cross-sections of the composite reinforced concrete span structure of bridges with a monolithic reinforced concrete roadway slab *IOP Conference Series: Materials Science and Engineering* (Bulgaria, Sozopol) volume 664 **10** 012014

Determining the dynamic characteristics of an existing timber belfry structure and impact on FEM modelling

T Kochański[1]
[1] Chair of Bridge, Metal and Timber Structures L-3, Cracow University of Technology, Warszawska 24 street, Cracow, *Poland*,

E-mail: Tomasz.kochanski@pk.edu.pl

Abstract. The paper studies the assessment of the dynamic characteristics of an existing historic timber belfry structure using FEM modelling and dynamic identification tests performed on a structure. The analysed structure undergoes geometrically measurement, including elements inspection, and on this basis the corresponding initial FEM model was built. The initial model including actual dead loads was computed to achieve dynamical and static calculations results, including eigenmodes and modal masses. On the belfry structure, dynamic identification tests were performed using accelerometers arranged in places of expected maximum amplitudes for belfry parts. The same initial model was calibrated according to the data achieved from computing in-situ identification tests (FFT spectrum analysis) and recalculated for static and dynamic results. Computed results from the initial and calibrated models were compared, presented and discussed. Brief summary includes conclusions from analysis and contribution to researches on belfries dynamics.

Keywords: timber, belfry, belltower, dynamics, FEM, dynamic identification tests

1. Introduction

Timber belfries were frequently built in the southern part of Poland but also, along the Carpathian Mountains. This kind of belfries are still not well identified structures in the field of structural mechanics and FEM modelling. The main purpose of the article is to determine actual structure and dynamic characteristics of the belfry in a FEM model using the in-situ dynamic identification tests.

1.1. History and localization of the belfry

The church with a belfry tower is located near the town centre of Skrzydlna, on sloping terrain, which makes the level of the belfry's tower foundation stand higher than the church's structure. According to the known history of the building, the first belfry structure was annexed to an existing church structure in the late 16th century. The structure was probably built in later periods (exact history and dates are unknown). It is located in front of the church's main front wall (a typical log timber construction) with main doors which creates an additional room – a vestibule in front of the building. Over the centuries, the belfry structure has undergone several renovations and rebuilding. Most of the modicifactions concerned its highest (bellroom with a turret) and lowest (ground beam, support) parts. The exact dates of the renovations and rebuilding are unknown [1][2].

IOP Conf. Series: Materials Science and Engineering 1141 (2021) 012044 doi:10.1088/1757-899X/1141/1/012044

Figure 1. Analysed belfry structure

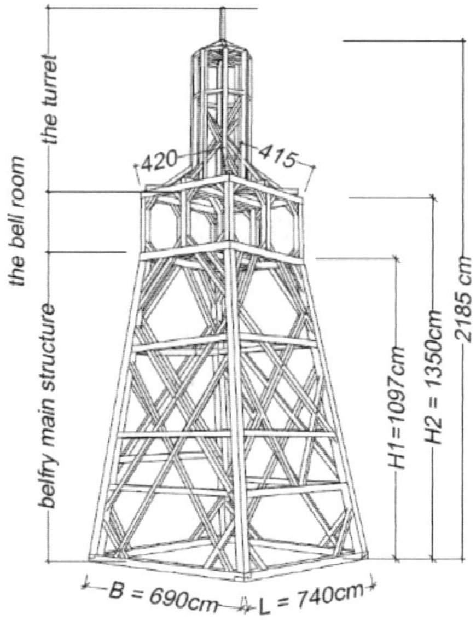

Figure 2. Characteristic dimensions of the structure

1.2. The main belfry structure

The main structure of the belfry presented in Fig.1 is composed of 4 main columns with varying cross-sections (from 29x30 to 38x40). The columns are supported (with additional knee bracing) on the ground beam which nowadays is rested on RC foundations (original foundations were simple stone-built footings) and connected with struts and beams along the height of the main structure. Those struts are crossing each other with lap jointing. At first sight, the main structure looks like a truss, but because the elements are not symmetrically connected, it works similarly to a frame – providing also bending moment. Originally, the analysed structure had both axis of symmetry, but due to progressive damages of the belfry's back wall, currently it has only one axis (symmetry of the side walls). This situation is caused by irregular subsidence of the whole structure in the past and wood rooting due to leakings, and this in time lead to disconnection of a strut and destruction of another one in the back wall. The elements are connected with each other with traditional, typical historical joints described in [3], which in some aspects are rigid types, but with limited work angle (dependent on connection loose) and limited resistance to tension. It needs to be mentioned, that the belfry has got 2 platform levels – the first above the vestibule creating ceiling and the second one creating a kind of a technical platform. Main dimensions of the structure are given in Fig.2, where H1 is height of the main structure.

1.3. The bell room and the turret

The bell room is built as independent space, supported on the main structure's columns by capping beams along the structure contour, however, the bell room is not the original one and it was built in the later periods (the exact dates are unknown). The bell room structure is composed of 8 posts (3 per each side) connecting two capping beams: upper (supporting the turret) and lower (rested on the main structure's columns). The structure is braced by knee braces arranged to connect upper and lower capping beams with posts. An additional king post was added in the centre of the structure, supported on purlins with knee bracing.

Figure 3. Interior – main belfry structure (platform)

Figure 5. Interior of the bell room and beams suspending formerly bells seen on lower parts on photography

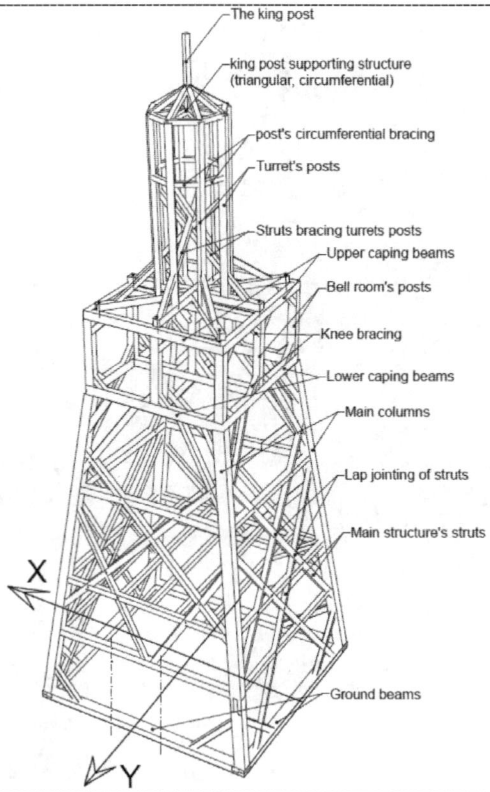

Figure 4. Description of the belfry structure

The turret is built as another structure composed of 8 posts each located on circumference, radial symmetry, placed on 8-star beam base forming the shape of a turret. Every two opposite posts and a single beam base are connected by a long strut forming a braced structure for each direction. Additionally, the posts were braced along the circle line with the horizontal elements in the connection levels. Tops of posts are connected with themselves with another horizontal bracing, also forming the structure of the king post. The explanation of the structure and localisations of the elements are given in Fig. 4. Additionally Fig. 3 and Fig. 5 present an overview of the belfry structure and the bell room.

1.4. Use of the belfry

Nowadays, the bells are dismantled, and the belfry serves as a historical attraction creating on the ground level a vestibule as the first room in the church.

2. FEM model building and assumptions

The whole structure of the building was measured including cross sections of elements, points of connections between the belfry elements, estimated characteristics of those connections, and claddings with their thickness (forming dead loads of the building, such as planks, roof steel panels and timber structure, according to [6]). Based on the measurements, the FEM model was built for a given structure. Belfry geometry, nodes (representing joints), and the relevance of the elements were converted to a 3D model which is the basis for the FEM calculation. In the first attempt, some assumptions were simplified (nodes were considered as pinned connection, pinned support on ground beam, excluding rigidity of a cladding). Tab. 1 shows the layout of characteristic fragments of the analysed belfry structure, including characteristic joints and problems specific for this structure according to the in-situ structural health check. Many aspects identified in this existing structure need to be validated in the model, as presented in Tab. 2.

Table 1. Joints – nodes. Presented the following structural specification of the existing belfry

 Overlapping joint between struts (Tab.2)	 Disconnection of the elements in pos. B (Tab.2)	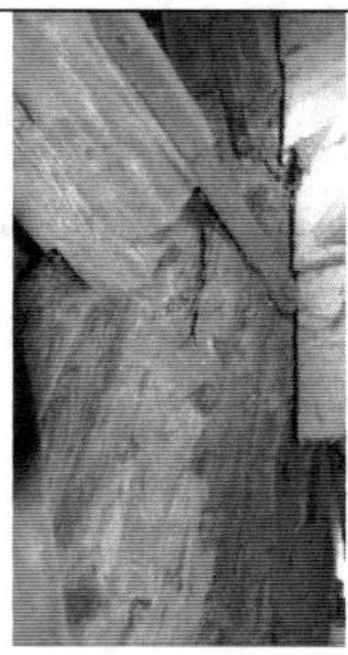 Damage of the main column pos. B (Tab.2)
 Knee braces and upper caping beam – corner of the bell room (Tab.3)	 Turret structure (posts and struts), view from 8-star beam level (Tab.3)	 Replaced destroyed fragment of the main column with a joint pos. A (tab.2) and knee bracing

Table 2. Modelled characteristics of each the main structure and the bell room structure sides

East side (nave entrance)	West side (church entrance)	North side (almost symmetrical to south side)
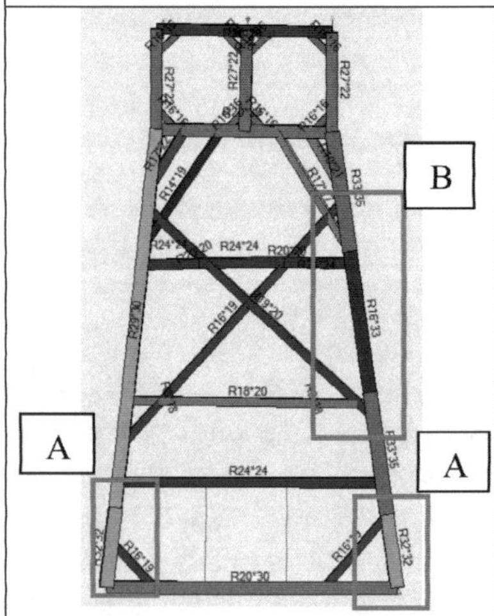	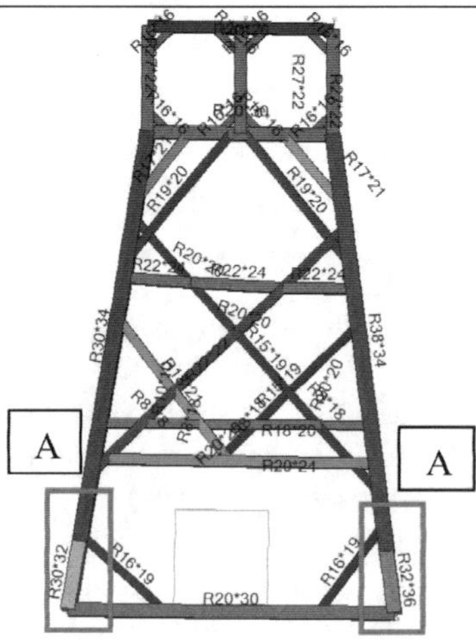	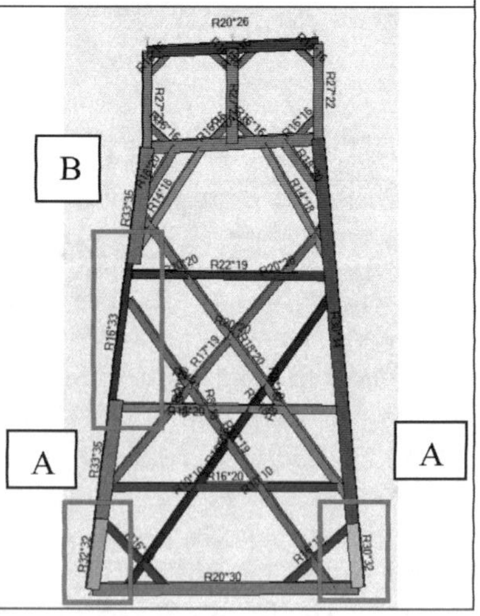

Table 3. Modelled characteristics of the bell room and the turret. Legend for the cross-section elements given in the lower right corner in cm. R stands for rectangular

The bell room	The turret with an 8-star beam base
	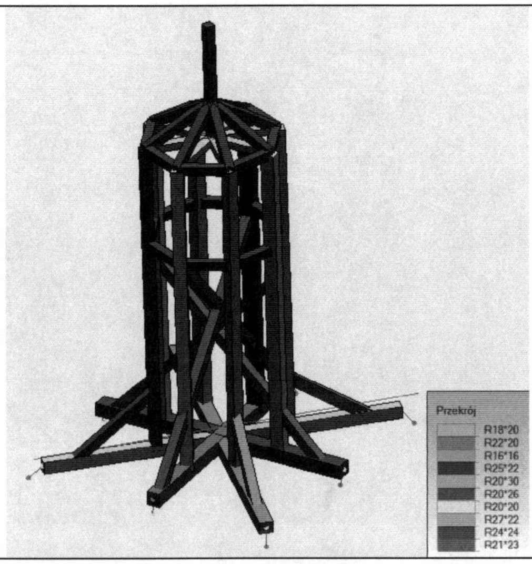

The structural layout of each wall presented in Tab. 2 consists of the main structure and the bell room structure. It is worth to pay attention to modelled deformation visible on each wall, which was mostly caused by uneven subsidence of the structure and rooting of the lower parts of the structure (modelled inclination of each wall on the basis of geodetic measurement). The destroyed lowest segments of the main columns together with the ground beams were already replaced with new ones which resulted in the creation of additional joints in the main columns (A). The present technical conditions of elements were considered in the model as a narrowing with eccentricity of a damaged main column (north-east one) (B). Tab. 3 shows the structural layout of the bell room and the turret with an 8-star beam. Red and green nodes above the caping beams and below 8-star beam base present connection between the bell room and the turret. According to [4] in the case of small displacements under actual dead loads, the connections between the elements are modelled as pinned connections because the static scheme are more common (most of connections are not tightly fitted and there are some significant imperfections). Tab. 2 and Tab. 3 segregate the belfry according to relations between elements and structures, for example, in terms of support, the turret structure is freely supported on the bell room structure, and the bell room is freely supported by caping beams on main columns (the bell room and the turret as separate structures consistently rest on each other). The representation of the existing structure principles in one calculation model is presented in Fig.6. The completed model underwent discretisation presented in Fig. 7. The materials' elastic modulus was generally assumed as for the timber strength class C20 (on the basis of structure's general visual wood characteristics).

IOP Conf. Series: Materials Science and Engineering 1141 (2021) 012044 doi:10.1088/1757-899X/1141/1/012044

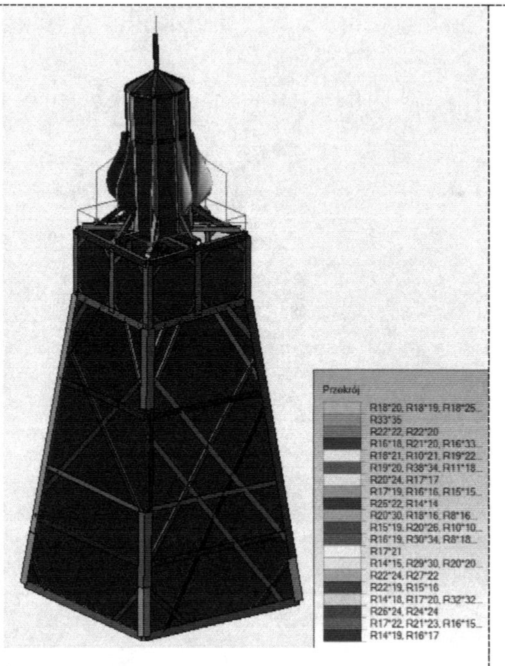

Figure 6. Completed model with panels representing planks, implemented dead loads

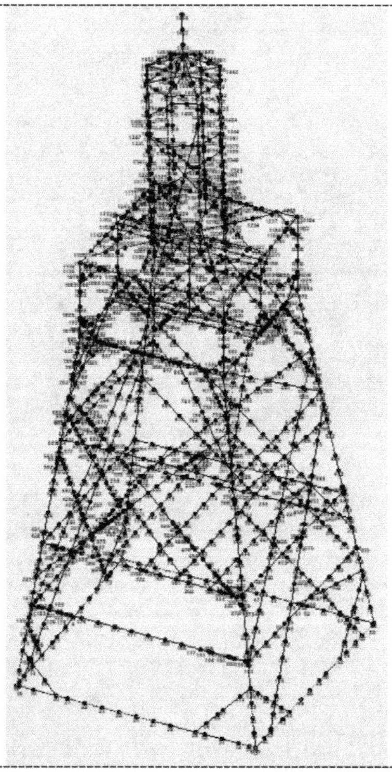

Figure 7. FEM model after discretisation

3. Initial FEM model analysis

An initial (with pinned connections) built-up model (Fig. 6) including dead loads (structure load, cladding load) was calculated to disclose belfry eigenmodes. Masses were distributed to nodes (Fig. 7), model computed and first 5 eigenmodes are presented in Tab. 4. The selected eigenmodes were chosen from a set containing over 50 eigenmodes based on the modal mass impact (5 with greatest modal masses).

Table 4. Computed belfry Eigenmodes

eigenmode 1	eigenmode 2	eigenmode 3	eigenmode 4	eigenmode 5
			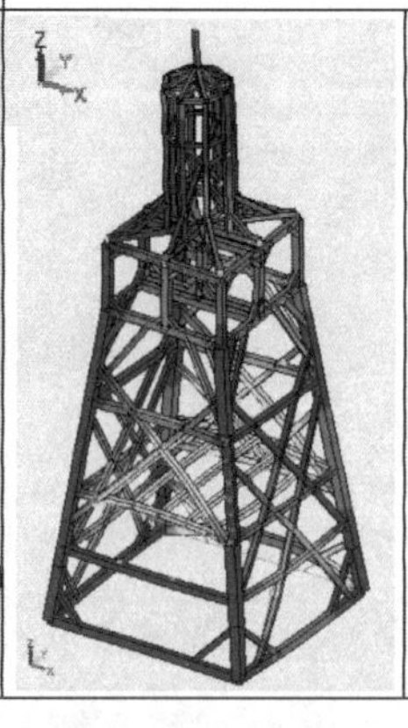	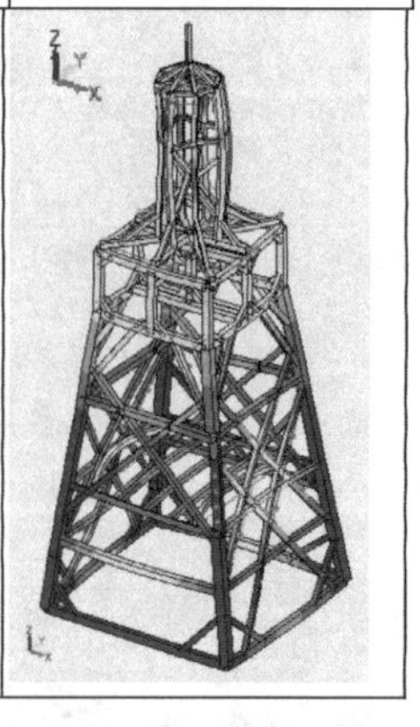

Table 5. Natural frequencies and oscillating masses (related to axis)

Eigen mode	Frequency [Hz]	Oscillating part of the structure	Contribution of oscillating mass	
			Related mass on axis X [%]	Related mass on axis Y [%]
1	**2.8**	**Turret, very limited oscillation of the capping beam**	**33**	0.2
2	3.1	Turret (local character)	0.7	**15**
3	**3.8**	**Turret, bell room, main structure**	**43**	0.2
4	6.3	Side walls (local character)	0.1	10
5	**7.0**	**Turret, bell room, side walls (limited)**	**0.4**	**14**

Eigenmode 3 received from the FEM model is related with a general structure response, while the 1, 2, 4 eigenfrequencies are related mostly with local responses. Eigenmode 5 may have impact on the upper parts of the structure (bell room structure, the turret and limited impact on side walls) and limited on the main structure (oscillation of struts). The way to describe the character of eigenmodes and their impact on the structure is to analyse oscillating masses related to each axis. After the analysis of the masses of each selected eigenmode, one may deduce that eigenmodes 1 and 3 have the greatest impact on the structure. Eigenmode 3, which has strictly global character, affects the whole structure. Eigenmodes 1 and 2 are characteristic for the turret structure (oscillation of the turret mass) of which eigenmode 1 shape is characteristic for cantilever vibration and have impact on the turret stability. Despite some asymmetry between opposite walls, specified eigenmodes have in this case directed character. For most important eigenmodes (no. 1, 3) vibration is strictly along the "x" axis. According to the initial FEM model, eigenfrequencies 1 and 3 would be suspected as potentially dangerous while using the bell and, for the safety purposes, those natural frequencies shouldn't be inducted during the usage of the belfry. The rest of the non-presented calculated eigenfrequencies have little impact on the structure (referring to the whole structure or its parts), impacting only individual elements, not considered in this article. For such a described model of the belfry structure, static calculation was conducted, and Tab. 6 presents the results which are characteristic values of internal forces caused by dead loads of the structure and cladding.

Table 6. Results of static calculations. Characteristic value of internal forces.

Main belfry structure

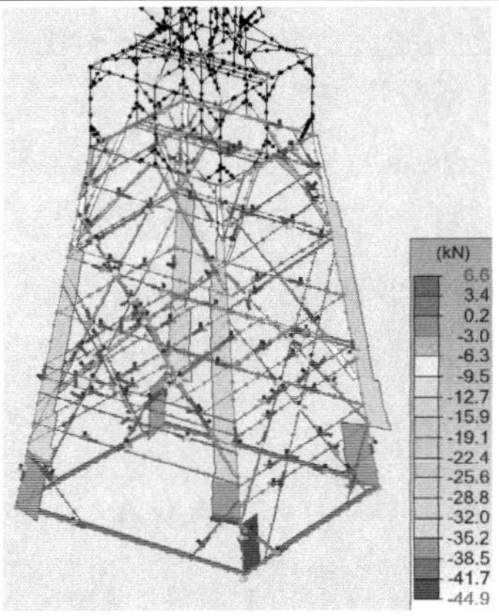

Axial force

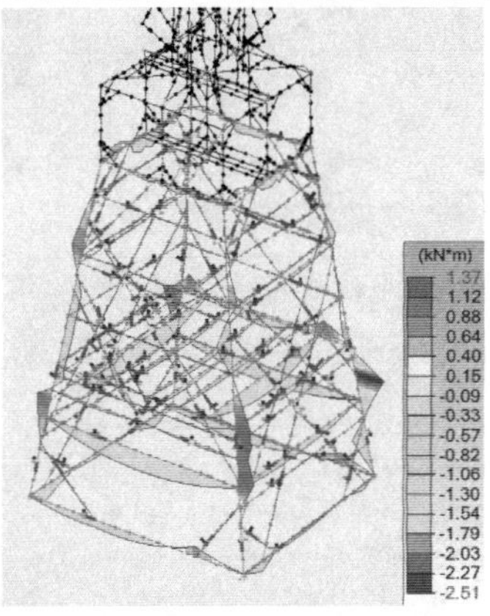

Bending moment (to the local "Y" axis)

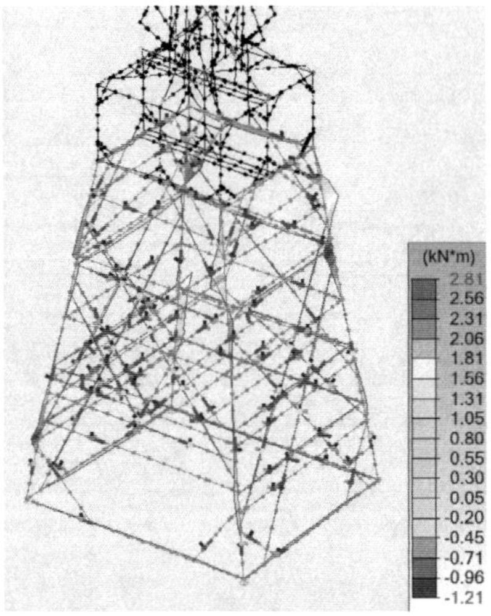

Bending moment (to the local "Z" axis)

Bell room

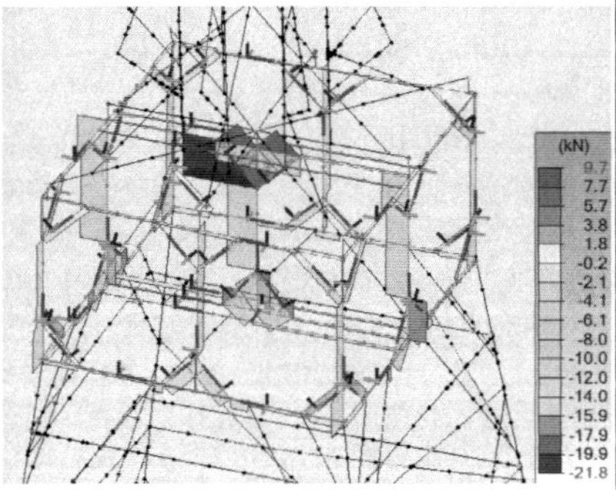

Axial force (view from below)

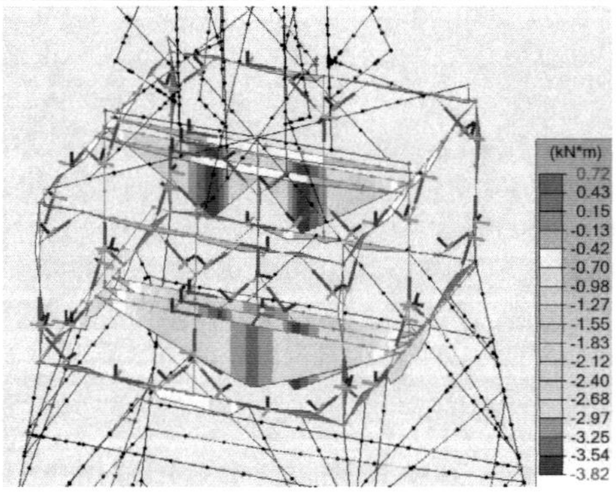

Bending moment (to the local "Y" axis)

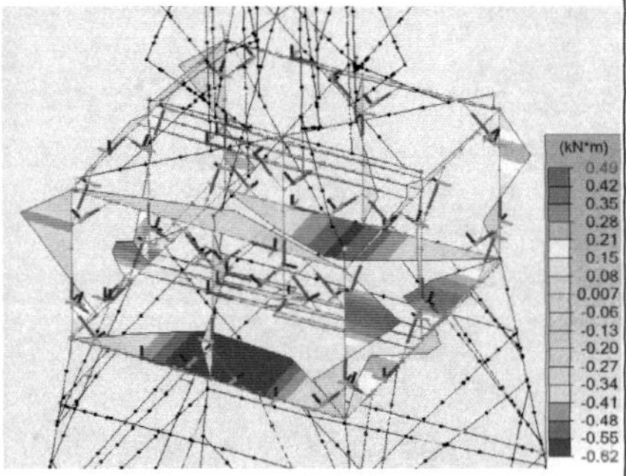

Bending moment (to the local "Z" axis)

Turret

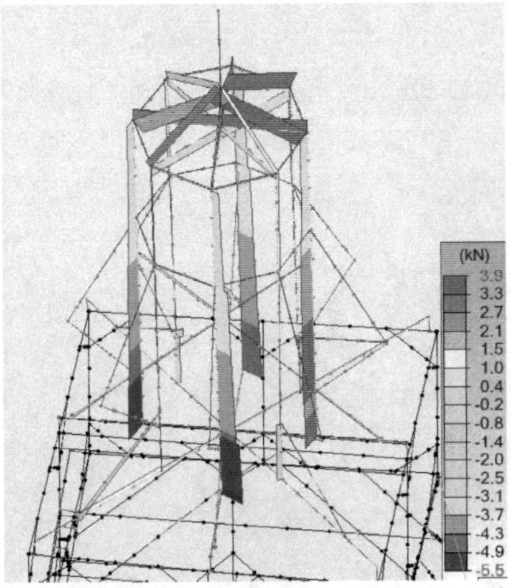

Axial force (view from below)

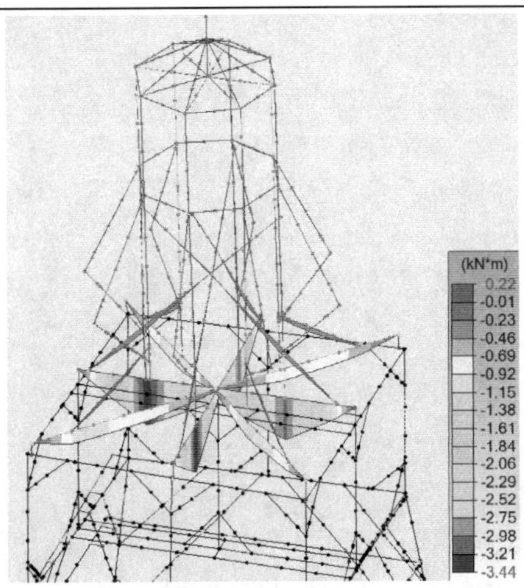

Bending moment (to the local "Y" axis)

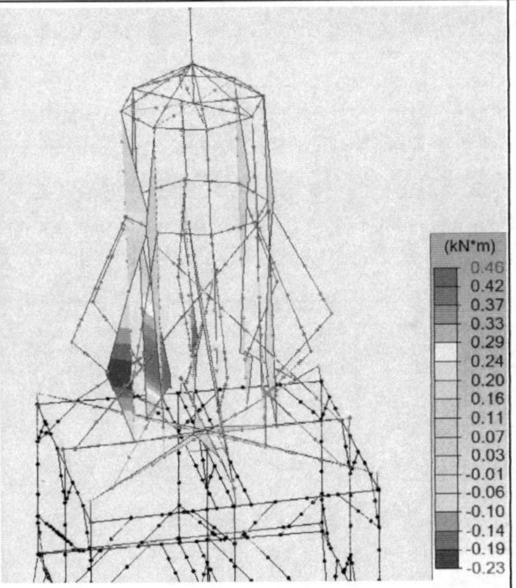

Bending moment to the local "Z" axis

The irregular distribution of the internal forces as presented in the Tab. 6, is a predictable behavior due to some imperfections between the walls of the main structure, monosymmetric character and reduction of the main column's cross-section (corrosion). The same situation was discovered in the elements of the bell room and turret, which was mostly caused by inclination of those structures and trapezoidal projection of the bell room (inclination and proportions obtained from geodetic measurements).

4. Dynamic identification tests

The author performed the dynamic identification tests using accelerometers for the main belfry structure, the bell room and also the turret (but the highest parts were inaccessible, so the measurement is not fully reliable for the turret). The first stage of testing includes inducting impulsive force to the structure and recording dynamical response (acceleration, Fig. 9) of the analysed structure with performing the FFT (Fast Fourier Tranformation) analysis. In the second stage, (if it was possible) frequencies received from the FFT analysis were inducted to confirm those results.

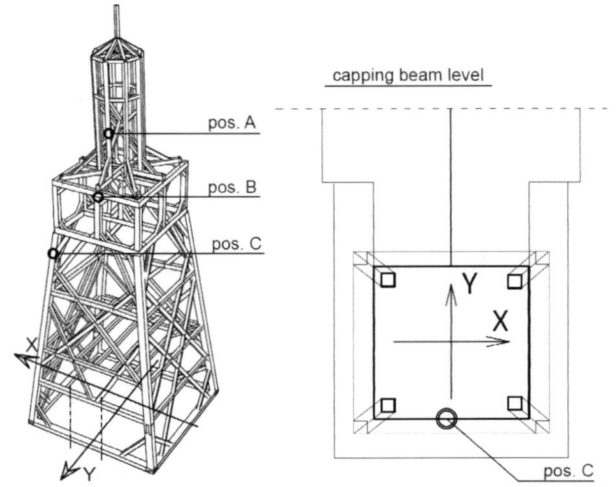

Figure 8. Accelerometers positions

As shown in Fig.8, the test was performed for the turret structure (pos. A), the second for the bell room at the level of capping beams (pos. B), and the third for the main belfry structure near the level of lower capping beams (pos. C). Those positions were determined by achieving possibly maximum amplitudes as well as the accessibility of those parts (that is why pos. A is not on the top of the turret).

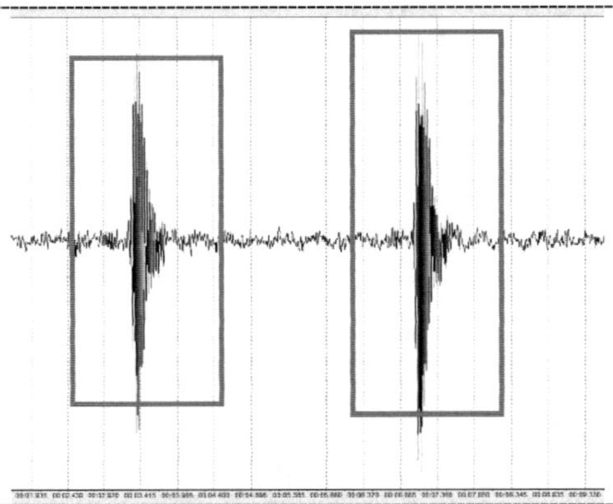

Figure 9. Sample of a recorded signal – dynamical responses to impulses of a capping beam, pos. B.

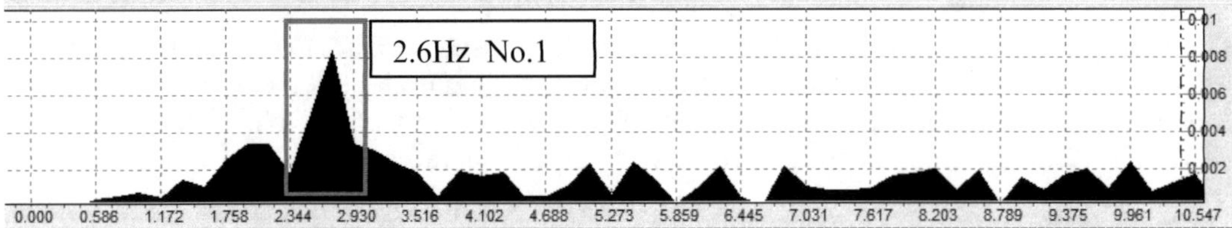

Figure 10. Example FFT spectrum analysis in pos. A. No.1 natural vibration frequency of the structure

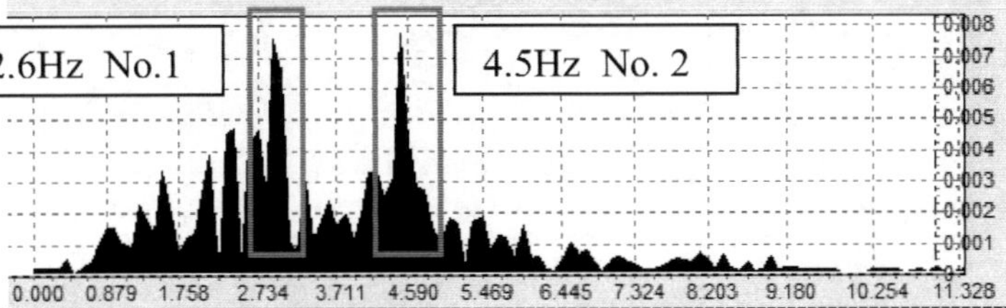

Figure 11. Example FFT spectrum analysis of a dynamical response of the main structure to impulses – natural vibration frequencies of the structure in pos. C. No.1 natural frequency 2.9 Hz, no.2 natural frequency 4.5 Hz specific for the main structure

Table 7. Dynamic identification tests results – natural frequencies

Position	Frequency No. [Hz]	Characteristic of the frequency
A	(No.1) 2.6	Main oscillation on the "x" axis, limited oscillation recorded on the "y" axis
B	(No.1) 2.7	Main oscillation on the "x" axis, limited oscillation recorded on the "y" axis
	(No.2, but limited amplitude) 4.5	
C	(No.1) 2.9	One direction of the oscillating. The "x" axis
	(No.2) 4.5	

Computed FFT presented in Fig.10 and Fig.11 shows disclosed natural frequency no.1 and no.2 Disclosed natural frequencies for positions A, B and C according to Tab.7 show convergence of the received natural frequencies for each tested position. This convergence suggests one eigenmode type, which is expected to be a fundamental eigenmode for the estimated cantilever scheme of this belfry structure. According to the results of the tests performed on different levels and parts of the structure, natural frequency fluctuates between 2.6 and 2.9 Hz. For pos. C (the main structure) the tests revealed another significant response of the structure with frequency about 4.5 Hz (No.2), and the amplitudes almost as high as 2.9Hz.

5. Comparison of the computed eigenfrequencies with the dynamic identification tests results

By comparing the results of tests in Tab. 7 with the computed results from the FEM model in Tab. 5, one may find potential paired eigenfrequencies. Such comparison is shown in Tab.8, below

Table 8. Comparison of eigenfrequencies

Computed results. Initial model			Identification tests [Hz] (Nat. freq. No.1 and No.2)			Convergence of frequencies [%]		
Eigenmode	Frequency [Hz], Axis	Oscillating part of the structure	The turret	The bell room	The main structure	The turret	The bell room	The main structure
1	**2.8, X**	**Turret, limited capping beam oscillation**	**2.6**	-	-	**93**	-	-
2	3.1, y	Turret, locally	No match					

IOP Conf. Series: Materials Science and Engineering 1141 (2021) 012044 doi:10.1088/1757-899X/1141/1/012044

3	**3.8, x**	**Turret, bell room, main structure**	**2.6**	**2.7**	2.9 and **4.5**	**68**	**71**	76 And **84**
4	6.3	Side walls	No match					
5	7.0, x	Turret, bell room, limited side walls oscillation	2.6	2.7	-	37	39	-

The comparison of the computed results with the in-situ test results demonstrates very high convergence between modelled and the turret natural frequencies for the eigenmode 1. However, considering behaviour of the computed bell room structure's eigenmode 1, the oscillation is very limited, whereas the same structure in the in-situ excitation tests (pos. B, Fig. 8) reaches similar amplitudes as the turret (pos. A). A comparison of the computed results for eigenmode 2 with the identification test results has no match - computed oscillation has only local impact on certain turret's elements which was not recorded during the tests. The analysis of the eigenmode 3 shows convergence with the recorded second natural frequency of the main structure which amounts to 4.5 Hz (Fig.11, no.2) and it may explain the second recorded significant frequency. Eigenmodes 4 and 5 will not be considered due to the lack of equivalence in records and insufficient convergence. The presented comparison shows that only the modelled turret structure converges towards the actual structure, whereas the bell room and the main structure should be modified.

6. Results of the model calibrations

The FEM model of the belfry structure was calibrated in accordance with the test results. Calibration was implemented by elastic releases of nodes positioned in joints. The main purpose of calibration was to achieve convergence between natural frequencies received during the in-situ tests and in FEM eigenmodes with simultaneous analysis of the eigenfrequency form. Best possible matching between the model and test results was achieved by calibration of bracing elements in the bell room and main structure's struts. The results of matching are shown in Tab 9.

Table 9. Convergence between the calibrated model and tests results

Computed results			Accelerometer tests [Hz]			Convergence of frequencies [%]		
Eigen mode	Frequency [Hz], Axis	Oscillating part of the structure	The turret	The bell room	The main structure	The turret	The bell room	The main structure
1	2.8, x	**Turret, Bell room's capping beam, main structure**	2.6	2.7	2.9	93	96	97
2	4.4, x	**Main structure, highest part of the turret**	-	-	4.5	-	-	98

Table 10. Comparison between matching eigenmodes of FEM model

Eigenmode	Calibrated FEM model	Initial FEM model
1		
2	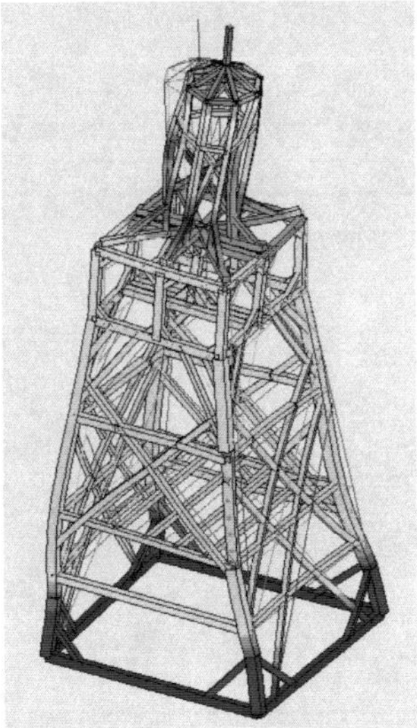	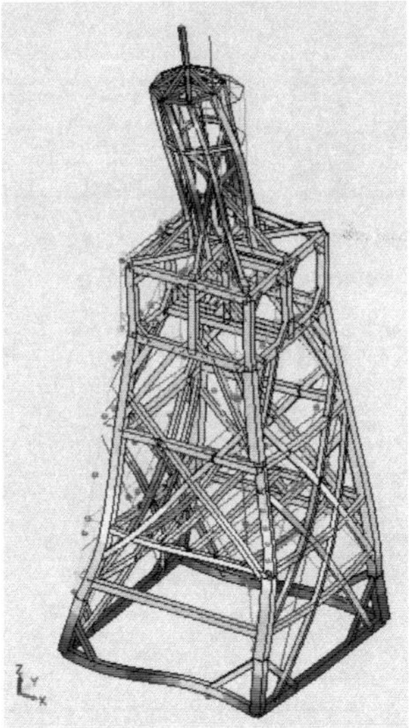

The comparison between the initial and calibrated models in Tab. 10 shows changes of eigenmodes' shapes. For eigenmode 1, the initial model shape shows oscillation only within the range of the turret structure, while the calibrated model oscillates within the range of the turret, the bell room and the main structure (oscillating part of the main structure has lower amplitude). In the case of eigenmode 1, both models give the same resultant frequency (2.8 Hz). Comparing the frequency results for eigenmode 2, there is a divergence between models. Calibrated model is much closer to No.2 natural frequency received in the dynamic identification test than initial one. For eigenmode 2, the initial model shows the same oscillations of the main structure and the bell room, while for the one calibrated oscillation of the bell room's capping beam is limited. Calibrated model shows that the turret's braced post oscillation is nearly zero (the same post where testing position A is set according to Fig.8). The calibrated model was calculated in terms of the impact of modal masses, and the results are presented in Tab. 11. Those results are compared to the initial model.

Table 11. Modal masses comparison

Calibrated FEM model			Initial FEM model			Difference in total modal mass [%]
Eigenmode	Related mass "x" axis [%]	Related mass "y" axis [%]	Eigenmode	Related mass "x" axis [%]	Related mass "y" axis [%]	
1	**47**	4	1	33	0	**+14**
2	34	0	2	**43**	0	-9

By analysing the difference in modal masses in Tab.11, one may conclude that calibration has animpact on dynamic characteristics by changing proportions of modal masses associated with eigenmodes. The analysis results show that the calibrated model in comparison with the initial one is more vulnerable to resonance in the case of eigenfrequency 1 due to higher modal mass than in the case of eigenfrequency 2. The situation for the eigenmode 2 is opposite, modal masses are greater in the case of the initial model. Comparing the character of both eigenmodes (according to Tab.10) it is visible that the results from both models differ within the range of the upper parts of the structure such as bell room and the main structure. It should be mentioned that the bell is located classically in the bell room or just below, inducing vibrations (energy) in a critical position, where high amplitudes are expected. In this case, calibration should not be skipped during the examination of the structure due to the ULS criteria according to [7]. It is worth to mention that resonance may occur with swinging bell's potential harmonic frequencies *f3* and *f5* [1], and those harmonics should be always checked while considering mantling a bell.

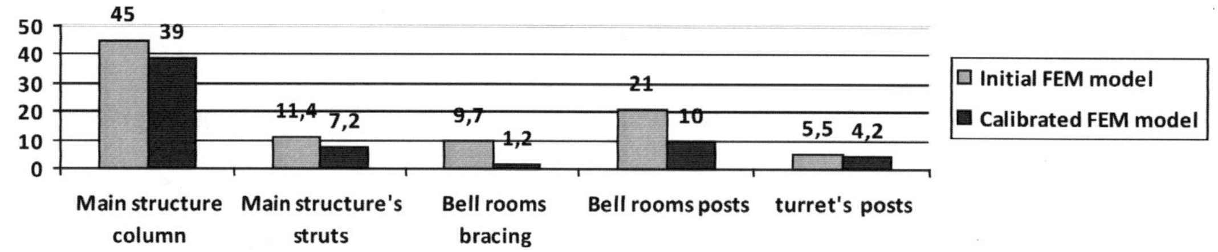

Figure 12. Axial forces comparison, maximum absolute values [kN]

In general, axial forces are more redistributed among the elements of the structure. The function of the bell room's bracing elements was reduced by given elastic releases, so forces (Fig.12) are very limited. In the initial model, some of the bell room posts' segments were tensioned, in the calibrated one there is only compression.

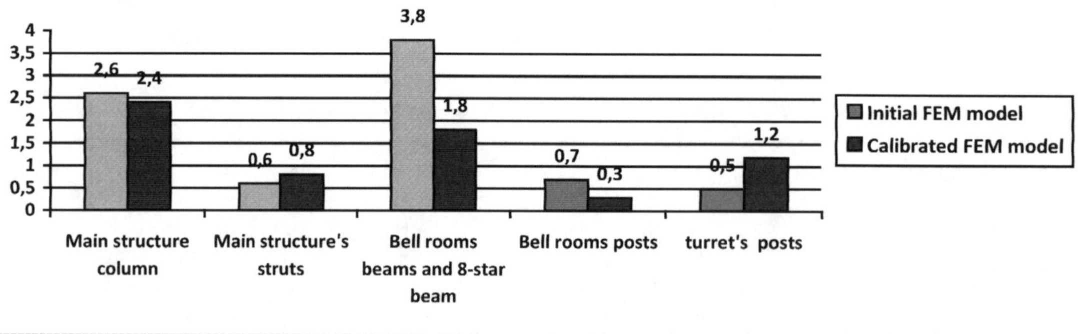

Figure 13. Bending moment, maximum absolute values, [kNm]

According to Fig. 12, there are no significant changes in momentum values except for the beam elements of the bell room and 8-star beam supporting the turret. In those elements, apart from the values, the distribution of moments has changed (mostly for the vertical bending). In general the values of the presented internal forces (Fig.12 and Fig. 13) calculated from the initial model are usually greater and do not consider redistribution as in the case of calibrated FEM model.

7. Conclusions

Building the calibrated model of such belfry structure is a complex task. It requires carrying out dynamical identification tests for as many points of the structure as possible (at least those which are the most characteristic). These points are supposed to give an approximated shape of the fundamental frequencies to achieve best results in calibration (based on the comparison between each tested point of the structure). The results of the tests show, that non calibrated model in this case did not give rational, convergence results. This indicates that its stiffness does not correspond to the actual, old structure of this belfry (initial model is characterised by overestimated stiffness of the elements and joints). The analysis of the related modal masses, which were computed from both models, shows that in the initial model the related mass is underestimated in case of the fundamental mode. The mentioned modal masses have impact on the behaviour of the belfry structure (forces from mass oscillation) during the resonance and they should be considered if a bell is mantled again. Static calculations for both models confirm the expected redistribution of forces among the structure elements. Currently, provided similar dynamic researches are mostly aimed at historical brick belfry towers [8,9] and modern reinforced concrete bell towers [10]. The results of the dynamic analysis presented in the this article can be used as a point of comparison with other types of belfries. The intention of the author is to provide further research to better recognize the dynamic of timber belfries based on the results from wide range of belfries. Such analysis is important to preserve the structure of historical timber belfries and to create a possibility for safe bell mantling/reinstalling.

8. References

[1] Forczek-Sajdak A 2020 *Historia kościoła filialnego pw. św. Mikołaja Biskupa w Skrzydlnej w świetle źródeł archiwalnych oraz najnowszych badań i prac konserwatorskich* (Studia jubileuszowe, 70 lat Wydziału Konserwacji i Restauracji Dzieł Sztuki Akademii Sztuk Pięknych im. Jana Matejki w Krakowie)
[2] Brykowski R and Kornecki M 1984 *Drewniane kościoły w Małopolsce Południowej,* Wrocław: Ossolineum
[3] Kopkowicz F 2009 *Ciesielstwo polskie,* Warszawa: Arkady
[4] Kochański T 2019 Dynamic influence of a swinging bell on a traditional wooden structure of the Lemko orthodox church belfry, Slovakia, *IOP Conf. Ser.: Mater. Sci. Eng.* **566** 012016)
[4] Kopkowicz F 2009 *Ciesielstwo polskie*, Warszawa: Arkady
[5] Gromysz K 2016 *Dynamika budowli. Obliczenia układów prętowych i o masach skupionych, Warszawa*, Wydawnictwo Naukowe PWN
[6] Eurocode 1: *Actions on structures* - Part 1-1: *General actions - Densities, self-weight, imposed ' loads for buildings*
[7] Eurocode 5: Design of timber structures- Part 1-1: General - Common rules and rules for buildings
[8] Lorenzo E, Canessa F, Chellini G, De Falco A, Resta C, Savelli E and Sevieri G. 2020 Modal identification of the San Francesco church in Pisa, Italy ed M Papadrakakis, M Fragiadakis and C Papadimitriou C, Athens *Proc. of EURODYN 2020. XI Internat.Conf. on Struct. Dyn.* Nov 23-26, pp 2386-99
[9] Azzara RM, Girardi M, Padovani C, Pellegrini D and Robol L FE 2020 Model updating of masonry towers: modeling and numerical issues ed M Papadrakakis, M Fragiadakis and C Papadimitriou C, Athens *Proc. of EURODYN 2020. XI Internat.Conf. on Struct. Dyn.* Nov 23-26, pp 2285-98
[10] Halicka A, Podgórski J, Bocian B, Jamińska P 2017 Analiza dynamiczna żelbetowej dzwonnicy wraz z oceną jej uszkodzeń korozyjnych, Poland, Międzyzdroje: *Inter. Conf. on Struct. Fail.*; 2017 May 22-26, pp 385-96

IOP Conf. Series: Materials Science and Engineering 1141 (2021) 012045 doi:10.1088/1757-899X/1141/1/012045

Comparative analysis on laboratory bearing tests for beams made of OSB3 and of solid timber

O A Poenaru (Coflea)[1], D N Isopescu[1], O Neculai[1]
[1] " Gheorghe Asachi" Technical University of Iasi-Romania, Faculty of Civil Engineering and Building Services, Blvd. Mangeron, No. 59A, 700050, Iasi, Romania

E-mail: oana-alexandra.poenaru@student.tuiasi.ro

Abstract. Current times impose the usage of less raw building and more recycled materials. Wood is a material that can be recycled and processed in the factory in order to create new materials. Due to its mechanical characteristics, OSB3 has been chosen to create glued beams, which are hybrid elements with a configuration similar to GLULAM. This paper presents an analysis of the results obtained in four points bending tests on real scale beams made of solid wood and of glued OSB3 lamellae. The current research intends to compare solid wood beams with OSB3 glued beams, from failure load and elastic behaviour points of view.

1. Introduction
The construction sector is currently one of the largest waste generators, but the environmental impact could be diminished with the help of the circular economy principles [1, 2].

To reduce the usage of raw building materials, it is necessary to develop new recycled ones. One of these raw materials is wood; a renewable resource that often comes from sustainable managed forests. It is highly sustainable to recycle wood that can be recovered from old furniture, formwork waste, pallets, packaging railways or bridges. Wood can also be recycled from raw products that cannot be used as they are in construction: branches, old chopped trees, trees with very small diameters. All these products can be processed in the mill in order to make wood chips, shavings and sawdust that are used to manufacture new recycled building materials [3].

Therefore, what if recycled materials, such as OSB – Oriented Strand Board could be used for structural elements? The current study is aimed to answer this question and focus on OSB3 material, which was chosen due to its bearing capacity and to the fact that it can be used outdoors, in humid environments. OSB3 lamellae were used to create glued beams, with a structure similar to GLULAM, which resulted in hybrid structural elements.

2. Production of OSB3 glued beams
The solid timber beams were made of fir, a softwood species.

The wood used to manufacture OSB3 is around 70% pine and spruce, the rest of 30% are species like: beech, birch, lime tree and willow. This ratio of softwood and hardwood species, in addition to the surface-to-core ratio and the pressing parameters determine the quality and characteristics of the final products. Other components of OSB3 are glue, water and paraffin wax emulsion [4, 5].

The hybrid structural elements have a lamellar structure and were manufactured from six OSB3 lamellae, each with a thickness of approximately 24 mm, glued with polyurethane adhesive for wood. To obtain high quality hybrid structural elements, a strict technological process was respected in the factory. For the OSB3 lamellae to have the same characteristics, the OSB3 was chosen from the same batch.

After delivery, the OSB3 lamellae were kept inside the factory for at least 24 hours for acclimatization. For the polyurethane adhesive to properly adhere to the OSB3 material, the OSB3 lamellae had their surface polished and individually calibrated. The calibration was executed by the Calibration and polishing machine, using a rubber lane with 100 granulation.

To manufacture a beam of 140 x 140 mm cross-section, six OSB3 lamellae were glued together using polyurethane adhesive. For this process, a Polyurethane Spray machine applied the polyurethane adhesive on the entire width and length of the first five OSB3 lamellae. The polyurethane adhesive was applied only on top of the first five OSB3 lamellae, the sixth lamella adhering to the fifth one. This assemble was then placed in the wood pressing machine. The applied pressure on the cross-section of the OSB glued element reached 64 bars, for a period of 60 minutes. To ensure the perfect conditions for this process, the temperature was between 10 – 25° C.

After the beams were removed from the wood press, they were stored for 8 hours. After 24 hours, the faces of the hybrid structural elements were finished using the Machine that processes on 4 sides. The dimensions for the resulted hybrid structural elements were of approximately 140 x 140 x 3000 mm (Figures 1 and 2).

Figure 1. Cross-section of the beam made of OSB3 lamellae

Figure 2. The view of beam made of OSB3 lamellae

3. Four-point bending tests on timber and OSB glued beams

3.1. Test specimens

To determine the static bending strength of timber and OSB3 glued beams, laboratory tests were carried out in accord to standards SR EN 14080 (july 2013) and EN 408:1995. The timber beams and the OSB3 glued beams were tested at the Faculty of Civil Engineering and Building Services, "Gheorghe Asachi" Technical University of Iasi. Experimental tests were made using the equipment of Zwick/Roell Material Testing Machine BP1-F1000SN.M11 [6, 7, 8, 9, 10].

Three solid timber beams and seven OSB3 glued beams were tested [11]. The first three OSB3 glued beams, GLO1, GLO2 and GLO3 were tested with the load, F, acting perpendicular on lamellae, Figure 3-a, and the other four beams, GLO4, GLO5, GLO6 and GLO7 were tested parallel to lamellae, Figure 3-b.

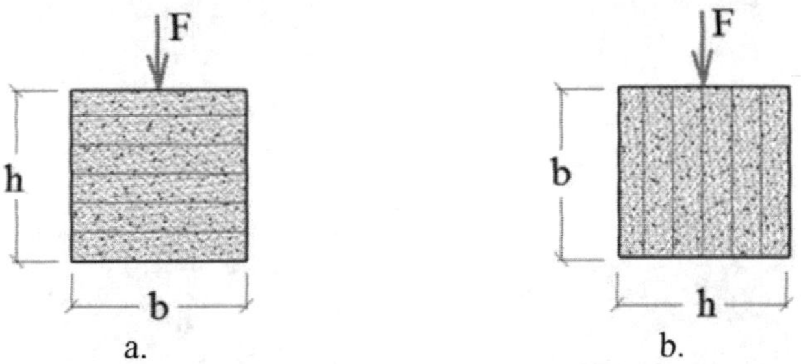

Figure 3. Direction of the force, F, applied on OSB3 glued beams:
a. perpendicular on lamellae, b. parallel to lamellae.

All beams had a prism shape, and their exact dimensions are presented in tables 1 and 2.

3.2. Testing procedure

The beams were subjected to four-point bending, hence the elements were simply supported on two pin supports of 150 mm width and the load was applied on two points, symmetrically placed at a distance a = L/3= 900 mm from the supports, where L is the beam span, Figure 4 [8, 9]. The load was applied continuously with constant test speed of 0.08 mm/sec.

Figure 4. Laboratory test equipment and beam sample.

3.3. Test results

Total failure was reached for all tested beams. The failing modes are depicted in Figures 5 and 6, for solid timber beams and OSB3 glued beams.

Figure 5. Failure mode of solid timber beams.

a. b.

Figure 6. Failure of OSB3 glued beam,
a. tested perpendicular on lamellae, b. tested parallel to lamellae.

Observing the figure above, it can be stated that failure occurred and remained in the tensile area for solid timber beams and for OSB3 glued beams, with the load applied perpendicular on lamellae. For the OSB3 glued beams tested parallel to lamellae, failure occurred on the entire width of the beams, thus splitting the element in two parts.

The load-displacement curves for all beams are depicted in Figures 7, 8 and 9.

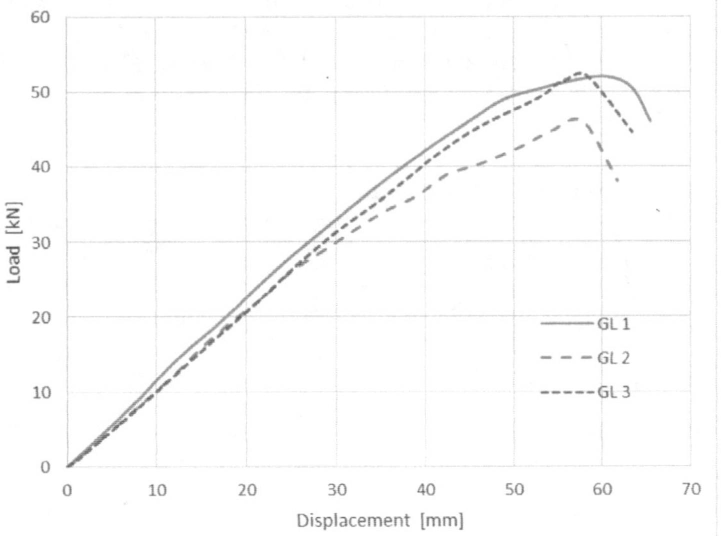

Figure 7. Load – displacement curves for solid timber beams.

As it can be observed for solid timber beams, their flexural behaviour is mostly elastic, up to a load of 40000 N, and then the beams start presenting plastic deformation, up to their failure at an average of 50025.56 N, with a displacement of 58.89 mm.

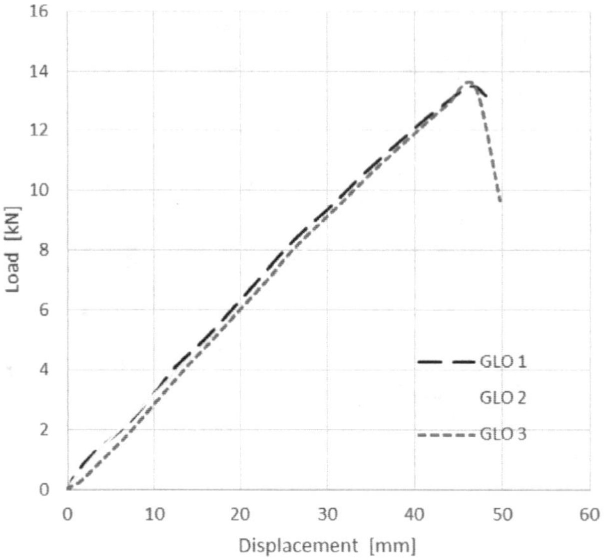

Figure 8. Load – displacement curves for OSB3 glued beams tested perpendicular on lamellae.

The OSB3 glued beams subjected to bending perpendicular to lamellae presented an elastic behaviour up to failure, which occurred at a load of 12800.66 N, with an average displacement of 45.37 mm, Figure 8. The different behaviour between GLO2 and the other beams can be explained due to the manufacturing procedure, thickness and uniformity of the adhesive layers.

The OSB3 glued beams subjected to bending parallel to lamellae also presented an elastic behaviour up to failure at a load of 11895.19 N with an average displacement of 37.24 mm, Figure 9.

The obtained results after testing the beams are presented in tables 1 and 2.

Table 1. Tests results for solid timber beams

Timber beam	b [mm]	h [mm]	L [mm]	F_{max}^{timber} [N]	w_{max}^{timber} [mm]
GL1	148	148	2700	51978.20	60.477
GL2	148	150	2700	45974.06	57.900
GL3	149	148	2700	52124.43	58.310
		Mean		50025.56	58.890

Table 2. Test results for OSB3 glued beams

Tested	OSB3 glued beam	b [mm]	h [mm]	L [mm]	F_{max}^{OSB3} [N]	w_{max}^{OSB3} [mm]
Perpendicular on lamellae	GLO1	138	143	2700	13490.61	46.400
	GLO2	139	143	2700	11392.82	42.890
	GLO3	138	145	2700	13518.50	46.809
		Mean			12800.66	45.380
Parallel to lamellae	GLO4	139	143	2700	11788.63	35.320
	GLO5	138	143	2700	12160.46	39.520
	GLO6	139	143	2700	10663.68	33.870
	GLO7	138	148	2700	12968.00	40.250
		Mean			11895.19	37.240

where: b is the width of beam; h is the height of beam cross-section; L is the length of beam; F_{max}^{timber} is the failure load of timber beam [N]; w_{max}^{timber} is the maximum displacement of timber beam, [mm], corresponding to the failure load; F_{max}^{OSB3} is the failure load of OSB3 glued beam, [N]; w_{max}^{OSB3} is the maximum displacement of OSB3 glued beam, [mm], corresponding to the failure load.

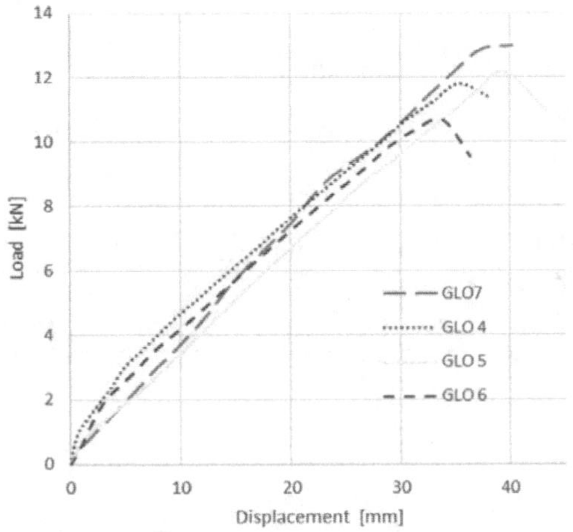

Figure 9. Load – displacement curves for OSB3 glued beams tested parallel to lamellae.

Analyzing the overall results, it can be stated that the OSB3 glued beams present similar values for failure load and displacements, and these values are around 24.7 % of the failure load of solid timber beams. Meanwhile, the displacements of OSB3 glued beams are approximately 70.13 % of the ones of solid timber beams. This indicates that, OSB3 glued beams in the presented configurations can be subjected to lower loads to attain deformations close to solid timber beams. Furthermore, looking at the load-displacement curves in Figure 10, for a load of 12347.925 N (which is the average failure load of OSB3 glued beams), the displacement for solid timber beams is approximately 10 mm. Comparing this displacement to the one of OSB3 glued beams of 41.3 mm, it results that OSB3 glued beams present deformations 413 % times greater than solid beams.

The curve shape of GLO4 beam is slightly different than the other, and this can be explained by certain irregularities in the manufacturing procedure, as the thickness or uniformity of the adhesive layers.

4. Evaluation of bending modulus of elasticity

The elasticity modulus of each beam was computed, $E_{m,g}$, using equation (1), according to NP 408-2004 [6]:

$$E_{m,g} = \frac{L^3(F_2-F_1)}{bh^3(w_2-w_1)}\left[\left(\frac{3a}{4L}\right)-\left(\frac{a}{L}\right)^3\right] \quad (1)$$

where: b is the beam width, [mm]; h is the beam height, [mm]; L is the distance between supports, meaning 2700 [mm]; F_1 is the load value of approximately $0.1F_{max}$, [N]; F_2 is the load value of approximately $0.4F_{max}$ [N]; w_1 is the displacement corresponding to load F_1 [mm]; w_2 is the displacement corresponding to load F_2, in mm; a is the distance between a loading position and the nearest support in a bending test, in this case 900 [mm].

Tables 3 and 4 present the results obtained for the bending elasticity modulus computation for each set of beams: solid timber beams, OSB3 glued beams subjected to bending perpendicular on lamellae and parallel to lamellae.

Table 3. Evaluation of elasticity modulus for solid timber beams

Timber beam	F_{max}^{timber} [N]	F_1 [N]	F_2 [N]	w_1 [mm]	w_2 [mm]	$E_{m,g}^{timber}$ [MPa]
GL1	54416.00	4330.80	14740.60	3.90	12.69	10350.90
GL2	42674.10	7921.00	15558.50	7.84	14.74	9287.93
GL3	50124.40	11080.80	21777.10	11.04	21.10	9227.49
			Mean			9622.09

The mean modulus of elasticity for solid timber beams is 9622.09 MPa, which corresponds to a strength class of C18, which is widely used for structural elements [10].

Table 4. Test results for OSB3 glued beams

Tested	OSB3 glued beam	F_{max}^{OSB3} [N]	F_1 [N]	F_2 [N]	w_1 [mm]	w_2 [mm]	$E_{m,g}^{OSB3}$ [MPa]
Perpendicular on lamellae	GLO1	13490.61	1273.47	5071.23	3.14	16.0	3076.87
	GLO2	11392.82	2440.35	5717.43	7.55	18.8	3004.38
	GLO3	12877.98	2014.43	5975.24	7.51	19.9	3193.12
			Mean $E_{m,g}^{OSB3,\perp}$				3091.46
Parallel to lamellae	GLO4	11788.63	1639.72	6641.34	2.1	16.7	3528.43
	GLO5	12160.46	1432.47	4759.70	3.34	14.7	3490.24
	GLO6	10663.68	2046.15	4294.50	3.38	10.3	3523.65
	GLO7	12968.00	1045.58	1880.55	2.48	4.86	3782.67
			Mean $E_{m,g}^{OSB3,II}$				3581.25

The average modulus of elasticity for OSB3 glued beams is 3091.46 MPa, for beams tested perpendicular on lamellae, and is 3581.25 MPa for beams tested parallel to lamellae. These values are about 3 times smaller than solid timber of strength class C18, which presents a standardised longitudinal modulus of elasticity of 9000 MPa [10].

5. OSB3 glued beams vs. timber beams of strength class C18

After analyzing the laboratory tests results, it has become clear that the tested OSB3 glued beams have low bearing capacity, as compared to solid timber beams of similar dimensions. Therefore, the OSB3 glued beams bearing capacity must be improved to match the solid timber beams resistance and rigidity characteristics for a case of strength class C18.

The maximum instantaneous deflection, $f_{max,inst}$, of a beam subjected to two concentrated loads equal to $F_{max}/2$, can be calculated according to equation (2), where E_m is the longitudinal modulus of elasticity, [MPa], and I_y is the moment of inertia of the cross-section, $b \cdot h^3/12$, in [mm^4] [7].

$$f_{max,inst} = \frac{23}{648} \cdot \frac{\frac{F_{max}}{2} l^3}{E_m \cdot I_y} \quad (2)$$

For OSB3 glued beams subjected to bending perpendicular on lamellae, the easiest method to improve the flexural behaviour is to increase the number of OSB3 lamellae, hence increasing the height of the glued beams.

The maximum instantaneous deflection was evaluated using equation (2), with $F_{max}=F_{max}^{timber}$, and $E_m=E_{m,g}^{OSB3,\perp}$, for various numbers of OSB3 lamellae, and the results are presented in table 5. The beam width is constant, of 138.33mm, which is the average between GLO1, GLO2 and GLO3.

Table 5. Evaluation of OSB3 glued beams displacement for various numbers of lamellae, for bending perpendicular on lamellae

Lamellae width, [mm]	Number of lamellae	Beam height, [mm]	Beam width, [mm]	Deflection, [mm]
24	6	144	138.33	164.20
	7	168		103.40
	8	192		69.28
	9	216		48.66
	10	240		35.47

Analysing Table 5, it can be observed that, for a load equal to the failure force of solid timber beams and the OSB3 longitudinal modulus of elasticity, the increase of lamellae number can provide a considerable reduction of deflection. For nine lamellae, the deflection for the OSB3 glued beams is smaller with 17.4% than that of solid timber beams, which is of 58.89 mm.

Therefore, increasing the number of lamellae to nine, resulting in a cross-section of 138.33 x 216 mm, the OSB3 glued beams present smaller deflections than solid timber beams of 148 x 149 mm cross-section of strength class C18.

For OSB3 glued beams subjected to bending parallel to lamellae, the increase of height can be done by choosing OSB3 lamellae with larger widths; some variations are presented in table 6.

The deflection was computed using equation (2) with $F_{max}=F_{max}^{timber}$ and $E_m=E_{m,g}^{OSB3,II//}$ and b=138.5[mm], which is the average between GLO4, GLO5, GLO6 and GLO7. For a height of OSB3 glued beams of 216 mm, the maximum deflection is 41.95 mm, which is 28,77% smaller than that of solid timber beams of class C18. The exact same maximum deflection of 58.89 mm as solid timber beams is met for OSB3 glued beams of 192.9 mm height.
Similar methods for improving flexural behaviour of OSB3 glued beams were detailed in their analytical study, where the focus was on the bearing capacity of hybrid structural elements such as columns and beams made of OSB3 lamellae [2].

Table 6. Evaluation of OSB3 glued beams displacement for various lamellae widths, for bending parallel to lamellae

Beam height, [mm]	Beam width, [mm]	Deflection, [mm]
144.0	138.33	141.58
168.0		89.16
192.0		59.73
192.9		58.89
216.0		41.90
240.0		41.95

6. Conclusions

The presented laboratory experiments evaluated the flexural behaviour of solid timber beams and OSB3 glued beams subjected to four-point bending, in an attempt to find new ways to use recycled wood products to replace standard structural elements.

Comparing the results of the timber beams with those of OSB3 glued beams, the modulus of elasticity for the glued beams is ≈ 33.47 % lower than the solid timber beams. In addition, the failure load of OSB3 glued beams is ≈ 24.7 % of the load of solid timber beams.

These results indicate that the cross-sections of OSB3 glued beams must be increased in order to produce structural elements with a flexural behaviour similar to solid timber beam. For OSB3 glued beams subjected to bending perpendicular to lamellae, by increasing the number of OSB3 lamellae from 6 to 9, the deflection of the glued beams became smaller with 17.4% than that of solid timber of strength class C18. Similarly, smaller deflections can be achieved for bending parallel to lamellae, by using OSB3 boards wider than 192.9 mm, resulting in OSB3 glued beams with bigger heights.

Other solutions that can improve the strength and rigidity characteristics for OSB3 glued beams experimented in the present study, can be the hybrid composition, in which there is a mixture of lamellae, most of them made of OSB3 and one or more to be made of materials with superior strength and rigidity. Examples include the use of flat lamellae made of fibre-reinforced polymeric composites placed in the tension area of the beam cross-section. In addition there are solutions with reinforcement made of independent bars made of steel or long-fibre-reinforced polymeric composites [12].

The experiments performed in the laboratory of the Faculty of Civil Engineering and Building Services from the "Gheorghe Asachi" Technical University of Iasi highlighted a new field for the use of OSB type products. Climate change and environmental degradation are an existential threat to Europe and the world. The studied elements can be the solution to this threat, being part of the development of a modern economy, competitive and efficient in terms of resource use.

7. References

[1] Poenaru (Coflea) O A and Isopescu D N 2018 *Eco-friendly products used in construction - results of the implementation of circular economy principles*, C65 International Conference, pp 41- 42

[2] Poenaru (Coflea) O A, Isopescu D N and Neculai O 202 The analytical methodology used to evaluate t*he bearing capacity for hybrid structural elements such as the column and the beam made from OSB3 lamellae*, Bulletin of the Polytechnic Institute of Iași, Construction. Architecture section, vol **3**

[3] Website *Circular economy*, https://en.wikipedia.org/wiki/Circular_economy

[4] Website *Egger*, https://egger.com

[5] Website *Oriented Strand board*, https://sciencedirect.com/topics/engineering/oriented-strand-board

[6] NP 005-2005, *Normativ proiectare pentru structuri din lemn*

[7] SR EN 408: 2004, *Structuri de lemn. Lemn masiv şi lemn lamelat încleiat. Determinarea anumitor proprietăţi fizice şi mecanice*, Asociaţia de Standardizare din România (ASRO), Bucureşti, România

[8] SR EN 1990–1–1: 2004, *Eurocod 0: Bazele proiectării structurilor*, Asociaţia de Standardizare din România (ASRO), Bucureşti, Romania

[9] SR EN 1995–1–2: 2004/AC:2006, *Eurocod 5: Proiectarea structurilor de lemn, Partea 1 – 2: Generalităţi – Proiectarea structurilor la foc*, Asociaţia de Standardizare din România (ASRO) Bucureşti, Romania

[10] SR EN 338: 2010, *Lemn pentru construcţii. Clase de rezistenţă*, Asociaţia de Standardizare din România (ASRO), Bucureşti, Romania, 2010.

[11] Neculai O and Isopescu D N 2017 *Timber structures. Building elements design*, Ed. Societăţii Academice Matei-Teiu Botez Iasi Romania

[12] Neculai O 2012 *Consolidarea elementelor din lemn folosind materiale compozite* (in Romanian), PhD thesis Technical University "Gheorghe Asachi" from Iaşi

IOP Conf. Series: Materials Science and Engineering 1141 (2021) 012046 doi:10.1088/1757-899X/1141/1/012046

Finite element analysis of square concrete columns strengthened with carbon fiber composites in different configurations

A N A Almagsoosi[1] and D P Georgescu[1]
[1] Department of Reinforced Concrete, Faculty of Civil Engineering/ Technical University of Civil Engineering in Bucharest (UTCB), Bd. Lacul Tei nr. 122 - 124, Sector 2, 020396, Bucharest, Romania

E-mail: alinajim_civil@yahoo.com

Abstract. This paper offers the findings of numerical study carried out to evaluate the behavior of low strength concrete columns in square section strengthen by Carbon Fiber Reinforced Polymer (CFRP) composites and subjected to axial compression load. In this study, a three dimensional non-linear finite element formulation model has been presented using ANSYS-16.1 (software computer program) to analyse low strength concrete columns in square section strengthen by CFRP composites in different configurations, to evaluate the improvement in the ultimate load capacities of columns as a result of CFRP confinement, and to study the impact of the most important parameters such as: compressive strength of concrete, thickness of the CFRP layers jacket/strips and the presence of steel reinforcement on the performance of th modelled columns subjected to compression load. A comparison has been made between the results obtained from the Finite Element Analysis (FEA) and experimental data. The results of the numerical FEA study showed that, external strengthening with CFRP composites are very effective in upgrading the ultimate load of square concrete columns. In addition, the findings revealed good agreement between the ANSYS and the experimental test results. The results obtained from the parametric study revealed that when the CFRP layers increased from 2 to 3 layers and from 2 to 4 layers the ultimate loads of the columns increased about (2% -10%) and (7%-15%) respectively. While the compressive strength showed significant effect on the upgrade of the ultimate load, when increase the compressive strength to 42MPa and to 60 MPa, the ultimate load increased to about 220% and to about 300% respectively. The presence of steel reinforcing did not show reasonable effect on the ultimate load.

1. Introduction

Columns behaviour are of critical importance for the performance and the safety of the concrete structures [1]. In the last two decades, it is commonly seen that there are require for strengthening and/or retrofitting of existing concrete structures and one of most common members are columns. There are many reasons could lead to damage the concrete structures like: the age of the concrete structures, design errors, amendments in the codes requirements or changing in the facility use, etc. External confinement of concrete columns with Fibre Reinforced Polymer (FRP) is an active technique can be used to upgrade the load carrying capacity and/or ductility of the concrete columns [2-9]. This technique has been studied theoretically through improving specific analysis procedures and experimentally on the performance of different substructures and structures.

The efficiency of the FRP jacket depends on several variables like: concrete type, shape and size of the cross-sectional area, corner radius for square and rectangular sections, section aspect ratio for non-circular sections, applied load condition and FRP stiffness (FRP type, number of layers and FRP configuration) [10].

Carlos and Silva [11], and Campione [12], observed that when increase the FRP layers resulting in increase of concrete strength. There are some studies conducted to investigate the most effective of FRP material type (aramid, glass, and carbon). Most of these studies indicate that the carbon fibre is the most effective material in providing strength and ductility [13-15].

Campione [12], observed from his study that ultimate strength decreased when the height of the column increased.

Sangeetha [16], carried out analytical study using FAE on a series of published experimental data of small scale of circular, square and rectangular columns strengthen by FRP. The study approved that the FEA can properly simulate the behaviour of confined concrete columns, moreover, the findings revealed that when strength of the columns increased due to the confinement by about (20-25%).

Tavio and Tat [17], approved that the actual stress-strain carves, upgrade in strength and ductility of confined reinforced normal strength rectangular concrete columns can be predicted through FEA modelling. This has been confirmed by comparison of the analytical results with experimental results.

Mazen and Zahraa [18], conducted a nonlinear FEA through using ANSYS computer program, three effective parameters including in the study: compressive strength of concrete, CFRP jacket thickness and radius of corners on the performance of RC columns having square section and strengthen by CFRP. This study made a comparison between the nonlinear results with published experimental results. The results from this study revealed that the CFRP jacket is very effective technique in upgrading the strength and the ductility of square columns, in addition, the study findings revealed good agreement between the FEA and the experimental results.

It's widely approved that less benefit gained from applying the external reinforcement (FRP) in rectangular/square columns with respect to the circular columns. In the circular columns, the FRP jacket offers a uniform confinement pressure to the expansion of the concrete core as a result of the axial loading, while columns with rectangular/square sections, the FRP confining pressure is focused at the section edges instead of over the whole perimeter; therefore, evaluating analytically the behaviour of CFRP-confined square concrete columns under compression load became an objective of this paper.

2. Finite element modelling of CFRP confined concrete columns

The FEA by using computer program ANSYS-16.1 is carried out using solid brick elements (SOLID65) to represent concrete and quadratic-order membrane shell (SHELL41) to model CFRP.

Solid-65 element is commonly used to model concrete members. This element possesses eight nodes, each node has three degrees of freedom in the nodal, x, y, and z-directions as illustrated in Figure 1 (a). The element is able to crack in tension, crushing in compression, and large deflection nonlinearities. This element also used to analyse concrete members without reinforcing steel bars.

SHELL41 shown in Figure 1 (b) is a three-dimensional element having membrane (in-plane) stiffness but no bending (out of plane) stiffness. It is intended for shell structures where bending of the elements is of secondary importance. The element has three degrees of freedom at each node: translations in the nodal x, y, and z- directions. The element has changeable thickness, stress stiffening, and enormous deflection [19].

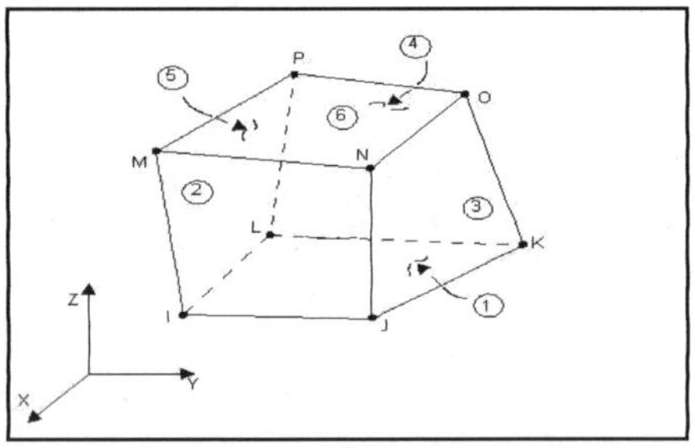

(a)

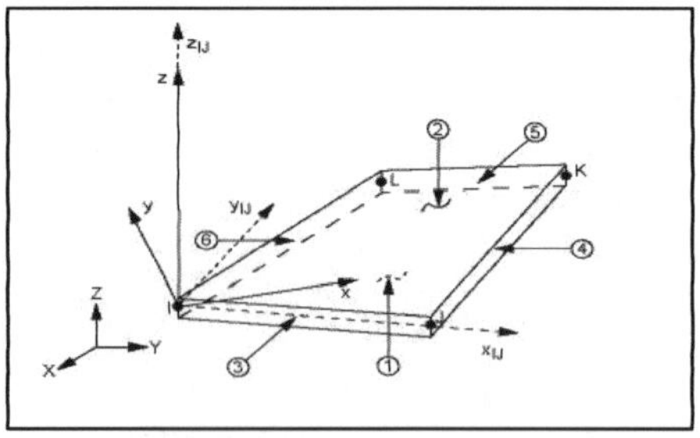

(b)

Figure 1. Finite element representation [19].

3. Properties of materials

FEA models for CFRP-confined concrete columns having square section are offered. In the first the material properties are specified, then specify the definitions of the material properties which are required to insert in computer software. Within this study, the computer program (ANSYS-16.1) is utilized to model the square concrete column and CFRP sheet. The nonlinear analysis is developed to simulate the nonlinear behaviour of the confined column. After whole model geometry definition, the material properties should be introduced. First, elastic behaviour of material is set. Hence, the elastic parameters such as: Young's modulus of concrete, E_c and Poisson's ratio, v, are inputted. From experimental results E_c is calculated as $E_c = 4700\sqrt{f_c'}$ where f_c' is given in MPa. The popular stress-strain relationship is used as illustrated in Figure 2 to make the uniaxial compressive simulation of the concrete column which is given by the following relationships [20].

$$f_c' = \frac{\varepsilon E_c}{1+\left(\frac{\varepsilon}{\varepsilon_o}\right)^2} \quad (1)$$

$$\varepsilon_o = \frac{2f_c'}{E_c} \quad (2)$$

$$E_c = \frac{f_c}{\varepsilon} \quad (3)$$

f_c the stress at any strain ε, ε is the strain at stress f_c, f_c' is the ultimate compressive stress and ε_o is the strain at the ultimate compressive strength f_c'

Poisson's ratio of concrete is assumed to be v_c=0.2. The CFRP behaviour is orthotropic (Figure. 3), the CFRP material is inputted as a linear elastic orthotropic material in the model. Indeed, it is necessary to introduce properties of the CFRP for each direction separately.

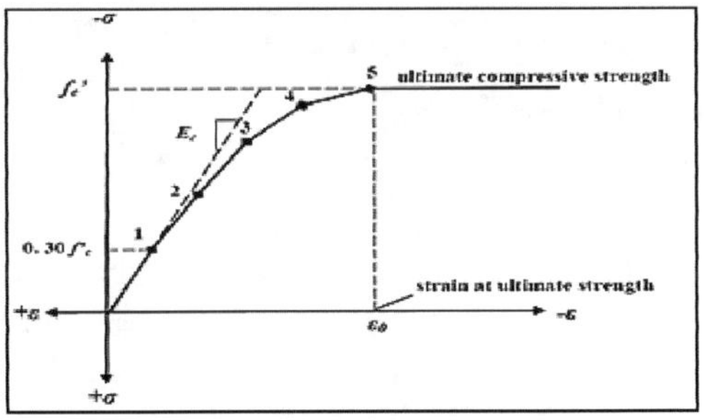

Figure 2. Stress-strain relationship for concrete under uniaxial compression.

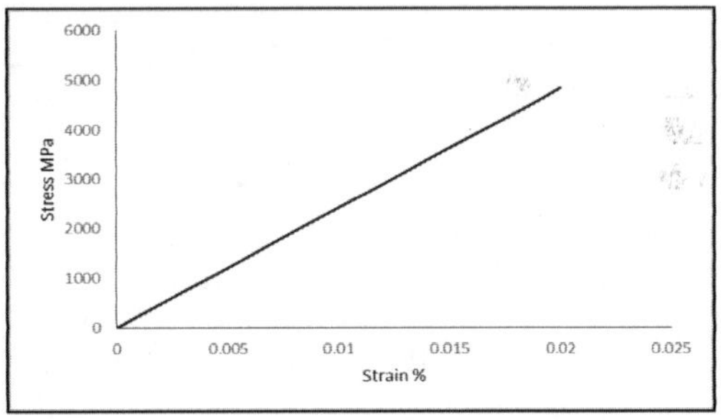

Figure 3. CFRP stress-strain relationship.

4. Numerical integration

An alternative scheme of numerical integration is required to carry out the numerical analysis because it is not possible to perform analytically the integration necessary to set up the element stiffness matrix. It is thus very necessary to nominate a suitable integration scheme that is both accurate and computationally efficient.

In the present work, the Gauss quadrature method is selected, since it has proved useful in finite element work [19]. The element stiffness matrix for brick element may be written in local element axes as:

$$I = \int_{-1}^{1}\int_{-1}^{1}\int_{-1}^{1} F(\xi,\eta,\zeta)\,d\xi\, d\eta\, d\zeta \quad (4)$$

Which can be calculated numerically as:

$$I = \sum_{i=1}^{p_1} \sum_{j=1}^{p_2} \sum_{k=1}^{p_3} W_i\, W_j\, W_k\, F(\xi_i, \eta_j, \zeta_k) \tag{5}$$

Where, p_1, p_2, and p_3 represent Gaussian points in the ξ, η, *and* ζ directions respectively. The function $F(\xi_i, \eta_j, \zeta_k)$ represents the matrix multiplication $([B]^T.[D].[B].|J|)$, generally the number of integration points is taken to be equal in the three directions: $p_1 = p_2 = p_3 = p$.

The integration rule used in this work is the 8 (2×2×2) points rule, Figure 4. Sampling point locations and weighting factors for 2×2×2 integration rule shown in table 1.

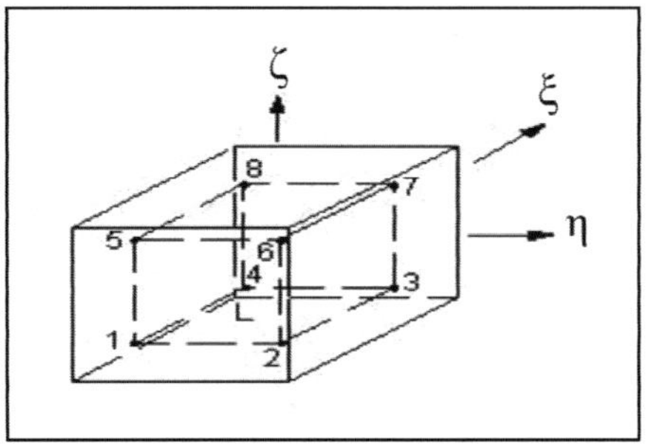

Figure 4. Integration points location for brick element [19].

Table 1. Sampling point locations and weighting factors for 2×2×2 integration rule [19].

Sampling Point	Load Coordinates			Weight
	ξ	η	ζ	
1-8	±0.57734	±0.57734	±0.57734	1

5. Nonlinear solution techniques

The solving of the nonlinear structural problems by using finite element discretion results in a set of nonlinear algebraic equations of the form:

$$[K]\{U\} = \{Fa\} \tag{6}$$

where:
[K]: stiffness matrix.
{U}: vector of nodal displacements.
{Fa}: vector of applied loads.

For linear elastic problems, equation (6) is used to seek out the answer of the unknown displacements *{U}*. In the case of nonlinear system, the stiffness matrix [K] is a function of the unknown displacements (or their derivatives). Then equation (6) cannot be exactly computed before determination of the unknown displacements {U}.

There are several techniques can be used to solve the non-linear equations (6); the basic techniques can be classified into:

1. *The incremental procedure:* load is applied in several small increments, and the structures is assumed to respond linearly within each increment with its stiffness recomputed based on the structural geometry and member end actions in the final of the previous load increment. This is a simple procedure, which requires no iterations, but errors are likely to accumulate after several increments unless very fine increments are used, Figure 5 (a).
2. *The iterative procedure:* the total load is applied in one increment at the first iteration, the out of balance forces are then computed and used in the next iteration to obtain progressively

improved solution. The iterative process is continued until the end of converging solution to be in balance, this internal load vector would equal the subjected load vector or within some tolerance. This process can be written as:

$$[K_i^T]\{\Delta U_i\}=\{F^a\}-\{F^{nr}\} \tag{7}$$

$$\{U_{i+1}\} = \{U_i\} + \{\Delta U_i\} \tag{8}$$

where:
$[K_i^T]$: stiffness matrix.
i: subscript representing the current equilibrium iteration.
$\{F^{nr}\}$: internal load vector.

This procedure fails to produce information regarding the intermediate stage of loading. For structural analysis including nonlinear (like plasticity), the solving method needs some average increments to be in balance to follow the correct load route. This could be gained by employing combined incremental-iterative method, Figure 5 (b).

3. *The combined Incremental-iterative procedure:* it can be used the combination of the incremental and iterative scheme. The load is subjected incrementally, and iterations are performed in order to obtain converged solution corresponding to the stage of loading under consideration, Figure 5 (c).

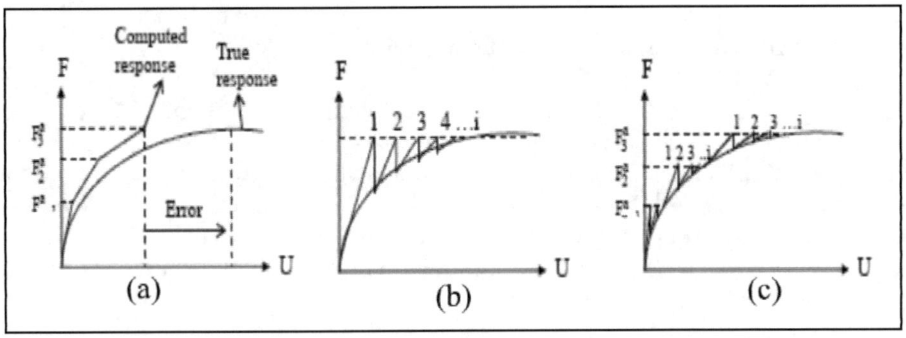

Figure 5. Basic technique to solve non-linear equations [21]:
(a) Incremental, (b) Iterative, (c) Incremental-Iterative.

ANSYS program implemented incremental-iterative solution procedures incorporated with Full and modified Newton-Raphson algorithm.

6. Finite element idealization

The structural behaviour of concrete columns having square section and strengthen by CFRP in different configurations is simulated depending on experimental test results [22] is presented in this study. Four column specimens have been chosen and analysed by using FEA. The columns verification study was carried out to check the validity of the analytical results with experimental test results, then carried out a parametric study to evaluate the influence of some effective parameter on behaviour of square concrete columns strengthen by CFRP composites.

6.1. Verification of the tested square concrete columns strengthen by CFRP composites

Four columns specimens were selected from experimental test data [22] to be analysed by FEA. All columns had the same cross section dimensions 300×300 and 1000 mm in height. The details of the tested columns are shown in table 2.

Table 2. Experimental results [22].

Specimen ID	f_c (MPa)	CFRP Strengthening Details	$P_{u,\ exp}$ (kN)
CP-Cnt 1	18	-	1830
CP-3	18	2 CFRP ply	2250
CP-5	19.3	2 CFRP ply+3 CFRP strip+10 CFRP anchors	2150
CP-7	18.8	1 CFRP ply+3 CFRP strip+10 CFRP anchors	2200

Note: The ultimate axial load values were the actual values obtained through an experimental test conducted by the author during the Ph.D. thesis' research.

From the table, it can be noticed that column specimen CP-5 has additional compressive strength and additional confinement (CFRP strips) than CP-3, but the ultimate load capacity was less than CP-3 (2150 kN vs 2250 kN), These unexpected values regarding to the ultimate capacities could happened because of the alternative arrangement of the anchors (not in the same position in each side). But on the other hand, the anchorage system improved based on the final failure mode of these confined columns where the failure didn't happen in the anchors zones, which was an aim of the research. In order to achieve a more effective confinement, the FRP strips be constrained on all sides through the use of anchors, thereby creating shorter distances which are confined between anchors.

These columns have been modelled using different element types with real constants and material properties. For these columns, the 1000 mm clear height of the columns was considered in the analysis and was divided into 3 divisions; first division is 50 mm in length, the second division is 900 mm in length (mid span) and the last division is 50 mm in length.

For control column CP-Cnt 1 (without CFRP), the FEA model consists of a total 12375 elements. For strengthen columns (with CFRP), the FEA model consists of a total 15675 elements for CP-3, 16590 for CP-5 and CP-7. The mesh and the boundary conditions used in the FEA for columns (control and strengthen) under concentric load are shown in Figure 6. The corner four nodes are restrained in the x direction for prevention solid shape to move of the idealized model. The distribution of the axial forces at the end elements of the model is shown in Figure 7. The middle nodes take full load and the nodes at the edges take half load, finally the nodes at the corners take quarter load as seen in Figure 8.

a) Unconfined column (b) Strengthen column

Figure 6. Mesh and boundary conditions of unconfined and confined columns

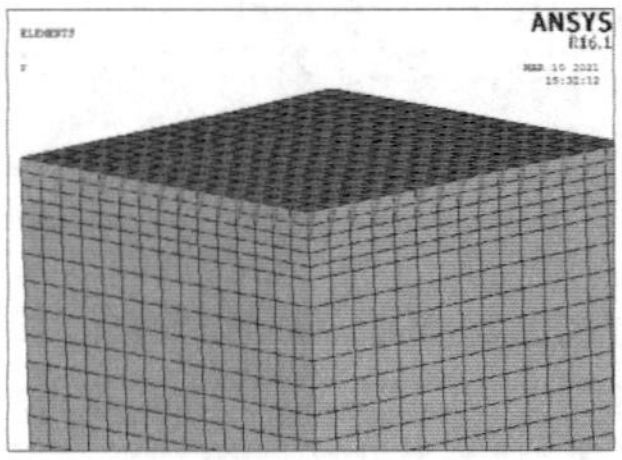

Figure 7. The distribution of the axial forces at the end.

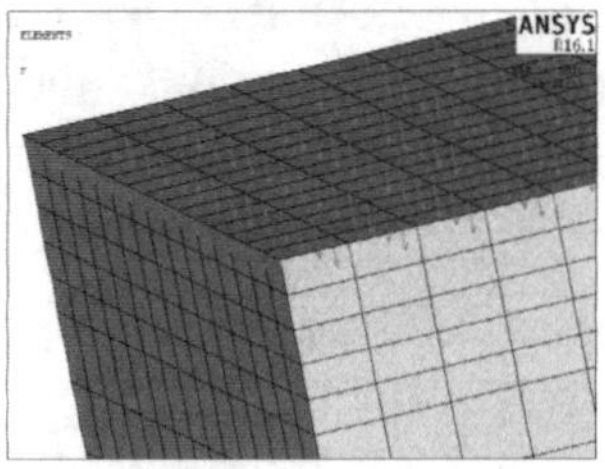

Figure 8. The nodes at the corners take quarter load.

The external nodes of concrete surfaces were used to represent the elements of CFRP element (in case of wrapping the column with full and partial jacketing). The both full and partial jacketing of CFRP elements were represented by using these nodes by connected to concrete elements or CFRP elements assuming full contact. In addition, the meshes size were bigger for the top and bottom part of the columns because of the load distribution. Figure 9 (a) and (b) shows the CFRP meshing.

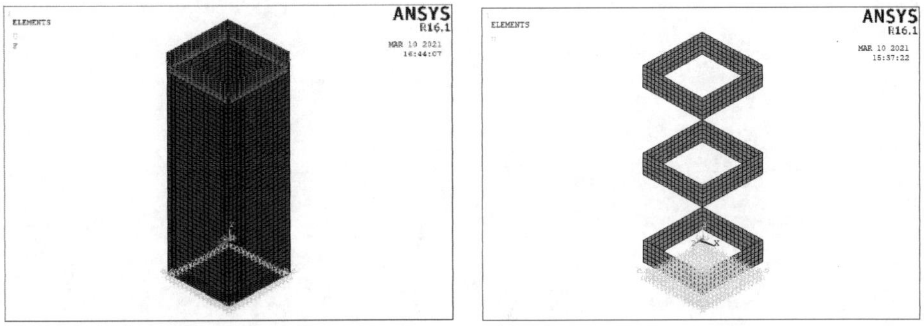

(a) (b)

Figure 9. CFRP meshing of jacket and strips: (a) CFRP meshing for the full confinement, (b) CFRP meshing for the partial confinement.

7. Results of the numerical analysis (ultimate load capacity)

The ultimate loads of the modelled columns were indicated by the state that the columns no longer can support additional load as indicated by the convergence failure of ANSYS program in failing to find a solution.

Table 3 shows the maximum failure load for each modelled columns obtained from the software computer program (ANSYS 16.1) and experimental tests. It can be observed from table 3 that the obtained ultimate load from the numerical analysis is a little less than the experimental ultimate load of square concrete columns strengthen by CFRP jackets/strips with least ratio $P_{u,Exp}$ / $P_{u,Ansys}$ was 2% for CP-3 and maximum ratio value was 9% for CP-Cnt 1 while the ratio CP-7 was 4%, except for CP-5 the ANSYS value was a little higher than the experimental value by 5%.

Table 3. Experimental and numerical (ANSYS) of ultimate load capacities.

Column ID	Ultimate Load kN		$P_{u,Exp} / P_{u,Ansys}$
	$P_{u,Exp}$	$P_{u,Ansys}$	
CP-Cnt 1	1830	1680	1.09
CP3	2250	2193	1.02
CP-5	2150	2260	0.95
CP-7	2200	2111	1.04

The axial load - deformation curves of columns CP-Cnt 1, CP-3 , CP-5 and CP-7, obtained from the numerical FEA along with the experimental curves reported by experimental data [22] presented and compared in Figure 10. These figures show good agreement between the experimental and finite element axial load-axial deformation results. In addition, complete curves have been obtained from ANSYS while the curves from the experimental were up to 75% of the P_u because mechanical dial gage used in reading the axial deformation during the experimental test. Figure 11 (a) – (i) shows the the nodal stress in the y direction at failure load of the modelled columns.

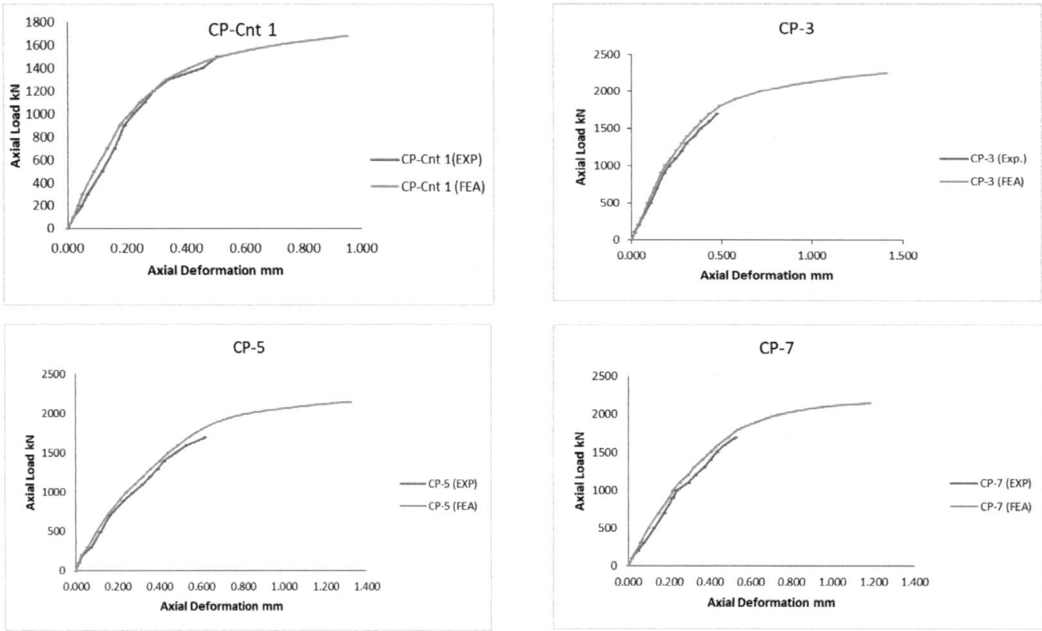

Figure 10. The axial load-axial deformation curves of columns CP-Cnt 1, CP-3 , CP-5 and CP-7 for both analytical and experimental.

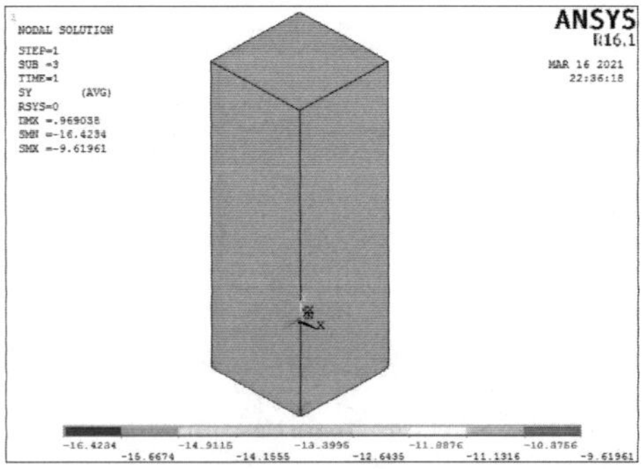

(a) Stresses of concrete of CP-Cnt 1

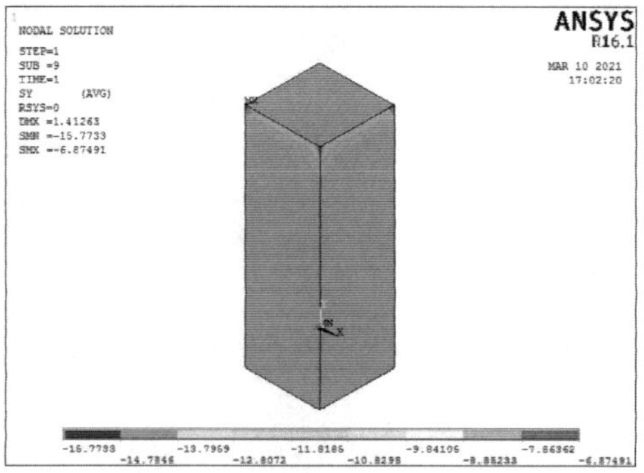

(b) Stresses of concrete of CP-3

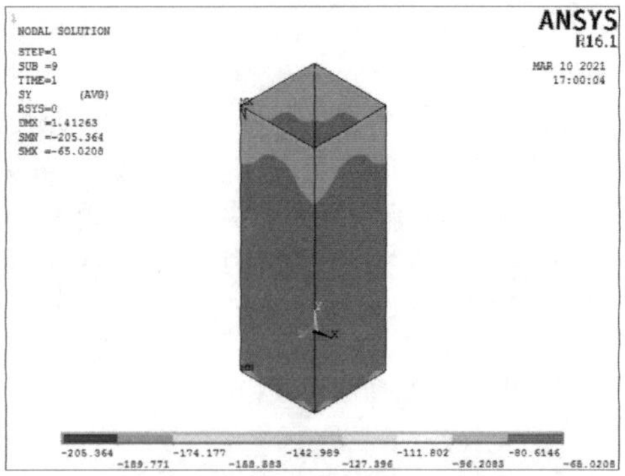

(c) Stresses of CFRP jacket of CP-3

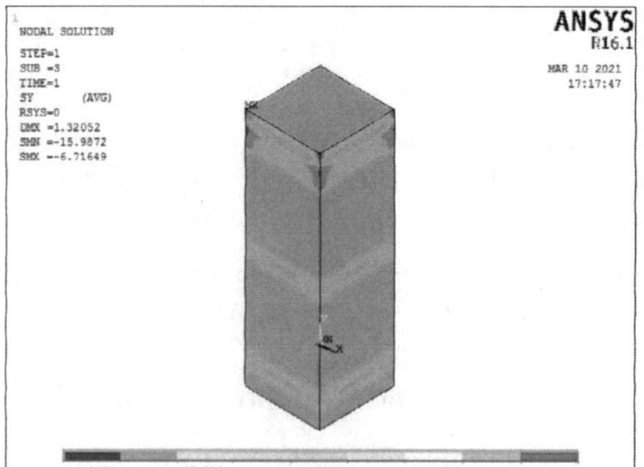

(d) Stresses of concrete of CP-5

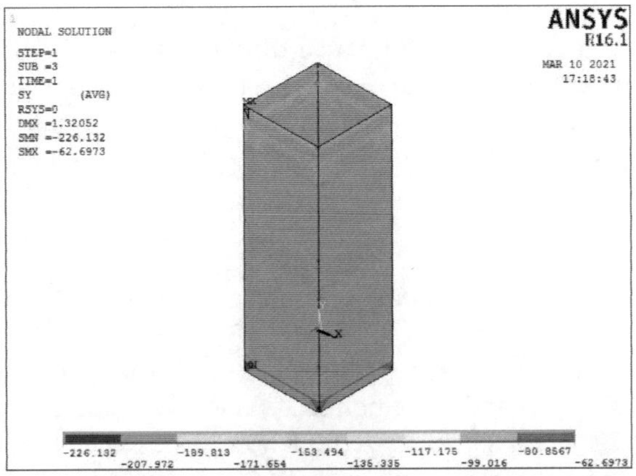

(e) Stresses of CFRP jacket of CP-5

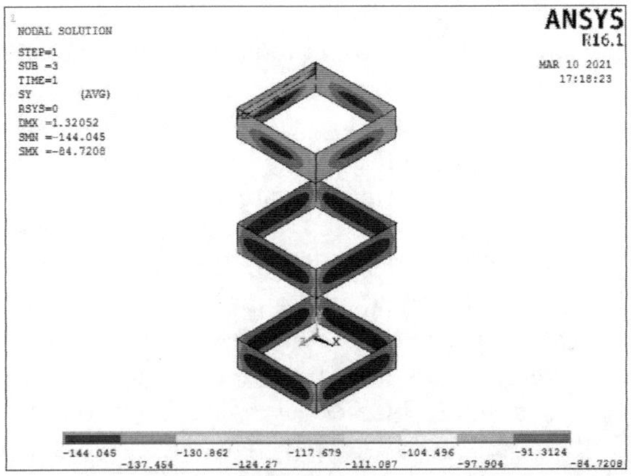

(f) Stresses of CFRP strips of CP-5

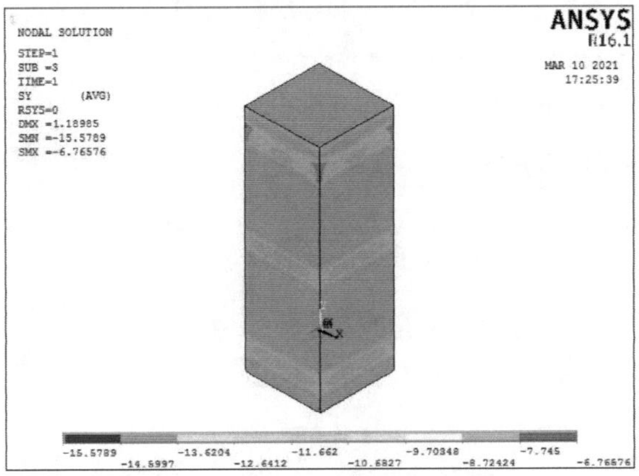

(g) Stresses of concrete of CP-7

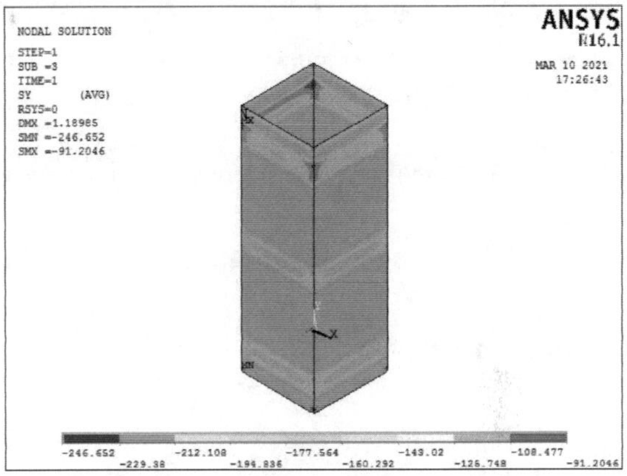

(h) Stresses of CFRP jacket of CP-7

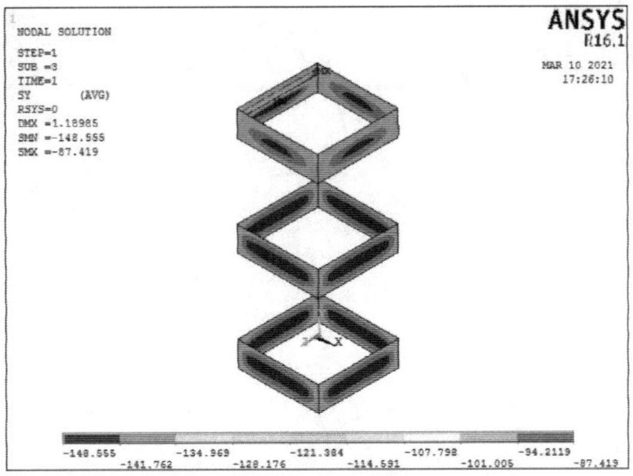

(i) Stresses of CFRP strips of CP-7

Figure 11. The nodal stress in the y direction at failure load of modelled columns

8. Parametric study

The evaluation to the most important parameters impact on the performance of concrete columns in square section that could not be performed during the experimental test were performed by using the numerical finite element ANSYS. In the parametric study, three parameters were included to evaluate their effect on the failure load (ultimate load) of the modelled columns. The parameters were included in this parametric study are:

8.1. Effect of number of CFRP layers

In this numerical study, the effect of CFRP jacket thickness applied on column specimens CP-3 and CP-5, the columns have been numerically analysed using different numbers of CFRP layers (full layers). The selected numbers of layers were 3 and 4, it is worth noting that in the experimental test, the amount of CFRP layers used for columns CP-3 was two and the amount of CFRP layers used for column CP-5 was also two but combined with CFRP strips and CFRP anchors. The comparison between the numerical maximum failure loads got from different amounts of CFRP layers (3 and 4 layers) of the tested columns are listed in table 4 and 5 respectively. It is obvious that the maximum load increases as the amount of CFRP layers increase. When the amount of CFRP layers increased from 2 to 3 layers as listed in table 4, the ultimate load increased by about 5% for both CP-3 and CP-5 respectively, while when the amount of CFRP layers increased from 2 to 4 layers as listed in table 5, the ultimate load increased by about 10% for both CP-3 and CP-5 respectively.
Figure 12 and 13 shows the concrete stresses and CFRP jacket distribution in the y direction at the maximum load of column specimen CP-3 with 3 and 4 CFRP layers respectively.

Table 4. Ansys ultimate load results with 2 and 3 CFRP layers.

Specimen ID	f_c (MPa)	Exp. Strengthening (No of Layers)	$P_{u,\,Ansys}$ (kN)	FEA Strengthening (No of CFRP layers)	$P_{u,\,Ansys}$ *(kN)*	$P_{u,Ansys}$ / $P_{u,Ansys}$
CP-3	18	2 CFRP ply	2193	3	2305	1.05
CP-5	19.3	2 CFRP ply+3 CFRP strip+10 CFRP anchors	2260	3	2375	1.05

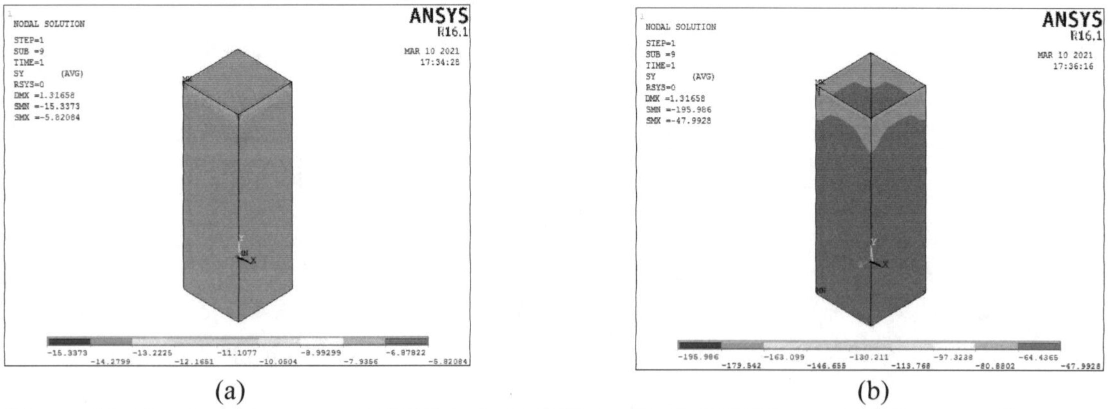

(a) (b)

Figure 12. Concrete stresses and CFRP jacket of CP-3 with 3-CFRP layers: (a) Stresses of concrete of CP-3 with 3-CFRP layers, (b) Stresses of CFRP jacket of CP-3 with 3-CFRP layers.

IOP Conf. Series: Materials Science and Engineering 1141 (2021) 012046 doi:10.1088/1757-899X/1141/1/012046

Table 5. Ansys ultimate load results with 2 and 4 CFRP layers.

Specimen ID	f_c (MPa)	Exp. Strengthening	$P_{u,\,Ansys}$ *(kN)*	FEA (ANSYS) Strengthening	$P_{u,\,Ansys}$ *(kN)*	$P_{u,Ansys}$ / $P_{u,Ansys}$
CP-3	18	2 CFRP ply	2193	4	2413	1.10
CP-5	19.3	2 CFRP ply+3 CFRP strip+10 CFRP anchors	2260	4	2475	1.10

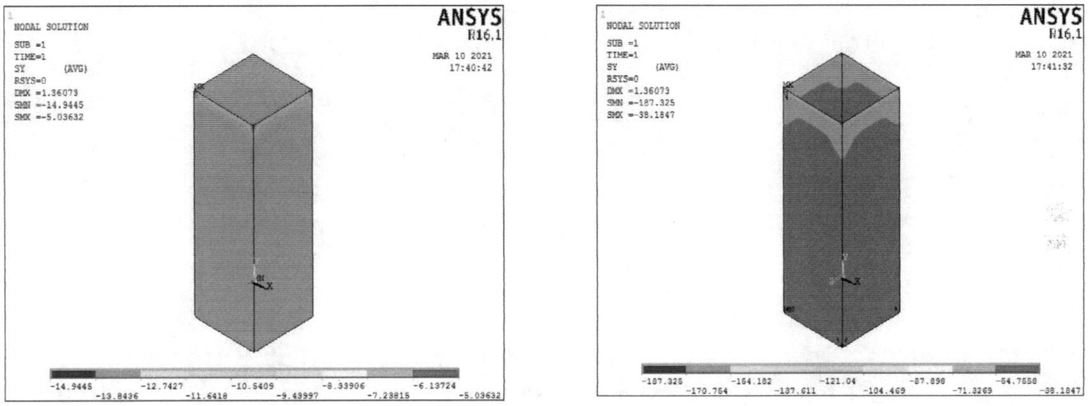

(a) (b)

Figure 13. Concrete stresses and CFRP jacket of CP-3 with 4-CFRP layers: (a) Stresses of concrete of CP-3 with 4-CFRP layers, (b) Stresses of CFRP jacket of CP-3 with 4-CFRP layers.

8.2. Effect of Concrete Compressive Strength

In this numerical study the selected values of the concrete strength (f_c) were 42 and 60 MPa for columns CP-3, CP-5 and CP-7. The numerical ultimate axial loads got from different values of concrete strength of the tested columns together with the experimental values are listed in table 6. It can be noted from the table that the concrete strength has a significant effect on ultimate axial loading. When the compressive strength of columns CP-3, CP-5 and CP-7 increased to 42 MPa, the ultimate load capacities increased by 222%, 221% and 223% respectively. When the concrete strength of columns CP-3, CP-5 and CP-7 increased to 60 MPa, the ultimate load capacities increased by 304%, 301% and 313% respectively. This effect was expected because the column (compression member) is essentially depends on the concrete compression strength as more as in this study the applied loads were pure compression (concentric load). Figure 14 shows the stresses of concrete, CFRP jacket and the CFRP strips of columns CP-3 for compressive strength 42 and 60 MPa.

Table 6. Experimental and numerical results of ultimate load with different compressive strength

Column ID	f_c *(MPa)* *(Exp.)*	P_{Exp} *(kN)*	f_c *(MPa)* *(ANSYS)*	$P_{u,Ansys}$ *(kN)*	$P_{u,Ansys}/P_{u,Exp}$	f_c *(MPa)* *(ANSYS)*	$P_{u,Ansys}$ *(kN)*	$P_{u,Ansys}/P_{u,Exp}$
CP3	18	2250	42	5004	2.22	60	6840	3.04
CP-5	19.3	2150	42	4770	2.21	60	6470	3.01
CP-7	18.8	2200	42	4914	2.23	60	6903	3.13

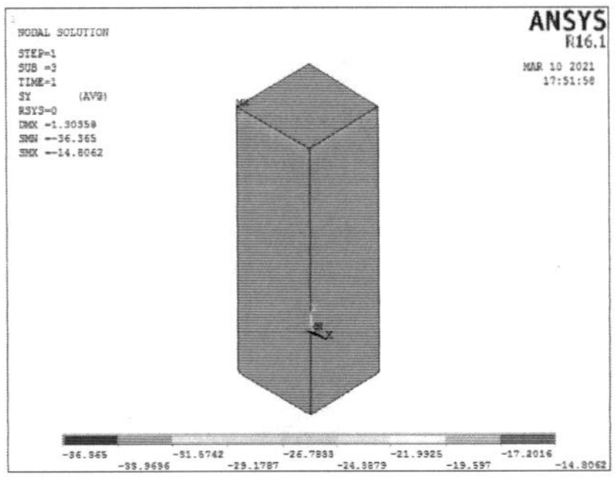

Stresses of concrete of CP-3 with f_c= 42 MPa

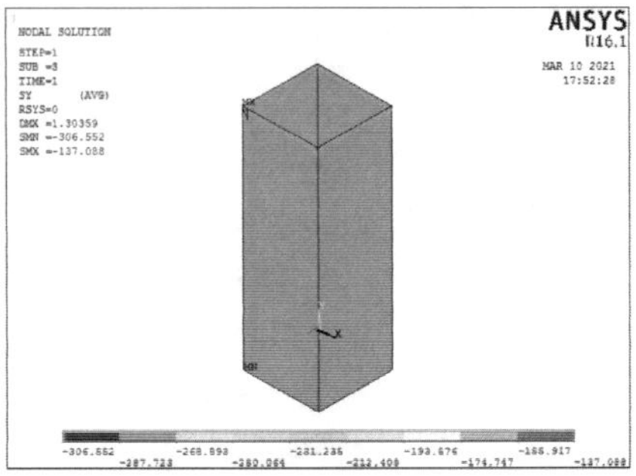

Stresses of CFRP jacket of CP-3 with f_c= 42 MPa

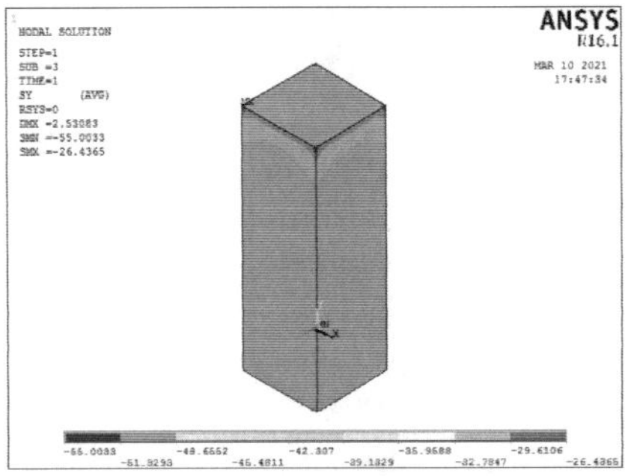

Stresses of concrete of CP-3 with f_c= 60 MPa

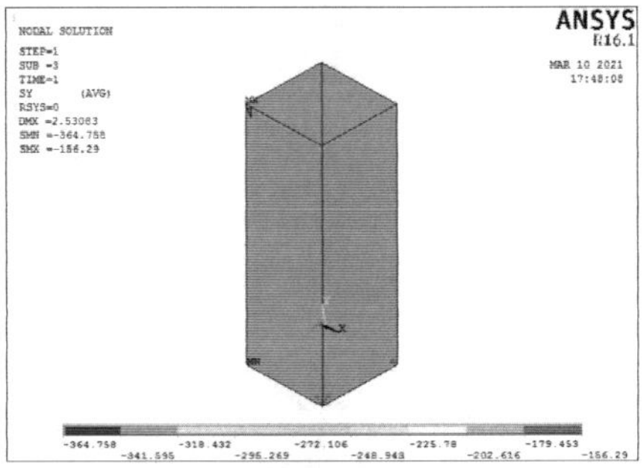

Stresses of CFRP jacket of CP-3 with f_c= 60 MPa

Figure 14. The stresses of concrete and CFRP jacket CP-3 with compressive strength 42 MPa

8.3. Presence of steel reinforcement

In order to evaluate the effectiveness of the presence of steel reinforcement on the maximum axial loading capacity of the tested concrete columns in square section strengthen by CFRP, assuming that the columns (CP- Cnt 1, CP-3, CP-5 and CP-7) having a steel reinforcing details (4 - Ø16 for main reinforcement and 7 Ø 10 for stirrups) as seen in Figure 15. All columns were modelled in finite element software ANSYS. In addition, three-dimensional spar element (LINK180) was used to represent the steel.

The numerical findings got from the finite element software ANSYS revealed that the reinforcement steel did not have a reasonable effect on the ultimate load capacity of the confined columns, the reason of that could be because the confined columns have been applied to compression loading and it have been approved previously that the concrete take the most action in compression than the steel, in addition to that the columns specimens have CFRP confinement which enhanced in the confinement instead of stirrups or spiral reinforcement. table 7 shows the ultimate load capacities of the modelled columns with and without steel reinforcement.

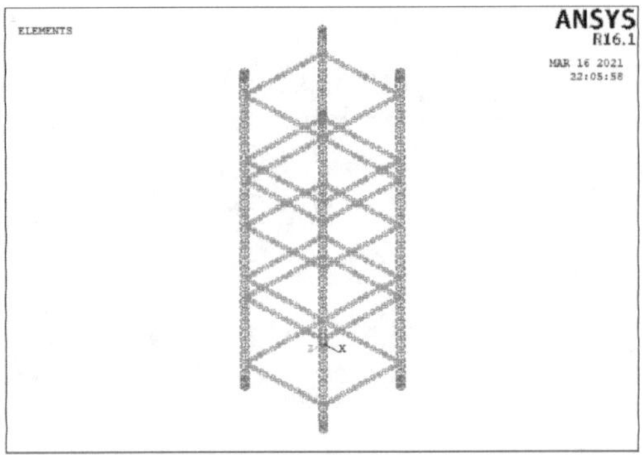

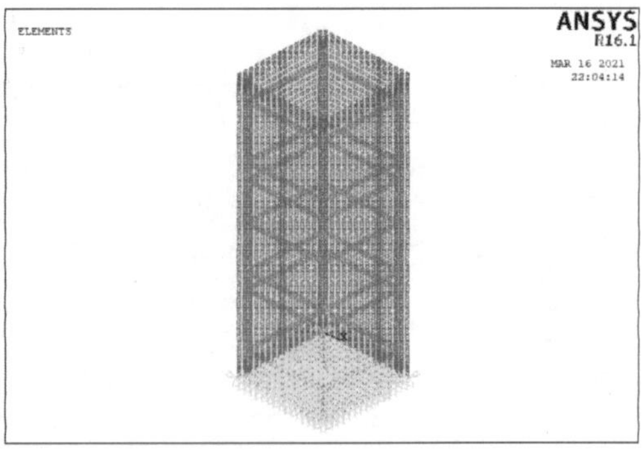

Figure 15. Reinforcement details in each modelled column.

Table 7. The ultimate load capacities with and without steel reinforcement

Column ID	Ultimate Load		$P_{u,Ansys}$ / $P_{u,Ansys}$
	Without reinforcing steel	With reinforcing steel	
	$P_{u,Ansys}$ (kN)	$P_{u,Ansys}$ (kN)	
CP-Cnt 1	1680	1900	1.13
CP3	2193	2315	1.05
CP-5	2260	2230	0.98
CP-7	2111	2190	1.03

9. Conclusion

Based on the analysis of comparison obtained from the analytical (ANSYS) results and experimental results it can be concluded that:

1. The numerical results showed that external strengthening with CFRP composites are very effective in upgrading the maximum compression loading capacity of the square concrete columns.
2. The numerical findings revealed good agreement between the ANSYS and the experimental test results.
3. The numerical results obtained from the parametric study revealed, when the CFRP layers increased from 2 to 3 and from 2 to 4 layers, the maximum compression loading of column specimen CP-3 increased by 5% and 10% respectively. While, when the CFRP layers increased from 2 to 3 and from 2 to 4 layers, the maximum compression loading of column specimen CP-5 increased by 5% and 10% respectively as listed in table 4 and 5.
4. The numerical results confirmed that the compressive strength showed significant effect on the upgrade of the ultimate load, when the concrete compression strength of modelled columns increased to 42MPa and to 60 MPa, the maximum compression loading of column specimens CP-3, CP-5 and CP-7 increased to about 220% and to about 300% respectively as shown in table 6. This effect was expected because the column is a compressive member and is essentially depends on the concrete compression strength as more as in this study the applied loads were pure compression (concentric load).
5. The numerical results revealed that the presence of steel reinforcing did not have a reasonable effect on the maximum axial loading of modelled confined columns. The reason of that could be because the columns were applied to pure compression load and it have been approved previously that the concrete take the most action in compression than the steel, in addition to that the columns specimens have CFRP confinement which enhanced in the confinement instead of stirrups or spiral reinforcement, while the unconfined column CP-Cnt 1 gained the higher increase in the ultimate load capacity by 13% due to the presence of steel reinforcement, which did not have external CFRP.

Acknowledgment

I would like to thank all those who have been instrumental in facilitating with this study, specifically the supervisor, Prof. Dr. Eng. Dan Paul Georgescu.

10. References

[1] Rocca R, Galati N and Nanni A 2008 Interaction diagram methodology for design of FRP-confined reinforced concrete columns. *Journal of Construction and Building Materials,* **23**, pp 1508–1520
[2] Lam L and Teng J G 2002 Strength models for fiber-reinforced plastic-confined concrete. *Journal of Structural Engineering* **128** (5), pp 612-623
[3] Mirmiran A and Shahawy M 1997 Behaviour of concrete columns confined by fiber

composites. *Journal of Structural Engineering, 123(5),* pp. 583-590.

[4] Nanni A and Bradford M 1995 FRP jacketed concrete under uniaxial compression. *Journal of Construction and Building Materials* **9** (2), pp 115-618

[5] Parvin A and Wang W 2001 Behaviour of FRP jacketed concrete columns under eccentric loading. *Journal of Composites for Construction* **5** (3), pp 146-152

[6] Parvin A and Jamwal A 2005 Effects of wrap thickness and ply configuration on composite-confined concrete cylinders. *Journal of Composite Structures* **67**, pp 437–442

[7] Pessiki S, Harries K, Kestner J, Sause R and Ricles J 2001 Axial behaviour of reinforced concrete columns confined with FRP jackets. *Journal of Composites for Construction* **5** (4), pp 237-245

[8] Rochette P and Labossie`re P 2000 Axial testing of rectangular column models confined with composites. *Journal of Composites for Construction* **4** (3), pp 129-136

[9] Xiao Y and Wu H 2000 Compressive behaviour of concrete confined by carbon fibre composite jackets. *Journal of Materials in Civil Engineering* **12** (2), pp 139-146

[10] Wu Y F and Wei Y Y 2010 Effect of cross-sectional aspect ratio on the strength of CFRP-confined rectangular concrete columns. *Journal of Engineering Structures* **32**, pp 32-45

[11] Chastre C and Silva G 2010 Monotonic axial behaviour and modelling of RC circular columns confined with CFRP. *Journal of Engineering Structures* **32**, pp 2268-2277

[12] Campione G 2006 Influence of FRP wrapping techniques on the compressive behaviour of concrete prisms. *Journal of Cement & Concrete Composites* **28**, pp 497–505

[13] Silva G 2011 Behaviour of square and circular columns strengthened with aramidic or carbon fibres. *Journal of Construction and Building Materials* **25**, pp 3222–3228

[14] Tastani S P and Pantazopoulou S J 2004 Experimental evaluation of FRP jackets in upgrading RC corroded columns with substandard detailing. *Journal of Engineering Structures* **26**, pp 817–829

[15] Sakisa T C, Karabinisa A I and Kiousisb P D 2007 FRP-confined concrete members: Axial compression experiments and plasticity modelling. *Journal of Engineering Structures* **29**, pp 1343–1353

[16] Sangeetha P 2007 Analysis of FRP wrapped concrete columns under uniaxial compression. *Journal of Scientific & Industrial Research* **66**, pp 235-242

[17] Tavio T and Tata A 2009 Predicting nonlinear behaviour and stress-strain relationship of rectangular confined reinforced concrete columns with ANSYS. *Journal of Civil Engineering Dimension* **11** (1), pp 23-31

[18] Mazen A and Zahraa K 2018 Nonlinear finite elements analysis of reinforced concrete columns strengthened with carbon fiber reinforced polymer (CFRP). *Journal of University of Babylon, Engineering Sciences* **26** (3), pp 1-23

[19] ANSYS, Inc. 2014 ANSYS fluent user's guide. Canonsburg, USA.

[20] Prakash D, Sundara K T and Sanjeeva T 1978 Equation for stress-strain curve of concrete confined in circular steel spiral. *Journal of Materials and Structures* **11** (5), pp 339-345

[21] Nyssen C 1981 An efficient and accurate iterative method, allowing large incremental steps, to solve elasto-plastic problems. *Journal of Computers and Structures* **13** (3), pp 63-71

[22] Almagsoosi A N A 2021 Experimental test on square concrete columns strengthened with CFRP composites in different configurations. Research report no.3 of the Doctoral Thesis, UTCB.

IOP Conf. Series: Materials Science and Engineering 1141 (2021) 012047 doi:10.1088/1757-899X/1141/1/012047

A meshfree formulation for CFD and linear elasticity problems

F R Saucedo-Zendejo[1]
[1]Centro de Investigación en Matemáticas Aplicadas, Universidad Autónoma de Coahuila, Camporredondo S/N, Saltillo, Coahuila, CP 25115, *México*

E-mail: fesaucedoz@uadec.edu.mx

Abstract. This paper documents the development and application of a meshless procedure for computational fluid dynamics and linear elasticity problems. The meshfree nature of this method gives the advantage of dealing with highly distorted domains and even fragmentations without the need of using computationally expensive remeshing approaches and with a very simple implementation. A description of the implementation of this method and also the solutions of some benchmark problems will be presented in order to demonstrate the potential of this formulation for dealing with problems in these fields and to introduce promising future fields of application.

1. Introduction

The ideal for solving many engineering problems is to obtain the best performance at the lowest possible cost. This goal can be reached with a deep knowledge and understanding of the nature that surrounds the problem to be solved. The numerical simulation, widely used nowadays in engineering, is a suitable technology to know in depth many physical phenomena since it provides a large amount of information that cannot be obtained through other methods. Furthermore, in many situations it is the most efficient, profitable, and powerful technique to achieve this purpose [1]. This tool involves the derivation of a mathematical model for the problem, which is frequently expressed in the form of partial differential equations (PDEs) whose analytical solution is generally not available, so numerical methods executed on a computer are used to approximate their solutions.

Numerical mesh-based methods such as the Finite Element Method (FEM), Finite Difference Method (FDM), Finite Volume Method (FVM) are universally accepted numerical techniques for solving the PDEs in different engineering problems, and that is why during the last years many numerical simulation tools are based on them. However, mesh-based methods are not suitable for problems associated with very large mesh deformations, moving discontinuities or problems requiring constant remeshing during the solution process, since it implies a high computational cost. To overcome some of the drawbacks arising when mesh-based methods are used, mesh-free methods have recently been proposed and developed as alternative [2].

The advantages of meshless methods over mesh-based methods are that they use a set of nodes to discretize the problem domain and its boundaries without requiring any information about the relationship between nodes such that they do not form an element mesh which lets to deal with problems involving discontinuities and deformations in the domain without the problems of mesh-based methods. Moreover, they have the flexibility to remove or add points whenever and wherever needed which promote the developing of adaptive schemes [3].

A truly Lagrangian strong-form meshfree method is the so called Finite Pointset Method (FPM) developed by J. Kuhnert [4]. It has proven to be far superior to traditional mesh-based and some other meshfree methods in several computational mechanics and engineering fields, such as flows on manifolds [5], additive manufacturing processes [6], fluid mechanics [7, 8], fluid structure interactions [9], free surface flows [10], radiative heat transfer [11], heat transfer with moving heat sources [12] and fluid flow coupled with heat transfer considering phase changes [13].It uses the weighted least-squares (WLSM) interpolation scheme to approximate the spatial derivatives [14]. It has many advantages over other methods since it is able to naturally and easily incorporate any kind of boundary conditions without requiring any special treatment or stabilization and it is really simple to implement.

Therefore, a novel meshless formulation based on the finite pointset method for linear elasticity problems is introduced in this work for computational fluid dynamics and linear elasticity problems. In order to get some insight on its performance we compute the numerical solution of some benchmark tests. The structure of the paper is as follows: Section 2 shortly describes the governing equations. Section 3 shows the main ideas behind the novel formulation of FPM for computational fluid dynamics and linear elasticity problems followed by the numerical tests presented in Section 4 with its corresponding results. Finally, some conclusions are given in the last section.

2. Mathematical model

Many physical phenomena such as linear elasticity, the resulting pressure Poisson and the velocity equations arising when the Chorin's projection method is applied in order to numerically solve the Navier-Stokes equations are governed by an elliptical partial differential equation whose general form reads

$$A_1\psi + B \cdot \nabla\psi + C_1\Delta\psi - F = 0 \quad (1)$$

with prescribed Dirichlet, Neumann ($\partial\psi/\partial n = \varphi$) or Robin ($A_2\psi + C_2\Delta\psi/\partial n = \varphi$) boundary conditions, where A_i, B, C_i, F and $\prod$ are known.

3. FPM formulation for general elliptic partial differential equations

Consider now the numerical solution of the elliptic partial differential equation (1). A novel FPM approach is proposed to numerically solve general elliptic PDEs and this is shortly introduced next. This approach is based on the FPM formulation proposed in [7] and considered in [10, 12, 13, 15], and it differs from it in the fact that it takes the discretization of the PDE directly in the sparse global linear system of equations and not in the local system of the weighted least squares solution. To give a general idea of this new approach, consider the numerical solution of the general elliptic PDE (1).

Let Ω be a given domain with boundary $\partial\Omega$ and suppose that the set of points $\mathbf{x}_1, \mathbf{x}_2, \ldots, \mathbf{x}_N$ are distributed with corresponding function values $\psi(\mathbf{x}_1)$, $\psi(\mathbf{x}_2)$, ..., $\psi(\mathbf{x}_N)$. The problem is to find an approximate value of ψ at some arbitrary location $\psi(\mathbf{x})$ using its discrete values at particles positions inside a neighbourhood of $\mathbf{x}$. To define the set of nodes and the neighbourhood of $\mathbf{x}$, a weight function $w(\mathbf{x} - \mathbf{x}_i)$ is introduced

$$w_i = w(x - x_i) = \{e^{-\gamma\|x-x_i\|^2/h^2}, \quad \frac{\|x-x_i\|}{h} \leq 1\ 0\ else \quad (2)$$

where h is the smoothing length, γ is a positive constant whose value is considered to be 6.5, and $\mathbf{x}_i$ is the position of the i–th point inside the neighbourhood.

The Taylor expansion of $\psi(\mathbf{x}_i)$ around a particle $\mathbf{x}$ is

$$\psi(\mathbf{x}_i) = \psi(\mathbf{x}) + \sum_{k=1}^{2} \frac{\partial\psi}{\partial x_k}(x_{k,i} - x_k) + \frac{1}{2}\sum_{j,k=1}^{3} \frac{\partial^2\psi}{\partial x_k \partial x_j}(x_{k,i} - x_k)(x_{j,i} - x_j) + e_{1,i}$$

for $i = 1, \ldots, m$, where m is the number of nodes that lie within the neighbourhood of $\mathbf{x}$.

In contrast to the approach studied in [7] where the linear system obtained from the Taylor's expansion is taken together with the general elliptic PDE (1) with the corresponding boundary conditions, in this approach and at this step only the linear system obtained from the Taylor's expansion is considered which in matrix form reads

$$e = Ma_\psi - b \quad (3)$$

where

$$M = \left(1\ h_{1,1}\ h_{2,1}\ \frac{1}{2}h_{1,1}^2\ h_{1,1}h_{2,1}\ \frac{1}{2}h_{2,1}^2\ 1\ h_{1,2}\ h_{2,2}\ \frac{1}{2}h_{1,2}^2\ h_{1,2}h_{2,2}\ \frac{1}{2}h_{2,2}^2\ \vdots\vdots\vdots\vdots\vdots\vdots\ 1\ h_{1,m}\ h_{2,m}\ \frac{1}{2}h_{1,m}^2\ h_{1,m}h_{2,m}\ \frac{1}{2}h_{2,m}^2\right) \tag{4}$$

with $h_{1,j} = (x_{1,j} - x_1)$ and $h_{2,j} = (x_{2,j} - x_2)$,

$$a_\psi = \left(\psi, \frac{\partial\psi}{\partial x_1}, \frac{\partial\psi}{\partial x_2}, \frac{\partial^2\psi}{\partial x_1{}^2}, \frac{\partial^2\psi}{\partial x_1\partial x_2}, \frac{\partial^2\psi}{\partial x_2{}^2}\right)^t \tag{5}$$

$$b = (\psi_1, \psi_2, \ldots, \psi_m)^t \tag{6}$$

$$e = (e_1, e_2, \ldots, e_m)^t \tag{7}$$

After a minimization of $\boldsymbol{ee}$ through the weighted least squares method, the unknown vector a_ψ is obtained as

$$a_\psi = (M^tWM)^{-1}(M^tW)b \tag{8}$$

where W is

$$W = [w(x - x_1)\ 0\ \cdots\ 0\ 0\ w(x - x_2)\ \cdots\ 0\ \vdots\ \vdots\ \ddots\ \vdots\ 0\ 0\ \cdots\ w(x - x_n)\,] \tag{9}$$

Thus, the general elliptic PDE (1) can be expressed in terms of $\mathbf{a}_\psi$ as

$$A_1 a_{\psi_1} + B_1 a_{\psi_2} + B_2 a_{\psi_3} + C_1(a_{\psi_4} + a_{\psi_6}) = F \tag{10}$$

while the Robin boundary condition reads

$$A_2 a_{\psi_1} + C_2 n_1 a_{\psi_2} + C_2 n_2 a_{\psi_3} = \varphi \tag{11}$$

If Q is defined as

$$Q = (M^tWM)^{-1} \tag{12}$$

and its kth– row is denoted by $\mathbf{q_k} = (q_{k,1}, q_{k,2}, \ldots, q_{k,6})$, the moving least squares solution reads

$$a_\psi = Q(M^tW)b, \tag{13}$$

and each element in $\mathbf{a}_\psi$ can be expressed as

$$a_{\psi_k} = \hat{a}_{i=1}^{m_j} w_{ij} p_{ki} \psi_{ij} \tag{14}$$

where m_j denotes the number of jth–particle neighbours within the shape function support, ψ_{ij} the unknown function value in the ith– neighbour particle position and p_{ki} is defined as

$$p_{ki} = q_{k,1} + q_{k,2}h_{1,i} + q_{k,3}h_{2,i} + q_{k,4}\frac{h_{1,i}^2}{2} + q_{k,5}h_{1,i}h_{2,i} + q_{k,6}\frac{h_{2,i}^2}{2}. \tag{15}$$

It couples each node with its neighbouring nodes yielding to the coupling between inner and boundary nodes. Therefore, the discretization of equation (1) under this setting reads

$$\sum_{i=1}^{m_j} w_{ij}[A_1 p_{1i} + B_1 p_{2i} + B_2 p_{3i} + C_1(p_{4i} + p_{6i})]\psi_{ij} = F \tag{16}$$

and if the particle is on a Robin boundary, (16) becomes

$$\sum_{i=1}^{m_j} w_{ij}(A_2 p_{1i} + C_2 n_1 p_{2i} + C_2 n_2 p_{3i})\psi_{ij} = \varphi \tag{17}$$

Since equation (16) is valid for j = 1,2,..,N, this can be arranged in a global sparse linear system $K\boldsymbol{\Psi} = \mathbf{f}$ which is solved with iterative methods. Thus, general elliptic PDEs such as (1) can be solved in this way, just adding appropriate entries in the systems of equations (16).

4. Numerical Examples

In this section the suitability and feasibility of this FPM formulation in order to solve computational fluid dynamics and linear elasticity problems will be evaluated.

4.1. *Computational fluid dynamics*

4.1.1. Filling Benchmark.

In order to demonstrate the suitability of this approach for computational fluid dynamics problem, the filling of a circular disk mould with a core in the center will be considered as a benchmark example since the experimental filling pattern is available in the literature [16]. The geometry of this example is shown in the figure 1. In this example, the problem domain was discretized with approximately 6000 points. The inlet velocity was taken as v = [0,0,18]t m/s. The pressure in all particles and the atmospheric pressure were considered as zero. The surface tension forces and the gravitational acceleration vector were neglected. The density and viscosity of the fluid were considered as ρ = 1000 kg/m3 and μ = 0.01 Pa·s, respectively. Finally, the smoothing length used in the simulation was h = 3.25 mm, a no- slip boundary condition was used at solid walls and the time step was chosen as Δt = 0.004 s.

The numerical results predicted with this formulation are compared with experimental results in figure 2. It indicates that this method perfectly reproduces the evolution of the domain over time, and also remains stable during the course of the filling simulation. Therefore, it reliable to simulate complex filling processes.

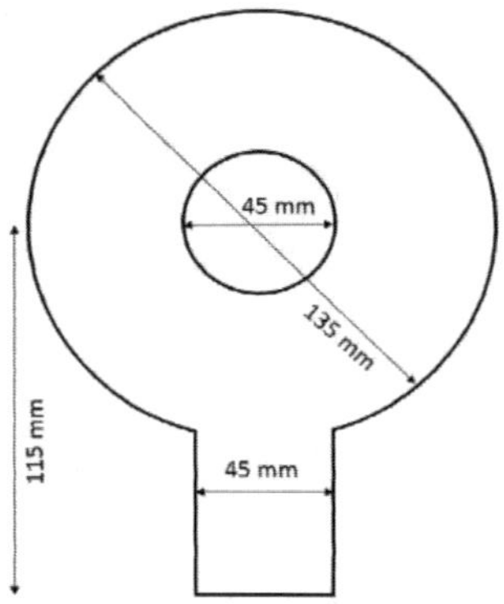

Figure 1. Geometry of the circular disk mould with a core in the center.

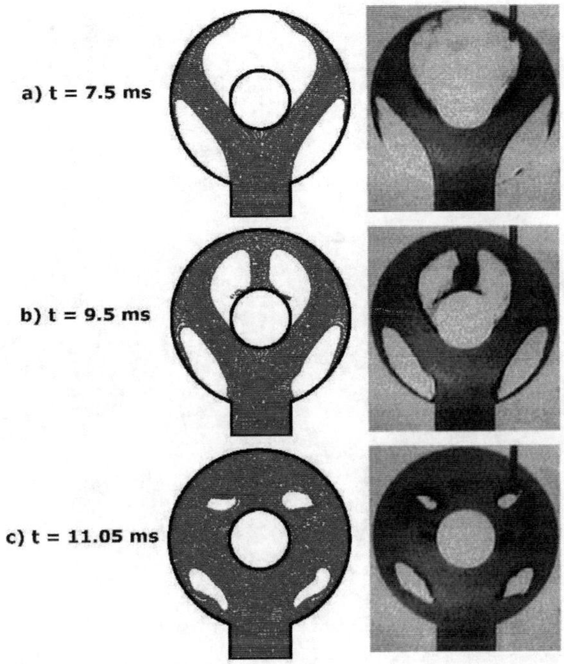

Figure 2. Filling profiles for the circular disk mould with a core in the center at different time steps. *First column:* FPM. *Second column:* experimental profile in [16].

4.1.2. Filling in casting.

The second numerical test considered here corresponds to the free surface flow problem of the filling of a pump cover. In this example, the problem domain was discretized with approximately 180000 points with a mean spacing of 0.0015 m. The inlet velocity was taken as $\mathbf{v} = [-0.1,0,0]^t$ m/s. The pressure in all particles as well as the atmospheric pressure were considered as zero. The surface tension forces and the gravitational acceleration vector were neglected. The density and viscosity of the fluid were considered as $\rho = 6964$ kg/m3 and $\mu = 0.0143$ Pa·s, respectively. These parameters correspond to the physical parameters of the molten cast iron. Finally, the smoothing length used in the simulation was h = 0.0045 m, a slip boundary condition was used at solid walls and the time step was chosen as $\Delta t = 0.004$ s.

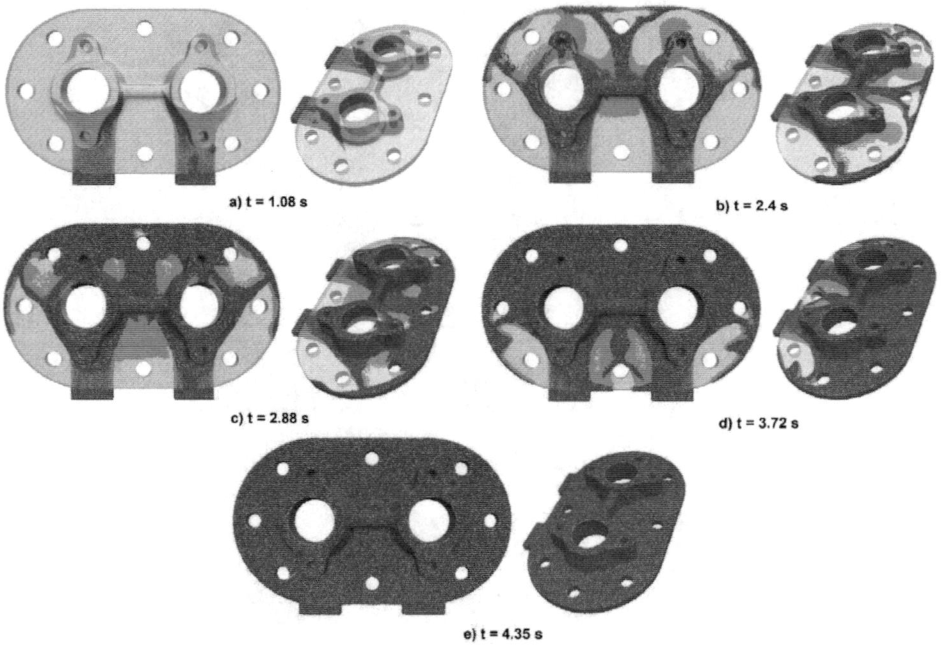

Figure 3. Free surface flow problems at different time steps.

Two perspective views of the filling patterns at different time steps are depicted in Figure 3. There, the picture on the left shows the view exactly from the top whilst the second one shows the view from the top at some angle to the right. As it is shown in this figure, the leading material is divided in four liquid fronts when it impacts the two annular central sections of the die. Two central jets partially merge forming a single liquid front which is split again when it impacts the central cylindrical obstacle. The emerging jets move backwards and starts filling the rear part of the mould. Splashing droplets and liquid fragmentations are visible in this part. The remaining two jets flow around the curved outsides of the die until they collide with the fronts coming from the rear and central parts of the mould. In the two annular central sections of the mould, the liquid flow up into the upper extensions. At this point the rear part of the mould is substantially filled and the fluid flow is towards the front part of the mould. Afterwards, almost all the mould cavity is filled and the biggest voids are principally behind two of the cylindrical obstacles near the inflow jets. They are uniformly filled until the filling process finishes.

These pictures show the robustness of this FPM formulation for dealing with complex computational fluid dynamics problems.

4.2. *Linear Elasticity*

Sections should be numbered as follows:

The numerical benchmark in linear elasticity corresponds to the very well known test example of a cantilever beam subjected to an end load, as geometrically shown in Figure 4. The beams has height D and a length L. A parabolic tangential load is considered on the right hand side and the left hand side is kept fixed. The parameters for this example are set as follows: L = 8 m, D = 1 m, E = 10 GPa, ν = 0.3 and P = 1000 Pa. The analytical solution for the stresses and displacements is found in [17] and reads

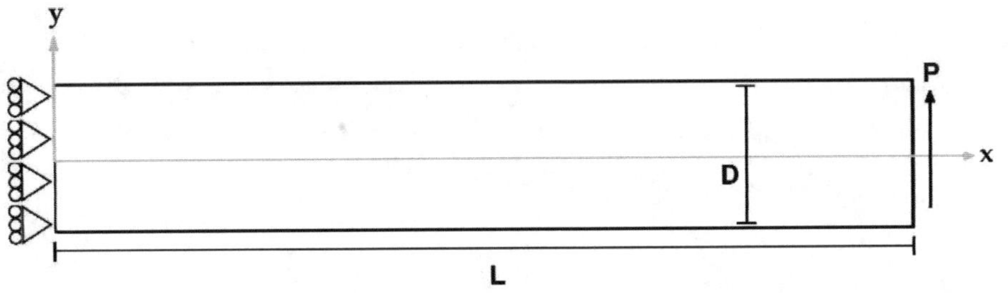

Figure 4. Geometrical description of the two-dimensional cantilever beam problem.

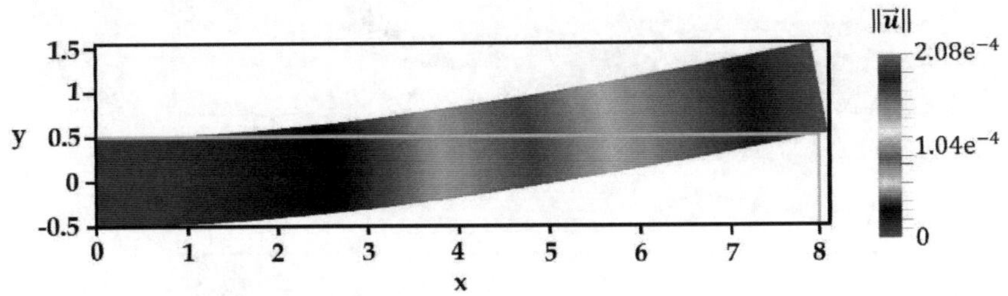

Figure 5. Displacement magnitude distribution

$$\sigma_{xx} = -\frac{P(L-x)y}{I}, \quad \sigma_{yy} = 0, \quad \sigma_{xy} = \frac{P}{2I}\left(\frac{D^2}{4} - y^2\right) \tag{18}$$

$$u = -\frac{Py}{6EI}\left[(6L-3x)x + (2+\nu)\left(y^2 - \frac{D^2}{4}\right)\right] \tag{19}$$

$$v = \frac{P}{6EI}\left[3\nu y^2(L-x) + \frac{D^2}{4}x(4+5\nu) + (3L-x)x^2\right] \tag{20}$$

where P is the total load force, E is the Young's modulus, ν is the Poisson's ratio and $I = \frac{1}{12}D^3$ is the moment of inertia around the horizontal axis. The displacements are imposed as essential boundary conditions on the left-hand side, traction free boundary conditions are defined on the top and bottom of the domain and traction boundary conditions on the right-hand side are prescribed. The corresponding numerical results regarding the displacement magnitude $||\vec{u}||$ and the stresses σ_{xx}, σ_{xy} and σ_v are shown in Figures 5 and 6, respectively, where the von Misses stress is defined as

$$\sigma_v = \sqrt{\sigma_{xx}^2 - \sigma_{xx}\sigma_{yy} + \sigma_{yy}^2 + 3\sigma_{xy}^2} \tag{21}$$

There the displacement has been scaled by a factor of $50e^3$ for visualization purposes and they were obtained considering a discretization with 1936 particles with a mean spacing of 0.0667 m and a smoothing length of 0.2 m leading to an average number of neighbouring particles of 28. These figures show smooth and stable stress patterns. The σ_{xx} and σ_v are compatible with the physical configuration. The approximating solution error regarding the obtained displacements and the corresponding stresses are shown in Figure 7.

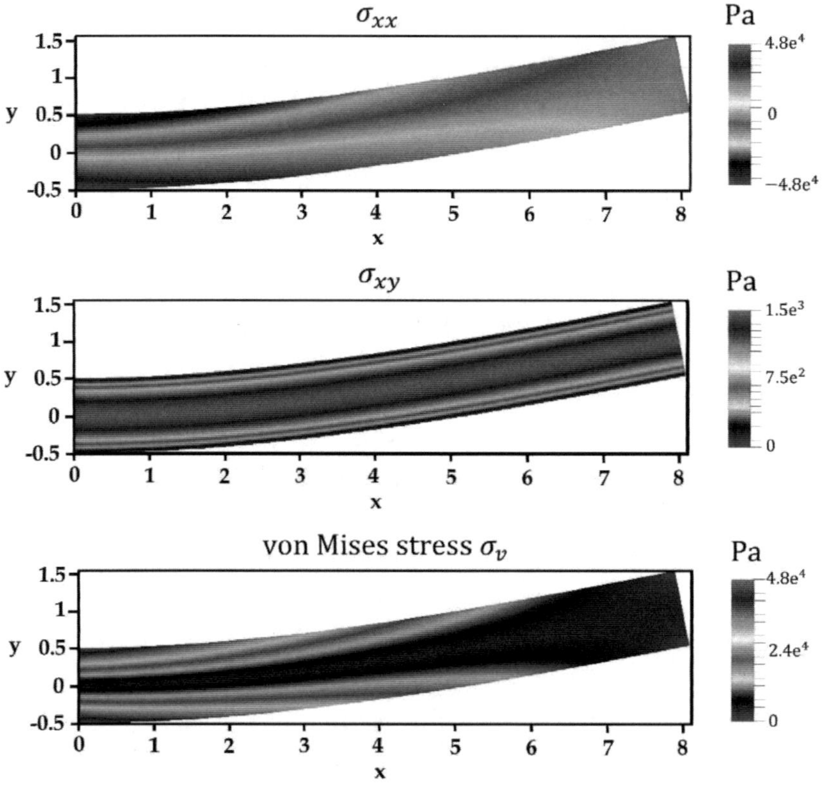

Figure 6. Stresses distributions.

These pictures indicate the robustness of this FPM formulation for the solution of this kind of problems since its numerical prediction matches very well with the exact solution, with a more numerically simple management of the boundary conditions.

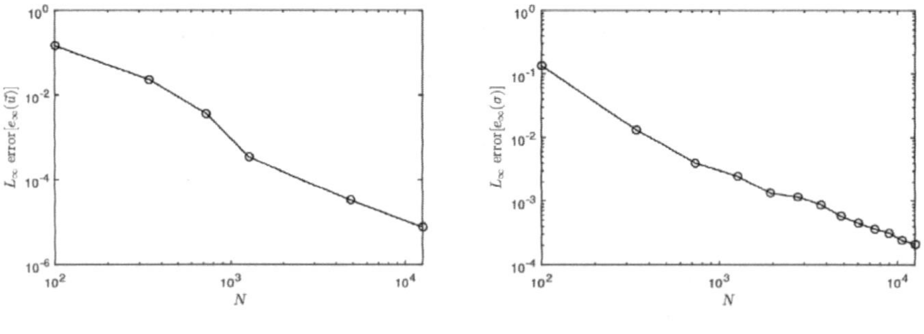

Figure 7. *Left:* displacement error with respect to the exact solution. *Right:* stress error with respect to the exact solution

5. Conclusions

Based on the numerical performance shown in the numerical examples we can conclude that the current approach is suitable and feasible for computational fluid dynamics and linear elasticity problems. It is stable and has enough accuracy in order to capture the physical behaviour observed in the governed processes. Since this formulation is a truly meshfree method it could be used for the study and analysis of complex problems involving high deformations and domain fragmentations with a great computational advantage since it does not need to compute any numerical quadrature and it does not need remeshing approaches. Further, it is able to naturally and easily handle any kind of boundary conditions without requiring any special treatment or stabilization and it is really simple to implement. Therefore, it could be a promising numerical tool for the simulation of engineering processes involving other phenomena described by elliptic partial differential equations as heat transfer. Consequently, it depicts a rich source of research opportunities.

6. References

[1] Kimatsuka A, Ohnaka I, Zhu J and Ohmichi T 2003 *International Journal of Cast Metals Research* **15** 149–152
[2] Chen J S, Hillman M and Chi S W 2017 *Journal of Engineering Mechanics* **143** 04017001
[3] Belytschko T, Krongauz Y, Organ D, Fleming M and Krysl P 1996 *Comput. Methods Appl. Mech. Engrg.* **139(1-4)** 3–47
[4] Kuhnert J 1999 *General smoothed particle hydrodynamics* Ph.D. thesis Technische Universitat Kaiserslautern
[5] Suchde P 2021 *International Journal for Numerical Methods in Fluids*
[6] Anedaf T, Abbes B, Abb`es F and Li Y 2018 *AIP Conference Proceedings* vol 1960 p 140002
[7] Tiwari S and Kuhnert J 2007 *Journal of Computational and Applied Mathematics* **203** 376–386
[8] Tiwari S and Kuhnert J 2002 *Revue Europeenne des Elements* **11:7-8** 965–987
[9] Peters C D, van der Spuy S J, Els D N and Kuhnert J 2018 *Journal of Fluids and Structures* **82** 173–197
[10] Saucedo-Zendejo F R 2020
[11] Wawrenczuk A, Kuhnert J and Siedow N 2007 *Comput. Methods Appl. Mech. Engrg.* **196** 4656–4671
[12] Resendiz-Flores E O and Saucedo-Zendejo F R 2015 *International Journal of Heat and Mass Transfer* **90** 239–245
[13] Resendiz-Flores E O and Saucedo-Zendejo F R 2018 *International Journal of Heat and Mass Transfer* **133** 13–21
[14] Tiwari S and Kuhnert J 2002 *Springer LNCSE: Meshfree methods for Partial Differential Equations* **26**
[15] Saucedo-Zendejo F R and Res´endiz-Flores E O 2020 *Computer Methods in Applied Mechanics and Engineering* **372** 113367
[16] Schmid M and Klein F 1995 *NADCA* 18. *International Die Casting Congress and Exposition* pp 93–99
[17] Timosenko S P and Goodier J N 1951 *Theory of elasticity* (McGraw-Hill)

IOP Conf. Series: Materials Science and Engineering 1141 (2021) 012048 doi:10.1088/1757-899X/1141/1/012048

Design and wind simulation of steel structures in a parametric environment

Á Kis
Technical University of Cluj-Napoca, Faculty of Civil Engineering, Cluj-Napoca, Romania

E-mail: Adam.Kis@student.utcluj.ro

Abstract. Nowadays, parametric design and various simulation methods are gaining ground in almost every engineering and creative profession. The paper investigates the practical applicability of the combination of these methods, by analysing a specific freeform structure with Grasshopper 3D and OpenFOAM and pointing out the differences between the results of a CFD simulation and the recommended methods of the Eurocode. The case study highlights the new perspectives that are opening up in the field of structural design, especially in the examination of wind effects.

1. Introduction

The ordinary practice of structural design is based on the use of many empirically introduced safety factors and statistical approaches, considering / applying different standards. As a result, the geometrical design of the buildings and the loads acting on them can often change significantly because of simplifications, especially if the geometry of the structure is freeform, which is practically impossible to accurately model and analyse with standard tools. However, detaching from reality can be a legitimate concern for a structural designer.

Although one of the main characteristics of today's architecture is the use of bold, yet fragile looking shapes. We are increasingly encountering so-called freeform structures, which provide structural design engineers tasks that require serious professionalism. Thoughtfully, but structural engineers must be able to respond to the needs mentioned above. However, a parametric environment helps the engineer to create a controllable model. A module called Grasshopper, integrated into CAD software called Rhino 3D, is perhaps the most flexible modelling tool of its kind [1]. Its own visual programming language allows for so-called algorithmic modelling, which is a key factor in increasing the efficiency of certain design processes.

Figure 1. Reception hall of Hungexpo Congress and Exhibition Centre Budapest. Structural model (left), architectural rendering (right).

During the design of the steel structure shown in Figure 1, beginning with the structural design and analysis of the model, through the detailing of the main structure and the secondary cladding structure, to the production and installation, parametric design was applied in every phase, showing an increased efficiency [2][3].

Wind simulation methods were also used to study the loads acting on the structure, which is the main topic of the study. However, it can be stated that generally these methods/tools, primarily the different types of simulations, in contrast to other industries, are far from widespread in the construction industry, especially not in the case of structural designers. However, after 2D and then 3D modelling techniques, this has perhaps the greatest untapped potential for the future of structural design.

2. Parametric design of the examined structure

In terms of its functionality, the examined building is the reception hall of the Budapest Congress and Exhibition Centre with office space and showrooms.

The uniqueness of the lattice façade is ensured by the adaptation to the three-level concrete support structure with different floor areas, while the special shape of the roof structure is ensured by the fact that it fits to a spherical surface.

In general, parametric design requires a well-structured set of rules (Figure 2). At the beginning, all the constraints and extremities have to be clarified, together with the basic conceptual principles, for instance the setting of the parameters. This is followed by data processing, which can mean dimensioning, modelling, or even detailing.

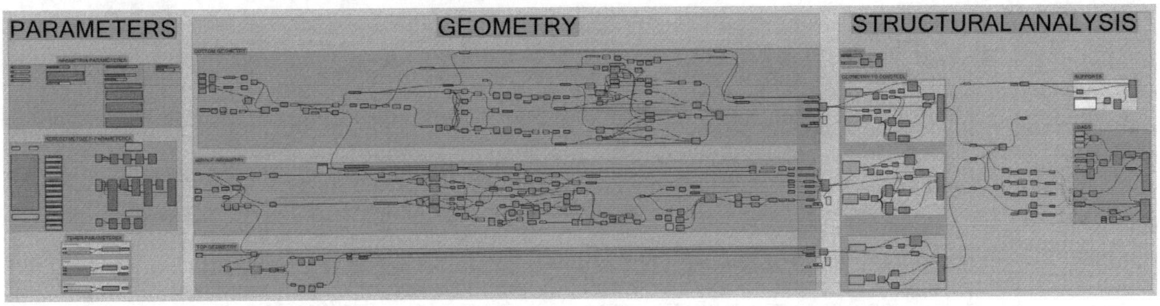

Figure 2. Example parametric algorithm in Grasshopper.

From an architectural point of view, the geometry of the designed steel structure is basically adjusted to the following fixed elements, so the whole structure should be arranged according to them (figure 3):

- The rotationally symmetrical edge curves, with constant heights, but different length in every level (blue)
- Rotationally symmetrical outer roof edge curve, with different heights along its length, due that fits to a spherical surface, (green)
- Horizontal and circular inner ring (orange).

IOP Conf. Series: Materials Science and Engineering 1141 (2021) 012048 doi:10.1088/1757-899X/1141/1/012048

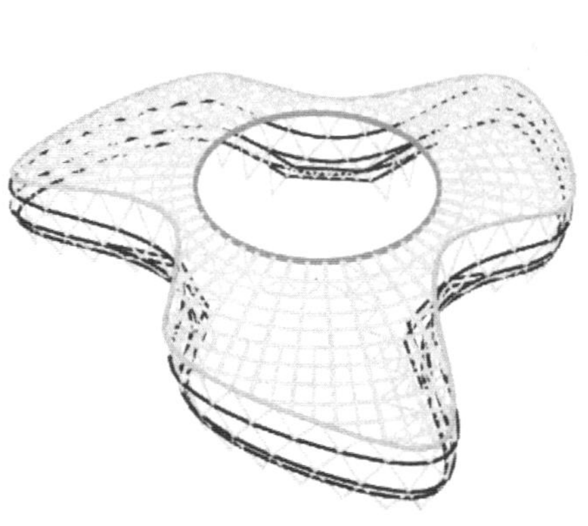

Figure 3. The final structural analysis model highlighted the architectural base geometry.

Therefore, for the structural designer the duty was to adapt to the abovementioned geometrical constraints, besides the load-bearing capacity, manufacturability and easy assembly. However, the primary task of the structural engineer was to develop the optimal load-bearing system, meaning the configuration and segmentation of the steel members, in addition to the global static inspections of the entire steel structure and the calculations of the joints. Several structural designs (figure 4 & 5) emerged, which were also examined in a parametric way [4] according to the following aspects: total weight, utilization, stability, deflection and node complexity (number of individual and uniform nodes).

After the examination, the final roof structure was chosen, where the outer and inner edge rings are connected by radial main girders, curved in their plane, acting as a traditional simply supported beam. The main girders are connected to each other by straight beams, resulting quadrilateral field. Certain fields are stiffened by braces and secondary radial beams.

Overall, working with Grasshopper during the design of the structure had huge advantages, as it made it possible to control the entire workflow, where the key is the coordination and interconnectivity with other software necessary for design (figure 6). In this workflow, Grasshopper, as a focal point, provided input data for preliminary design (Karamba plug-in), global structural analysis and design (ConSteel), connection design (IDEA Statica), detailing (Tekla Structures) and to perform wind simulations (OpenFOAM).

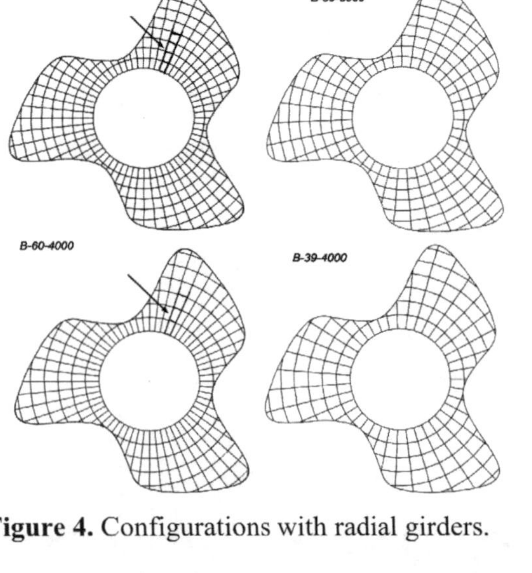

Figure 4. Configurations with radial girders.

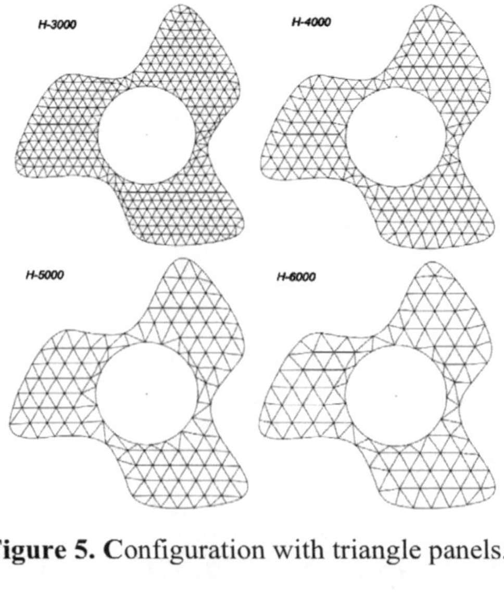

Figure 5. Configuration with triangle panels.

IOP Conf. Series: Materials Science and Engineering 1141 (2021) 012048 doi:10.1088/1757-899X/1141/1/012048

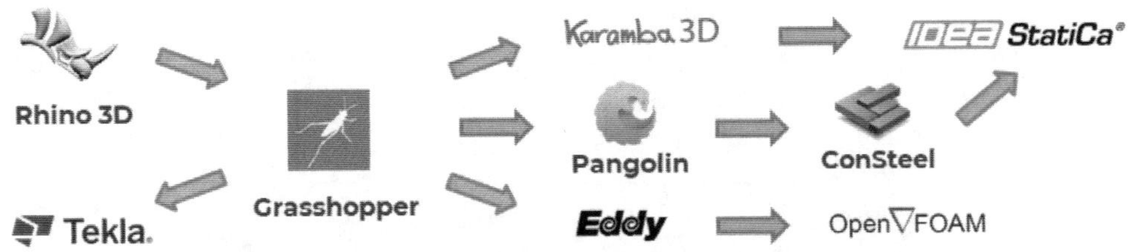

Figure 6. Parametric design workflow

3. Wind loads according to Eurocode

The effect of the wind acting on a structure can be calculated from the average wind speed and the fluctuating wind speed generated by the turbulence around the building [5]:

$$V(z,t) = v_m(z) + v(z,t) \tag{1}$$

However, EN-1991-1-4 [6] defines the wind-generated effects as simplified surface pressures for the calculation of wind loads acting on structures, considering the shape of the structure, its location, the roughness of the terrain, etc. Thus, the wind loads can be considered as a quasi-static pressure equal to the effect of the turbulent wind with maximum speed.

$$q_p(z) = \frac{1}{2}\rho v_p(z)^2 \tag{2}$$

The effective surface pressure instead (pressure or suction), can be determined as the result of the product of the peak pressure (at "z" height above the ground) and the pressure coefficients belonging to the building considered to be rigid:

$$w_e = q_p(z_{ref}) \cdot c_{pe} \tag{3}$$

In the case of the structure already presented, the main challenge, considering the methods of the codes, is to determine the appropriate pressure coefficients. For the sake of simplicity, the study is limited to the determination of the external pressure factors, in which case the code basically covers several regular shapes: flat roofs, different pitched roofs, vaulted roofs, domes, etc. Due to the relatively low roof slope of the structure, the pressure coefficients for a flat roof with curved eaves, or even for a monopitched roof, can be "falsely" used. For the sake of safety, by assuming the pressure factors corresponding to a roof with a 5° slope, the pressure distribution on the structure shown in figure 7 can be assumed.

Of course, this assumption is highly debatable for several reasons. It is more logical to consider the structure under study as a dome (due that it fits to the surface of a sphere). In this case, the standard gives 3 characteristic values in a relatively cumbersome way (figure 8), depending on the eaves height, roof height and diameter of the building.

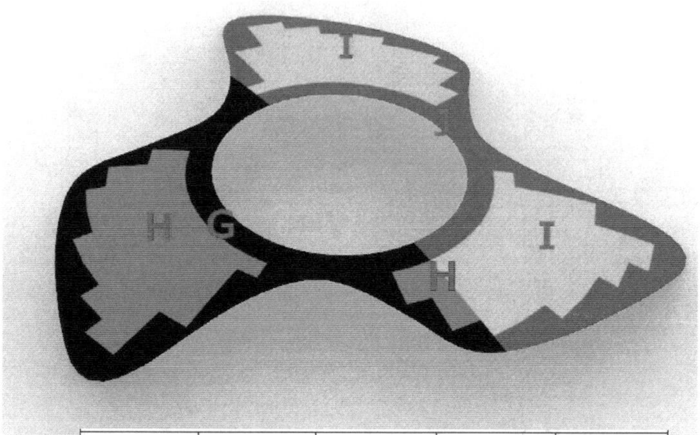

	F		G		H		I		J	
	$c_{pe,10}$	$c_{pe,1}$	$c_{pe,10}$	$c_{pe,1}$	$c_{pe,10}$	$c_{pe,1}$	$c_{pe,10}$	$c_{pe,1}$	$c_{pe,10}$	$c_{pe,1}$
5°	−1,7	−2,5	−1,2	−2,0	−0,6	−1,2	−0,6		+0,2	
	+0,0		+0,0		+0,0				−0,6	

Figure 7. Pressure coefficients on the roof. 1st case.

Along the sectioning circles created by planes parallel to the wind direction, the values between A, B and C is estimated by linear interpolation, thereby the following values shown in figure 9 can be taken into consideration.

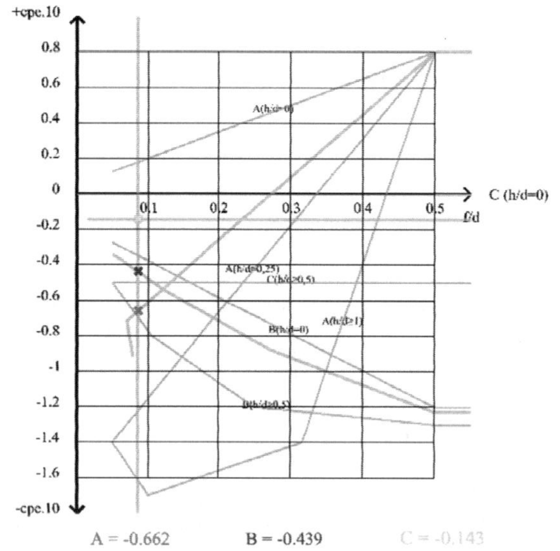

Figure 8. Function evaluating the pressure coefficients. 2nd & 3rd case.

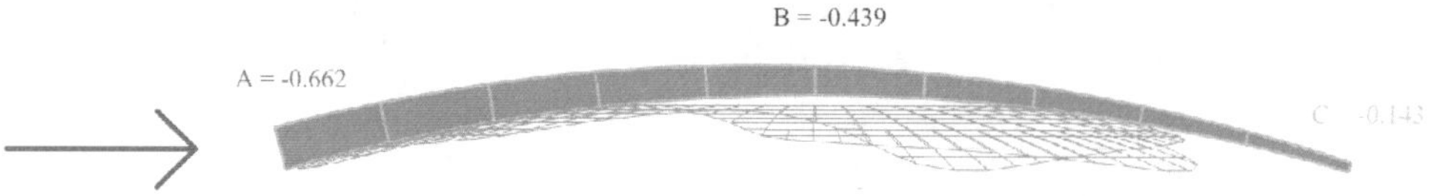

Figure 9. Pressure coefficient diagram on the roof.

However, due to the complexity in plan, this is also debatable. If the roof structure is considered by

IOP Conf. Series: Materials Science and Engineering 1141 (2021) 012048 doi:10.1088/1757-899X/1141/1/012048

examining the pressure coefficients for a circle written around the entire surface (figure 10), then maximum suction is assumed only at the nearest points of the windward side:

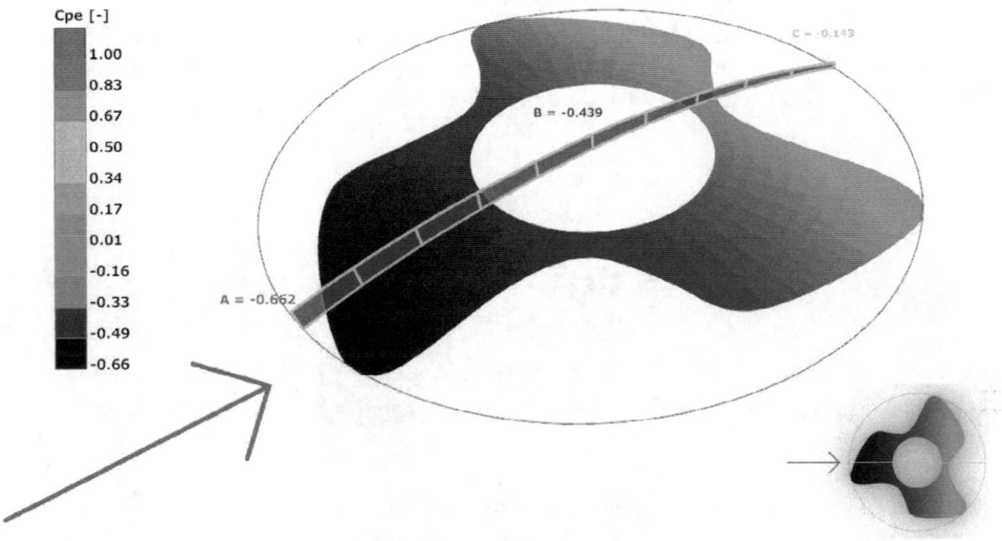

Figure 10. Pressure coefficients on the roof, 2nd case.

It is more reasonable to divide the structure into several segments, assuming more circles written around these sub-surfaces, so the pressure coefficients are different, as shown in figure 11. In this case, substructures with different diameters must be assumed.

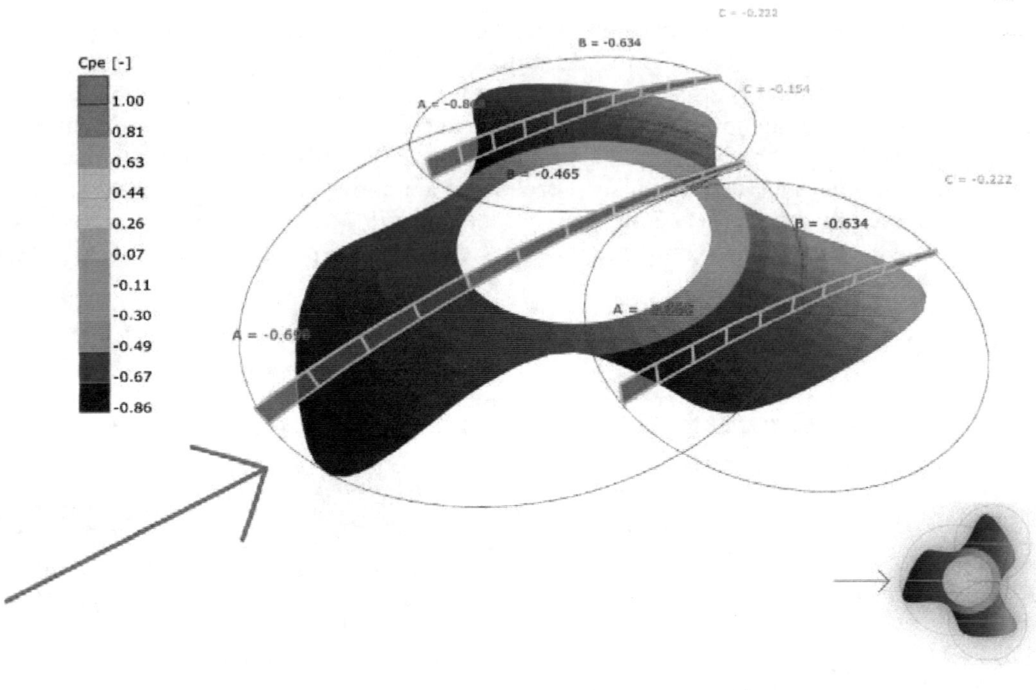

Figure 11. Pressure coefficients on the roof, 3rd case.

IOP Conf. Series: Materials Science and Engineering 1141 (2021) 012048 doi:10.1088/1757-899X/1141/1/012048

The determination of the dominant wind direction is also a topic to be examined in terms of the dimensioning of the entire structure, which means at least 6 wind directions. However, the present study only covers the case when the closest panels to the windward side are also the lowest in terms of roof height, as this has proved to be the governing case especially from the point of view of the roof structure. This is also confirmed by figure 12, which shows the smaller pressure values obtained in a similar way to the previous cases, with a wind direction rotated by 60 °:

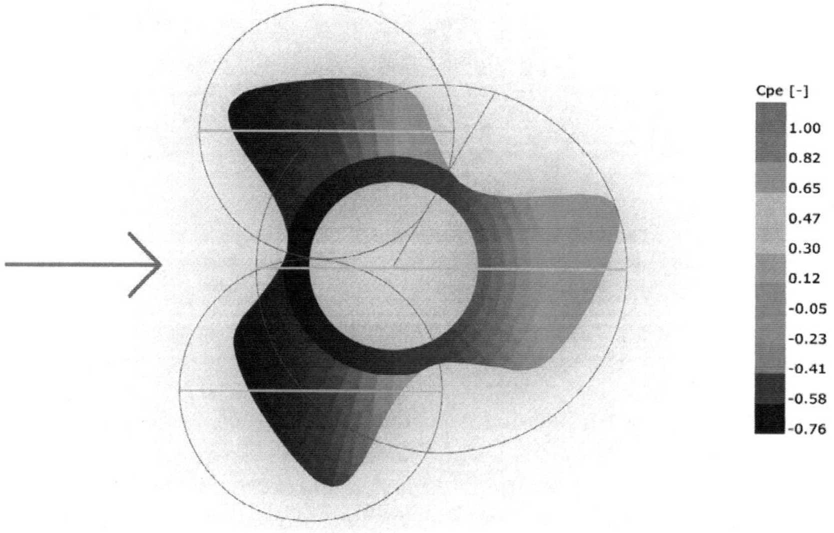

Figure 12. Pressure coefficients on the roof, 3rd case, wind acting rotated with 60°.

In the case of wind loads acting on the façade, they can also be separated to different wind zones, only according to a specific, intuitive principle, as the standard only provides guidance for vertical and rectangular walls. It is obvious that the separation of a minimum of 3 types of zones is justified: windward, leeward, and suctioned side zones, perpendicular to the wind direction. The generated panels of the cladding can be considered as direct load-transfer surfaces. By dividing each panel in half, the created triangles are defining planes, each having a normal vector of which it can be decided that the wind direction is more "perpendicular" or more "parallel" to the given panel, according to the angle defined by the two vectors, comparing to 45 ° (figure 13):

	A		B		C		D		E	
h/d	$c_{pe,10}$	$c_{pe,1}$	$c_{pe,10}$	$c_{pe,1}$	$c_{pe,10}$	$c_{pe,1}$	$c_{pe,10}$	$c_{pe,1}$	$c_{pe,10}$	$c_{pe,1}$
5	−1,2	−1,4	−0,8	−1,1	−0,5		+0,8	+1,0	−0,7	
1	−1,2	−1,4	−0,8	−1,1	−0,5		+0,8	+1,0	−0,5	
≤ 0,25	−1,2	−1,4	−0,8	−1,1	−0,5		+0,7	+1,0	−0,3	

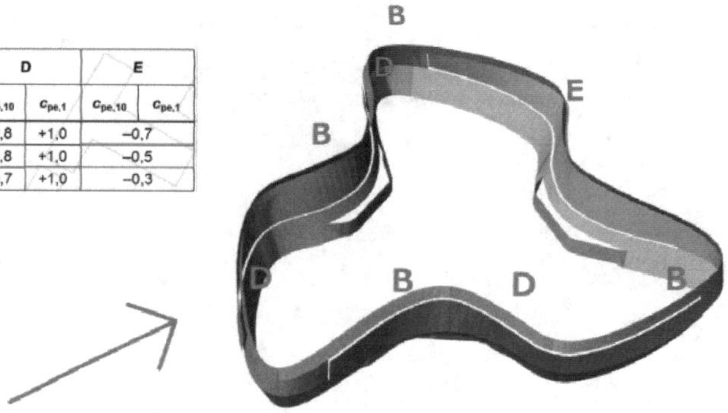

Figure 13. Pressure coefficients on the façade.

4. Wind Simulation

Computational Fluid Dynamics (CFD) is a branch of fluid mechanics, which is the basis of widely used simulation methods. It is used in several fields, such as aerodynamics, natural sciences, weather simulations etc. There are several computer tools / methods for different air, liquid, heat, or gas flow analyses.

OpenFOAM [7] (Open Source Field Operation and Manipulation) is a free C ++ toolkit suitable for this type of numerical fluid dynamics calculations (figure 14). In essence, it allows applications to be made and executed to simulate different fluid behaviours and their environments. There are basically two main types of applications that can be created. Solver-type applications are suitable for solving specific solid or liquid body mechanical problems, while utilities can be used for data management.

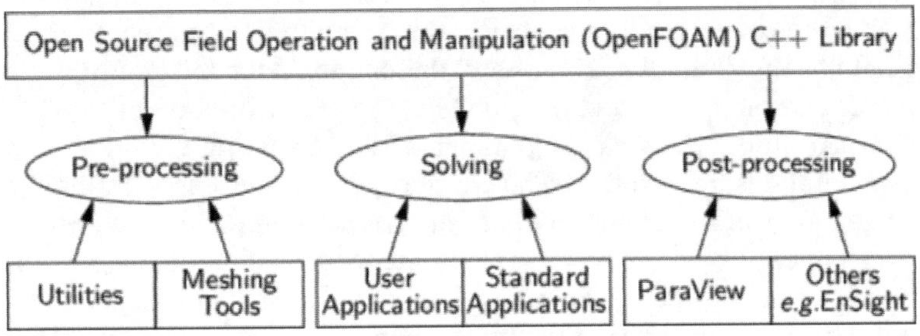

Figure 14. Overview of OpenFOAM Toolbox.

To prepare a CFD-based simulation, the following main steps can be identified:

- Enter input parameters (wind speed, vp (z) corresponding to the peak value of the wind pressure, wind direction, roughness length z0, boundary conditions [8], etc.)
- Preliminary mesh generation processes
- Selection of turbulence model
- Solver selection

The most important task before effective wind simulation is to decompose the investigated building and the environment into finite elements. The mesh generation must meet certain criteria to provide a valid and consequently accurate solution or result. The basic or pre-liminary mesh generation can be performed manually, which in the present study was done according to the principles developed within Grasshopper, to obtain the resultant surface pressure values in the ap-propriate positions when it comes to convert them to loads (figure 15). Furthermore, in this way we can control the complexity of the geometry and consequently to simplify the partial differential equations to be solved but keeping in mind the expected physical behaviour.

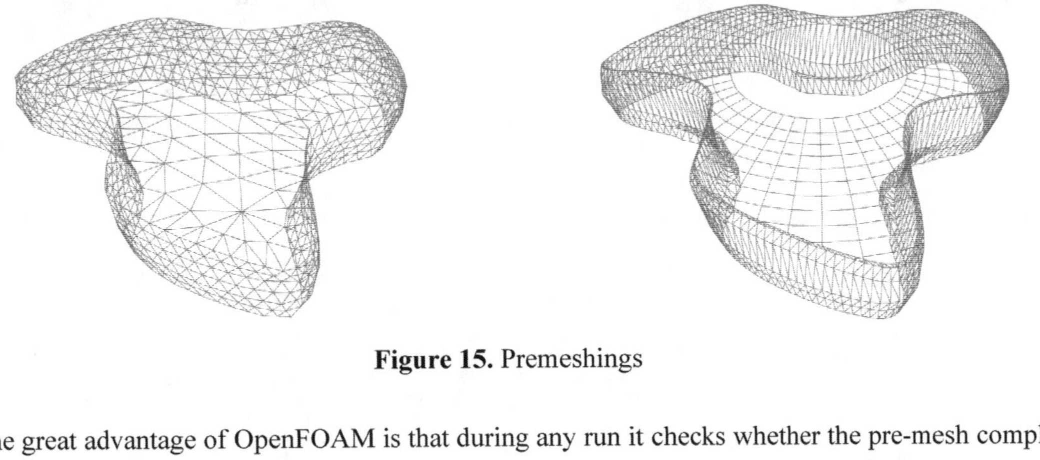

Figure 15. Premeshings

The great advantage of OpenFOAM is that during any run it checks whether the pre-mesh complies

IOP Conf. Series: Materials Science and Engineering 1141 (2021) 012048 doi:10.1088/1757-899X/1141/1/012048

with the validity constraints, corrects it in case of an error, creating the final mesh of the structure based on the given desired mesh sizes, also decomposing the environment to so-called blocks.

The different turbulence models and solving algorithms, all assume the media and the objects within it according to specific principles. The most used [9], also utilised in this case solves the partial differential equations of the kinetic energy and the energy distribution rate (k-ε model), while as a solver, Simple-Foam (SIMPLE = Semi -Implicit Method for Pressure Linked Equations) was selected. In this case, the simulation assumes the following conditions:

- Incompressible, rigid bodies
- Turbulent flow
- There is no physical time, quasi-static pressure

OpenFOAM can also be controlled through Grasshopper. Thus, simulation models can be developed on a much more user-friendly interface, which can be part of a parametric system. The most used extensions are Butterfly (Ladybug Tools) and the one used for this study, Eddy [10]. Both plug-ins toolbar contains the commands needed to set up a complete simulation.

Performing numerical fluid dynamics calculations and determining wind loads by simulation is not commonplace in structural design practice. Thus, a preliminary study is warranted on a geometrically regular building (rectangular in floor plan, with a flat roof structure and with vertical walls). Based on this, the "enveloping" intention of the values recommended by the standard is clearly obvious, the separated zones assume the maxi-mum pressure values uniformly, while the simulation shows the real pressure change at the surface, considering the more intense local effects (figure 16).

Interestingly, from a structural point of view, if the building is considered as a frame structure, as a typical steel hall structure, the difference in the behaviour of the structure can be well demonstrated based on the two methods. Regarding the displacements, with 50-80% higher values can occur in the case of the methods proposed by the codes for specific nodes (column end, main girder midpoint). If the structure is to be dimensioned, utilization differences of about 10-20% can be found in favour of the simulation method. The extent of the deviations depends on how the given pressure values obtained at the specific points (at the corners of the finite elements) is interpreted, namely how it is converted to surface loads in a structural analysis. In the case of a structural model, the most obvious is to define load-bearing surfaces (like the zones recommended by the standard) in which uniformly or variably distributed area loads can be defined, with an intensity given by the average or eventually the maximum of the corresponding pressure values (figure 17).

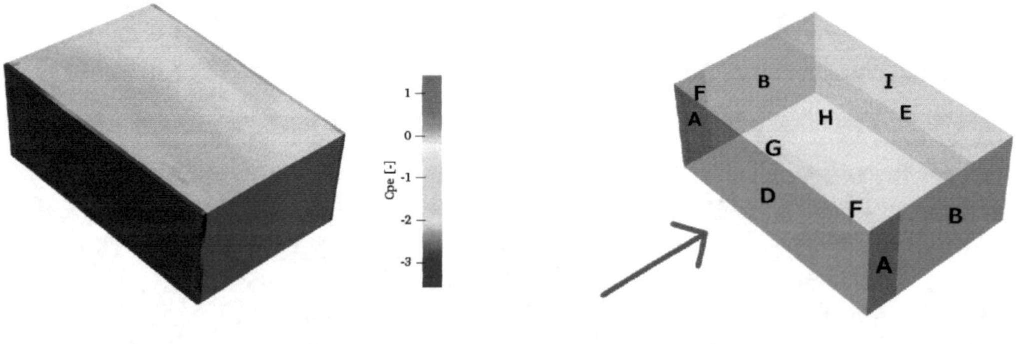

Figure 16. Pressure coefficients according to the simulation (left) and the code recommendations (right).

Figure 17. Wind pressures as surface loads in a structural analysis model (ConSteel).

Using the simulation methodology of the simple hall structure, the results of the examined freeform structure showed a more significant deviations when examining extreme values of the extracted pressures (maximum suction and pressure). In addition, the mesh generation parameters of the different models were recorded, based on which OpenFOAM performed the final mesh generation, and the time required to run the given simulation. Based on this, it can be shown that the "density" of the mesh has major impact when examining local pressure concentrations and, of course, also in terms of analysis duration. Figure 18 shows the results of the last, most dense mesh, in accordance with Table 1, which summarizes the results of four highlighted simulations.

Figure 18. Wind profile (left) and the pressure coefficients according to the simulation (right).

Table 1. The results of the simulations

Block size [m]	FE Size [m]	Min Cpe [-]	Max Cpe [-]	Meshing [min]	Simulation [min]	Post-processing [min]
10	1	-2,02	0,57	2	4	3
10	0,5	-2,26	0,43	4	5	3
5	0,5	-1,75	0,46	10	35	5
2	0,5	-5,72	0,83	16	50	12

5. Conclusion

Based on the study, the real potential of parametric design can be observed, mainly due to its versatility. The use of a parametric methodology can be an advantage in virtually any design process, but it is important to emphasize that setting up such a system is a time-consuming task. Aware of this, adequate experience is required to estimate the degree of efficiency that can be implemented. In addition, it has become worth-while to keep up with the developers behind different parametric design applications, because there is a constant progress with the emergence of new tools, making easier to meet the need for automation. In fact, if the routine in everyday tasks is recognized any work can be made extremely efficient with little effort. In justified cases, it is also worthwhile to design such a parametric system for performing complex tasks, especially if the goal is optimization (since Grasshopper was basically developed for this purpose [11]).

However, the topic of wind simulation is much more complex. Experience basically shows that although CFD can be used to perform wind load tests for industrial use, its reliability is highly questionable. Mainly due to the fact that it requires a serious fluid dynamics expertise and in case of improper use, contradictory results can be easily obtained. Nevertheless, in justified cases, a simulation built in a parametric environment can re-place the methods proposed by the codes, especially if the only way is to use the values of similar geometries, which can lead to a false interpretation, thereby the structure could be dimensioned improperly. In the case of the presented structure, it can be stated that quite different pressure values can be evaluated based on the methods proposed by the codes. The most critical zones of the structure are the roof edges under suction at windward sides, where the values of the pressure coefficients determined in three different ways according to the standard are -0.66, -0.86, -1.2, respectively. On the other hand, based on the simulation, it can be shown more precisely that in the case of local pressure concentrations coefficients higher than the above values (around -2) can be evaluated, but as required by the standard.

Of course, it is not entirely advisable to rely on these data, but it can certainly give a general picture or guidance to the structural engineer in determining the pressure distribution and thus the wind loads on the building.

The golden mean between the codes and computer wind simulation may be the laboratory wind tunnel studies, but here, too, some of the obstacles need to be highlighted. The primary consideration is also the expertise and technology, which is provided in a few cases, especially in our region, in the case of building wind simulation. As it is a relatively rare need, because few serious investments justify such studies, there is no expertise or even a protocol for data communication between the laboratory and the client in the case of such a task. Another important consideration for a wind tunnel test is the amount of time spent, which in a few cases catches up with the rapid pace of the market. Because building a wind simulation model alone can take up to weeks, especially if a modular model is needed to simulate different construction phases. Thereafter, the evaluation of the measurements and the results can be a similarly long process, so that the duration of a complete experiment can take up to 3-6 months, while the full global structural design of a structure, which can be considered complex, can be completed in just few months. Thus, in many cases, there is no data available during the initial design phase, to design more economically by overwriting the code recommendation with wind simulation results. Therefore, in the early phase of designing a freeform structure, it is worth considering the use of wind simulation methods. This means that more studies and experiments are needed to carry out regarding CFD analysis for civil engineering problems, to develop reliable and validated simulation methodologies.

6. References

[1] Akos G, Parsons R 2006 *Foundations. The grasshopper primer third edition*, https://issuu.com/pabloherrera/docs/mode_lab_grasshopper_primer_third_e/3
[2] Szalai J, Juhász M I, Kis Á 2020 *Eingangshalle Hungexpo – parametrische Entwurfs, Berechnungs- und Nachweismethoden*. Ernst&Sohn Special, BIM – Building Information Modeling, 11.2020/A61029, 106-112, https://issuu.com/ernst-und-

sohn/docs/bim2020_web/1?ff
[3] Kis Á 2020 *Proiectarea structurilor metalice si simularea actiunii vantului asupra lor intr-un mediu parametric* (in Romanian) dissertation thesis, UTCN
[4] Arendt E 2020 *Szabadformájú rácsos acél felületszerkezet paraméteres vizsgálata* (in Hungarian) Diplomamunka, Budapesti Műszaki Egyetem
[5] Văcăreanu R 2012 *Cod de proiectare. Bazele proiectării şi acţiuni asupra construcţiilor. Acţiunea vântului (Revizuire NP 082-04. Comentarii, recomandări de proiectare şi exemple de calcul)*, Universitatea Tehnică de Construcţii Bucureşti
[6] MSZ EN 1991-1-3:2005: *Eurocode 1: A tartószerkezeteket érő hatások,* 1-4. rész: Általános hatások, Szélhatás
[7] OpenFOAM – The open source CFD toolbox, https://www.openfoam.com/ (downloaded: 17.02.2021.).
[8] Holmes, J D 2007 *Wind Loading on Structures*, S. Press, Ed., New York, USA: Taylor & Francis e-Library
[9] Nagy-György T, Boros J, Crişan A 2020 Egy szélérzékeny tetőszerkezet számítási érdekességei. Computational curiosities of a wind-sensitive roof structure. *XXIV Conference on Civil Engineering and Architecture* – ÉPKO http://ojs.emt.ro/index.php/EPKO/article/view/246/191
[10] Eddy – Airflow and Microclimate Simulations for Rhino and Grasshopper https://www.eddy3d.com/ (downloaded: 17.02.2021).
[11] Evolutionary Principles applied to Problem Solving https://ieatbugsforbreakfast.wordpress.com/2011/03/04/epatps01/ (downloaded: 17.02.2021).

AUTHOR INDEX

Institute of Physics
Dirac House, Temple Back
Bristol BS1 6BE UK

ISSN: 1757-8981
ISBN 978-1-7138-4618-5

9 781713 846185

CURRAN ASSOCIATES INC.
proceedings.com